T0344503

Automotive Ethernet

Third Edition

Learn about the latest developments in Automotive Ethernet technology and implementation with this fully revised third edition. Including 20% new material and greater technical depth, coverage is expanded to include

- Detailed explanations of the new PHY technologies 10BASE-T1S (including multidrop) and 2.5, 5, and 10GBASE-T1
- Discussion of EMC interference models
- Descriptions of the new TSN standards for automotive use
- More on security concepts
- An overview of power saving possibilities with Automotive Ethernet
- Explanation of functional safety in the context of Automotive Ethernet
- An overview of test strategies
- The main lessons learned

Industry pioneers share the technical and non-technical decisions that have led to the success of Automotive Ethernet, covering everything from electromagnetic requirements and physical layer technologies, QoS, and the use of VLANs, IP, and service discovery, to network architecture and testing. *The* guide for engineers, technical managers, and researchers designing components for in-car electronics, and those interested in the strategy of introducing a new technology.

Kirsten Matheus is a communications engineer who is responsible for the in-vehicle networking strategy at BMW and who has established Ethernet-based communication as a standard technology within the automotive industry. She has previously worked for Volkswagen, NXP, and Ericsson. In 2019 she was awarded the IEEE-SA Standards Medallion "For vision, leadership, and contributions to developing automotive Ethernet networking."

Thomas Königseder is CTO at Technica Engineering, supporting the smooth introduction of Ethernet-based systems for automotive customers. At his former employment at BMW, Thomas was responsible for launching the first serial production car with an Ethernet connection in 2008. He paved the path to today's Automotive Ethernet by enabling the first automotive Ethernet physical layer for series production start in 2013.

Automotive Ethernet

Third Edition

KIRSTEN MATHEUS
BMW AG

THOMAS KÖNIGSEDER
Technica Engineering GmbH

CAMBRIDGE
UNIVERSITY PRESS

University Printing House, Cambridge CB2 8BS, United Kingdom

One Liberty Plaza, 20th Floor, New York, NY 10006, USA

477 Williamstown Road, Port Melbourne, VIC 3207, Australia

314–321, 3rd Floor, Plot 3, Splendor Forum, Jasola District Centre, New Delhi – 110025, India

79 Anson Road, #06–04/06, Singapore 079906

Cambridge University Press is part of the University of Cambridge.

It furthers the University's mission by disseminating knowledge in the pursuit of education, learning, and research at the highest international levels of excellence.

www.cambridge.org
Information on this title: www.cambridge.org/9781108841955
DOI: 10.1017/9781108895248

First edition published 2015
Second edition published 2017
Third edition published 2021

A catalogue record for this publication is available from the British Library.

ISBN 978-1-108-84195-5 Hardback

Contents

Preface to the Third Edition

By the time we were working on the third edition of this book in 2020, Automotive Ethernet had expanded far and wide. All major car manufacturers had Automotive Ethernet in series production cars or were in preparation of their series production start. The physical layer technologies on the road were (in order of introduction) IEEE 100BASE-TX, 100BASE-T1, 1000BASE-T1, and 1000BASE-RH. Furthermore, the IEEE had just published automotive suitable physical layer specifications for 10 Mbps, 2.5 Gbps, 5 Gbps, and 10 Gbps, and was starting the specification work for 10 Gbps plus for optical as well as electrical transmission. Sharing the medium between more than two users had been reintroduced with 10BASE-T1S, and enhancements to this so-called multidrop technology were also being developed.

On layer two, the Time Sensitive Networking (TSN) standardization had completed a number of new standards to extend Quality-of-Service to time-critical control traffic (important features for automated driving) and was well into the specification of a dedicated Automotive TSN Profile. The OPEN Alliance had more than 400 members, and the complete ecosystem had matured with good supporting solutions from tools and test houses to cables and connectors that regularly met at three well established conferences around the world: the IEEE-SA Ethernet&IP@Automotive Technology Day (at worldwide different locations), the Automotive Ethernet Congress (in Munich), and the Nikkei Automotive Ethernet Seminar (at different locations in Japan).

So, all good?

All promising (but not quite there). The technological base has been made available, but true automotive networks are still only at their beginning. They are closely coupled to the shift from hardware-defined cars to software-defined cars, and also here the industry is, with exceptions, only just starting. So, while the industry is expert in physical layer technologies, electromagnetic compatibility, and hardware costs, the chances and choices the protocol layers offer are less explored.

This book has therefore been amended with a description of all new developments within TSN and the physical layer (and because the physical layer chapter would otherwise have become too large, we split it into three chapters: automotive environment, physical layer technologies, and power). We also enhanced the protocol sections above layer two. These layers are what (can) make all the difference. They are what allows the network to support distributed computing and to explore different choices in the EE architecture.

Furthermore, we added ten important lessons learned. These were generated not only from our own experiences at BMW, but also from what we observed in the industry in general. As Thomas left BMW and joined Technica Engineering, we were in the fortunate position of having broader insights to share on a general basis. We are sure the lessons learned can make a huge difference to those who are still beginning to explore the potential of Automotive Ethernet and to those who are wondering what is going wrong.

As with every new technology, it takes time and experience to find the most suitable way to adopt a technology. Automotive Ethernet offers plenty of choices, and car manufacturers must decide on the most suitable path for themselves. We are looking forward to accompanying and supporting the process. In general, we would like to thank all who are making Automotive Ethernet happen on a daily basis. For this third edition, we would like to thank all who answered many of the smaller and not so small questions that came up during the process of writing. In particular, we would like to thank (in alphabetical order):

- Piergiorgio Beruto, Canovatech. Without Piergiorgio, the IEEE 10BASE-T1S standard would not have been as ground-breaking as it is. He reviewed the 10BASE-T1S section of this edition and provided viable background information.
- Stefany Chourakorn, BMW, who, as a new adopter of IEEE 10BASE-T1S was able to point out those aspects we forgot to explain, but which help tremendously to understand the technology.
- Brian Petersen, Ethernovia, for proof-reading the entire book. With his background, he ironed out some inaccuracies and gave fresh insights from the perspective of someone who knows what networking can be. The reader will greatly benefit from his suggestions and corrections (including those with respect to the English language).
- Lars Völker, Technica Engineering. Lars is one of the key people and early contributors to Automotive Ethernet as such. Lars decisively shaped SOME/IP(-SD) and with that contributed a decisive piece of Automotive Ethernet and all the possibilities it offers to automotive networking, as it allows the integration of modern communication paradigms within the existing automotive infrastructure. Thank you for your contributions to the protocol chapter, especially the reworking of the security section.
- George Zimmerman, CMS consulting, for reviewing the MultiGBASE-T1 and Energy Efficient Ethernet sections and for providing viable background information on both.
- Helge Zinner, Continental. Helge did not only review the complete protocol chapter, but also did a lot of the groundwork that helped structure the new TSN specifications.

Last but not least, I, Kirsten, would like to thank BMW for supporting the work on the book and for giving me the opportunity to make a difference. Thomas, now CTO of Technica Engineering, will simply continue to always make a difference.

Preface to the Second Edition

In September 2011, Automotive Ethernet was still at its very beginning. BMW was far and wide the only car manufacturer seriously interested. In 2011, BMW had been in production with 100BASE-TX for diagnostics and flash updates for three years, and decided to go into production with what is now called 100BASE-T1 in its new surround view system in 2013.

In September 2011, strong doubts still had the upper hand. The main concern was that transmitting Ethernet packets at 100M bps over a single Unshielded Twisted Pair (UTP) cable would not be possible under the harsh automotive electromagnetic compatibility (EMC) conditions. Another concern was the missing ecosystem. At the time there was only one supplier of the transceiver technology, Broadcom, who had no prior experience with the written and unwritten requirements of the automotive industry. Additionally, BMW was only just starting to involve the supporting industry of test institutions, tool vendors, software houses, etc.

For an outsider, September 2011 was thus a time of uncertainty. From the inside, however, it was the time in which the foundations for the success of Automotive Ethernet were being laid and in which we ensured that the right structural support was in place. In the background we were finalizing the framework of the OPEN Alliance, NXP was in full speed evaluating its chances as a second transceiver supplier, BMW was preparing to congregate the industry at the 1st Ethernet&IP@Automotive Technology Day, while first discussions on starting the next generation standardization project, 1000BASE-T1, concurred.

One of my, Kirsten Matheus', many jobs at the time was to interest more semiconductor vendors in Automotive Ethernet. In September 2011 this meant getting them to negotiate a licensing agreement with Broadcom, one of their competitors, while the market prospects were still foggy. In one of the discussions I had, an executive manager explained to me, in detail, why this was out of question, based on the following experience.

In the past, he had worked for another semiconductor company that was addressed by a powerful customer to be the second supplier for a proprietary Ethernet version (just like 100BASE-T1 was proprietary when it was still BroadR-Reach and not yet published in the OPEN Alliance). This customer offered significantly higher volumes than BMW ever could, and it was even in the position to technically support them with interoperability and other technical questions, which they did not expect BMW to be capable of. They invested and developed a respective Ethernet PHY product.

However, shortly after, the IEEE released an Ethernet specification for the same use case. This IEEE version was seen to be technologically inferior. However, it had one technical advantage over the proprietary technology they had invested in: It was backwards compatible to previously designed IEEE Ethernet technologies. The IEEE technology prevailed, whereas the solution they had invested in never gained any serious traction. In consequence, they would not again invest in a technology that was not a public standard. The prospect of the OPEN Alliance acting as an organization that ensured transparency in respect to licensing and technical questions did not make any difference to them.

Today, five years later, in 2016, we know that if that semiconductor company had invested in 100BASE-T1/BroadR-Reach in 2011, their business prospects today would be excellent. Not only because the technology persevered but also because they would have been early in the market. Was the executive all wrong in his saying it needs to be a public standard? I do not know.

Many things happened in the meantime. Based on the experiences with BroadR-Reach/100BASE-T1, what BMW had wanted to start with became doable: Transmitting 100 Mbps Ethernet over unshielded cables during runtime using 100BASE-TX PHYs. This solution, sometimes called Fast Ethernet For Automotive (FEFA), was based on a public IEEE standard. For BMW it came too late. But many (most) other car manufacturers had not taken any decision yet. For a while, it was not so sure whether the "proprietary" (but licensed) BroadR-Reach would succeed in the market or the tweaked "public" 100BASE-TX.

Well, today we know: BroadR-Reach made it. But, in the meantime, it has also become a public standard, called IEEE 802.3bw or 100BASE-T1. Only three weeks after handing in the manuscript of the first edition for this book, a respective Call for Interest (CFI) successfully passed at IEEE 802.3. The IEEE released a "BroadR-Reach compliant" specification as an IEEE 802.3 standard in October 2015. Maybe BroadR-Reach would have succeeded also without IEEE's blessing. Who knows? The fact is, the IEEE standardization made life easier. It erased the topic of technology ownership from the discussions.

And it was a main motivator to write this second edition. The now publicly available 100BASE-T1 and BroadR-Reach specifications allowed us to go into detail. The reader will thus find a significantly extended PHY chapter. This section now includes a detailed explanation of the 100BASE-T1 technology as well as the 1000BASE-T1 technology, whose standardization has also been completed in the meantime. While the description of the 100BASE-T1 technology includes experiences made while implementing and using the technology, the 1000BASE-T1 description includes the methodology used behind developing a technology in the case of an unknown channel – something new and also useful for potential future development projects.

Furthermore, the PHY chapter now has a distinct power supply section. Specifications on wake-up and Power over Dataline (PoDL) have been released in the meantime and need context. Additionally, power supply impacts the EMC behavior. This influence on Automotive Ethernet is also described. On the protocol layer, there are new developments with respect to Time Sensitive Networking which have been included in the protocol chapter. Furthermore, the security section has been extended significantly. Last, but not least, we have updated all chapters with the latest developments and insights.

Like the first edition, this edition would also not have happened without the support of the colleagues who make Automotive Ethernet happen on a daily basis. For this second edition we would like to extend our gratitude to (in alphabetical order):

- Karl Budweiser, BMW, who had the (mis)fortune to start working at BMW just at the right time to proofread the PHY section.
- Thomas Hogenmüller, BOSCH, who did not contribute directly to this book, but who successfully dared to drive the standardization of BroadR-Reach at IEEE, and without whom the main reason for writing this second edition might not have happened.
- Thomas Lindner, BMW, who dissected the BroadR-Reach/100BASE-T1 technology and was thus able to contribute vital insights to the 100BASE-T1 description. The reader will benefit a lot from his scrutiny.
- Brett McClellan, Marvell, who answered many questions on the 1000BASE-T1 specification and helped in understanding the technology.
- Mehmet Tazebay, Broadcom, who, as the key designer of BroadR-Reach/ 100BASE-T1 and 1000BASE-T1, has not only provided the basis for what happened in Automotive Ethernet as such, but who also answered many questions.
- Michael Ziehensack, Elektrobit, who supported with insights to the security section.
- Helge Zinner, Continental, who relentlessly counterread the complete second edition and made it a significantly more consistent and precise book than it would have been without him.

Last, but not least, we would like to thank BMW for supporting our work on the book and for giving us the opportunity to make a difference.

Preface to the First Edition

On November 11, 2013, I, Kirsten Matheus, attended a celebration of 40 years since the invention of Ethernet at an IEEE 802 plenary meeting. During the celebration, Robert Metcalfe, David Boggs, Ronald Crane, and Geoff Thompson were honored as the pioneers of Ethernet. If I had to name the people without who Automotive Ethernet would not have happened as it did, I would name Thomas Königseder, technical expert at BMW and co-author of this book, and Neven Pischl, EMC expert at Broadcom.

It all started in 2004, when Thomas received the responsibility for speeding up the software flash process for BMW cars. With the CAN interface used at the time, flashing the 1 Gbyte of data anticipated for 2008 would have required 16 hours to complete. After careful evaluation, Thomas chose and enabled the use of standard 100 Base-TX Ethernet for this purpose. Thus, in 2008, the first serial car with an Ethernet interface, a BMW 7-series, was introduced to the world.

This was only a small beginning though. The problem was that the EMC properties of standard 100Base-TX Ethernet were not good. So the technology was usable with cost competitive unshielded cables only when the car was stationary in the garage for the specific flash use case. To use 100Base-TX also during the runtime of the car would have required shielding the cables, and that was too expensive.

Yet, Thomas was taken with the effectiveness of Ethernet-based communication and therefore investigated ways to use 100Base-TX over unshielded cables. He identified the problem, but could not solve it. So in 2007 he contacted various well-established Ethernet semiconductor suppliers to work with him on a solution. (Only) Broadcom responded positively, and engineers from both companies evaluated the BMW 100Base-TX Ethernet EMC measurements. Then, in January 2008, it happened: BMW performed EMC measurements with boards the Broadcom engineer Neven Pischl had optimized using a 100 Mbps Ethernet PHY variant Broadcom had originally developed for Ethernet in the First Mile and which Broadcom engineers had further adapted for the automotive application. The very first measurements ever performed at a car manufacturer with this technology were well below the limit lines and yielded better EMC performance results than even the existing FlexRay!

This was when Automotive Ethernet was born. Without having had this technology available at the right time, without proving that 100 Mbps can be transmitted over unshielded twisted pair (UTP) cables in the harsh automotive EMC environment, none of all the other exciting, complementary, futuristic, and otherwise useful developments in the field would have happened. BMW would likely be using Media Oriented Systems Transport (MOST) 150 and be working on the next speed grade of MOST, together with the rest of the industry.

Naturally, from the discovery of a solution in 2008 to the first ever introduction of the UTP Ethernet in a serial car, a BMW X5, in 2013 and to establishing Automotive Ethernet in the industry was and is a long run. Thomas and I would therefore like to thank all those who helped to make this happen up till now, and those who are today fervently preparing the bright future Ethernet has in the automotive industry, inside and outside of BMW, with a special mention of Stefan Singer (Freescale), who, among other things, established the first contact between BMW and Broadcom. Using Ethernet for in-car networking is a revolution, and it is an unparalleled experience to be able to participate in its development.

This book explains the history of Automotive Ethernet in more detail and, also, how Automotive Ethernet can technically be realized. We would like to thank all those who supported us with knowhow and feedback in the process of writing this book. First, we thank Thilo Streichert (Daimler), who made it his task to review it all, and who saved the readers from some of the blindness that occurs to authors having worked on a particular section for too long. Then there are (in alphabetical order): Christoph Arndt (FH Deggendorf), Jürgen Bos (Ericsson, EPO), Karl Budweiser (TU München), Steve Carlson (HSPdesign), Bob Grow (RMG Consulting), Mickael Guilcher (BMW), Robert von Häfen (BMW), Florian Hartwich (BOSCH), Thomas Hogenmüller (BOSCH), Michael Johas Teener (Broadcom), Michael Kaindl (BMW), Oliver Kalweit (BMW), Ramona Kerscher (FH Deggendorf), Matthias Kessler (ESR Labs), Max Kicherer (BMW), Yong Kim (Broadcom), Rick Kreifeld (Harman), Thomas Lindenkreuz (BOSCH), Thomas Lindner (BMW), Stefan Schneele (EADS), Mehmet Tazebay (Broadcom), Lars Völker (BMW), Ludwig Winkel (Siemens), Helge Zinner (Continental). Last, but not least, we would like to thank BMW for supporting our work on the book.

Abbreviations and Glossary

#	Number of
1PPoDL	One Pair Power over Data Line (name of IEEE 802.3 study group)
2D	Two-Dimensional
3B2T	Three Bits to Two Ternary conversion
3D	Three-Dimensional
4B3B	Four Bits to Three Bits conversion
4D	Four-Dimensional
4D-PAM5	Four-Dimensional Five-Level Pulse-Amplitude Modulation
AAA2C	Avnu sponsored Automotive Avb gen 2 Council (name of an Avnu initiative to gauge and channel interest for TSN; shifted to IEEE P802.1DG)
AAF	AVTP Audio Format (part of TSN)
ACK	Acknowledgment
ACL	Access Control List
ACR-F	Attenuation to Cross talk Ratio at Far end
ACR-N	Attenuation to Cross talk Ratio at Near end
ADAS	Advanced Driver Assist System
ADC or A/D	Analog to Digital Converter
ADSL	Asymmetric Digital Subscriber Line
AEC	Automotive Electronics Council (name of US based organization that standardizes the qualification of electronic components)
AFDX	Avionics Full-Duplex Switched Ethernet
AFEXT	Alien Far End Cross Talk (part of EMC)
AGC	Adaptive Gain Control
AH	Authentication Header (part of IPsec)
AIDA	AutomatisierungsInitiative der Deutschen Automobilhersteller (Automation Initiative of German Automobile Manufacturers)
ALOHA	Hawaiian greeting (name for the multiple user access method developed at the University of Hawaii)
AM	Amplitude Modulation
AMIC	Automotive Multimedia Interface Corporation (discontinued early automotive initiative to standardize multimedia interfaces for automotive use)

Amp. or AMP	Amplifier
ANEXT	Alien Near End Cross Talk (part of EMC)
ANSI	American National Standards Institute
API	Application Programming Interface
APIX	Automotive PIXel link (name for a proprietary SerDes interface)
ARINC	Aeronautical Radio, Inc. (a company founded in 1929 that is known for its ARINC standards, since 2018 part of Collins Aerospace [1])
ARP	Address Resolution Protocol (used with IPv4)
ARPANET	Advanced Research Projects Agency Network (discontinued predecessor of the Internet)
ASA	Automotive SerDes Alliance
ASIC	Application Specific Integrated Circuit
ASIL	Automotive Safety Integrity Level (part of functional safety/ ISO 26262)
ASN	Avionics Systems Network
ATM	Asynchronous Transfer Mode (telecommunications protocol used at layer two)
ATS	Asynchronous Traffic Shaping (part of TSN)
AUTOSAR	AUTomotive Open System Architecture (organization dedicated to the development of software development standards in the automotive industry)
AV, A/V	Audio and Video
AVB	Audio Video Bridging (early name of a set of IEEE standards enabling QoS for Ethernet-based communication)
AVBgen1	First generation of IEEE AVB standards
AVBgen2	Second generation of IEEE AVB standards, renamed TSN
Avnu	Includes the AV for Audio Video and also means road in Creole [2] (organization to industrialize AVB/TSN)
AVS	Audio Video Source
AVTP	AVB Transport Protocol (part of IEEE 1722)
AWGN	Additive White Gaussian Noise
AXE	Name of Ericsson's digital telephone exchange/switching product
B	Billion
BAG	Bandwidth Allocation Gap (part of AFDX)
BCI	Bulk Current Injection (part of EMC)
BER	Bit Error Rate
BLW	BaseLine Wander correction
BM	Bus Minus (FlexRay terminology)
BMCA	Best Master Clock selection Algorithm (part of TSN)
BP	Bus Plus (FlexRay terminology)
BPDU	Bridge Protocol Data Unit
BSD	Berkeley Standard Distribution or Berkeley Software Distribution (operating system based on early Unix)
C2C	Car-to-Car communication

C2X	Car-to-anything communication
CA	Coupling Attenuation (part of EMC)
CaaS	Car-as-a-Service
CAGR	Compound Annual Growth Rate (constant rate of growth over a time period CAGR = $(\text{Volume}_{t2}/\text{Volume}_{t1})^{(1/(t2 - t1))} - 1$)
CAN	Controller Area Network
CAN FD	CAN with Flexible Data rate
CC	Communication Controller (part of FlexRay)
CCITT	Comité Consultatif International Téléphonique et Télégraphique (renamed ITU-T in 1993 [3])
CD	Compact Disc
CDM	Charged Device Model (part of ESD)
CE	Consumer Electronics or Carrier Ethernet (the latter is a marketing name for extensions to Ethernet for the telecommunications industry)
CFI	Call for Interest (part of the IEEE 802.3 process to establish new standardization projects)
CIA	Confidentiality, Integrity, and Availability (part of security)
CIDR	Classless Inter-Domain Routing (part of IPv4)
CISPR	Comité International Spécial des Perturbations Radioélectriques (International Special Committee on Radio Interference, belongs to IEC)
CM	Common Mode
CMC	Common Mode Choke
cmd	command
CML	Current Mode Logic (one technical principle to realize SerDes interfaces)
COL	COLlision (signal needed with CSMA/CD Ethernet)
COTS	Commercial-Off-The-Shelf
CPU	Central Processing Unit
CRC	Cyclic Redundancy Check (a form of channel coding used to detect and sometimes correct errors in a transmission)
CRF	Clock Reference Format (part of IEEE 1722)
CRS	CaRrier Sense (signal needed with CSMA/CD Ethernet)
CSMA/CD	Carrier Sense Multiple Access with Collision Detection
CSN	Coordinated Shared Network
CW	Continuous Wave
D^2B	Domestic Digital Bus
DAC or D/A	Digital to Analog Converter
DAS	Driver Assist Systems or Driver ASsist
DC	Direct Current or Daisy Chain
DDS	Data Distribution Service (name for a middleware)
DEC	Digital Equipment Corporation
DEI	Drop Eligible Indicator (part of the 802.1Q header)
DFE	Decision Feedback Equalizer

DHCP	Dynamic Host Configuration Protocol (used with IP)
DIX	DEC Intel Xerox (name for the early Ethernet promoter companies)
DLL	Data Link Layer
DLNA	Digital Living Network Alliance
DM	Differential Mode
DMA	Direct Memory Access
DME	Differential Manchester Encoding
DMIPS	Dhrystone Million Instructions Per Second
DMLT	Distinguished Minimum Latency Traffic
DNS	Domain Name System (part of IP)
DoIP	Diagnostic over IP
DoS	Denial of Service
DPI	Direct Power Injection (part of EMC)
DRM	Digital Rights Management
DSP	Digital Signal Processor
DSQ 128	Double SQuare constellation, 2-times 16 discrete levels of PAM16 mapped on a 2-dimensional checkerboard (one variant of Ethernet signaling)
DTLS	Datagram Transport Layer Security
DUT	Device Under Test
EADS	European Aeronautic Defence and Space company (Airbus is a division of EADS)
EAP	Extensible Authentication Protocol (part of IEEE 802.1x)
ECN	Explicit Congestion Notification (part of IP)
ECU	Electronic Control Unit
EE or E/E	Electric Electronic
EEE	Energy-Efficient Ethernet (defined in IEEE 802.3af)
EFM	Ethernet in the First Mile (defined in IEEE 802.3ah)
EIA	Electronic Industries Alliance (US-based standards and trade association that ceased operations in 2011, standardized – among other – inexpensive wiring used with Ethernet [4])
ELFR	Early Life Failure Rate (part of AEC-Q100 qualification)
ELTCTL	Equal Level Transverse Conversion Transfer Loss (part of EMC)
EMC	ElectroMagnetic Compatibility
EMD	Electronic Master Device
EME	ElectroMagnetic Emissions
EMI	ElectroMagnetic Immunity (in other documents sometimes also used for ElectroMagnetic Interference!)
EMS	Electro Magnetic Susceptibility (more common: EMI)
EPO	European Patent Office
EPON	Ethernet Passive Optical Network (part of EFM)
ESD	ElectroStatic Discharge or End Stream Delimiter (the latter is explained with 100BASE-T1)
Eth.	Ethernet

Euro NCAP	European New Car Assessment Program (a European car safety performance assessment program)
EWSD	Elektronisches Wählsystem Digital (Electronic Digital Switching System/Electronic World Switch Digital, telephone exchange system discontinued in 2017 [5])
FBAS	FarbBildAustastSynchron signal (analog video signal format, English equivalent is CVBS: Color, Video, Blanking, and Synchronous Signal)
FCC	Federal Communications Commission
FCDM	Field induced Charge Device Model (part of ESD)
FCS	Frame Check Sequence (CRC at the end of an Ethernet packet)
FEC	Forward Error Correction
FEFA	Fast Ethernet For Automotive
FEXT	Far End Cross Talk
FFE	Feed Forward Equalizer
FIFO	First In First Out
FlexRay	Name for a serial, deterministic and fault tolerant fieldbus for automotive use
FOT	Fiber Optical Transmitter
FPD	Flat Panel Display
fps	Frames per second
FRAND	Fair, Reasonable And Non-Discriminatory (the European equivalent of RAND)
FTZ	Forschungs- und Transfer Zentrum (research and transfer center, part of the University of Applied Science in Zwickau, Germany)
GB	Giga bytes (i.e., 2^{30} bytes)
Gbps	Giga bits per second (i.e., 10^9 bits per second)
GDP	Gross Domestic Product
GENIVI	(name of an automotive industry alliance dedicated to open source software in the in-vehicle infotainment. GENIVI is a word construct taken from Geneva, the international city of peace, in which apparently the concept of GENIVI was publicly presented for the first time, and In-Vehicle Infotainment [6])
GEPOF	Gigabit Ethernet over Plastic Optical Fiber (1000BASE-RH defined in IEEE802.3bv)
GMII	Gigabit Media Independent Interface
GND	GrouND
GOF	Glass Optical Fiber
GPS	Global Positioning System
gPTP	Generalized Precision Time Protocol (part of TSN)
h	Hour
H.264	(Name for MPEG-4 Part 10 or Advanced Video Coding, video compression standard of ITU-T)
HB	HeartBeat

HBM	Human Body Model (part of ESD)
HD	High Definition
HDCP	High-bandwidth Digital Content Protection
HDMI	High-Definition Multimedia Interface (proprietary audio/video interface)
HE	High End
HF	High Frequency
hi-fi	High Fidelity (term used to refer to high-quality reproduction of sound in the home, invented in 1927 [7])
HMI	Human Machine Interface
HPF	High Pass Filter
Hres	Horizontal RESolution
HS CAN	High Speed CAN
HSE	High Speed Ethernet (Industrial Ethernet variant of the Fieldbus Foundation)
HSFZ	High Speed Fahrzeug Zugang (BMW term for first High Speed Car Access supporting Ethernet)
HSM	Hardware Security Module
HTTP	HyperText Transfer Protocol (loads website into a browser)
HU	Head Unit (main infotainment unit inside the car, former radio)
I^2C	Inter-Integrated Circuit (referred to also as I-two-C or IIC, used especially for intra PCB communication)
I^2S	Inter-IC Sound (referred to also as Integrated Interchip Sound, or IIS, used especially for connecting digital audio devices on PCB)
IANA	Internet Assigned Numbers Authority (oversees global IP address allocation)
IC	Integrated Circuit
ICMP	Internet Control Message Protocol (part of IP)
ID	IDentifier, IDentification
IDL	Interface Definition Language or Interface Description Language
IEC	International Electrotechnical Commission (headquarter in Geneva)
IEEE	Institute of Electrical and Electronics Engineers (headquarter in New York)
IEEE-RA	IEEE Registration Authority
IEEE SA	IEEE Standards Association
IET	Interspersing Express Traffic (see IEEE 802.3br)
IETF	Internet Engineering Task Force (releases standards especially for the TCP/IP protocol suite)
IFE	In-Flight Entertainment
IGMP	Internet Group Management Protocol (part of IP)
IL	Insertion Loss or Attenuation (part of channel definition)
IMAP	Internet Message Application Protocol (part of IP)
infotainment	INFOrmation and enterTAINMENT

INIC	Intelligent Network Interface Controller (used for MOST to control the higher layers of the ISO/OSI layering model)
I/O	Input/Output
IoT	Internet of Things
IP	Industrial Protocol or Internet Protocol
IPC	InterProcess Communication
IPG	InterPacket Gap (follows every Ethernet packet)
IP(R)	Intellectual Property (Rights)
IPsec	Internet Protocol SECurity
IRQ	Interrupt ReQuest (part of the OA-SPI)
ISI	InterSymbol Interference
ISO	International Organization for Standardization (headquarters in Geneva)
IT	Information Technology
ITU-T	International Telecommunication Union – Telecommunications standardization sector (headquarters in Geneva)
IVN	In-Vehicle Networking
JASPAR	Japan Automotive Software Platform and ARchitecture
JPEG	Joint Photographic Experts Group (standardized in ISO/IEC 10918-1, CCITT Recommendation T.81, describes different methods for image compression)
K-Line	Name for a single-ended, RS-232 similar technology standardized in ISO 9141-2
kbps	Kilo bits per second (i.e., 10^3 bits per second)
LAN	Local Area Network
LCL	Longitudinal Conversion Loss (part of EMC)
LCTL	Longitudinal Conversion Transmission Loss (part of EMC)
LED	Light Emitting Diode
LFSR	Linear Feedback Shift Register
Lidar	LIght Detection And Ranging (method for measuring distances using laser light)
LIN	Local Interconnect Network (single ended automotive bus)
LLC	Logical Link Control (ISO/OSI layer 2)
LLDP	Link Layer Discovery Protocol (used with IET)
LPF	Low Pass Filter
LPI	Low Power Idle (part of EEE)
LS CAN	Low Speed CAN
LVDS	Low Voltage Differential Signaling (physical principle for SerDes interfaces often used synonymously for SerDes)
MAAP	MAC Address Acquisition Protocol (for dynamic allocation of multicast addresses with IEEE 1722)
MaaS	Mobility-as-a-Service (from owning to using)
MAC	Media Access Control
MB	Mega Bytes (i.e., 2^{20} bytes)

Mbps	Mega bits per second (i.e., 10^6 bits per second)
MCL	Mode Conversion Loss (part of EMC)
MDC	Management Data Clock (used with the Ethernet PHY management)
MDI	Media Dependent Interface
MDIO	Management Data Input/Output
MEF	Metro Ethernet Forum (combines of marketing and specification work for connectivity services, was originally dedicated to Carrier Ethernet/telecommunication only)
MGbps	MultiGigabit per second
MHL	Mobile High-definition Link (evolution of HDMI)
MIB	Management Information Base (IEEE 802.3 standardization project)
MIDI	Musical Instrument Digital Interface manufacturers association (standard to connect electronic instruments)
MII	Media Independent Interface
min	Minutes
Mio	Millions
MIPI	Mobile Industry Processor Interface (develops specifications for the mobile eco-system)
MIPS	Million Instructions Per Second
MISRA	Motor Industry Software Reliable Association
MJPEG	Motion JPEG (video compression format)
MLB	Media Local Bus (interface to INIC specified for MOST)
MLD	Multicast Listener Discovery
MM	Machine Model (part of ESD)
MMRP	Multiple MAC Registration Protocol (used with TSN)
MoCa	Multimedia over Coax
MOST	Media Oriented Systems Transport (automotive bus system)
MOST Co	MOST Cooperation (organization that industrialized MOST)
MP3	MPEG-1 Audio Layer III (MPEG 1 Part 3) or MPEG-2 Audio Layer III (MPEG-2 Part 3)
MPEG	Moving Picture Experts Group (sets standards for audio/video compression and transmission)
MPEG2-TS	MPEG No. 2-Transport Stream (one of the formats of MPEG)
MPLS	Multi-Protocol Label Switching (used, e.g., within telecommunication networks)
MQS	Micro Quadlock System (type of connector common in the automotive industry)
MSE	Mean Square Error
Msps	Mega symbols per second, equals MBaud
MSRP	Multiple Stream Reservation Protocol (part of TSN)
MVRP	Multiple VLAN Registration Protocol (used with TSN)
μC	MicroController
n/a	Not available or not applicable
NACK	Negative ACK (packet was not received as expected)

NAT	Network Address Translation (part of IP)
NBI	Narrow Band Interference (part of EMC)
NC	Numerically Controlled
NDP	Neighbor Discovery Protocol (used with IPv6)
NEXT	Near End Cross Talk (part of EMC)
NFV	Network Function Virtualization (used with SDN in the telecom industry)
NIC	Network Interface Controller
NM	Network Management
nMQS	Nano MQS (smaller version of the MQS connector)
NRO	Number Resource Organization (protects the unallocated IP numbers)
NRZ	Non Return to Zero (two level signaling)
ns	Nanoseconds
OA-SPI	OPEN Alliance Serial Peripheral Interface (defined by the OPEN Alliance for the 10BASE-T1S MACPHY)
OA3p	OPEN Alliance 3-pin interface (for 10BASE-T1S transceivers)
OABR	Open Alliance BroadR-Reach (sometimes also referred to as UTSP Ethernet or as simply as BroadR-Reach, now IEEE 100BASE-T1)
OAM	Operation, Administration, and Management (side channel for those purposes available with many transceiver specifications)
OBD	OnBoard Diagnostic (automotive interface for diagnosis)
OCF	Open Connectivity Foundation
OEM	Original Equipment Manufacturer (in the automotive industry often used as a synonym for car manufacturer)
OPEN	One Pair EtherNet alliance (SIG founded to support the Automotive Ethernet eco-system)
OS	Operating System
OSEK	"Offene Systeme und deren Schnittstellen für die Elektronik im Kraftfahrzeug" ("Open systems and their interfaces for electronics in automobiles" is a consortium that describes an OS suitable for embedded systems)
OSI	Open Systems Interconnection (used in ISO/OSI layering)
OTN	Optical Transport Network
P2MP	Point-to-MultiPoint (refers to a form of sharing a medium)
P2P	Point-to-Point (represents a medium that is not shared; can, in another context, also mean Peer-to-Peer)
PAM	Pulse Amplitude Modulation
PAMx	x-level Pulse Amplitude Modulation
PAN	Personal Area Network
PC	Personal Computer
PCB	Printed Circuit Board
PCS	Physical Coding Sublayer
PD	PhotoDiode or Powered Device

PFS	Perfect Forward Secrecy
PHY	Physical Layer (refers to the physical signaling and media, layer one of the ISO/OSI layering model)
PLC	Programmable Logic Controller or Power Line Communication
PLCA	Physical Layer Collision Avoidance (renamed from Physical Layer Carrier Access) (method organizing medium access for shared 10BASE-T1S)
PLL	Phase-Locked Loop
PLS	PhysicaL Signaling sublayer (used with the MAC)
PMA	Physical Medium Attachment
PMD	Physical Medium Dependent (additional sublayer needed in case of optical transmission)
PoC	Power over Coaxial cabling
PoDL	Power over DataLine (often used for transmission of power over single pair technologies but is actually independent from the number of pairs needed)
PoE	Power-over-Ethernet (refers directly to the implementation described in IEEE 802.3af focusing on 2 pair 100Base-TX Ethernet, was later incorporated as clause 33 into the revision document IEEE 802.3-2005)
POF	Polymeric/Plastic Optical Fiber
PoMD	Power over MultiDrop (used for power over a 10BASE-T1S multidrop segment)
PON	Passive Optical Network
POSIX	Portable Operating System Interface
PPM	Parts Per Million, sometimes also called Defects Per Million (DPM)
PS-ACR-F	Power Sum Attenuation to Cross talk Ratio at Far end (part of EMC)
PS-ACR-N	Power Sum Attenuation to Cross talk Ratio at Near end (part of EMC)
PS-NEXT	Power Sum for Near End Cross Talk (part of EMC)
PSA	Peugeot Société Anonyme
PSAACRF	Power Sum for Alien Attenuation to Cross talk Ratio at Far end (part of EMC)
PSANEXT	Power Sum for Alien Near End Cross Talk (part of EMC)
PSD	Power Spectral Density
PSE	Power Sourcing Equipment
PSTN	Public Switched Telephone Network
PTP	Precision Time Protocol (IEEE 1588-2002, part of TSN)
PTPv2	PTP version 2 (IEEE 1588-2008, part of TSN)
QM	Quality Management
QoS	Quality of Service
RAND	Reasonable and Non-Discriminatory
RARP	Reverse Address Resolution Protocol (part of IPv4)
RDMA	Remote Direct Memory Access

RF	Radio Frequency
RFC	Request For Comment
RFI	Radio Frequency Interference
RfQ	Request for Quote
RGB	Red Green Blue (analog video transmission based on transmitting one color per cable)
RIR	Regional Internet Registry (administers and registers IP addresses)
RL	Return Loss or echo (part of channel definition)
RMII	Reduced Media Independent Interface
RoCE	RDMA over Converged Ethernet
ROM	Read Only Memory
RPC	Remote Procedure Call
RS	Reconciliation Sublayer
RS-232	Binary, serial interface first introduced by the EIA in 1962
RS-FEC	Reed Solomon Forward Error Correction
RSE	Rear Seat Entertainment
RSTP	Rapid STP
RTP	Real-time Transport Protocol (part of TSN)
RTPGE	Reduced Twisted Pair Gigabit Ethernet (study group name for IEEE 1000BASE-T1)
RTPS	Real-Time Publish Subscribe (used with DDS)
Rx / RxD	Receiver ingress
S-parameter	Scattering parameter
SA	Screening Attenuation (part of EMC)
SAE	Society of Automotive Engineers (US-based industry association)
SD	Service Discovery
SD-DVCR	Standard Definition Digital Video Cassette Recorder (one of the formats supported with IEEE 1722)
SDH	Synchronous Digital Hierarchy (technology for core telecommunications networks)
SDN	Software Defined Networks
SecOC	SECure Onboard Communication (AUTOSAR specification for security)
SEIS	Sicherheit in Eingebetteten IP-basierten Systemen (Security in Embedded IP-based Systems, early Germany-based research project that addressed Ethernet in automotive use)
Semicond.	Semiconductor(s)
SER	Symbol Error Rate
SerDes	SERializer DESerializer (SerDes links are sometimes also called "pixel links," "High Speed Video links," or – incorrectly – "LVDS")
SFD	Start Frame Delimiter (part of an Ethernet packet)
SG	Study Group
SIG	Special Interest Group
SL	StripLine (part of EMC)

SMTP	Simple Mail Transfer Protocol (first protocol for transporting emails)
SNR	Signal-to-Noise Ratio
SOA	Service Oriented Architecture
SoC	System on Chip
SOME/IP	Scalable service-Oriented MiddlewarE over IP
SONET	Synchronous Optical NETworking (technology for core telecommunications networks)
SOP	Start of Production
SPI	Serial Peripheral Interface
SQI	Signal Quality Indicator
SR	Stream Reservation (part of TSN)
SRP	Stream Reservation Protocol (part of TSN)
SRR	Substitute Remote Request (part of CAN)
SSD	Start Stream Delimiter (part of the Ethernet packet)
SSL	Secure Sockets Layer (replaced by TLS)
SSO	Standard Setting Organization
STP	Shielded Twisted Pair or Spanning Tree Protocol
SUV	Service or Sport Utility Vehicle
SVS	Surround View System
SW	SoftWare
TAS	Time Aware Shaping (part of TSN)
tbd	to be defined
TC	Technical Committee
TCI	Tag Control Information (part of the IEEE 802.1Q header)
TCL	Transverse Conversion Loss (part of EMC)
TCM	Trellis Coded Modulation
TCP	Transmission Control Protocol
TCTL	Transverse Conversion Transfer Loss (part of EMC)
TDM	Time Division Multiplexing (also used as a synonym for circuit switched networks)
TEM	Transversal ElectroMagnetic wave (part of EMC)
TF	Task Force
TIA	Telecommunications Industry Association or TransImpedance Amplifier
TLS	Transport Layer Security
TLV	Type Length Value or Tag Length Value (discussed with SOME/IP)
TO	Transmit Opportunity (part of 10BASE-T1S)
TP	Twisted Pair or Transport Protocol
TSMC	Taiwan Semiconductor Manufacturing Company
TSN	Time Sensitive Networking
TTL	Time-To-Live (part of IP)
Tx / TxD	Transmitter Egress
UBAT	Battery Voltage
UBS	Urgency Based Scheduler (part of TSN)

UDP	User Datagram Protocol
UDS	Unified Diagnostic Services
UNECE	United Nations Economic Commision for Europe
UNFCCC	United Nations Framework Convention on Climate Change
UNH-IOL	University of New Hampshire InterOperability Lab
UNI	User Network Interface
Unix	Derived from Uniplexed Information and Computing Service (UNICS)
UPnP	Universal Plug and Play
USB	Universal Serial Bus
USP	Unique Selling Proposition, Unique Selling Point
UTP	Unshielded Twisted Pair
UTSP	Unshielded Twisted Single Pair (if combined with Ethernet, this often also refers to OABR)
UWB	Ultra Wide Band (IEEE 802.15.4a)
VAN	Vehicle Area Network
VCC	Pin for IC voltage supply
VCIC	Video Communication Interface for Cameras (ISO 17215)
VDA	Verband der Automobilindustrie (German Association of the Automotive Industry)
VDD	Pin for IC voltage supply
VDE	Verband Deutscher Elektrotechniker (Association for Electrical, Electronic & Information Technologies based in Germany)
VID	VLAN Identifier
VIN	Vehicle Identification Number
VL	Virtual Link
VLAN	Virtual LAN
VLSM	Variable Length Subnet Mask (used with IP)
VoIP	Voice over IP
Vpp	Volts peak to peak
Vres	Vertical RESolution
WAN	Wide Area Network
WiFi	Marketing name invented by the WiFi Alliance for IEEE 802.11 enabled WLAN products, often synonymously used for WLAN [8])
WLAN	Wireless LAN
WPAN	Wireless PAN
WRAN	Wireless Regional Area Network
WUP	Wake-Up Pattern (part of CAN partial networking)
WUR	Wake-Up Request (part of CAN partial networking)
WWH-OBD	World Wide Harmonized OnBoard Diagnostics
www	World Wide Web
xMII	any of the many MII variants
xor	either or (exclusive or)
XTALK	Crosstalk (part of the channel definition and EMC)

References

[1] Wikipedia, "ARINC," May 10, 2020. [Online]. Available: https://en.wikipedia.org/wiki/ARINC. [Accessed May 26, 2020].

[2] R. Kreifeld, *Email correspondence*, 2013.

[3] ITU, "Welcome to the History of ITU Portal," 2020 (continuously updated). [Online]. Available: www.itu.int/en/history/Pages/Home.aspx. [Accessed May 26, 2020].

[4] Wikipedia, "Electronic Industries Alliance," May 8, 2020. [Online]. Available: https://en.wikipedia.org/wiki/Electronic_Industries_Alliance. [Accessed May 26, 2020].

[5] Wikipedia, "EWSD," April 1, 2020. [Online]. Available: https://en.wikipedia.org/wiki/EWSD#cite_note-1. [Accessed May 7, 2020].

[6] GENIVI Alliance, "GENIVI FAQ," July 22, 2013. [Online]. Available: www.genivi.org/sites/default/files/GENIVI_FAQ_072213.pdf. [Accessed May 26, 2020].

[7] Hartley Loudspeakers, "A Brief History," 2013. [Online]. Available: www.hartleyloudspeakers.com/new_page_1.htm. [Accessed October 30, 2013, no longer available].

[8] Wikipedia, "WiFi," May 23, 2020. [Online]. Available: http://en.wikipedia.org/wiki/Wi-Fi. [Accessed May 27, 2020].

Timeline

1965	AT&T installs the world's first electronic telephone switch (special purpose computer) in a local telephone exchange [1].
1968	Invention of Programmable Logic Controllers (PLCs) [2].
1969	AT&T employees at Bell Labs develop the operating system Unix, which eventually enabled distributed computing with remote procedure calls and the use of remote resources. For antitrust reasons, AT&T was neither allowed to sell Unix nor to keep it to itself. In consequence, they shipped it to everyone interested [3].
1969 Apr. 7	The RFC 1 is published [4]. It discusses the host software for ARPANET's switching nodes. ARPANET represents one of the world's first operational packet switching networks [5].
1969 Oct. 29	The first ARPANET link is established between University of California, Los Angeles, and Stanford Research Institute [6].
1971 Nov. 3	Publication of the first *UNIX Programmer's Manual* [7].
By 1973	Unix was recoded in C (it was first developed in [an] Assembly language) [8]. This greatly enhanced Unix' portability to different hardware and further incited its distribution.
1973	The International Electrotechnical Commission (IEC) creates a technical committee (TC77) to specifically handle questions of electromagnetic compatibility [9].
1973 May 22	First documentation of Ethernet as an idea in a memo from Robert Metcalfe at Xerox PARC [10]. At that time, Xerox PARC was selling the first personal computer workstations (called "Xerox Alto") and had invented the first laser printers [11]. Metcalfe was working on a solution for data transmission between these products and the early Internet.
1973 Oct.	Unix was presented publicly to the Fourth Association for Computer Machinery on Operating System Principles [3].
1973 Nov. 11	First Xerox internal demonstration of Ethernet [10].
1974 Dec.	Release of the "Specification of Internet Transmission Control Program," RFC 675 [12], which was a monolithic specification that covered both network (Internet Protocol, IP) and connection (Transmission Control Protocol, TCP) protocols. It was initiated by

	the Defense Advanced Research Projects Agency (DARPA), influenced by early networking protocols from Xerox PARC, and refined by the Networking Research Group of the University of Stanford [13].
1975	Honeywell and Yokogawa introduce the first distributed computer control systems for industrial automation [14].
1975 Mar. 31	Xerox files a patent application listing Robert Metcalfe, David Boggs, Charles Thacker, and Butler Lampson as inventors of Ethernet [15].
1976 Jul.	First paper published on Ethernet [16].
1977	The ISO formed a committee on Open System Interconnection (OSI) [17]. Somewhat later a group from Honeywell Information Systems presented their seven-layer model to the ISO OSI group [18].
1978 Mar. 9	The Computer System Research Group of the University of California, Berkeley, released its first Unix derivative, the Berkeley Software Distribution (BSD) [19].
1978 Apr. 1	ARINC publishes the first ARINC 429 communication standard for avionic equipment [20].
1979 Jun.	ISO publishes the OSI layering model [18].
1979 Jun. 4	Metcalfe founds 3Com to build Ethernet competitive products and convinces DEC, Intel, and Xerox (referred to as DIX) to use and promote Ethernet as a standard for their products [10, 21].
1979–82	Next to 3Com, several start-up companies were founded that built Ethernet products. The most successful ones in the mid-1980s were Ungermann-Bass (U-B), Interlan, Bridge Communications, and Excelan [21].
1980 Feb.	IEEE starts the 802 project to standardize LANs [21].
1980 May	The DIX group joins the IEEE 802 project and offers Ethernet for adoption while still working on it [21].
1980 Aug. 29	The User Datagram Protocol (UDP) was published as RFC 768 [22].
1980 Sep. 30	Publication of the first version of the so-called DIX Standard (from DEC/Intel/Xerox) on Ethernet. Operating at 2.94 Mbps, it was able to support 256 devices [23].
1980 Dec.	IEEE 802 LAN effort was split into three groups: 802.3 for CSMA/CD (Ethernet), 802.4 for Token Bus (for the factory automation vendors), and 802.5 for Token Ring (driven by IBM) [21].
1981 Mar.	3Com shipped its first 10 Mbps Ethernet 3C100 transceiver [24].
1981 Sept.	With the fourth version the Transmission Control Protocol (TCP) and the Internet Protocol are published in separate documents, RFC 793 [25] and RFC 791 [26].
1982 Aug.	Simple Mail Transfer Protocol (SMTP) is published as RFC 821 [27].
1982 Sep.	3Com ships the first Ethernet adapter for IBM PCs [10].
1982 Nov.	The second version of the DIX Ethernet Standard is published [28].

1983	IEEE publishes 802.3 10BASE-5 for 10 Mbps over thick coax cable [29].
1983	The trade press names at least 21 companies either developing or manufacturing Ethernet products: The five startups (3Com, U-B, Interlan, Bridge Communications, and Excelan), eight computer manufacturers (DEC, HP, Data General, Siemens, Tektronix, Xerox, ICL, and NCR), and seven chip manufacturers (Intel, AMD, Mostek, Seeq, Fujitsu, Rockwell, and National Semiconductors), all fiercely competing [21].
1983	BOSCH starts a company internal project to develop CAN [30].
1984 Jan. 1	AT&T monopoly is broken up, existing installed telephone wiring is usable by competing companies for their services [1].
By 1985	Approximately 30,000 Ethernet networks have been installed, connecting at least 419,000 nodes [21].
1985	IEEE publishes 802.3 10BASE-2 for 10 Mbps over thin coax cable [29].
1986	Market introduction of Token Ring, quickly gaining momentum as it is able to use telephone wires, is more reliable, and easier to trouble shoot [21].
1987	200 vendors of Ethernet equipment counted [21].
Mid-1987	SynOptics (Xerox spinout) shipped the first (proprietary) 10 Mbps Ethernet version for telephone wire. Even if this solution was proprietary, it proved the feasibility [21].
1987 Dec.	BMW introduces the first car with a communication bus for diagnostic purposes.
1988	The all-electronic fly-by-wire system is introduced into commercial airplane service (on the Airbus A320) [31].
1989 Oct.	Publication of the TCP/IP Internet Protocol (IP) suite as "Requirements for Internet Hosts – Communication Layers," RFC 1122 [32] and "Requirements for Internet Hosts – Application and Support," RFC 1123 [33].
1989–90	The World Wide Web is invented at CERN [34].
1990 Sep.	IEEE 802.3 ratified 10BASE-T [29] (with some effort, as various proprietary solutions had evolved [21]). Ethernet had won the battle against competing technologies, by adapting to market realities and shifting from coax to twisted pair cabling [10].
1991	TIA publishes TIA-568. It describes an inexpensive and easy to maintain UTP structured wiring plant. This includes the definition of pin/pair assignments for eight-conductor 100-Ohm balanced twisted-pair cabling for wires in 8P8C/RJ-45 eight-pin modular connector plugs and sockets [35].
1992	The first cars using CAN roll off the assembly line at Mercedes Benz [30].

1993	IEEE 802.3 releases 10BASE-F, its first of a large number of optical versions [29].
1994 Jun.	Initial release of the first automotive quality specification for integrated circuits AEC-Q100 [36].
1995	The first commercial VoIP product allows real-time, full-duplex voice communication over the Internet using 1995 available hardware and bandwidth [37].
1995	IEEE 802.3 releases 100BASE-TX (-T4, -FX) including auto-negotiation [29].
1995	The ISO/IEC publishes a backwards compatible MPEG-2 Audio (MPEG-2 Part 3) specification – commonly referred to as MP3 – with additional bit and sample rates [38].
1995 Jun.	IETF releases the IPv4 specification "Requirements for IP Version 4 Routers," RFC 1812 [39].
1995 Aug.	IETF releases the first IPsec specification, RFC 1825 [40].
1995 Dec.	IETF release the first specification for IPv6 as RFC 1883 [41].
1996 Feb. 14	The Windows 95 Service Pack-1 includes Explorer 2.0 (i.e., built-in TCP/IP networking) [13, 42, 43].
1996 May	HTTP/1.0 is published as RFC 1945 [44].
1997	IEEE 802.3 releases 802.3x full-duplex and flow control [29].
1997 Apr.	The Fieldbus Foundation funds the project to develop the "High Speed Ethernet (HSE)" Industrial Ethernet version [45].
1998	IEEE 802.1 publishes the IEEE 802.1D-1998 revision that incorporates IEEE 802.1p with new priority classes [46] and IEEE 802.1Q, which enables VLANs [47].
1998	IEEE 802.3 releases 802.3ac, which extends the maximum frame size to 1522 bytes, in order to allow 802.1Q VLAN information and 802.1p priority information to be included ("Q-tag") [29].
1998	Founding of the LIN consortium by Audi, BMW, Daimler, Volkswagen, Volvo, Freescale (erstwhile Motorola), and Mentor Graphics (erstwhile Volcano) [48].
1998 Sep. 10	Founding of the MOST corporation by BMW, Daimler, Oasis (now Microchip), and (Harman) Becker [49].
1998 Dec.	IETF publishes the "Internet Protocol, Version 6 (IPv6) Specification," RFC 2460 [50].
1999	IEEE 802.3 releases the 1000BASE-T specification 802.3ab [29].
1999 May	Napster launches and significantly simplifies MP3 music sharing. It was closed in February 2001 [51].
2000 May	Boeing delivers its first 747-400 with an advanced flight deck display system that uses the Rockwell Collins-developed, Ethernet-based Avionics Systems Network (ASN) as a communication system [52].
2000 Dec. 31	IEC adopts its IEC 61158 standard on fieldbusses. It contains no less than 18 variants. The Ethernet-based variants HSE, EtherNet/IP, and ProfiNet represent three of them [53].

2000	Freescale (formerly Motorola, now NXP), NXP (formerly Philips), BMW, and DaimlerChrysler (today again Daimler) found the FlexRay Consortium [54].
2001 Oct.	DaimlerChrysler (today again Daimler) introduces LIN as the first car manufacturer [55].
2001 Nov.	The first (BMW) car with MOST25 bus and an LVDS-based SerDes goes into production.
2002 Nov.	Release of the IEEE 1588 PTP standard, which had been initiated a few years earlier by Agilent Technologies [56].
2003	IEEE 802.3 releases the first Power over Ethernet (PoE) specification (IEEE802.3af) [29].
2003	The AUTOSAR consortium is founded by BMW, BOSCH, Continental, DaimlerChrysler (today Daimler), Siemens VDO (today Continental), and Volkswagen [57].
2003 Jun. 10	Release of the ARINC Specification 664 Part 2 "Ethernet Physical and Data-Link Layer Specification" [58].
2003 Nov.	LIN 1.3 is published [48].
2004	Start of investigations at BMW to use Ethernet as an in-vehicle networking technology.
2004 Feb.	The Metro Ethernet Forum releases the first of a number of standards for the deployment of Carrier Ethernet [59].
2004 Jul.	IEEE 802.3 passes a CFI on "Residential Ethernet" and starts a respective SG, i.e., the Audio Video Bridging (AVB) activities [60].
2004 Sep.	IEEE 802.3 releases the first Ethernet in the first Mile (EFM) specification (IEEE 802.3ah) [29].
2005 Apr. 27	First flight of the A380 using an AFDX network for its avionics system, see e.g. [61, 62].
2005 Jun. 27	Publication of the ARINC 664 Part 7 specification on "Avionics full-duplex switched Ethernet (AFDX) network" [58].
2005 Nov. 21	The AVB activities are shifted from IEEE 802.3 to IEEE 802.1 [63].
2006	IEEE 802.3 releases the 10GBASE-T specification (IEEE 802.3an) [29].
2006 Feb.	First cars with built-in USB interface for connecting consumer devices are being sold [64, 65].
2006 Aug. 18	IEEE 802.1 releases the 802.1AE specification, also known as MACsec [66].
2006 Nov.	BMW has the first car with a FlexRay bus in production [67].
2007	Toyota introduces the first car with MOST50 [68].
2007 Jul. 20	IEEE 802 confirms the renaming of the 802.3 group from "CSMA/CD (Ethernet)" to "Ethernet" [69].
2008 Jan.	First automotive EMC measurements of Broadcom's BroadR-Reach, today referred to as IEEE 100BASE-T1 Ethernet, at BMW.
2008 Oct.	SOP of the BMW 7 series using 100BASE-TX unshielded as a diagnostic interface and using 100BASE-TX shielded for the communication between HU and RSE [70].

2009	The development of FlexRay is completed. The work in the FlexRay Consortium is terminated [71] and the specifications are transferred to ISO 17458.
2009 Mar.	The GENIVI Alliance is founded by BMW, Delphi, General Motors, Intel, Magneti Marelli, PSA Peugeot Citroën, Visteon, and Wind River [72].
2009 Aug. 25	The AVnu Alliance is founded by Broadcom, Cisco, Harman, Intel, and Xilinx [73].
2009 Dec. 7	AUTOSAR 4.0 is published and provides means to support Diagnostics over IP (DoIP), i.e., Ethernet communication-based diagnosis and software flashing via IP and UDP [74].
2010	IEEE 802.3 releases 802.3az on Energy Efficient Ethernet (EEE) [29].
2010 Jan	First informal discussion among various car manufacturers and FTZ on UTSP Ethernet [75].
2010 Mar.	BMW internal decision on using Broadcom's BroadR-Reach (which later became IEEE 100BASE-T1) Ethernet for the next surround view system [75].
2011 Jan.	First discussion between Broadcom, NXP, and BMW on founding the OPEN Alliance [75].
2011 Jan. 31	The IANA assigns the last available blocks of IPv4 addresses to the Regional Internet Registries (RIR) [76]. This means that there are no longer any IPv4 addresses available for allocation from the IANA to the five RIRs.
2011 Mar.	BMW internal decision on using BroadR-Reach/100BASE-T1 Ethernet for the infotainment domain [75].
2011 Aug. 8	The FlexRay Consortium is officially dissolved.
2011 Oct. 15	ISO publishes the DoIP standard part 1 [77].
2011 Nov. 9	NXP, Broadcom, and BMW start the OPEN Alliance. In the same month C&S, Freescale (now NXP), Harman, Hyundai, Jaguar Land Rover, and UNH-IOL join [78].
2011 Nov. 9	NXP announces the development of a BroadR-Reach/100BASE-T1 Ethernet compliant PHY [79].
2011 Nov. 14	First Ethernet&IP@Automotive Technology Day at BMW in Munich [80].
2011 Sep. 30	The IEEE ratifies and publishes the last of its "Audio Video Bridging (AVB)" standards (IEEE 802.1BA) [81].
2012 Feb.	The Metro Ethernet Forum publishes a suite of specifications as Carrier Ethernet 2.0 [82].
2012 Mar. 15	Call for Interest (CFI) passes for Reduced Twisted Pair Gigabit Ethernet (RTPGE, later called 1000BASE-T1) at IEEE 802.3 [83].
2012 Jun.	ISO publishes the DoIP standard part 2 [84].
2012 Sep. 19	Second Ethernet&IP@Automotive Technology Day, hosted by Continental in Regensburg [85].

2012 Sep.	Audi starts the production of its first car with a MOST150 network [86].
2012 Nov.	IEEE renames the AVB activities as Time Sensitive Networking (TSN) [87].
2012 Nov. 15	CFI passes for "distinguished minimum latency traffic in a converged traffic environment," later called Interspersing Express Traffic (IET)/ IEEE802.3br, at IEEE 802.3 [88] after it had failed its first attempt on March 12 [89].
2013 Jan.	Start of RTPGE/1000BASE-T1 task force at IEEE 802.3 [88].
2013 Jul.	The LIN standardization is seen as completed. The LIN specifications are transferred to ISO 17987 [90] and the LIN Consortium is dissolved.
2013 Jul. 16	CFI passes for Power over Data Line (PoDL) at IEEE 802.3 [91].
2013 Sep.	SOP of the BMW X5 using BroadR-Reach/100BASE-T1 Ethernet for connecting the cameras to the surround view system [75].
2013 Sep. 25	Third Ethernet&IP@Automotive Technology Day, hosted by BOSCH in Stuttgart [92].
2013 Nov.	Acceptance of Interspersing Express Traffic (IET)/IEEE 802.3br Task Force at IEEE 802.3 [93] after failing the attempt in July [94].
2014 Jan.	Start of PoDL Task Force at IEEE 802.3 [95]
2014 Mar. 20	CFI for 1 Twisted Pair 100 Mbps Ethernet (1TPCE) PHY at IEEE 802.3, i.e., the transfer of BroadR-Reach to the IEEE standard 100BASE-T1 [96].
2014 Mar. 20	CFI for Gigabit Ethernet over Plastic Optical Fiber, later named 1000BASE-RH, at IEEE 802.3 [97].
2014 Mar. 31	AUTOSAR Version 4.1 is published and supports TCP, Service Discovery (SD), and the connection to the MAC and PHY layers (including BroadR-Reach/100BASE-T1) [98].
2014 Jun. 9	The OPEN Alliance has more than 200 members [99].
2014 Sep.	Start of 100BASE-T1 Task Force at IEEE 802.3 [100].
2014 Oct. 23	IEEE-SA (4th) Ethernet&IP@Automotive Technology Day, hosted by General Motors in Detroit [101] and organized by IEEE-SA.
2015 Jan.	Start of GEPOF/1000BASE-RH Task Force at IEEE 802.3 [102] after failing to move into Task Force in July [103].
2015 May 12	Publication update of the Automotive Ethernet AVB specification [104].
2015 Sep.	SOP of 7-series BMW using 100BASE-T1 Ethernet as system bus to connect a variety of ECUs [75].
2015 Oct. 14	Among other car manufacturers, Volkswagen and Jaguar Land Rover publicly announce the use of BroadR-Reach/100BASE-T1 Ethernet in their cars [105].
2015 Oct. 26	Publication date of 100BASE-T1 specification by IEEE [106].
2015 Oct. 27	Fifth Ethernet&IP@Automotive Technology Day, hosted by Jaspar in Yokohama [107] and organized by Nikkei BP.

2015 Dec. 12	The United Nations Framework Convention on Climate Change (UNFCCC) adopts the so-called Paris agreement. Its goal is to limit global warming to below 2°C (ideally to 1.5°C) above preindustrial level [108]. This leads to stringent CO_2 targets for the car industry.
2016 Jan.	ISO starts Project 21111 Part 1 and 3 on "Road vehicles – In-vehicle Gigabit Ethernet system" with focus on specifications to support the optical Gbps Ethernet standard 1000BASE-RH [109, 110].
2016 Mar. 4	A significantly amended IEEE 1722 specification is published [111].
2016 Mar. 22	OPEN Alliance has more than 300 members [112].
2016 Jun.	The ISO registers ISO 21806 in order to accommodate the completed MOST specifications at ISO.
2016 Jun. 30	Publication date of the 1000BASE-T1 specification by IEEE [113].
2016 Jun. 30	Publication date of the Interspersing Express Traffic (IET) specification by IEEE [114].
2016 Jul. 28	CFI passes at IEEE 802.3 in order to establish a study group to investigate the standardization of a 10 Mbps Ethernet for use in automotive and industrial applications [115].
2016 Sep. 20	IEEE-SA (6th) Ethernet&IP@Automotive Technology Day, hosted by Renault in Paris [116] and organized by IEEE-SA.
2016 Sep.	The ISO project 21111 is renamed from "Road vehicles – In-vehicle Gigabit Ethernet system" to "Road vehicles – In-vehicle Ethernet system" in order to be able to comprise future Automotive Ethernet support specifications for different PHY technologies. The original parts 1 and 3 are split into part 1 to part 4, with the new parts 1 and 2 containing information that is applicable to all Automotive Ethernet PHY variants.
2016 Nov. 10	IEEE 802.3 agrees on requesting to move the 10 Mbps PHY activity for industrial and automotive applications to Task Force [117]. This effort receives the number IEEE 802.3cg and the two PHYs developed are called 10BASE-T1S and 10BASE-T1L.
2016 Nov. 10	CFI passes at IEEE 802.3 in order to establish a study group to investigate the standardization of a multi-Gbps Ethernet for use in the automotive industry [118].
2017 May 22	First Task Force meeting of IEEE 802.3ch [119].
2017 Oct. 31	IEEE-SA (7th) Ethernet&IP@Automotive Technology Day, hosted by US Car in San Jose [120] and organized by IEEE-SA.
2018 Jul. 9	Initiation slides for starting the development of a Time Sensitive Networking (TSN) Automotive Profile for Automotive at IEEE 802.1 [121].
2018 Oct. 8	IEEE-SA (8th) Ethernet&IP@Automotive Technology Day, hosted by JLR in London [122] and organized by IEEE-SA.
2019 Feb. 8	Approval of the Project Authorization Request for the Time-Sensitive Networking Profile for Automotive In-Vehicle Ethernet Communications, IEEE P802.1DG [123].

2019 Mar. 14 — CFI presentation at IEEE 802.3 in order to establish a Study Group to investigate the standardization of a $>$10G Automotive electrical Ethernet PHY [124]. CFI passes [125].

2019 Jul. — The OPEN Alliance has more than 400 Members [126].

2019 Jul. 18 — CFI passes at IEEE 802.3 in order to establish a study group to investigate the standardization of multidrop enhancements for 10BASE-T1S [127].

2019 Jul. 18 — CFI passes IEEE 802.3 in order to establish a study group to investigate the standardization of a $\geq$10G optical Automotive Ethernet PHY [127, 128].

2019 Sep. 24 — IEEE-SA (9th) Ethernet&IP@Automotive Technology Day, hosted by Ford in Detroit [129] and organized by IEEE-SA.

2019 Nov. 19 — Approval date of the IEEE 802.3cg 10BASE-T1S and 10BASE-T1L specification [130].

2019 Nov. 25 — As the last of the five RIRs the PIRE NCC responsible for Europe, the Middle East, and Central Asia has run out of IPv4 addresses [131]. The pool for original IPv4 addresses has thus been completely exhausted, and new assignments are only possible should IPv4 addresses be recovered.

2020 Jun. 24 — First meeting of the greater than 10Gb/s Electrical Automotive Ethernet PHYs Task Force [132].

2020 Jul. 14 — First meeting of the Multi-Gigabit Optical Automotive Ethernet Task Force [133].

2020 Sep. 15 — The IEEE-SA holds the IEEE-SA Ethernet&IP@Automotive Technology Day as a virtual event.

References

[1] AT&T, "Milestones in AT&T History," 2004. [Online]. Available: www.thocp.net/companies/att/att_company.htm. [Accessed May 6, 2020].

[2] A. Dunn, "The Father of Invention: Dick Morley Looks Back on the 40th Anniversary of the PLC," September 12, 2008. [Online]. Available: www.automationmag.com/855-the-father-of-invention-dick-morley-looks-back-on-the-40th-anniversary-of-the-plc/. [Accessed May 6, 2020].

[3] M. Lasar, "The UNIX Revolution – Thank You, Uncle Sam?," arstechnica, July 19, 2011. [Online]. Available: http://arstechnica.com/tech-policy/2011/07/should-we-thank-for-feds-for-the-success-of-unix/. [Accessed May 6, 2020].

[4] S. Crocker, "Host Software," April 7, 1969. [Online]. Available: http://tools.ietf.org/html/rfc1. [Accessed May 6, 2020].

[5] Wikipedia, "ARPANET," May 4, 2020. [Online]. Available: http://en.wikipedia.org/wiki/ARPANET. [Accessed May 6, 2020].

[6] C. Sutton, "Internet Began 35 Years Ago at UCLA with First Message ever Sent between Two Computers," September 2, 2004. [Online]. Available: http://web.archive.org/web/20080308120314/http://www.engineer.ucla.edu/stories/2004/Internet35.htm. [Accessed May 6, 2020].

[7] Bell Labs, "Unix Programmer's Manual," Wikipedia, November 3, 1971. [Online]. Available: www.bell-labs.com/usr/dmr/www/1stEdman.html. [Accessed May 6, 2020].
[8] D. M. Ritchie, "The Evolution of the Unix Time-sharing System," September 1979. [Online]. Available: www.bell-labs.com/usr/dmr/www/hist.pdf. [Accessed May 6, 2020].
[9] D. E. Möhr, "Was ist eigentlich EMV? – Eine Definition," not known. [Online]. Available: www.emtest.de/de/what_is/emv-emc-basics.php. [Accessed May 6, 2020].
[10] R. M. Metcalfe, "The History of Ethernet," December 14, 2006. [Online]. Available: www.youtube.com/watch?v=g5MezxMcRmk. [Accessed May 6, 2020].
[11] C. E. Surgeon, *Ethernet: The Definite Guide*, Sebastopol, CA: O'Reilly, 2000, February.
[12] V. Cerf and Y. Dalal, "Specification of Internet Transmission Control Program," December 1974. [Online]. Available: https://tools.ietf.org/html/rfc675. [Accessed May 6, 2020].
[13] Wikipedia, "Internet Protocol Suite," Wikipedia, April 27, 2020. [Online]. Available: http://en.wikipedia.org/wiki/Internet_protocol_suite. [Accessed May 6, 2020].
[14] S. Djiev, "Industrial Networks for Communication and Control," (likely) July 2009. [Online]. Available: https://data.kemt.fei.tuke.sk/SK_rozhrania/en/industrial%20networks.pdf. [Accessed May 6, 2020].
[15] R. M. Metcalfe, D. R. Boggs, C. P. Thacker and B. W. Lampson, "Multipoint Data Communication System (with Collision Detection)." U.S. Patent 4,063,220, March 31, 1975.
[16] D. Boggs and R. Metcalfe, "Ethernet: Distributed Packet Switching for Local Computer Networks," *Communications of the ACM*, vol. 19, no. 7, pp. 395–405, July 1976.
[17] A. L. Russel, "OSI: The Internet That Wasn't," July 30, 2013. [Online]. Available: http://spectrum.ieee.org/computing/networks/osi-the-internet-that-wasnt. [Accessed May 6, 2020].
[18] W. Stallings, "The origin of OSI," 1998. [Online]. Available: http://williamstallings.com/Extras/OSI.html. [Accessed May 6, 2020].
[19] Wikipedia, "Berkeley Software Distribution," Wikipedia, May 3, 2020. [Online]. Available: http://en.wikipedia.org/wiki/Berkeley_Software_Distribution. [Accessed May 6, 2020].
[20] Avionics Interface Technologies, "ARINC 429 Protocol Tutorial," Avionics Interface Technologies, date unknown. [Online]. Available: http://aviftech.com/files/2213/6387/8354/ARINC429_Tutorial.pdf. [Accessed July 14, 2013, no longer available].
[21] U. v. Burg and M. Kenny, "Sponsors, Communities, and Standards: Ethernet vs. Token Ring in the Local Area Networking Business," *Industry and Innovation*, vol. 10, no. 4, pp. 351–375, December 2003.
[22] J. Postel, "User Datagram Protocol," August 29, 1980. [Online]. Available: http://tools.ietf.org/html/rfc768. [Accessed May 6, 2020].
[23] Digital Equipment Corporation, Intel Corporation, Xerox Corporation, "The Ethernet, A Local Area Network. Data Link Layer and Physical Layer Specifications, Version 1.0," September 30, 1980. [Online]. Available: http://ethernethistory.typepad.com/papers/EthernetSpec.pdf. [Accessed May 6, 2020].
[24] Wikipedia, "Ethernet," Wikipedia, April 23, 2020. [Online]. Available: http://en.wikipedia.org/wiki/Ethernet. [Accessed May 6, 2020].
[25] Information Sciences Institute University of Southern California, "Transmission Control Protocol," September 1981. [Online]. Available: http://tools.ietf.org/html/rfc793. [Accessed May 6, 2020].

[26] Information Sciences Institute University of Southern California, "Internet Protocol," September 1981. [Online]. Available: http://tools.ietf.org/html/rfc791. [Accessed May 6, 2020].

[27] J. Postel, "Simple Mail Tranport Protocol," August 1982. [Online]. Available: http://tools.ietf.org/html/rfc821. [Accessed May 6, 2020].

[28] Digital Equipment Corporation, Intel Corporation, Xerox Corporation, "The Ethernet, A Local Area Network. Data Link Layer and Physical Layer Specifications, Version 2.0," November 1982. [Online]. Available: http://decnet.ipv7.net/docs/dundas/aa-k759b-tk.pdf. [Accessed May 6, 2020].

[29] Wikipedia, "IEEE 802.3," Wikipedia, April 28, 2020. [Online]. Available: http://en.wikipedia.org/wiki/IEEE_802.3. [Accessed May 6, 2020].

[30] CiA, "History of CAN Technology," 2019 (contiuously updated). [Online]. Available: www.can-cia.de/can-knowledge/can/can-history/. [Accessed May 6, 2020].

[31] Condor Engineering, *AFDX Protocol Tutorial*, Santa Barbara, CA: Condor Engineering, 2005.

[32] R. Braden, "RFC 1122: Requirements for Internet Hosts – Communication Layers," October 1989. [Online]. Available: http://tools.ietf.org/pdf/rfc1122.pdf. [Accessed May 6, 2020].

[33] R. Braden, "RFC 1123: Requirements for Internet Hosts – Application and Support," October 1989. [Online]. Available: http://tools.ietf.org/pdf/rfc1123.pdf. [Accessed May 6, 2020].

[34] P. Barford, "The World Wide Web," University of Wisconsin, September 11, 2008. [Online]. Available: http://pages.cs.wisc.edu/~pb/640/web.ppt. [Accessed May 6, 2020].

[35] Wikipedia, "TIA/EIA-568," Wikipedia, March 1, 2020. [Online]. Available: https://en.wikipedia.org/wiki/TIA/EIA-568. [Accessed May 6, 2020].

[36] Automotive Electronics Council, "AEC History," not known. [Online]. Available: www.aecouncil.com/AECHistory.html. [Accessed May 6, 2020].

[37] iLocus, "The 10 That Established VoIP," September 2007. [Online]. Available: https://de.scribd.com/document/275142588/The-10-That-Established-VoIP. [Accessed May 24, 2020].

[38] ISO, *ISO 13818-3:1995 – Information Technology – Generic Coding of Moving Pictures and Associated Audio Information – Part 3: Audio*, ISO, Geneva: ISO, 1995.

[39] F. Baker, "Requirements for IP Version 4 Routers," June 1995. [Online]. Available: http://tools.ietf.org/pdf/rfc1812.pdf. [Accessed May 6, 2020].

[40] R. Atkinson, "Security Architecture for the Internet Protocol," August 1995. [Online]. Available: http://tools.ietf.org/html/rfc1825. [Accessed May 6, 2020].

[41] S. Deering and R. Hinden, "Internet Protocol, Version 6 (IPv6)," December 1995. [Online]. Available: http://tools.ietf.org/html/rfc1883. [Accessed May 6, 2020].

[42] Wikipedia, "Windows 95," Wikipedia, May 5, 2020. [Online]. Available: http://en.wikipedia.org/wiki/Windows_95. [Accessed May 6, 2020].

[43] Microsoft, "Verfügbarkeit von Microsoft Windows 95 Service Pack 1-Komponenten," February 14, 2006. [Online]. Available: https://support.microsoft.com/de-de/kb/142794. [Accessed June 23, 2013, no longer available].

[44] T. Berners-Lee, R. Fielding and H. Frystyk, "HTTP/V1.0," May 1996. [Online]. Available: http://tools.ietf.org/html/rfc1945. [Accessed May 6, 2020].

[45] Contemporary Controls, "Dick Caro – One of the Most Influential Persons in the Field of Industrial Networking – Part 2," 2006. [Online]. Available: www.ccontrols.com/pdf/Extv6n3.pdf. [Accessed May 6, 2020].

[46] IEEE Computer Society, *802.1D-1998 – IEEE Standard for Local Area Network MAC (Media Access Control) Bridges*, New York: IEEE, 1998.

[47] IEEE Computer Society, *802.1Q-1998 – IEEE Standards for Local and Metropolitan Area Networks: Virtual Bridged Local Area Networks (VLANs)*, New York: IEEE, 1998.

[48] D. Marsh, "LIN Simplifies and Standardizes In-Vehicle Networks," *EDN*, pp. 29–40, April 28, 2005.

[49] A. Grzemba, *MOST, the Automotive Multimedia Network; from MOST 25 to MOST 150*, Poing: Franzis Verlag GmbH, 2011.

[50] S. Deering and R. Hinden, "Internet Protocol, Version 6 (IPv6) Specification," December 1998. [Online]. Available: http://tools.ietf.org/pdf/rfc2460.pdf. [Accessed May 6, 2020].

[51] T. Lamont, "Napster: the Day the Music Was Set Free," February 24, 2013. [Online]. Available: www.theguardian.com/music/2013/feb/24/napster-music-free-file-sharing. [Accessed May 6, 2020].

[52] J. W. Ramsey, "Boeing's 767-400ER: Ethernet Technology Takes Wing," May 1, 2000. [Online]. Available: www.aviationtoday.com/av/commercial/Boeings-767-400ER-Ethernet-Technology-Takes-Wing_12652.html. [Accessed May 6, 2020].

[53] M. Felser and T. Sauter, "Standardization of Industrial Ethernet – the Next Battlefield?," *Proc. IEEE 5th International Workshop on Factory Communication Systems*, pp. 413–421, September 22, 2004.

[54] U. Niggli, "FlexRay, Eine Übersicht über die neue Datenbus-Generation für den Automobil-Bereich," January 2006. [Online]. Available: https://prof.hti.bfh.ch/uploads/media/flexray.pdf. [Accessed September 24, 2013, no longer available].

[55] D. Müller, D. Sommer and S. Stegemann, "Local Interconnect Network (LIN)," November 13, 2009. [Online]. Available: http://prof.beuth-hochschule.de/uploads/media/LIN-Bus_MuellerSommerStegemann.pdf. [Accessed September 17, 2013, no longer available].

[56] Hirschmann, "Precision Clock Synchronization, The Standard IEEE 1588," 2008. [Online]. Available: www.belden.com/docs/upload/precision_clock_synchronization_wp.pdf. [Accessed October 28, 2013, no longer available].

[57] F. Leitner, "AUTOSAR AUTomotive Open System ARchitecture," 2007. [Online]. Available: www.inf.uni-konstanz.de/soft/teaching/ws07/autose/leitner-autosar.pdf. [Accessed October 9, 2013, no longer available].

[58] AIM GmbH, "AFDX Workshop," March 2010. [Online]. Available: www.afdx.com/pdf/AFDX_Training_October_2010_Full.pdf. [Accessed May 6, 2020].

[59] Metro Ethernet Forum, "MEF Technical Specifications," Metro Ethernet Forum, 2020 (continuously updated). [Online]. Available: www.mef.net/carrier-ethernet/technical-specifications. [Accessed May 6, 2020].

[60] R. Brand, S. Carlson, J. Gildred, S. Lim, D. Cavendish and O. Haran, "Residential Ethernet, IEEE 802.3 Call for Interest," July 2004. [Online]. Available: http://grouper.ieee.org/groups/802/3/re_study/public/200407/cfi_0704_1.pdf. [Accessed May 6, 2020].

[61] AIM GmbH, "What Is AFDX," 2016. [Online]. Available: www.afdx.com/. [Accessed May 6, 2020].

[62] J. Henley, "At Seven Storeys High, £6bn to Develop and With a Wingspan of 80 metres – the A380 Makes Its Maiden Flight," *The Guardian*, www.theguardian.com/business/2005/apr/28/theairlineindustry.travelnews1, April 27, 2005.

[63] IEEE 802.3, "IEEE 802.3 Residential Ethernet Study Group Homepage," January 10, 2006 (closed). [Online]. Available: http://grouper.ieee.org/groups/802/3/re_study/. [Accessed May 6, 2020].

[64] just-auto, "USA: Cars to Get USB Ports Soon – CD Players to Become Obsolete?," October 31, 2005. [Online]. Available: www.just-auto.com/news/cars-to-get-usb-ports-soon-cd-players-to-become-obsolete_id84660.aspx. [Accessed May 6, 2020].

[65] bookofjoe, "Microsoft Windows Automotive – World's First Production Car with a Factory-installed USB Port," March 12, 2006. [Online]. Available: www.bookofjoe.com/2006/03/microsoft_windo.html. [Accessed May 6, 2020].

[66] IEEE Computer Society, *802.1AE-2006 – IEEE Standard for Local and Metropolitan Area Networks: Media Access Control (MAC) Security*, New York: IEEE, 2006.

[67] F. Völkel, "BMW X5: Erstes Auto mit FlexRay Datenbus," August 17, 2006. [Online]. Available: www.tomshardware.de/bmw-x5-flexray-datenbus.testberichte-1546.html. [Accessed September 15, 2013, no longer available].

[68] H. Schoepp, "In-Vehicle Networking Today and Tomorrow," January 17, 2012. [Online]. Available: www.automotive-eetimes.com/en/in-vehicle-networking-today-and-tomorrow.html?cmp_id=71&news_id=222902008. [Accessed August 28, 2013, no longer available].

[69] IEEE, IEEE 802 LMSC, "Agenda & Minutes (Unconfirmed) – IEEE 802 LMSC Executive Committee Meeting (updated 24 September 2007)," July 20, 2007. [Online]. Available: www.ieee802.org/minutes/jul2007/Minutes%20-%20Friday%20July%2020%202007.pdf. [Accessed May 6, 2020].

[70] K. Matheus, "OPEN Alliance – Stepping Stone to Standardized Automotive Ethernet," in 2nd Ethernet & IP @ Automotive Technology Day, Regensburg, 2012.

[71] K. R. Avinash, P. Nagaraju, S. Surendra and S. Shivaprasad, "FlexRay Protocol Based an Automotive Application," *International Journal of Emerging Technology and Advanced Engineering*, vol. 2, no. 5, pp. 50–55, May 2012.

[72] B. Wuelfing, "CeBIT 2009: BMW and Partners Found GENIVI Open Source Platform," March 3, 2009. [Online]. Available: www.linuxpromagazine.com/Online/News/CeBIT-2009-BMW-and-Partners-Found-GENIVI-Open-Source-Platform. [Accessed May 6, 2020].

[73] Business Wire, "AVnu Alliance Launches to Advance Quality of Experience for Networked Audio and Video," August 25, 2009. [Online]. Available: www.businesswire.com/news/home/20090825005929/en/AVnu-Alliance-Launches-Advance-Quality-Experience-Networked. [Accessed May 6, 2020].

[74] AUTOSAR, "Requirements on Ethernet Support in AUTOSAR, Release 4.0, Revision 1," December 7, 2009. [Online]. Available: www.autosar.org/fileadmin/user_upload/standards/classic/4-0/AUTOSAR_SRS_Ethernet.pdf. [Accessed May 13, 2020].

[75] K. Matheus, "Structural Support for Developing Automotive Ethernet," in *3*rd Ethernet & IP @ Automotive Technology Day, Leinfeld-Echterdingen, 2013.

[76] Number Resource Organization, "Free Pool of IPv4 Address Space Depleted," February 3, 2011. [Online]. Available: www.nro.net/news/ipv4-free-pool-depleted. [Accessed May 6, 2020].

[77] ISO, *ISO 13400-1:2011 – Road vehicles – Diagnostic Communication over Internet Protocol (DoIP) – Part 1: General Information and Use Case Definition*, Geneva: ISO, 2011.

[78] PR Newswire, "Broadcom, NXP, Freescale, and Harman Form OPEN Alliance Special Interest Group," November 9, 2011. [Online]. Available: www.prnewswire.com/news-releases/broadcom-nxp-freescale-and-harman-form-open-alliance-special-interest-group-133514928.html. [Accessed May 6, 2020].

[79] PHYS ORG, "NXP Develops Automotive Ethernet Transceivers for In-Vehicle Networks," November 9, 2011. [Online]. Available: https://phys.org/news/2011-11-nxp-automotive-ethernet-transceivers-in-vehicle.html. [Accessed May 6, 2020].

[80] I. Riches, "BMW 1st Ethernet & IP @ Automotive Techday – Momentum Achieved," November 14, 2011. [Online]. Available: www.strategyanalytics.com/strategy-analytics/blogs/automotive/powertrain-body-chassis-safety/powertrain-body-chassis-and-safety/2011/11/16/bmw-1st-ethernet-ip-@-automotive-techday—momentum-achieved. [Accessed May 6, 2020].

[81] Wikipedia, "Audio Video Bridging," Wikipedia, April 29, 2020. [Online]. Available: http://en.wikipedia.org/wiki/Audio_Video_Bridging. [Accessed May 6, 2020].

[82] P. Marwan, "Carrier Ethernet 2.0 ist an den Start gegangen," February 28, 2012. [Online]. Available: www.zdnet.de/41560473/carrier-ethernet-2-0-ist-an-den-start-gegangen/. [Accessed May 6, 2020].

[83] S. Carlson, T. Hogenmüller, K. Matheus, T. Streichert, D. Pannell and A. Abaye, "Reduced Twisted Pair Gigabit Ethernet Call for Interest," March 2012. [Online]. Available: www.ieee802.org/3/RTPGE/public/mar12/CFI_01_0312.pdf. [Accessed May 6, 2020].

[84] ISO, *ISO 13400-2:2012 – Road Vehicles – Diagnostic Communication over Internet Protocol (DoIP) – Part 2: Transport Protocol and Network Layer Services*, Geneva: ISO, 2012.

[85] C. Wirth, "Die Netzwerktechnologie Ethernet verspricht eine neue Zukunft im Auto," January 10, 2013. [Online]. Available: www.oth-regensburg.de/fakultaeten/informatik-und-mathematik/nachrichten/einzelansicht/news/die-netzwerktechnologie-ethernet-verspricht-eine-neue-zukunft-im-auto.html. [Accessed May 6, 2020].

[86] MOST Cooperation, "MOST150 Inauguration in Audi A3," *MOST Informative*, no. 8, p. 2, October 2012.

[87] IEEE. IEEE 802.1, "802.1 Plenary –11/2012 San Antonio Closing," November 2012. [Online]. Available: www.ieee802.org/1/files/public/minutes/2012-11-closing-plenary-slides.pdf. [Accessed May 6, 2020].

[88] IEEE. IEEE 802.3, "Approved Minutes IEEE 802.3 Ethernet Working Group Plenary Grand Hyatt, San Antonio, TX USA," November 12–15, 2012. [Online]. Available: www.ieee802.org/3/minutes/nov12/minutes_1112.pdf. [Accessed May 6, 2020].

[89] IEEE. IEEE 802.3, "Approved Minutes IEEE 802.3 Ethernet Working Group Plenary Hilton Waikoloa Village, Waikaloa, HI USA," March 12–15, 2012. [Online]. Available: www.ieee802.org/3/minutes/mar12/minutes_0312.pdf. [Accessed May 6, 2020].

[90] Consortiuminfo, "Local Interconnect Network Consortium (LIN)," July 8, 2015. [Online]. Available: www.consortiuminfo.org/links/linksdetail2.php?ID=428. [Accessed May 6, 2020].

[91] D. Dwelley, "Power over Data Line, Call for Interest," July 16, 2013. [Online]. Available: www.ieee802.org/3/1PPODL/public/jul13/CFI_01_0713.pdf. [Accessed May 6, 2020].

[92] Hanser Automotive, "Rückblick: 3rd Ethernet & IP @ Automotive Technology Day," *Hanser Automotive Sonderheft Autmotive Networks,* November 2011.

[93] IEEE. IEEE 802.3, "Approved Minutes IEEE 802.3 Ethernet Working Group Plenary Dallas, TX," November 11–14, 2013. [Online]. Available: www.ieee802.org/3/minutes/nov13/1113_minutes.pdf. [Accessed May 6, 2020].

[94] IEEE. IEEE 802.3, "Approved Minutes IEEE 802.3 Ethernet Working Group Plenary Geneva, CH," July 15–18, 2013. [Online]. Available: www.ieee802.org/3/minutes/jul13/minutes_0713.pdf. [Accessed May 6, 2020].

[95] IEEE. IEEE 802.3, "IEEE 802.3bu 1-Pair Power over Data Lines (PoDL) Task Force Public Area," IEEE, November 2, 2016. [Online]. Available: www.ieee802.org/3/bu/public/index.html. [Accessed May 6, 2020].

[96] T. Hogenmüller, S. Abbenseth, S. Buntz, K. M. Albert Kuo and H. Zinner, "Call for Interest for 1 Twisted Pair 100 [C] Mbit/s Ethernet," March 2014. [Online]. Available: www.ieee802.org/3/cfi/0314_2/CFI_02_0314.pdf. [Accessed May 6, 2020].

[97] C. Pardo, T. Lichtenegger, A. Paris and H. Hirayama, "Gigabit over Plastic Optical Fibre; Call for Interest," March 2014. [Online]. Available: www.ieee802.org/3/GEPOFSG/public/CFI/GigPOF%20CFI%20v_1_0.pdf. [Accessed May 6, 2020].

[98] AUTOSAR, "Specification of Ethernet Interface, Release 4.1., Revision 3," March 31, 2014. [Online]. Available: www.autosar.org/fileadmin/user_upload/standards/classic/4-1/AUTOSAR_SWS_EthernetInterface.pdf. [Accessed May 13, 2020].

[99] OPEN Alliance, "World's Leading Auto Makers, Tier Ones and Tech Companies Partner to Drive Wide Scale Adoption of Automotive Ethernet," June 9, 2014. [Online]. Available: http://opensig.org/news/press-releases/. [Accessed May 6, 2020].

[100] IEEE. IEEE 802.3, "IEEE 802.3bw 100BASE-T1 Task Force Public Area," IEEE, August 17, 2015. [Online]. Available: www.ieee802.org/3/bw/public/index.html. [Accessed May 6, 2020].

[101] IEEE-SA, "2014 IEEE-SA Ethernet & IP @ Automotive Technology Day," October 23, 2014. [Online]. Available: www.facebook.com/events/1533916296821800/. [Accessed May 6, 2020].

[102] IEEE. IEEE 802.3, "IEEE 802.3bv Gigabit Ethernet Over Plastic Optical Fiber Task Force Public Area," IEEE, November 30, 2016. [Online]. Available: www.ieee802.org/3/bv/public/index.html. [Accessed May 6, 2020].

[103] IEEE. IEEE 802.3, "Approved Minutes IEEE 802.3 Ethernet Working Group Plenary San Diego, CA, USA," July 15–18, 2014. [Online]. Available: www.ieee802.org/3/minutes/jul14/0714_minutes.pdf. [Accessed May 6, 2020].

[104] G. Bechtel, B. Gale, M. Kicherer (Turner) and D. Olsen, *Automotive Ethernet AVB Functional and Interoperability Specification Revision 1.4*, Beaverton: Avnu, 2015.

[105] IEEE Computer Society, *802.3bw-2015 – IEEE Standard for Ethernet Amendment 1 Physical Layer Specifications and Management Parameters for 100 Mb/s Operation over a Single Balanced Twisted Pair Cable (100BASE-T1)*, New York: IEEE-SA, 2015.

[106] Wikipedia, "Paris Agreement," April 30, 2020. [Online]. Available: https://en.wikipedia.org/wiki/Paris_Agreement. [Accessed May 9, 2020].

[107] Nikkei BP, "5th annual IEEE-SA Ethernet&IP@Automotive Technology Day," October 27–28, 2015. [Online]. Available: http://techon.nikkeibp.co.jp/info/ieee/regist/en.html. [Accessed June 26, 2016, no longer available].

[108] OPEN Alliance, "Automotive Ethernet Hits the Road in Wide Range of New Vehicles," October 15, 2015. [Online]. Available: http://opensig.org/news/press-releases/. [Accessed May 6, 2020].

[109] F. Horikoshi, *NWIP on Road Vehicles – In-vehicle Gigabit Ethernet System – Part 1: General Requirements of Gigabit Ethernet System and Physical and Data-link Layer Requirements of Optical Gigabit Ethernet System*, Geneva: ISO, 2015.

[110] F. Horikoshi, *NWIP on Road Vehicles – In-vehicle Gigabit Ethernet System – Part 3: General Requirements and Test Methods of Optical Gigabit Ethernet Components*, Geneva: ISO, 2015.

[111] IEEE Computer Society, *1722–2016 – IEEE Standard for a Transport Protocol for Time-Sensitive Applications in Bridged Local Area Networks*, New York: IEEE-SA, 2016.

[112] OPEN Alliance, "Non-Profit Alliance Reached a New Milestone: over 300 Members," March 22, 2016. [Online]. Available: http://opensig.org/news/press-releases/. [Accessed May 6, 2020].

[113] IEEE Computer Society, *802.3bp-2016 – IEEE Standard for Ethernet Amendment 4: Physical Layer Specifications and Management Parameters for 1 Gb/s Operation over a Single Twisted Pair Copper Cable*, New York: IEEE-SA, 2016.

[114] IEEE Computer Society, *802.3br-2016 – IEEE Standard for Ethernet Amendment 5: Specification and Management Parameters for Interspersing Express Traffic*, New York: IEEE-SA, 2016.

[115] L. Winkel, M. McCarthy, D. Brandt, G. ZImmerman, D. Hoglund, K. Matheus and C. DiMinico, "10 Mb/s Single Twisted Pair Ethernet Call for Interest," July 26, 2016. [Online]. Available: www.ieee802.org/3/cfi/0716_1/CFI_01_0716.pdf. [Accessed May 24, 2020].

[116] IEEE-SA, "2016 IEEE-SA Ethernet & IP @ Automotive Technology Day," 2016. [Online]. Available: www.aconf.org/conf_70994.2016_IEEE-SA_Ethernet_&_IP_@_Automotive_Technology_Day.html. [Accessed May 6, 2020].

[117] G. Zimmerman, "IEEE 802.3 10 Mb/s Single Twisted Pair Ethernet (10SPE) Study Group Closing Report," November 10, 2016. [Online]. Available: www.ieee802.org/3/minutes/nov16/1116_10M_stp_close_report.pdf. [Accessed May 6, 2020].

[118] S. Carlson, H. Zinner, K. Matheus, N. Wienckowski and T. Hogenmüller, "CFI Multi-Gig Automotive Ethernet PHY," November 9, 2016. [Online]. Available: www.ieee802.org/3/ad_hoc/ngrates/public/16_11/20161108_CFI.pdf. [Accessed May 6, 2020].

[119] IEEE, IEEE 802, "IEEE 802.3 Multi-Gig Automotive Ethernet PHY Task Force," IEEE, March 14, 2020 (continuously updated). [Online]. Available: www.ieee802.org/3/ch/public/index.html. [Accessed May 6, 2020].

[120] IEEE-SA, "2017 IEEE-SA Ethernet & IP @ Automotive Technology Day," October 31, 2017. [Online]. Available: www.strategyanalytics.com/strategy-analytics/webinars-and-events/events/meet-strategy-analytics/2017/10/31/default-calendar/2017-ieee-sa-ethernet-ip-@-automotive-technology-day. [Accessed March 6, 2020].

[121] M. Potts, "Ethernet TSN and Security IEEE 802.1 Automotive Specification Profiles," July 9–13, 2018. [Online]. Available: www.ieee802.org/1/files/public/docs2018/new-autoprof-mpotts-tsn-automotive-profile-0718-v1.pdf. [Accessed December 7, 2019].

[122] IEEE-SA, "2018 IEEE-SA Ethernet & IP @ Automotive Technology Day," IEEE, October 8, 2018. [Online]. Available: https://standards.ieee.org/events/automotive/presentations-2018.html. [Accessed November 23, 2019].

[123] IEEE, IEEE 802, "P802.1DG – TSN Profile for Automotive In-Vehicle Ethernet Communications," 2019 (continuously updated). [Online]. Available: https://1.ieee802.org/tsn/802-1dg/. [Accessed December 7, 2019].

[124] S. Carlson, C. Mash, C. Wechsler, H. Zinner, O. Grau and N. Wienckowski, "10G+ Automotive Ethernet Electrical PHYs Call for Interest," March 12, 2019. [Online]. Available: www.ieee802.org/3/cfi/0319_1/CFI_01_0319.pdf. [Accessed December 7, 2019].

[125] IEEE, IEEE 802.3, "Approved Minutes IEEE 802.3 Ethernet Working Group Plenary Vancouver, Canada," March 11–14, 2019. [Online]. Available: www.ieee802.org/3/minutes/mar19/0319_minutes.pdf. [Accessed December 7, 2019].

[126] D. Martini, *Meeting Minutes of the OPEN Alliance Steering Committee*, Irvine, CA: OPEN Alliance, July 2019.

[127] IEEE, IEEE 802.3, "Approved Minutes IEEE 802.3 Ethernet Working Group Plenary Vienna, Austria," July 15–18, 2019. [Online]. Available: www.ieee802.org/3/minutes/jul19/0719_minutes.pdf. [Accessed December 7, 2019].

[128] C. Pardo, H. Goto, T. Nomura and B. Grow, "Call for Interest Automotive Optical Multi Gig PHY Consensus Presentation," July 2019. [Online]. Available: www.ieee802.org/3/cfi/0719_1/CFI_01_0719.pdf. [Accessed February 3, 2020].

[129] IEEE-SA, "2019 IEEE-SA Ethernet & IP @ Automotive Technology Day," May 2, 2019. [Online]. Available: https://beyondstandards.ieee.org/event/get-ready-for-ethernet-ip-automotive-technology-day-2019/. [Accessed November 23, 2019].

[130] IEEE Computer Society, *802.3cg-2019 – IEEE Standard for Ethernet Amendment 5: Physical Layer and Management Parameters for 10 Mb/s Operation and Associated Power Delivery over a Single Balanced Pair of Conductors*, New York: IEEE-SA, 2019.

[131] RIPE NCC, "The RIPE NCC Has Run Out of IPv4 Addresses," November 25, 2019. [Online]. Available: www.ripe.net/publications/news/about-ripe-ncc-and-ripe/the-ripe-ncc-has-run-out-of-ipv4-addresses. [Accessed April 19, 2020].

[132] IEEE, IEEE 802.3, "Homepage of the IEEE 802.3 Greater than 10 Gb/s Electrical Automotive Ethernet PHYs Task Force," 2020 (continuously updated). [Online]. Available: www.ieee802.org/3/cy/public/jun20/index.html. [Accessed September 7, 2020].

[133] IEEE, IEEE 802.3, "Homepage of the IEEE 802.3 Multi-Gigabit Optical Automotive Ethernet Task Force," 2020 (continuously updated). [Online]. Available: www.ieee802.org/3/cz/public/index.html. [Accessed September 6, 2020].

1 A Brief History of Ethernet (from a Car Manufacturer's Perspective)

1.1 From the Beginning

In 1969, employees at AT&T/Bell Labs developed the first version of Unix. The original intention was to aid the company's internal development of software on and for multiple platforms. Unix had an unpretentious beginning. However, it evolved to be a very widespread and powerful operating system that facilitated distributed computing. Distributed computing requires communication, which is why the success of Unix is closely linked to the history of Ethernet. One important reason for the success of Unix was that, for antitrust reasons, AT&T was neither allowed to sell Unix nor to keep the intellectual property to itself [1]. As a consequence, Unix – in source code form – was shared with everybody interested.

It was especially, but not only, embraced by universities, and the community that evolved provided the basis for the computing environment we are used to today and in which Ethernet has its place. At a time when computing was dominated by large, proprietary, and very expensive mainframe computers few people had access to, Unix created a demand for Local Area Networking (LAN) while at the same time providing an affordable, common platform for developing Unix [2]. As one example, a group at the University of California, Berkeley, created a Unix derivative known as the Berkeley Software Distribution (BSD). It was first released in 1978 and its evolutions became as established as the "BSD-style license" attached to it [3]. Another example is the Transmission Control Protocol (TCP). The first version of this, published in 1974, was implemented for Unix by Stanford University by 1979 [4]. Later, in 1989, the then up-to-date TCP/IP (Industrial Protocol) code for Unix from AT&T was placed in the public domain and thus significantly helped to distribute the TCP/IP Internet Protocol Suite [5].

The advent of Unix represents an important milestone in the early days of computing. It coincides with a time when a significant number of public as well as proprietary research projects were initiated to investigate methods to interchange data locally and at higher speeds than could be provided for by the telephone system [6]. One of the most momentous projects was the one at Xerox PARC. Xerox needed a solution for data transmission between its first personal computer workstations (called "Xerox Alto"), its laser printers, and the early Internet. Thus, Ethernet was invented (1973), patented (1975) [7], and published (1976) [8].

The general opinion (see e.g. [9]) is that the foundation of Ethernet's later success was laid almost as early in time as this, because of the following two choices:

1. **Opening the technology to others:** At the time, it was common for computer companies to try to bind customers to their products by using proprietary technologies or at least restricting competition with the licensing policy of their patents. Xerox held the patents on Ethernet, but there seems to have been an early understanding that they would profit more from the network effects of a widely deployed Ethernet than from selling the technology itself.[1] Seven years after the invention, on September 30, 1980, Xerox published the "DIX Standard" for Ethernet [10] jointly with the Digital Equipment Corporation (DEC) and Intel. They also offered the technology for adoption to the Institute of Electrical and Electronics Engineers (IEEE) 802 group, very shortly after the group had been founded.[2] With several competing technologies being proposed and followed up, it was by no means evident that Ethernet would prevail. But it did, and one of the reasons attributed to this is that Xerox followed a relaxed licensing policy while not trying to dominate the standardization effort [6]. In the authors' view, such an approach substantially supports the proliferation and success of a standard. Unfortunately, it was an attitude as uncommon then as it is today.
2. **Limiting the technical solution to the task at hand:** Ethernet addressed, and still does address, the communication mechanisms needed on the lower 1½ layers of the ISO/OSI layering model only (see also Figure 1.4 in Section 1.2.1), at a time when the ISO/OSI layering model had yet to be completed. It provided a container that gets a packet through a network with multiple participants, but is as independent from the application layer as possible [11]. When new communication systems are being designed today, there is still a tendency to define all layers, often even without adhering to their strict separation. What supposedly allows for "simplifications" and provides the advantages of complete control over the whole communication stack generally makes the system less flexible and less adaptable to future, and hence unknown, requirements. Indeed, Ethernet's adaptability has proven itself to the extent that it has now been successfully introduced in a completely different physical and application environment: Automotive.

In the years that followed, the IEEE became the host for the development of Ethernet. In 1983, IEEE 802.3 published the first of many Ethernet Standards, 10BASE-5 for 10 Mbps over thick coax cable [12]. In the same year, at least 21 companies were mentioned in the trade press to be developing and/or manufacturing Ethernet products [6]. When on January 1, 1984 the AT&T monopoly ended, the existing installed telephone wiring became usable for competing services and applications [13], and a whole new range of possibilities opened to the networking world. Thus, in 1987, SynOptics, a Xerox spin-off, was the first company to prove the feasibility of transmitting Ethernet at 10 Mbps over telephone wire–type cables with a proprietary Ethernet product [6]. The IEEE ratified the respective 10BASE-T standard in September 1990. Because of the many other proprietary versions of Ethernet that had evolved in the meantime, standardization of 10BASE-T was not obvious and required some effort. Nevertheless, when successful, it sealed the victory over other networking technologies in the market [14]. Shortly after, an optical Ethernet version was developed and published as 10BASE-F in 1993.

Meanwhile, the world around Ethernet did not stand still, but continued to provide means and create demands for networking. Various evolutions of TCP and IP were developed. In October 1989, the Internet Engineering Task Force (IETF) published the complete set of protocols in the TCP/IP Internet protocol suite [15, 16]. As mentioned, the success of TCP/IP was fueled by AT&T's public domain implementation of TCP/IP on Unix [5]. In 1991, the Telecommunications Industry Association or TransImpedance Amplifier (TIA) published a standard for inexpensive Unshielded Twisted Pair (UTP) wiring: TIA/EIA-568. Even today, it is hard to imagine an Ethernet network without the 8P8C/RJ-45 connector described in that standard. The World Wide Web was launched in 1994 [17], and the IETF released a specification for IPv4 routers in June 1995 [18]; the Windows 95 Service Pack-1, released on February 14, 1996, automatically included the Microsoft Internet Explorer 2.0 (i.e., built-in TCP/IP networking), bringing the Internet to the masses [19]. Internet Explorer had been available before but had to be purchased separately.

Subsequently, the IEEE amended and enhanced Ethernet, proving Ethernet's adaptability. First, IEEE 802.3 added, and continues to add, new speed grades. Figure 1.1 gives an overview of the increase in data rates for copper and fiber optical channels, in comparison to when similar speed grades were developed for the automotive industry. The largest data rates envisioned today are 40 Gbps for transmission for twisted pair cables and 400 Gbps for optical communication. Figure 1.2 gives an overview of all Ethernet Physical Layer (PHY) variants developed or under development. It is notable that many of the new developments no longer simply increase the previous data rate by a factor of 10, but that the market diversified with many in-between speed grades.

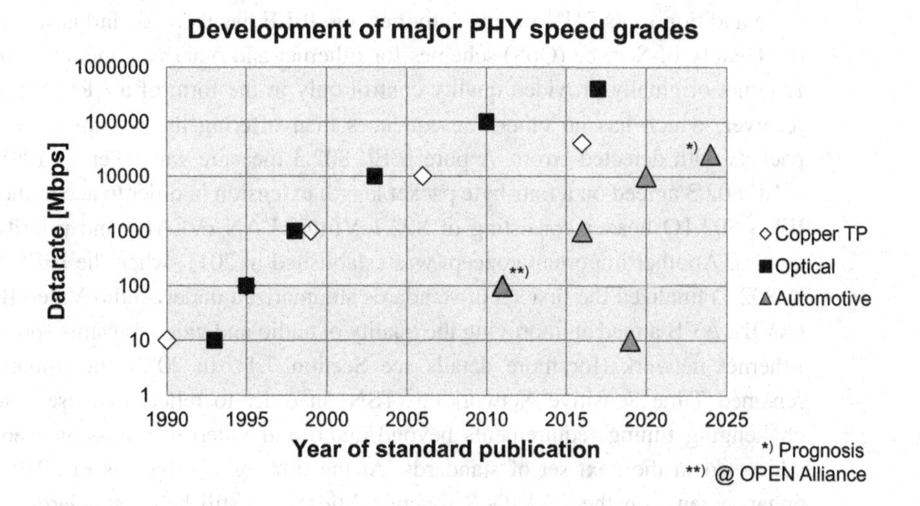

Figure 1.1 Timeline of major PHY speed increases in comparison with automotive speed grades

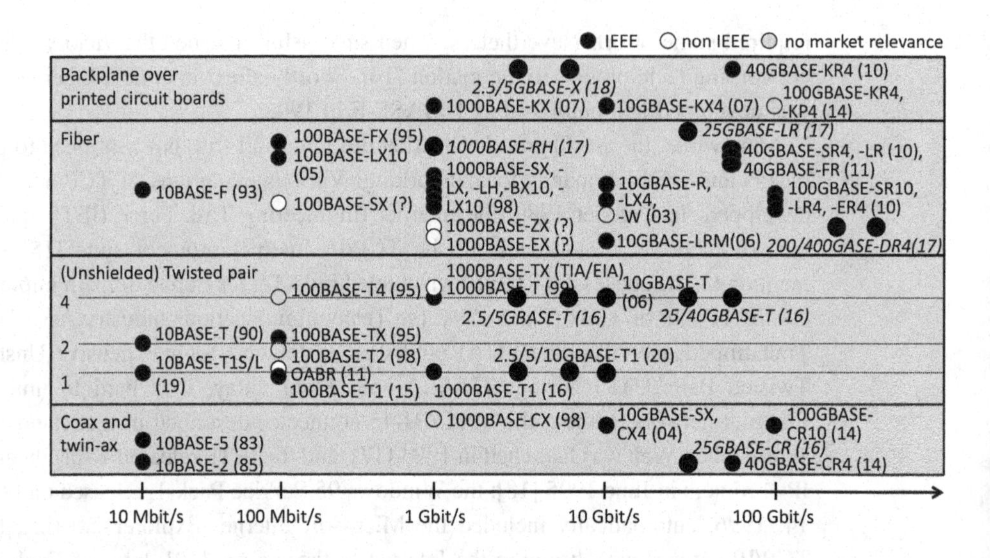

Figure 1.2 2016 overview on Ethernet PHY variants. The (expected) year of release is in brackets, when known [L. Völker, Modified from non-public document, 2011]

Over time, IEEE 802.3 added new functionality and use cases. In 1997, IEEE 802.3 enabled full-duplex communication and flow control to replace the shared media approach prevailing until then. In 1995, IEEE 802.3 added autonegotiation. In 2003, it added Power over Ethernet (PoE), and, in 2010, Energy Efficient Ethernet (EEE). The new use cases include Ethernet in the First Mile (EFM, 2004, see also Section 1.2.2), Ethernet over copper backplane (2007), and finally, in 2013, automotive. The start of work in the automotive space is marked by the establishment of an IEEE 802.3 task force to develop a Reduced Twisted Pair Gigabit Ethernet (RTPGE) (see also Section 5.3.1).

In addition to its PHY-related activities, the IEEE has worked, and is still working on, Quality-of-Service (QoS) schemes for Ethernet and other management functions. Ethernet originally provided quality control only in the form of a CRC check at the receiver, which has no other consequences than offering the possibility to discard packets with detected errors. A pure IEEE 802.3 measure was taken in 1998, when IEEE 802.3 agreed on a four-byte packet length extension in order to accommodate an IEEE 802.1Q header consisting of 802.1 Virtual LAN (VLAN) and priority information. Another important concept was established in 2011, when the IEEE (mainly in 802.1) finalized the first set of standards summarized under Audio Video Bridging (AVB). AVB aimed at improving the quality of audio and video transmissions over an Ethernet network (for more details see Section 7.1). In 2012, the initiative was renamed Time Sensitive Networking (TSN) in order to reflect that use cases with challenging timing requirements beyond audio and video transmission were being addressed in the next set of standards. At the time of writing this in 2020, further enhancements on the AVB/QoS functionalities were still being standardized under TSN at IEEE 802.1.

1.2 The Meaning of "Ethernet"

The term "Ethernet" was first used in 1973, the name referring to the "luminiferous ether" believed by nineteenth-century physicists to be a passive medium between the Sun and Earth, which allows electromagnetic waves to propagate everywhere. The coax used for the inventors' communication system was equally passive and they also intended their data packets to go everywhere [14].

Nevertheless, at first the IEEE did not officially adopt the name (although, unofficially, it did). As an open standards body, the IEEE did not want to give the impression of favoring any company in particular. Despite the fact that Xerox had relinquished their trademark on the name, IEEE 802.3 was instead called "Carrier Sense Multiple Access with Collision Detection (CSMA/CD)" [11]. The official renaming of the IEEE 802.3 efforts into "Ethernet" did not happen until 2007 [20].

As a result, in the various application fields and industries, the name "Ethernet" is used with different meanings, some of which have little in common with what is specified in IEEE 802.3. The following sections thus provide an outline on how different industries use (the term) "Ethernet."

1.2.1 Ethernet in IEEE

Ethernet is standardized in IEEE 802.3 (see Figure 1.3). This comprises the complete PHY and those parts of the Data Link Layer (DLL) that are technology-specific, like the packet format and the medium access method chosen (see also Figure 1.4). Various other aspects also in the IEEE standards (e.g., in IEEE 802.1) affect the implementation of an Ethernet-based communication system. While being relevant, these standards are applicable to all technologies addressed in 802 and are therefore

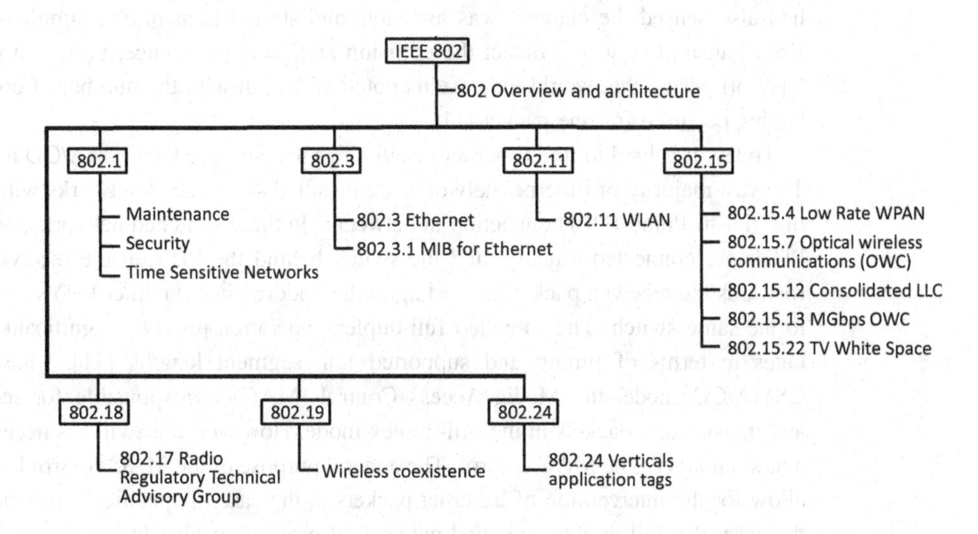

Figure 1.3 IEEE 802 standardization groups active in 2019 [22]

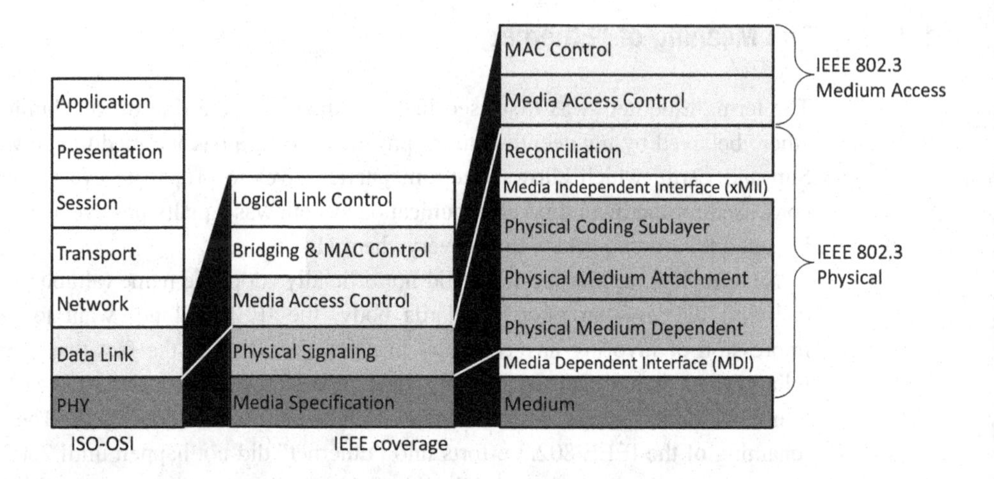

Figure 1.4 Ethernet in IEEE (e.g., [21])

not "IEEE Ethernet"-specific. This is the same for the Logical Link Control (LLC), whose standardization has been concluded in IEEE 802.2 and whose task is to harmonize various methods of medium access toward the network layer [11, 21].

One of the main inventions of the original Ethernet was sharing the media with the help of a CSMA/CD mechanism. CSMA/CD was based on the ALOHA method, which had been developed at the University of Hawaii a few years earlier as a multiuser access method and which more or less simply proposed retransmissions in case collisions were detected [14]. In the case of CSMA/CD, this was enhanced by additionally establishing whether the channel was occupied prior to the start of a transmission. Only if the channel is sensed available is the transmitter allowed to send its packet. Nevertheless, even in this case, collisions can occur, e.g., when another unit had also sensed the channel was available and started transmitting simultaneously. Both transmitters would detect the collision and, in consequence, go into a random back-off period that would increase its potential length with the number of collisions having occurred for one packet [11].

Today, it is hard to find Ethernet installations that still use the CSMA/CD method.[3] The vast majority of Ethernet networks are installed as switched networks with a type of Point-to-Point (P2P) connection in between.[4] In these switched networks, only two PHYs are connected directly, and the switch behind the PHY in the receiving unit forwards the received packets according to their addressing via other PHYs connected to the same switch. The so-called full-duplex[5] operation provides significant advantages in terms of timing and supported link segment lengths [11]. Like in the CSMA/CD mode, the Media Access Control (MAC) is responsible for receiving and transmitting packets in the full-duplex mode. However, the switches necessitated a new sublayer: The MAC control. The general purpose of the MAC control layer is to allow for the interception of Ethernet packets in the case of specific requirements. In the case of a full-duplex switched network, it prevents packet losses in cases where more data is received than can be forwarded, i.e., the MAC control layer enables flow

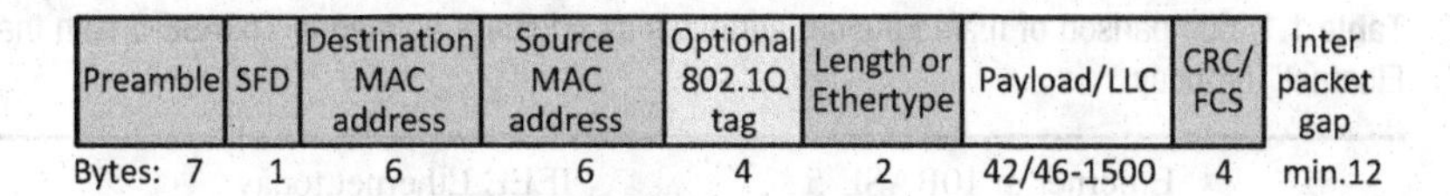

Figure 1.5 Elements of an Ethernet frame/packet

control to prevent packet losses in the case of network congestion, while allowing for limited resources in terms of the buffering and switching bandwidth. The MAC control layer provides the mechanisms that determine when packets may be sent [21].

The most pronounced and stable element of Ethernet is the Ethernet frame/Ethernet packet (see Figure 1.5). The packet starts with a preamble and the Start Frame Delimiter (SFD). These were originally introduced to help synchronize incoming data in the case of CSMA/CD operation. Starting with 100BASE-TX, more complex signal encoding was introduced, which allows for the deployment of special symbols to detect the beginning and end of a packet. So, when CSMA/CD is not used, preamble and SFD are not needed. They are, nevertheless, kept for backward compatibility reasons.

Each Ethernet interface is assigned a unique 48-bit serial number, often referred to as the "MAC address" or the "hardware address."[6] Following the preamble, every packet contains information on where the packet is to be sent and where the packet comes from, using the respective MAC addresses. End node MACs initially only examine the destination address to evaluate whether a packet is intended for this end node (as direct/uni-, multi-, or broadcast). If the packet's destination address matches, the packet is read completely; if the addresses do not match, the packet is ignored. Switch nodes evaluate both the destination address, deciding which port to send the packet to, and the source address, remembering for future incoming packets on which port to find the addressee with that address. This means that there normally is a learning period after start-up in a switched Ethernet network.

The next four bytes represent an optional IEEE 802.1Q tag. The first two bytes identify that this indeed is an 802.1Q header. The remaining two provide the Tag Control Information (TCI) and are divided into three bits for the priority information according to the 802.1p standard, one bit representing the Drop Eligible Identifier (DEI), and 12 bits for the VLAN identifier, which specifies to which VLAN the packet belongs [23]. VLANs represent an important concept for partitioning a physical LAN into various logical domains at layer two (see also Section 7.2).

The next field indicates either the length of the packet or the Ethertype. The Ethertype states what type of data to expect in the payload in respect to the higher layers. It covers content like IP (v4 or v6) or certain AVB packets, but also various proprietary types that have accumulated over time. Ethernet was designed to be a container for whatever data needs to be transmitted; for example, several of the Industrial Ethernet variants – e.g., Profinet, EtherCat, Sercos, Powerlink, High-Speed Ethernet (HSE) – have their own Ethertype (see also Section 1.2.3). The IEEE 802.1Q identifier has the Ethertype 0x8100. A list of Ethertypes is maintained by the IEEE [24]. When the field represents the packet's length, its value is a number

Table 1.1 Comparison of main Ethernet attributes as originally defined in 10BASE-5 with the "IEEE Ethernet" of today

	Ethernet in 10BASE-5	IEEE Ethernet today
Packet	26+12 bytes overhead, 46–1,500 bytes payload	Optional 4 bytes for 802.1Q tag added
Medium access	CSMA/CD	Full-duplex switches with flow control
QoS	Best-effort traffic without acknowledgments	Additionally, 802.1Q, TSN possible
Signaling	Manchester encoding	Various, e.g., PAM-2/3/4/5, DSQ128, NRZ
Media	Coax	TP, fiber, backplane, Twinax

equal to or less than 1500 (see next paragraph). In this case, the IEEE 802.3 LLC protocol can be used to identify the type of data that is being transmitted.

The payload has a minimum size of 42 bytes when the 802.1Q tag is present, and 46 bytes when it is not.[7] If the data being sent is shorter than the minimum payload, then the remaining bytes of the payload are filled with padding. The maximum payload length is 1,500 bytes. Note that the payload represents user data only from a layer two perspective. Various headers from other layers, like the IP or User Datagram Protocol (UDP) headers, will further reduce the bytes available for the actual application.

Finally, the packet is terminated with a Cyclic Redundancy Check (CRC) called the Frame Check Sequence (FCS). The FCS is 32 bits long and checks the integrity of the various bits of the packet (other than preamble and SFD). Following the packet there must be an InterPacket Gap (IPG) of a minimum of 12 bytes. With a fully loaded payload this means that the header/payload efficiency is over 97%.

Table 1.1 provides an overview of the main components of Ethernet and how they have changed over time. As has been visualized in Figure 1.2, Ethernet has been developed for various media, and almost all but the original one are being addressed today. As a result of higher data rates and advancements in signal processing, the physical signaling has changed with the media and has also been standardized in various forms. The original media access mechanism CSMA/CD is seldom used today. Nevertheless, the principle that Ethernet performs no quality control in form of acknowledgments or retransmits, as well as its "container" function, has been kept. If needed, retransmits have to be initiated on higher layers. Likewise, the Ethernet packet has remained almost unchanged, with only the addition of the optional 802.1Q header.

1.2.2 Ethernet in Telecommunications

The telecommunication providers laid many of the foundations for today's Information Age. Among other things, they were involved in the development of computers, in how to use them (see Section 1.1), and in physically connecting companies and households to the (telecommunication) network, which is still fundamentally important in our networked society today. It thus is surprising that in respect to the adoption of Ethernet and IP the telecommunication industry seems belated and struggling to keep up [25]. However, for a very long time, enabling voice communication between two parties at different physical locations was the main service provided by telecommunications companies. From the invention of telephony in the middle of the nineteenth century [26] to acknowledging the need for changes, the twenty-first century had to arrive [27]. By now, 2020, these changes have become more than just adaptations. Telecommunication providers are in active transition to completely abandoning the original, voice-oriented principles of communication – i.e. circuit-switched/Time Division Multiplex (TDM) communication – for packet-switched IP traffic [28–31]. To explain the developments that led to this and the relationship to Ethernet, a (simplified) distinction is made between the following communication areas: The user domain, the access technologies, and the core telecommunication networks (see also Figure 1.6).

For years, the focus of telecom providers was on improving voice communication. They invested in better voice quality and better coverage. They enabled more simultaneous calls and more connections between the continents and became more cost efficient with improved automation in call switching. The carrier switches that manage the calls between two different subscribers represented a key element in the network and thus were first to go digital. AT&T introduced the first digital carrier switch in 1965 [13]. Ericsson and Siemens developed their commercial and very successful digital switching products AXE and "Elektronisches Wählsystem Digital" (EWSD) in

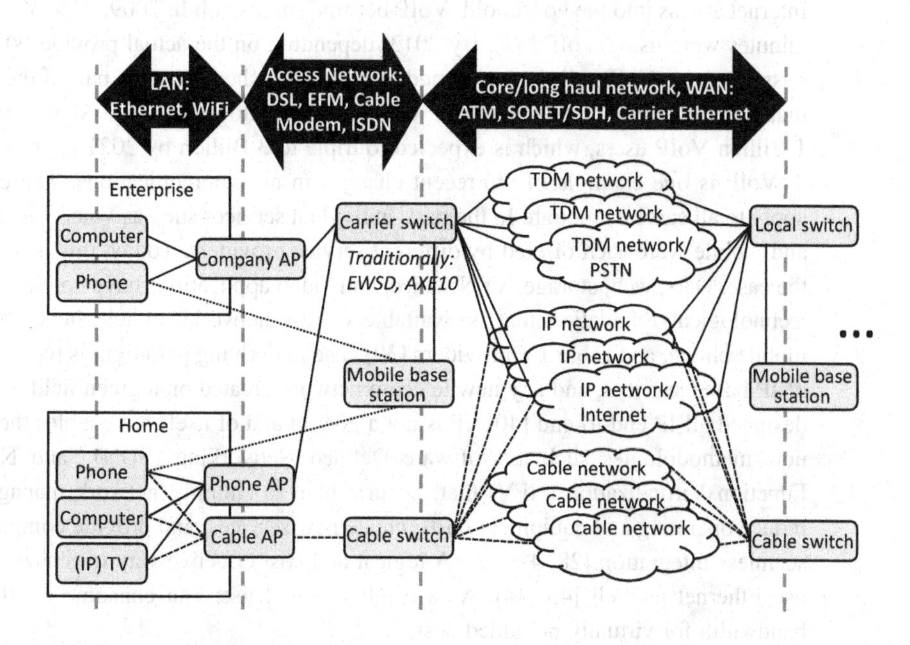

Figure 1.6 Simplified diagram of the relation between the different elements in telecommunications

the late 1970s [32, 33]. All the while, on the subscriber side the communication stayed analog.

The Information Age was thus not driven by the digitization of the Public Switched Telephone Network (PSTN) but by the US military and the computer industry and their desire to share data, which eventually led to the establishment of the Internet. These are the important milestones in this process:

1. The development and publication of IPv4 and TCP as RFCs in 1981 [34, 35].
2. The already mentioned break-up of the AT&T monopoly on January 1, 1984 [13].
3. The creation of the IETF in 1986.
4. The invention of the World Wide Web in 1989/90 [17].
5. The decommissioning of the military run Advanced Research Projects Agency Network (ARPANET).
6. The deregulation of the Internet for commercial use in 1990.

The Internet changed everything. Not only the way people work, live, and communicate but also the paradigm that telephony/voice communication needs to be circuit switched. The application that proved this was Voice over IP (VoIP). There had been very early experiments that showed that voice could be transmitted with data packets. However, the real start of VoIP was in 1995, when the first commercial product allowed real-time, full-duplex voice communication [36]. VoIP first took root in international calls, where the cost savings for users were most significant. The technology improved with the rapidly increasing number of calls and their management in the network became more powerful. With the proliferation of broadband Internet access into the household, VoIP became successful. In 2009, 25% of all voice minutes were using VoIP [37]. By 2013, depending on the actual provider(s), it was possible that a VoIP call was converted to circuit switched for some part of the way or that a regular circuit-switched call was changed into VoIP. The year 2017 saw 1 Billion VoIP users, which is expected to triple to 3 Billion by 2021 [38].

VoIP is one example of the recent changes in telecommunications. The changes apply to all services though. In the past, individual services such as voice, video, data, and mobile were each offered by different service providers. Today, this is no longer the case. Data, web, storage, VPN, voice, and video applications have converged on a technological foundation that is available on the networks of telephone, Internet, mobile phone, and cable TV providers [39]. The underlying paradigm is the provision of IP-based services, and any new telecom network created on a green field would be designed all-IP end-to-end [40]. IP is not a goal in and of itself but enables the use of new methodologies such as Software Defined Networking (SDN) and Network Function Virtualization (NFV) that, in turn, help to simplify network management, reduce operating and equipment costs, create new services, and provide complete and seamless integration [28, 41, 42]. A logical and cost effective way to realize IP is to use Ethernet as well [43, 44]. As a result, the end user can consume hundredfold bandwidth for virtually no added costs.

To deploy Ethernet at the user level (see also Figure 1.6) is simple. Ethernet was primarily designed for enterprise LANs and from there it spread into homes.

Enterprise LANs and consumer devices (like PCs, printers, etc.) thus represent the original Ethernet market. The developments specific to telecommunications are "Ethernet in the First Mile (EFM)" for the access network and "Carrier Ethernet"[8] for the core network.

EFM is an IEEE 802.3 effort, which resulted in the IEEE 802.3ah standard released in September 2004. EFM focuses on two aspects relevant for access networks: Longer link distances and additional diagnostic monitoring capabilities. The latter is necessary as – and this is different in the case of LANs – the service provider is likely to be located at a significant physical distance from the subscriber's potential problem with the network [21]. The "Far-end Operation, Administration, Management (OAM)" of the standard's objectives thus includes remote failure indication, remote loopback, and link monitoring.

For the physical links, optical transmission offers higher capacities and longer reaches. As there are nevertheless significant regional differences [21], the EFM specification describes 14 different PHYs. These are divided into three main categories: Optical Point-to-Point (P2P) for 10 km@100 Mbps/1 Gbps, optical Point-to-Multipoint (P2MP) for 10/20 km@1 Gbps (enhanced to 10 Gbps with IEEE 802.3av in 2009), and electrical ("copper") P2P for 0.75 km@10 Mbps or 2.7 km@2 Mbps. The P2MP version, also called "Ethernet Passive Optical Network (EPON)," was added for efficiency reasons, albeit requiring a special adaptation in the (multipoint) MAC Control. It either eliminates the need to lay long individual links per user or the need for a complex switch [45]. Of the many Passive Optical Network (PON) variants, the 10 Gbps EPON is expected to hold a revenue share of about 20% in 2022 [46]. Optical broadband access experienced a huge growth since the first edition of this book. It grew from 17% of the worldwide 624 million broadband connections in 2012 [47] to 65% of the worldwide 1.084 billion broadband connections in 2019 [48].

"Carrier Ethernet" uses native Ethernet links (typically high-speed fiber) to build the physical Wide Area Network (WAN) infrastructure as a replacement for the traditional carrier technologies like Optical Transport Network (OTN), Synchronous Optical NETworking (SONET)/Synchronous Digital Hierarchy (SDH). IP routers are used to convey packets across the network. On the other hand, traditional carrier technologies (SONET, etc.) have all been adapted to convey IP packets. Ethernet can be tunneled on top of IP to provide end-to-end Ethernet services for organizations that need to have remote networks operate as a single, large layer two network.

To achieve the needed scalability, resiliency, and manageability in the core network a variety of layer two principles from IEEE 802.1 and 802.3 can be adopted. Examples are VLANs, prioritization, the OAM defined for EFM, MAC addressing, switching, link aggregation, as well as variants of the Spanning Tree Protocol [43]. Another important concept is label switching. The IETF is publishing specifications for Multi Protocol Label Switching (MPLS), which allows adding a label to an IP packet with a predefined tunnel, that is, a predefined path that routes a packet in accordance to the rules of the network technology the packet is going over [43].

The Metro Ethernet Forum (MEF), an organization dedicated to the deployment of Carrier Ethernet, provides a suite of specifications suitable for the implementation of [49] and certification for Carrier Ethernet. "Carrier Ethernet CE 2.0," launched by the MEF at the beginning of 2012 [50], further improves the options available to providers when implementing Carrier Ethernet by increasing the number of services and their manageability over interconnected provider networks. Market research companies see Carrier Ethernet growing continuously alongside IP traffic in the network (see e.g. [51]).

With respect to the automotive industry, the layout of Carrier Ethernet seems to have little in common with the requirements in cars, where the links are short – even compared with standard LAN Ethernet – and where the traffic is predictable. However, a number of technologies developed for Carrier Ethernet are also useful in automotive applications, e.g., perfect QoS, fault detection, time synchronization, resiliency, security. When not looking at in-vehicle networking but at the car as an additional node in the networked world, the all-prevailing IP determines the services the car needs to be connected to. Various automotive-specific applications in terms of diagnostics, etc. are envisioned and the telecommunications industry has to have the capacity to support these, in most cases wirelessly.

1.2.3 Ethernet in Industrial Automation

Communication in industrial automation is generally structured hierarchically (see also Figure 1.7). The lowest level of communication happens between sensor or actuator and the low-level controller [52, 53]. The amount of data transmitted with every cycle can consist of a few bits only. Nevertheless, the communication needs to be cost efficient and the response time short. Cycle times for tasks like motion control can be much less than 1 ms, with a synchronization accuracy within 1 μs [54]. At a

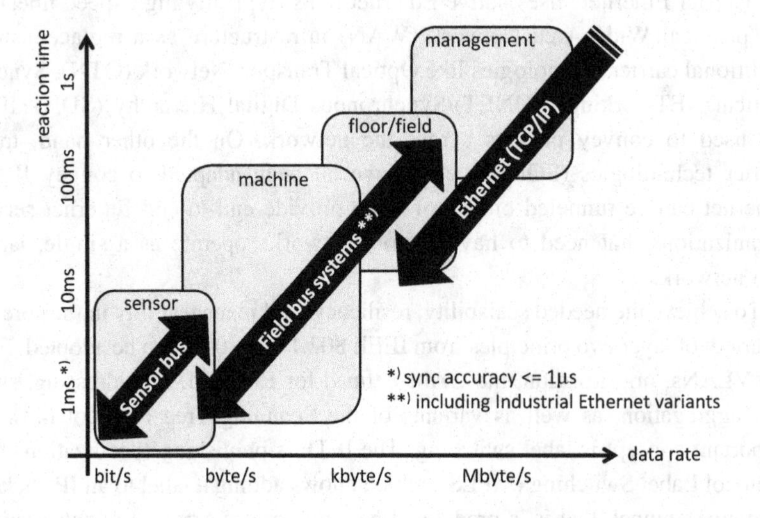

Figure 1.7 Hierarchical approach to factory communication [54, 63]

machine level, more intelligent field devices like I/O stations, operator panels, and Programmable Logic Controllers (PLCs) exchange data. For most tooling machines or remote Input/Output (I/O) a response time of below 10 ms is required. At the floor (or "field") level, automation and operator stations communicate with PCs. A response time of 100 ms is sufficient for activities like process monitoring and, thus, most processes in process automation and building control [54]. Often the floor level is subdivided into smaller "cells" and larger "areas." This allows the separation of critical from not-so-critical cells and, in the case of issues, enables them to be isolated as well as repaired without affecting the whole operation. At the highest management levels, orders, reports, quality statistics, etc. are handled. The requirements for the reaction time here are less critical, while the packet size and amount of data increase.

The process of industrialization is the foundation of wealth in occidental society. Hence, right from the beginning of the industrial revolution, efforts have been made to improve and optimize production processes. Naturally, the possibilities of computerization were explored from the early days and the foundations for hierarchical communication were laid in the early 1970s. After the 1960s had brought a number of inventions impacting industrial manufacturing – mini computers, robots, computer/Numerically-Controlled (NC) machines, and especially PLCs – there was a need for efficient communication between the units as well as the possibility for decentralizing their control [55]. Decentralization improved the quality and availability of process observation and control. Additionally, it unburdened the central computer. Furthermore, it removed the need to use a star topology, and thus reduced the amount of cabling [56]. The first commercially available distributed computer control systems were introduced by Honeywell and Yokogawa in 1975 [57].

The rest of industry followed and in the 1980s every company in the automation business seemed to have developed their own "fieldbus"[9] system in order to support the respective communication in manufacturing plants. The large number of fieldbus variants (more than 50 [58]) nevertheless did not appeal to customers. In the case of technical problems, manufacturing plant owners need immediate access to replacements – potentially from a different vendor – to minimize the risk and impact of downtimes. In response to this need, suppliers published their specifications [59], which helped to establish fieldbus systems in industrial automation. All the way until 2017, fieldbus connected nodes represented the majority of new networking nodes in industrial plants [60]. Efforts toward standardization were made. The outcome of those efforts is, however, a double-edged sword: When the International Electrotechnical Commission (IEC) finally adopted its IEC 61158 standard on 31 December 2000, it contained no less than 18 variants [54]. To have perfectly fitting solutions for different use cases with some interoperability was obviously more important and more advantageous than to have a single solution that covers all [61]. The respective standardization efforts in IEEE (802.4) were disbanded in 2004 [62].[10]

Fieldbuses can fulfill very small reaction time requirements (see also Figure 1.7). Investigations into the use of fieldbus technologies showed that it is advantageous to use a single technology [64]. Nevertheless, many publications mention the additional use of separate sensor busses for cost reasons (e.g., [63]). On top, the standard

Ethernet and TCP/IP are used to integrate the management level, which makes it three technologies at minimum. The desire for seamless communication over all hierarchy levels and parts of the production process for complexity and cost reasons is easy to understand, and this made Ethernet, being part of the system anyway, an obvious choice. The trend toward the Internet of Things (IoT) added to the suitability of Ethernet and IP [62].

Standard Ethernet and TCP/IP nevertheless started out being non-deterministic, with reaction times above 100 ms, though there have always been simple means to reduce this, such as using UDP instead of TCP or restricting the traffic load in local sections of the network. Taking these steps can reduce Ethernet's reaction time to 10 ms or less, covering a significant number of applications in industrial automation [53].

To make Ethernet (even) more suitable for real-time applications and to fulfill various additional requirements on robustness, functional safety, high availability, and security combined with low latency, "Industrial Ethernet"[11] solutions were developed. Figure 1.8 shows the different concepts behind them. In the simplest case, a protocol specifically catering for time-critical use cases is used on the application layers ("Industrial 1"). The next option ("Industrial 2") is to have the time-critical traffic bypass the IP and TCP/UDP layers and to directly communicate with the data link layer. With this, the reaction time can apparently be shortened down to 1 ms [53]. This bypass concept is also used in IEEE 802.1 AVB and is described in more detail in Section 7.1.

In the last variant depicted ("Industrial 3") the data link layer is redefined in order to accommodate the real-time requirements directly in the MAC. This implies the most significant changes that might even affect the implementation in hardware down to the PHY. Even though the aviation industry does not reuse any of the Industrial Ethernet variants for the communication between avionics systems in an aircraft (see also Section 1.2.4), Figure 1.8 depicts the "Aviation" structure for comparison purposes.

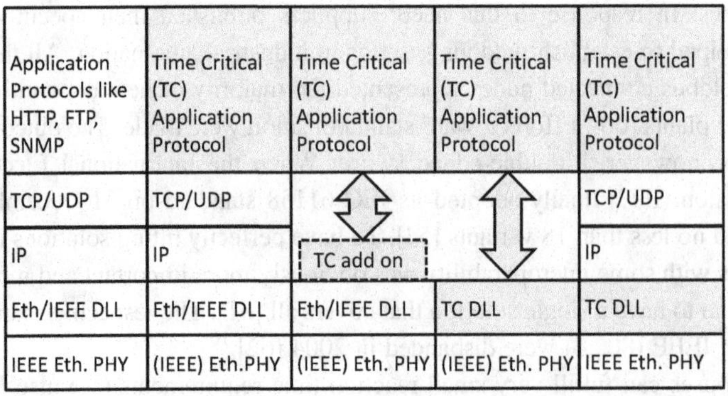

IT	Industrial 1	Industrial 2	Industrial 3	Aviation
Application Protocols like HTTP, FTP, SNMP	Time Critical (TC) Application Protocol	Time Critical (TC) Application Protocol	Time Critical (TC) Application Protocol	Time Critical (TC) Application Protocol
TCP/UDP	TCP/UDP			TCP/UDP
IP	IP	TC add on		IP
Eth/IEEE DLL	Eth/IEEE DLL	Eth/IEEE DLL	TC DLL	TC DLL
IEEE Eth. PHY	(IEEE) Eth.PHY	(IEEE) Eth. PHY	(IEEE) Eth. PHY	IEEE Eth. PHY

Figure 1.8 Conceptual real-time variants in Industrial Ethernet [53, 54, 65]

Table 1.2 Market share and volume of industrial networking nodes between 2011 and 2015 [67]

Technology	Number of installed nodes in industrial networks in 2011 (millions)	Projected number of installed nodes, 2015 (millions)	Average number of new nodes per year (millions)	Expected growth, CAGR (%)
Ethernet TCP/IP	3.46	5.40	0.49	11.4
Industrial Ethernet	3.81	6.42	0.65	14.0
Overall Ethernet	7.26	11.82	1.14	12.9
Overall fieldbus	24.04	33.28	2.31	8.5
Overall	31.30	45.10	3.45	9.6

The Aviation structure is, in the end, just another version of "Industrial 3." One of the basic principles behind almost all Industrial Ethernet versions is that the "IT" part of the communication is used for best-effort traffic and that Standard Ethernet hardware is used for the PHY. Note that special variants of cabling and connectors are generally always used with Industrial Ethernet, for robustness in the physically harsh environment of industrial manufacturing [54].

The authors are not aware of any interest being expressed by the automotive industry in adopting Industrial Ethernet's non-standard parts of the DLL when discussing the introduction of Ethernet.

Industrial Automation not only uses a large number of fieldbus variants but also various incompatible types of Industrial Ethernet. Twenty-nine versions of real-time Ethernet are listed in Laboratory of Process Data Processing of Reutlingen University [66], and seven are mentioned with respect to their market share in the Industrial Ethernet Book [67]. From an automotive perspective, this is surprising, because many incompatible networking technologies result in additional costs and overheads. Even if the costs for the networking technologies are of lower priority in industrial automation than in the automotive industry, it is of high priority for industrial automation customers to be able to obtain replacement units from different vendors to carry out any necessary maintenance and repair work. If all vendors use the same networking technologies, this should also be easier to achieve. Potentially, the industrial automation customers are too diverse, and it is only possible in smaller groups, like the Automation Initiative of German Automobile Manufacturers (AIDA), to request uniform solutions.

Table 1.2 shows example market data for industrial automation networking technologies comparing the industrial networking nodes installed in 2011 with the nodes installed in 2015. The market shift from fieldbus solutions toward Ethernet solutions is clearly visible in the installed nodes (as well as in the new nodes [60]). Also, the diversity in the solutions is decreasing: 2019 market research lists fewer technologies than comparable 2015 research (compare e.g., [68] with [69]). However, a clear shift toward "standard Ethernet with TCP/IP" or even just toward one Industrial Ethernet standard is not visible, also not when comparing the 2019 data [69].

There are efforts ongoing to change this situation. In order to be able to use IEEE standardized Ethernet better in Industrial Automation, various activities have been and are still being initiated and supported in IEEE. For one, requirements from Industrial Automation were the main driver for the new standard activities of TSN in IEEE 802.1 (see also Section 5.1.4). Then IEEE 802.3 concluded in 2016 its specification on Interspersing Express Traffic (IET)\IEEE802.3br [70], a provision to further reduce latency in an Ethernet network by interrupting the transmission of lower priority packets by high priority ones.

Last but not least, the combined interest of stakeholders in industrial automation and the automotive industry led to the standardization of a 10 Mbps Ethernet PHY at IEEE 802.3 in order to be able to extend the Ethernet network also to those domains, which require lower data rates [71]. This effort – officially started in 2016 – resulted in the finalization of the IEEE 802.1cg standard for a long reach and a short reach PHY in 2019 [72]. The long reach PHY, 10BASE-T1L, supports link lengths of up to 1 km, and therefore targets the industrial automation environment. The short reach PHY, 10BASE-T1S, is usable in both the automotive industry as well as industrial automation industries. Stakeholders from industrial automation also drove the effort to enhance the capabilities of 10BASE-T1S further in a new IEEE 802.3 project [73].

Together with the TSN efforts, all this, at least theoretically, allows Ethernet in industrial automation to cover the vast majority of use cases with standard IEEE Ethernet solutions.

1.2.4 Ethernet in Aviation

In a passenger plane, four areas of communication can be distinguished by their different networking requirements: (1) Communication between the avionics systems; (2) operational communication for avionics system and cabin communication (maintenance, configuration, ...); (3) air–ground communication; and (4) passenger communication as part of the in-flight entertainment (IFE). The IFE (4) and maintenance and configuration (2) can potentially use a variety of existing networking technologies from other industries, including Ethernet, and will not be discussed further. The air–ground communication (3) is not relevant in the context of this book. Even though (3) is special in the sense that it requires a frequency band to use and worldwide harmonization, standardization has long been realized. As early as 1929, when commercial air traffic emerged, the Aeronautical Radio, Inc. (ARINC) was established for this purpose and started with coordinating the air–ground communication.

The communication area with very specific requirements and of interest for "Aviation Ethernet" is the communication between avionics systems (1). Today, the ARINC also hosts the development and publication of various standards relevant in this area [74]. One of the most commonly used ARINC specifications for the communication between avionics systems is the ARINC 429, which was first released in April 1978 [74, 75]. ARINC 429 allows for the simplex, i.e., unidirectional communication at either 12.5 kbps or 100 kbps over STP cabling with a word size of 32 bits overall,

19 of which represent the data area. One transmitter can be connected to up to 20 receivers. Nevertheless, all receivers wanting to respond to the transmitter require a separate ARINC 429 link. Star, bus, or mixed topologies are in principle all possible, but as transmitter and receiver need to be directly connected (per transmission direction) any slightly more complex communication needs quickly lead to a significant amount of wiring. As commercial planes heavily rely on their avionics systems – the first all-electronic fly-by-wire system was introduced into commercial airplane service in 1988 [76] – the cost and weight of the wiring is considerable.

There were thus several reasons to introduce Ethernet-based communication systems in aviation: Allow for larger content per packet, allow for higher data rates, and allow for a modular architecture with standardized components (Integrated Modular Avionics [77]) and flexibility in the network in order to optimize aircraft design. It was of interest to be able to share resources, support the increased interdependencies of avionics systems, reduce hardware costs through the use of less wiring and, by basing the system on Commercial-Off-The-Shelf (COTS) technologies, reduce cabling weight and allow for the integration/reuse of existing communication technologies outside the plane (especially IP and UDP), while at the same time addressing real-time requirements (see e.g. [78]).

Boeing was the first to introduce an avionics system with Ethernet-based communication adapted to the rigors of the environment [79]. The first Boeing 747-400 with the respective advanced flight deck display system was delivered in May 2000 [78]. Boeing also introduced Ethernet in the Boeing 777 for non-critical systems at less than 10 Mbps. Airbus developed the Avionics Full-Duplex Switched Ethernet (AFDX), which saw its maiden flight with the commercial aircraft A380 in April 2005. AFDX was published in two ARINC standards: ARINC 664 part 2 "Ethernet Physical and Data-Link Layer Specification" and ARINC 664 part 7 "Avionics Full-Duplex Switched Ethernet (AFDX) network." AFDX (see also Figure 1.8) relies on standard IEEE 802.3 Ethernet PHYs, but uses them with cabling suitable for the specific environment (which in the case of planes means shielded cables with special temperature resilience). The DLL is changed and adapted, in order to achieve the necessary timing and reliability requirements.

In AFDX, logical communication is based on so-called Virtual Links (VLs) that restrict the communication depending on an identifier. VLs do not define the LAN inside which the traffic can go unrestricted, as is the case for VLANs. They define, as the name indicates, the communication partners for every communication in the system. The MAC addressing fields in the AFDX packets are used in a specific way to accommodate this. AFDX thus (also) allows emulation of the communication defined for ARINC 429, while at the same time profiting from the flexibility and reduced cabling of a switched Ethernet network. An additional control is being performed at every receiver by evaluating the VL identifier and the assigned bandwidth of incoming packets against the preconfigured values [80]. For the bandwidth allocation a Bandwidth Allocation Gap (BAG) is defined per VL. The BAG restricts the amount of traffic that can be transmitted in a specific interval and thus has a sort of traffic shaping function.

Another property of AFDX is the support of redundancy. Redundancy is achieved by the use of sequence numbers and the installation of two parallel links for every physical connection and every VL. At the receiver only the first packet to arrive with a specific sequence number is processed. Should a redundant packet arrive at a later time, it is ignored [80]. Furthermore, careful system design assesses upfront whether the network – with all latencies, jitters, and buffer sizes – will allow the expected communication to pass without congestion related delay or loss.

The described efforts show that also the aviation industry is interested in the (cost) advantages that can be realized with the standardization of non-differentiating functions like networking and the reuse of solutions successful in other industries. As an industry, it nevertheless has extremely harsh safety and security requirements: Avionics systems require a "Design Assurance Level" A [81], with less than one error in 10 billion hours airborne time [82], and the respective certification from the authorities. This sometimes impedes the reuse of technologies[12] and is the main reason why the overlap with, e.g., Industrial Ethernet is very small. At the same time, of the industries discussed in the context of Ethernet in this chapter, aviation is the one with the smallest number of manufacturers (see also Table 1.3). This can make standardization easier, but as every manufacturer is used to being a very powerful customer, it can make them less likely to compromise. As a consequence, all of the aircraft manufacturers listed in October 2019 as using Ethernet-based ARINC 664 part 7 solutions (Airbus, ATR, Boeing, Bombardier, Comac, Irkut, Learjet, Sukhoi, plus the helicopter vendor AgustaWestland [83]) are likely to use a specific adaptation of the standard.

1.2.5 Automotive Ethernet

The previous sections have presented the various uses of Ethernet in different industries. Every industry has kept some parts of the original IEEE Ethernet but has amended or changed others according to the particular needs and the structure of the respective market.

Automotive relates to the other industries in various ways. First, the automotive industry is one of the biggest customers in industrial automation [84]. Second, the car is going to be another node in the global telecommunication network. While the first is of no concern for this book, the second relates to Automotive Ethernet in that seamless connectivity between the car and "the rest of the world" provides an attractive outlook into relevant future use cases. From a technical viewpoint, Automotive Ethernet is yet something different.

The focus of Automotive Ethernet is on in-vehicle networking (IVN), i.e., the communication between the various Electronic Control Units (ECUs) inside the car. For this, the automotive industry would like to reuse as much existing technology as possible, over all protocol layers. At the same time, the industry produces close to 80 million cars per year (e.g., 2017 [85]), with every car containing an increasing number of ECUs that need to be connected. This means that the market volume for Ethernet transceivers is big enough to justify the development of a special PHY in

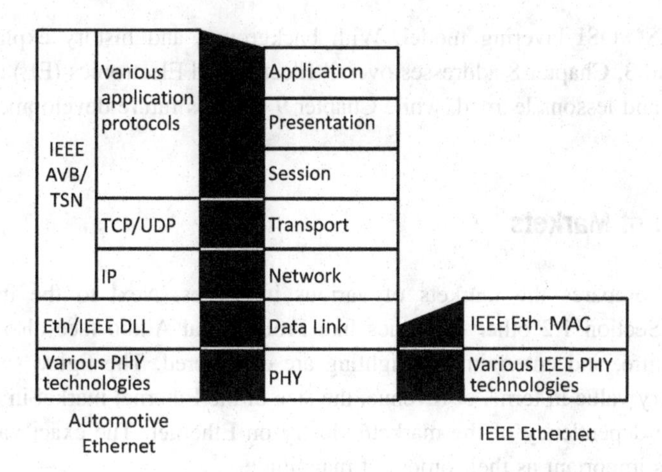

Figure 1.9 Simplified protocol stack of Ethernet-based communication in the automotive industry ("Automotive Ethernet") in comparison

order to meet the automotive requirements while reducing the costs to a level that makes the technology attractive for the industry. At the time of this writing, five Ethernet PHY technologies usable in the automotive industry had come out of this effort (in order of completion): 100BASE-T1, 1000BASE-T1, 1000BASE-RH, 10BASE-T1S, and 2.5, 5, and 10GBASE-T1 (see also Chapters 3, 4, and 5). While this can justify calling those PHY technologies "Automotive Ethernet," we think it is not sufficient and that Automotive Ethernet entails more.

Figure 1.9 shows the simplified ISO/OSI layer model of "Automotive Ethernet." The descriptions of the use of Ethernet in other industries indicated it: It is almost impossible to limit the explanations to just the PHY and the MAC. Automotive Ethernet – or more correctly Ethernet-based communication in the automotive industry – covers all layers of the ISO/OSI layering model. A very important motivation for the industry to use Ethernet is that protocols for all levels are available and that the industry can select the appropriate solution for each layer. Additionally, the clearly defined interfaces between the layers allow the development of new protocols for individual layers while reusing protocols for the rest. Just as IP and Ethernet provide the basis for fundamental changes in the telecommunications market, Ethernet holds a similar potential in the automotive industry. The physical layer technologies are just one aspect of it. The real change comes from new possibilities in the realm of software development, E/E-architecture and the like that have more to do with the higher layers of the communication system.

It is the goal of this book to explain how using Automotive Ethernet affects all the different aspects of in-vehicle networking. Chapters 4 and 5 explain impacts related to the Physical Layer, including different PHY variants. Chapter 7 discusses the protocols used on layers two to seven: Section 7.1 for AVB, Section 7.2 for VLANs, Section 7.3 for IP, Section 7.4 for the application layers (i.e., middleware), and Section 7.5 for security. But, this book shows also that Automotive Ethernet goes beyond the ISO/OSI layering model. With background and history explained in Chapters 2 and 3, Chapter 8 addresses overall Electric and Electronics (EE) architecture, tooling, and lessons learned, while Chapter 9 outlines future developments.

1.3 Comparison of Markets

Figure 1.10 compares the markets of various industries. Next to the industries discussed in Section 1.2 other industries like Professional Audio (see also Section 7.1), architecture, and entertainment lighting are considered. The figure shows the overall industry value in terms of revenue, the size of the Ethernet market in terms of ports, and the dependency of the market/industry on Ethernet. The exact values are thereby not as important as their orders of magnitude.

It can be seen that, with respect to Ethernet ports, there are two groups: Those with clearly less than 10 million (Mio) ports per year and those with more than 100 Mio ports per year. Naturally, those markets with smaller volumes rather reuse/have to reuse the PHY hardware, as it is more difficult to justify the development of a specific PHY technology.[13] They have to make sure that industry specific adaptations can be realized with help of software and special Application Specific Integrated Circuits (ASICs). For Industrial Automation and Aviation, e.g., significant changes are therefore realized on MAC layer and above.

This is different for the markets with large volumes. Telecommunications, e.g., is a prime application area for the optical PHY solutions that have been specifically developed to cater to the long range requirements of the industry. At the same time,

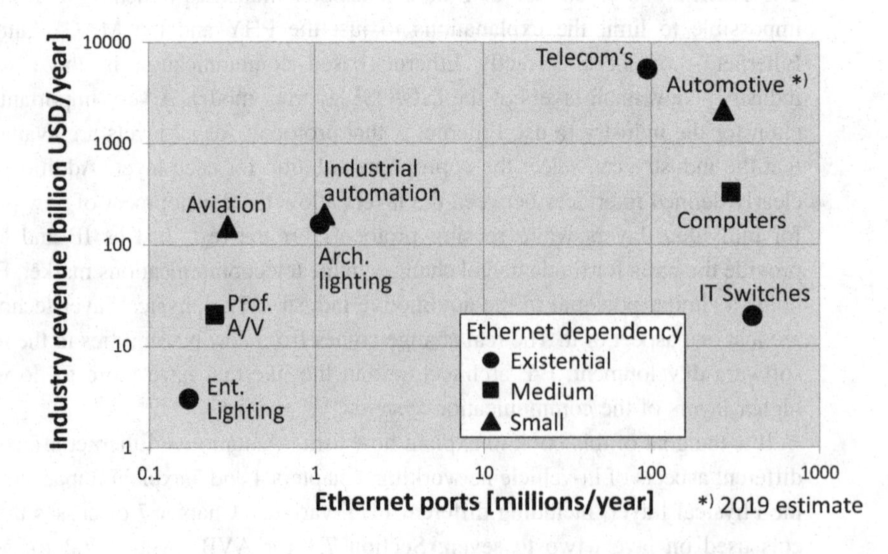

Figure 1.10 Ethernet ports, overall market revenue, and dependency of the market on Ethernet for different industries (mainly year 2012 data, see Table 1.3 for references)

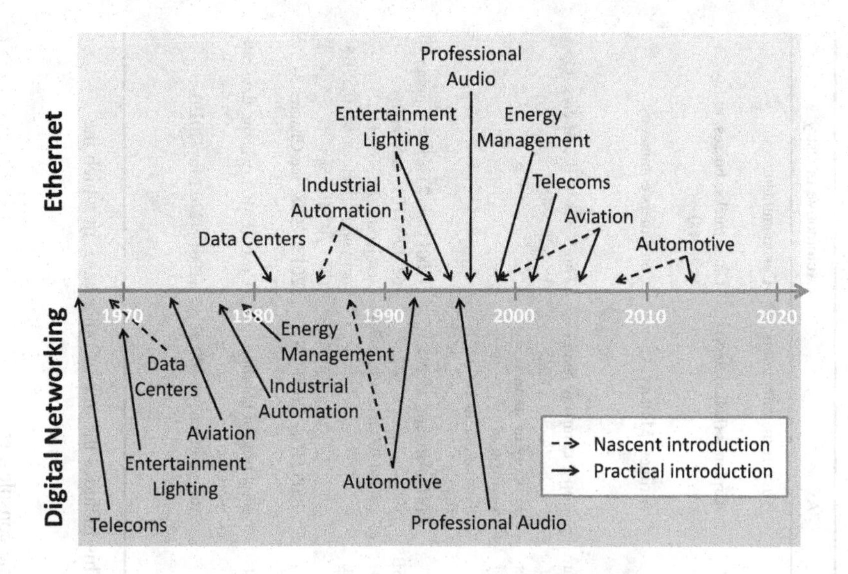

Figure 1.11 Timeline of introducing digital networking versus introducing Ethernet in different industries

Carrier Ethernet requires specific management functions, which are realized with IEEE 802.1 protocols or others found in the specifications of the MEF. The automotive industry also justifies the development of specific PHYs that meet the cost and performance targets, while Computers and IT Switches represent the original market for IEEE 802.3 PHY specifications anyway.

Figure 1.11 depicts when different industries introduced digital networking compared to when they introduced Ethernet. Evidently, the automotive industry was the latest industry to introduce Ethernet and that was about 20 years after starting with digital networks. When comparing this with other industries, timing does not seem to matter though. For some industries like telecommunications or aeronautics this gap is even larger. Other industries (e.g. data centers) thrived directly with Ethernet.

Table 1.3 gives an overview of some of the specific properties of the various Ethernet-using industries. As can be seen, the specifics of these industries are quite diverse. The differences start with the type of product for which Ethernet is being used. It can be a consumer good/service or an investment good to produce consumer goods and services. The end-products' values can vary between a few hundred to a few million dollars. The volume in which it is produced can vary between a few thousand to hundreds of millions per year. Next, the market for the Ethernet parts might be centralized around very few customers or consist of a multitude that sell the end-product. Some are heavily dependent on Ethernet (like switches), for others it is a small part of their product (like in cars or planes). The most astounding observation thus is, that, despite the differences in requirements and circumstances, Ethernet is a solution that works for all.

Table 1.3 Main properties of different Ethernet market segments

	Data centers	Consumer products	Telecommunications	Industrial automation	Aviation	Automotive industry
Implementer*	LAN equipment manufacturer	Consumer electronics manufacturers	Communication providers (carriers)	Industrial Automation suppliers	Aircraft manufacturers	Car manufacturers
Product	LAN equipment	(Networked) CE like computers, printers, ...	Communication services	Industrial automation equipment	Airplanes (helicopters)	Cars (trucks, busses, utility vehicles)
Customer	All entities/ companies with a LAN	Consumers/ companies	Consumers/companies	All manufacturers, building operators	Airlines, Military, ...	Consumers/companies
Industry structure	Top 1 holds >60%, Top 5 share >80% of Ethernet LAN switches [86]	Top 5 share >50% of the computer** market [87], but tablets are taking over [88]	Top 12 share 50% of mobile subscribers [89]; overall 798 mobile network operators [90]	~Top 11 companies share 50% of the market, ~50 share 90% of the market [53]	Top 2 company share >60% (incl. military, duopoly in some segments), 10 manufacturers >90% [91]	Top 5 sell >50% of cars, top 18 sell >90%
Market 2012, revenue in Billion USD (B$) or pieces	[64] 19.8B$ Ethernet LAN switches	[92] Computer sales 329B$ [93] microprocessors for PCs 40B$; (@ 352.7Mio PCs [87])	[94] Telecom service 2200B$; [51] carrier Eth. equip. 34B$; (@ 6.8B mobile, 1.2 B. fixed line subscribers, 750 Mio Internet households [95])	[84] Indus. automation supplier 200B$; [96] 75B$ of which electronics; [97] 2.6B$ wireline industrial networking	[91] Aircrafts 150B$; [98] avionics 6.3B$ (@ ~1500 aircrafts [99])	[100] Cars 2130B$, supplier 631B$; [101] automotive semicond. 25.5B$; [102] networking semicond. 0.55B$ (@ ~70 Mio cars)
Year of Ethernet intro	1981 [6]	1981 [6]	2000 [28]	~2000 [59] (1985 [103])	2005 [104] (2000 [78])	2013 (2008, see Chapter 3)
Key Eth. Requirements	Original Ethernet use	Original Ethernet use	Long reach, management, QoS	Short response time and reliability	Real time, reliability, weight	Costs (EMC), weight, data rate
Ethernet market in ports	>400 Mio (2012) [105]	Rough estimate***: 300 Mio, but decreasing	95 Mio (2017) [51].	~1.14 Mio (2012) [67]	Rough estimate: 300 T/y	[106] >270 Mio (2019)

* "Implementer" is the one most likely to drive the decisions on the networking technology used. This includes the one or other case, in which the implementer's customer makes the choice.

** Computers represent just one of a variety of products so that overall there are more vendors.

*** Most desk-based and mobile PCs [87], share of DSL routers for homes [47], some printers, game consoles, etc.

Notes

1 In 1980, Robert Metcalfe, the author of the Xerox internal memo that first mentioned Ethernet, presented what became the so-called Metcalfe's Law. This states that the value of a telecommunications network increases proportionally with the square of the number of compatible communication devices [110]. It can thus be expected that the thinking behind this was used to promote the idea of a networking technology that can be used by more than one company to build products.

2 During the celebration of 40 years of Ethernet at the IEEE 802 plenary meeting in November 2013 in Dallas, Robert Metcalfe stated that it had been on their lawyers' advice that the DIX group opened up the standard and later offered it to the IEEE. Restricting the technology to DEC, Intel, and Xerox only would have violated antitrust laws.

3 The same CSMA/CD mechanism is still used in environments where the channel is always shared, such as in the wireless communication of IEEE 802.11 WLANs (sometimes disconcertingly referred to as "wireless Ethernet"). Some sharing (but not CSMA/CD) was reintroduced with Ethernet in the First Mile (EFM), for which link distances of up to 20 km were defined. In one use case, this included the possibility of sharing the link without switches in order to allow for cost-efficient installation (see also Section 1.2.2). Full sharing capabilities were reintroduced with IEEE 802.3cg/10BASE-T1S. This uses a mechanism called PHY Level Collision Avoidance (PLCA) that makes use of some of the CSMA/CD principles but is much more bandwidth-efficient on the shared, so-called multidrop, link (see also Section 5.4.3.2). After initial resistance in IEEE 802.3, PLCA became so interesting that a separate study group for multidrop enhancements was set up in 2019 [107].

4 First and foremost, Point-to-Point (P2P) communication identifies a direct communication between two units. Depending on the communication aspect under discussion (ISO/OSI layer, physical versus logical, etc.), this can have completely different meanings. Historically, the expression P2P was used for a data link with exactly two end points that were not even part of a network and on which no data or packet formatting was performed [108]. Generally, such communication was based on an RS-232 interface or similar, which coincidentally is also something the car industry considered in the very early days of in-vehicle networking (see also Section 2.2.1). At the other end of the spectrum, P2P also denotes the communication between two end nodes in the one-to-one communication of a telephone call, independent of the physical network layout. While the RS-232–like communication means P2P on the physical layer, in the case of a telephone call it is P2P only at the application layer.

In the case of Ethernet communication, P2P can mean the communication between two PHYs in a switched network on the physical layer (layer one). No matter how many hops there are between two end units or how many end units there are for each hop, exactly two nodes are physically connected P2P. It is not P2P on layer one when Ethernet uses a shared medium with, e.g., CSMA/CD. However, P2P in Ethernet can also be used to identify unicast communication, where a packet is addressed from one sender to exactly one receiver (in contrast to multicast or broadcast, where one packet is addressed to multiple or all receivers). Unicast then identifies a P2P communication at MAC level (layer two); independent of whether the data is transmitted over a switched or shared network. This extends to higher layers such as IP or the applications used on top of the Ethernet basis.

In the context of the in-vehicle networking discussed in this book, we use P2P to denote the communication on layer one and the lower parts of layer two. P2P requires that there be a direct physical connection between exactly two units and that those two units be unambiguously identified as the next communication partners, so that they do not have to directly share the bandwidth with anybody else. This is the case in a solely switched Automotive Ethernet network but is also the case for a one-to-one communication over, e.g., LVDS/pixel links/SerDes links or USB (see also Sections 2.2.6 and 2.2.7). This is clearly different from the case of LIN and CAN where more than two nodes can be physically connected and the medium is shared. This is also different from MOST or FlexRay where, often, only two units are directly physically connected. Nevertheless, this is at layer one only. MOST and FlexRay share the data rate of the link between all participants on the network.

5 The term "full-duplex" and its counterpart "half-duplex" are also somewhat ambiguous, as the terms can mean different things depending on the context and the ISO/OSI layer considered. In principle, "full-duplex communication" refers to communication in which transmission and reception can happen simultaneously, and "half-duplex communication" refers to a communication in which either transmission or reception is possible.

If a medium is shared, regardless of whether it is Ethernet using CSMA/CD or any other sharing system, it is always half-duplex. Only one unit can transmit, while all other units connected listen. If more than one unit transmits at the same time on the shared link, as would be possible for two units in a full-duplex system, a collision occurs and the transmission is unsuccessful. It is therefore correct to say that CSMA/CD Ethernet operation is half-duplex. However, at IEEE 802.3 the terms CSMA/CD Ethernet and half-duplex are often used synonymously, while half-duplex also applies to different sharing methodologies, e.g., the multidrop PLCA mechanism of 10BASE-T1S or even the physical medium realization of 100BASE-TX.

The situation is also complex for full-duplex operation as it is necessary to consider both the PHY technology and the ISO/OSI layer. For example, 100BASE-TX uses one pair of wires for transmission and one pair for reception. On each wire pair the traffic is always uni-directional, i.e., there is no simultaneous transmission or reception on one pair. However, on the MII interface between PHY and MAC level, or even in the processing of the PHY, transmission and reception can happen simultaneously. At IEEE 802.3, 100BASE-TX is therefore a full-duplex technology, even if it is "dual simplex" at PHY level.

In the automotive industry, the terminologies are slightly different. "Sharing" the media means a "bus system" is used. If it is not a bus system that is used, it is either a "switched system" (meaning Automotive Ethernet) or some "direct P2P" technology such as SerDes or USB. The most relevant aspects for differentiation are whether a media is shared by more than two units or not and, if it is shared, how. In the context of this book, the terms half-duplex and full-duplex are thus used with care. They will not be a direct placeholder for indicating if Ethernet is operating in CSMA/CD mode or not but for cases in which the (in)capability to simultaneously receive and transmit at PHY level on the same medium is relevant.

6 MAC addresses do not identify only Ethernet interfaces but are also deployed by interfaces of other IEEE communication technologies and of technologies standardized by other organizations. Examples of the latter are FDDI or ATM. The MAC address originates from the original Xerox Ethernet addressing scheme. Today, MAC address assignments are coordinated by IEEE-RA [109].

7 This is also a legacy from CSMA/CD operation, where packets needed to be long enough in order to process a collision.

8 The extension "Carrier" originates in the jargon for telecommunications network providers: "Common Carriers." "Carrier Ethernet," as such, is more the name for a market segment than one specific technology.

9 It seems that the name "fieldbus" was used for a while before meaning and definitions were added to it [56]. Most simply, a fieldbus connects units "in the field," i.e., units distributed on a factory floor, which can have dimensions large enough to be fieldlike.

10 Industrial automation represents a safety critical environment with challenges different from what can be found in data centers. So, when the computerization and local networking reached the automation industry, other aspects needed to be standardized in addition to the (LAN/fieldbus) communication technology. The programming of the PLCs needed to be standardized (IEC 61131). The installation of the communication networks was (and still is) a challenge (IEC 61918). Not only is the environment not always friendly with respect to temperature, vibration, dirt, acids, etc., it is also necessary to keep wiring changes to a minimum when the line is changed to produce a different product. Other topics like redundancy (e.g., IEC 62439), the development of safety critical systems (e.g., IEC 61508), functional safety (e.g., IEC 61511), security for industrial automation and control systems (e.g., IEC 62443), parameterization, and diagnosis are also important.

11 "Industrial Ethernet" is an expression so commonly used that it seems to defy the need for an unambiguous definition. It generally refers to the case where some or even all elements of Ethernet as defined in IEEE 802 are (re)used in an industrial environment for tasks directly related to the manufacturing process, fulfilling exactly those additional requirements for robustness, high availability (e.g., ring redundancy), functional safety, cyber security, etc. The expression was apparently introduced with "Profinet" [L. Winkel, email correspondence, August 22, 2013].

12 It was, e.g., a major concern for the automotive industry when standardizing FlexRay that its reuse in aviation could lead to liability issues for the car industry.

13 The efforts to standardize a 10 Mbps Ethernet version for industrial automation (10BASE-T1L at IEEE 802.3 [71] seem to contradict the statement that, for markets with small volumes, it is not worthwhile to develop a special PHY. There are two aspects to consider. For one, the value per port can, of course, make a difference. The special qualification needed for industrial adds to the value of the product. Second, Ethernet in industrial is seeing a significant upward trend [60]. Having the right PHY can additionally accelerate this and thus increase the market size. However, even a threefold increase of the market for Industrial Ethernet would still be small in comparison to markets of other industries. However, in the end, it is up to the silicon vendors to decide about market opportunities.

References

[1] M. Lasar, "The UNIX Revolution – Thank You, Uncle Sam?" *arstechnica*, July 19, 2011. [Online]. Available: http://arstechnica.com/tech-policy/2011/07/should-we-thank-for-feds-for-the-success-of-unix/. [Accessed May 6, 2020].

[2] B. Montague, "Why You Should Use a BSD Style License for Your Open Source Project," freebsd, May 17, 2013. [Online]. Available: www.freebsd.org/doc/en/articles/bsdl-gpl/index.html. [Accessed May 7, 2020].

[3] Wikipedia, "Berkeley Software Distribution," Wikipedia, May 3, 2020. [Online]. Available: http://en.wikipedia.org/wiki/Berkeley_Software_Distribution. [Accessed May 6, 2020].

[4] V. G. Cerf, "Final Report of the Stanford TCP Project," April 1, 1980. [Online]. Available: ftp://ftp.rfc-editor.org/in-notes/ien/ien151.txt. [Accessed May 7, 2020].

[5] Wikipedia, "Internet Protocol Suite," Wikipedia, April 28, 2020. [Online]. Available: http://en.wikipedia.org/wiki/Internet_protocol_suite. [Accessed May 6, 2020].

[6] U. von Burg and M. Kenny, "Sponsors, Communities, and Standards: Ethernet vs. Token Ring in the Local Area Networking Business," *Industry and Innovation*, vol. 10, no. 4, pp. 351–375, December 2003.

[7] R. M. Metcalfe, D. R. Boggs, C. P. Thacker, and B. W. Lampson, "Multipoint Data Communication System (with Collision Detection)," US Patent 4,063,220, March 31, 1975.

[8] D. Boggs and R. Metcalfe, "Ethernet: Distributed Packet Switching for Local Computer Networks," *Communications of the ACM*, vol. 19, no. 7, pp. 395–405, July 1976.

[9] S. Carlson, "Fast, Simple, Reliable and Cheap: Ethernet at 40," in 3rd Ethernet & IP @ Automotive Technology Day, Leinfeld-Echterdingen, 2013.

[10] Digital Equipment Corporation, Intel Corporation, Xerox Corporation, "The Ethernet, A Local Area Network. Data Link Layer and Physical Layer Specifications, Version 1.0," September 30, 1980. [Online]. Available: http://ethernethistory.typepad.com/papers/EthernetSpec.pdf. [Accessed May 6, 2020].

[11] C. E. Surgeon, *Ethernet: The Definite Guide*, Sebastopol, CA: O'Reilly, February 2000.

[12] Wikipedia, "IEEE 802.3," Wikipedia, April 28, 2020. [Online]. Available: http://en.wikipedia.org/wiki/IEEE_802.3. [Accessed May 6, 2020].

[13] AT&T, "Milestones in AT&T History," 2004. [Online]. Available: www.thocp.net/companies/att/att_company.htm. [Accessed May 6, 2020].

[14] R. M. Metcalfe, "The History of Ethernet," December 14, 2006. [Online]. Available: www.youtube.com/watch?v=g5MezxMcRmk. [Accessed May 6, 2020].

[15] R. Braden, "RFC 1122: Requirements for Internet Hosts – Communication Layers," October 1989. [Online]. Available: http://tools.ietf.org/pdf/rfc1122.pdf. [Accessed May 6, 2020].

[16] R. Braden, "RFC 1123: Requirements for Internet Hosts – Application and Support," October 1989. [Online]. Available: tools.ietf.org/pdf/rfc1123.pdf. [Accessed May 6, 2020].

[17] P. Barford, "The World Wide Web," University of Wisconsin, September 11, 2008. [Online]. Available: http://pages.cs.wisc.edu/~pb/640/web.ppt. [Accessed May 6, 2020].

[18] F. Baker, "RFC 1812: Requirements for IP Version 4 Routers," June 1995. [Online]. Available: http://tools.ietf.org/pdf/rfc1812.pdf. [Accessed May 6, 2020].

[19] Microsoft, "Verfügbarkeit von Microsoft Windows 95 Service Pack 1-Komponenten," February 14, 2006. [Online]. Available: https://support.microsoft.com/de-de/kb/142794. [Accessed November 16, 2016, no longer available].

[20] IEEE 802 LMSC, "Agenda & Minutes (Unconfirmed) – IEEE 802 LMSC Executive Committee Meeting (updated 24 September 2007)," July 20, 2007. [Online]. Available: www.ieee802.org/minutes/jul2007/Minutes%20-%20Friday%20July%2020%202007.pdf. [Accessed May 6, 2020].

[21] W. W. Diab and H. M. Frazier, *Ethernet in the First Mile*, New York: IEEE, 2006.

[22] IEEE 802, "IEEE 802 LAN/MAN Standards Committee," IEEE, 2019 (continuously updated). [Online]. Available: www.ieee802.org/. [Accessed December 19, 2019].

[23] IEEE Computer Society, *802.1Q-2011 – IEEE Standard for Local and Metropolitan Area Networks – Media Access Control (MAC) Bridges and Virtual Bridge Local Area Networks*, New York: IEEE, 2011.

[24] IEEE, "Ethertype," constantly updated. [Online]. Available: http://standards.ieee.org/develop/regauth/ethertype/eth.txt. [Accessed May 7. 2020].

[25] J. J. Río, A. v. Maltzahn, and J. Ong, "The Future of Telecoms: New Models for a New Industry," February 2012. [Online]. Available: https://leslivresblancs.fr/livre/filieres-specialisees/tic/future-telecoms-new-models-new-industry. [Accessed May 7, 2020].

[26] Wikipedia, "History of the Telephone," May 4, 2020. [Online]. Available: http://en.wikipedia.org/wiki/History_of_the_telephone. [Accessed May 7, 2020].

[27] US Department of Transportation, "Telecommunications Handbook for Transportation Professionals: The Basics of Telecommunications: Chapter 1. Telecommunication

Basics," 3 January 3, 2005. [Online]. Available: www.ops.fhwa.dot.gov/publications/telecomm_handbook/chapter1.htm. [Accessed May 7, 2020].

[28] S. Cherry, "The End of the Public Phone Network," December 19, 2012. [Online]. Available: http://spectrum.ieee.org/podcast/telecom/internet/the-end-of-the-public-phone-network. [Accessed May 7, 2020].

[29] Wikipedia, "Geschichte des Telefonnetzes," April 7, 2020. [Online]. Available: https://de.wikipedia.org/wiki/Geschichte_des_Telefonnetzes. [Accessed May 7, 2020].

[30] J. Mitchell, "100 Years in the Making: The End of PSTN and ISDN Networks," Claranet Limited, February 22, 2019. [Online]. Available: https://insight.claranet.co.uk/improving-communications/the-end-of-pstn-and-isdn. [Accessed December 20, 2019].

[31] A. Kuch, "Telekom: ISDN Abschaltung dauert noch bis 2020," Congstar, June 29, 2019. [Online]. Available: www.teltarif.de/telekom-isdn-abschaltung-2020/news/77131.html. [Accessed December 20, 2019].

[32] J. Armstrong, "A History of Erlang," May 21, 2009. [Online]. Available: http://web.archive.org/web/20090521100504/http://www.cs.chalmers.se/Cs/Grundutb/Kurser/ppxt/HT2007/general/languages/armstrong-erlang_history.pdf. [Accessed May 7, 2020].

[33] Wikipedia, "EWSD," April 1, 2020. [Online]. Available: http://en.wikipedia.org/wiki/EWSD/EWSD. [Accessed May 7, 2020].

[34] Information Sciences Institute University of Southern California, "Internet Protocol," September 1981. [Online]. Available: http://tools.ietf.org/html/rfc791. [Accessed May 6, 2020].

[35] Information Sciences Institute University of Southern California, "Transmission Control Protocol," September 1981. [Online]. Available: http://tools.ietf.org/html/rfc793. [Accessed May 6, 2020].

[36] iLocus, "The 10 That Established VoIP," September 2007. [Online]. Available: https://de.scribd.com/document/275142588/The-10-That-Established-VoIP. [Accessed May 24, 2020].

[37] B. Injaz, "How Does VoIP work," 2013. [Online]. Available: http://de.slideshare.net/badr212/how-does-voip-work. [Accessed May 7, 2020].

[38] N. Gibert, "71 Key VoIP Statistics: 2019 & 2020 Data Analysis & Market Share," FinancesOnline, probably 2019. [Online]. Available: https://financesonline.com/voip-statistics/. [Accessed 20 December 2019].

[39] Wikipedia, "Next Generation Network," May 6, 2020. [Online]. Available: http://de.wikipedia.org/wiki/Next_Generation_Network. [Accessed May 7, 2020].

[40] R. Bennett and S. Nurenberg, "Transforming a Modern Telecom Network – From All-IP to Network Cloud," in *Building the Network of the Future*, Boca Raton, FL: CRC Press, 2017, pp. 9–24.

[41] US Department of Transportation, "Telecommunications Handbook for Transportation Professionals: The Basics of Telecommunications: Chapter 10. The Future," 2005. [Online]. Available: www.ops.fhwa.dot.gov/publications/telecomm_handbook/chapter10.htm. [Accessed May 6, 2020].

[42] J. Donovan and K. Prabhu, "The Need for Change," in *Building the Network of the Future*, Boca Raton, FL: CRC Press, 2017, pp. 1–8.

[43] C. G. Gruber and A. Autenrieth, "Carrier Ethernet Transport in Metro and Core Networks," September 28, 2008. [Online]. Available: www.netmanias.com/ko/post/cshare/5063. [Accessed May 6, 2020].

[44] M. Ghodrat, "Carrier Ethernet Improves IP Service Delivery," October 2008. [Online]. Available: www.lightwaveonline.com/articles/print/volume-25/issue-10/technology/carrier-ethernet-improves-ip-service-delivery-54889962.html. [Accessed May 7, 2020].

[45] G. Pesavento, "Ethernet in the First Mile; Point to Multipoint; Ethernet Passive Optical Network Tutorial," July 9, 2001. [Online]. Available: www.ieee802.org/3/efm/public/jul01/tutorial/pesavento_1_0701.pdf. [Accessed May 7, 2020].

[46] A. Diana, "Next-Gen PON $7B Market by 2022," August 1, 2018. [Online]. Available: www.broadbandworldnews.com/author.asp?section_id=548&doc_id=745018. [Accessed December 21, 2019].

[47] POINT topic, "World Broadband Statistics," June 2012. [Online]. Available: http://point-topic.com/wp-content/uploads/2013/02/Sample-Report-Global-Broadband-Statistics-Q2-2012.pdf. [Accessed May 7, 2020].

[48] POINT topic, "World Fixed Broadband Statistics – Q2 2019," June 2019. [Online]. Available: http://point-topic.com/free-analysis/world-fixed-broadband-statistics-q2-2019/. [Accessed December 21, 2019].

[49] Metro Ethernet Forum, "MEF Technical Specifications," Metro Ethernet Forum, 2020 (continuously updated). [Online]. Available: www.mef.net/carrier-ethernet/technical-specifications. [Accessed May 6, 2020].

[50] P. Marwan, "Carrier Ethernet 2.0 ist an den Start gegangen," February 28, 2012. [Online]. Available: www.zdnet.de/41560473/carrier-ethernet-2-0-ist-an-den-start-gegangen/. [Accessed May 6, 2020].

[51] Infonetics, "Carrier Ethernet Market to Top $39B by 2017," May 10, 2013. [Online]. Available: www.businesswire.com/news/home/20130510005674/en/Infonetics-Carrier-Ethernet-Market-Top-39B-2017. [Accessed May 7, 2020].

[52] W. Lawrenz, *CAN Controller Area Network Grundlagen und Praxis*, 2nd edition, Heidelberg: Hürthig, 1997.

[53] L. Larsson, "Fourteen Industrial Ethernet Solutions Under the Spotlight," September 2005. [Online]. Available: www.iebmedia.com/index.php?id=4811&parentid=74&themeid=255&hft=28&showdetail=true&bb=1&PHPSESSID=38a2saj9cvll4kaguj5m14c4j5. [Accessed May 7, 2020].

[54] M. Felser and T. Sauter, "Standardization of Industrial Ethernet – the Next Battlefield?," *Proc. IEEE 5th International Workshop on Factory Communication Systems,* pp. 413–421, September 22, 2004.

[55] S. Djiev, "Industrial Networks for Communication and Control," July 2009 (likely). [Online]. Available: https://data.kemt.fei.tuke.sk/SK_rozhrania/en/industrial%20networks.pdf. [Accessed May 6, 2020].

[56] T. Sauter and M. Wollschlaeger, "Feldbussystems – Historie, Eigenschaften und Entwicklungstrends," *Informationstechnik und Technische Informatik*, vol. 42, no. 4, pp. 7–16, 2000.

[57] Wikipedia, "Distributed Control System," November 11, 2020. [Online]. Available: http://en.wikipedia.org/wiki/Distributed_control_system. [Accessed November 28, 2020].

[58] G. Färber, "Feldbus-Technik heute und morgen," *Automatisierungstechnische Praxis,* no. 11, pp. 16–36, 1994.

[59] M. Felser and T. Sauter, "The Fieldbus War: History or Short Break between Battles?," *4th IEEE International Workshop on Factory Communication Systems*, August 28, 2002.

[60] T. Carlsson, "Industrial Ethernet Is Now Bigger Than Fieldbuses," February 16, 2018. [Online]. Available: www.anybus.com/about-us/news/2018/02/16/industrial-ethernet-is-now-bigger-than-fieldbuses. [Accessed December 21, 2019].

[61] Contemporary Controls, "Dick Caro – One of the Most Influential Persons in the Field of Industrial Networking – Part 2," 2006. [Online]. Available: www.ccontrols.com/pdf/Extv6n3.pdf. [Accessed May 6, 2020].

[62] IEEE 802, "Re: [802SEC] Disbanding of 802.4 and 802.9," IEEE 802, July 30, 2004. [Online]. Available: http://ieee802.org/secmail/msg05489.html. [Accessed May 7, 2020].

[63] Schneider Electric, "Chapter 9: Industrial Networks," [Online]. Available: www.schneider-electric.hu/documents/automation-and-control/asg-9-industrial-networks.pdf. [Accessed July 22, 2013, no longer available].

[64] B. Vogel-Heuser, Thomas Tauchnitz, Wilfied Schmieder, Sven Seintsch, Manfred Diets, Frans van Laak, Harry van Rijt, Jürgen George, and Thomas Kasten, "*FuRIOS: Fieldbus and Remote I/O System Comparison*," Mannheim: Pepperl&Fuchs GmbH, 2002.

[65] J.-P. Thomesse, "Fieldbus: Industrial Network Real-Time Network," 2005. [Online]. Available: http://etr05.loria.fr/slides/vendredi/ETR2005_Thomesse.pdf. [Accessed July 22, 2013].

[66] Laboratory of Process Data Processing of Reutlingen University, "Information about Real-Time Ethernet in Industry Automation," Laboratory of Process Data Processing of Reutlingen University, 2008. [Online]. Available: www.realtime-ethernet.de/. [Accessed May 7, 2020].

[67] Industrial Ethernet Book, "The World Market for Industrial Ethernet," April 2012. [Online]. Available: www.iebmedia.com/index.php?id=8595&parentid=74&themeid=255&hpid=2&showdetail=true&bb=1&appsw=1&sstr=world_market. [Accessed May 7, 2020].

[68] Automation.com, "EtherNet/IP Leads Industrial Network Market Shares," February 10, 2016. [Online]. Available: www.automation.com/automation-news/industry/ethernetip-leads-industrial-network-market-shares. [Accessed December 21, 2019].

[69] A. Jacobsen, "Industrial Network Market Shares 2019 According to HMS," May 7, 2019. [Online]. Available: www.hms-networks.com/news-and-insights/news-from-hms/2019/05/07/industrial-network-shares-2019-according-to-hms. [Accessed December 21, 2019].

[70] IEEE Computer Society, *802.3br-2016 – IEEE Standard for Ethernet Amendment 5: Specification and Management Parameters for Interspersing Express Traffic*, New York: IEEE-SA, 2016.

[71] L. Winkel, M. McCarthy, D. Brandt, G. ZImmerman, D. Hoglund, K. Matheus, and C. DiMinico, "10Mb/s Single Twisted Pair Ethernet Call for Interest," July 26, 2016. [Online]. Available: www.ieee802.org/3/cfi/0716_1/CFI_01_0716.pdf. [Accessed May 24, 2020].

[72] IEEE Computer Society, *802.3cg-2019 – IEEE Standard for Ethernet Amendment 5: Physical Layer and Management Parameters for 10 Mb/s Operation and Associated Power Delivery over a Single Balanced Pair of Conductors*, New York: IEEE-SA, 2019.

[73] P. Jones, C. Jones, G. Zimmerman, D. Brandt, D. Temblay, and L. Yseboodt, "10SPE Multidrop Enhancements Call for Interest Consensus Presentation," July 2019. [Online]. Available: www.ieee802.org/3/cfi/0719_03/CFI_03_0719.pdf. [Accessed December 21, 2019].

[74] Avionics Interface Technologies, "ARINC 429 Protocol Tutorial," Avionics Interface Technologies, date unknown. [Online]. Available: http://aviftech.com/files/2213/6387/8354/ARINC429_Tutorial.pdf. [Accessed July 14, 2013, no longer available].

[75] J. Wittmer, "ARINC 429 Ein Avionik Feldbus der zivilen Luftfahrt," Berner Fachhochschule, January 23, 2007. [Online]. Available: https://prof.hti.bfh.ch/uploads/media/ARINC_429.pdf. [Accessed July 14, 2013].

[76] Condor Engineering, *AFDX Protocol Tutorial*, Santa Barbara, CA: Condor Engineering, 2005.

[77] C. B. Watkins and R. Walters, "Transitioning from Federated Avionics Architectures to Integrated Modular Avionics," *Trans. of Digital Avionics Systems Conference*, October 21, 2007.

[78] J. W. Ramsey, "Boeing's 767-400ER: Ethernet Technology Takes Wing," May 1, 2000. [Online]. Available: www.aviationtoday.com/av/commercial/Boeings-767-400ER-Ethernet-Technology-Takes-Wing_12652.html. [Accessed May 6, 2020].

[79] L. Buckwalker, *Avionics Databusses*, 3rd edition, Leesburg: avionics.com, 2005.

[80] AIM GmbH, "AFDX Workshop," March 2010. [Online]. Available: www.afdx.com/pdf/AFDX_Training_October_2010_Full.pdf. [Accessed May 6, 2020].

[81] RTCA, *Software Considerations in Airborne Systems and Equipment Certification, DO-178C*, Washington: RTCA, 2011.

[82] Joint Aviation Administration, "ADVISORY MATERIAL JOINT – AMJ 25.1309," 1998. [Online]. Available: www.mp.haw-hamburg.de/pers/Scholz/materialAFS/AMJ25.1309-Change15.pdf. [Accessed September 14, 2013, no longer available].

[83] Wikipedia, "Avionics Full-Duplex Switched Ethernet," Wikipedia, October 7, 2019. [Online]. Available: http://en.wikipedia.org/wiki/AFDX. [Accessed December 21, 2019].

[84] Credit Suisse, "Global Industrial Automation," August 14, 2012. [Online]. Available: https://doc.research-and-analytics.csfb.com/docView?language=ENG&source=emfromsendlink&format=PDF&document_id=994715241&extdocid=994715241_1_eng_pdf&serialid=hDabUewpvOqQcRiLxK7rxIQJZZ8TPLDrYHs47S97OOI%3d. [Accessed May 7, 2020].

[85] European Automotive Manufacturer Association (ACEA), "World Passenger Car Production," June 21, 2019. [Online]. Available: www.acea.be/statistics/article/world-passenger-car-production. [Accessed December 21, 2019].

[86] IDC, "IDC's Worldwide Quarterly Ethernet Switch and Router Tracker Shows Sequential Improvement in Both Markets," March 1, 2013. [Online]. Available: www.businesswire.com/news/home/20130301005268/en/IDCs-Worldwide-Quarterly-Ethernet-Switch-Router-Tracker. [Accessed May 9, 2020].

[87] Gartner, "Gartner Says Declining Worldwide PC Shipments in Fourth Quarter of 2012 Signal Structural Shift of PC Market," January 14, 2013. [Online]. Available: www.bloomberg.com/press-releases/2013-01-14/gartner-says-declining-worldwide-pc-shipments-in-fourth-quarter-of-2012-signal-structural-shift-of-pc-market. [Accessed May 7, 2020].

[88] NPD Displaysearch, "Tablet PC Market Forecast to Surpass Notebooks in 2013," prweb, January 7, 2013. [Online]. Available: www.prweb.com/releases/The_NPD_Group/NPD_DisplaySearch/prweb10295949.htm. [Accessed May 7, 2020].

[89] Wikipedia, "List of Mobile Network Operators," August 13, 2013. [Online]. Available: http://en.wikipedia.org/wiki/List_of_mobile_network_operators. [Accessed August 14, 2013, update available].

[90] Telecoms Networks, "MNO Directorio 2013 now Available," Telecoms Networks, 2013. [Online]. Available: www.telecomsnetworks.com/. [Accessed May 7, 2020].

[91] JEC composites, "Aircraft Market Forecast," April 19, 2011. [Online]. Available: www.jeccomposites.com/news/composites-news/aircraft-market-forecasts. [Accessed May 7, 2020].

[92] Statistic Brain, "Computer Sales Statistics," August 24, 2012. [Online]. Available: www.statisticbrain.com/computer-sales-statistics/. [Accessed September 10, 2013, update available].

[93] IDC, "Worldwide PC Microprocessor Revenues in 2013 to Rise 1.6% Compared to 2012," TechPowerUp, January 16, 2013. [Online]. Available: www.techpowerup.com/178838/worldwide-pc-microprocessor-revenues-in-2013-to-rise-1-6-compared-to-2012. [Accessed May 7, 2020].

[94] Send2Press Newswire, "Worldwide Telecommunications Industry Revenue to Reach $2.2 Trillion in 2013, says Insight Research Corp.," January 28, 2013. [Online]. Available: www.send2press.com/newswire/Worldwide-Telecommunications-Industry-Revenue-to-Reach-2-2-Trillion-in-2013-says-Insight-Research-Corp_2013-01-0128-002.shtml. [Accessed May 7, 2020].

[95] ITU, "ICT Facts & Figures," 2013. [Online]. Available: www.itu.int/en/ITU-D/Statistics/Documents/facts/ICTFactsFigures2013-e.pdf. [Accessed May 7, 2020].

[96] IMS Research, "Industrial Automation," 2013. [Online]. Available: www.imsresearch.com/research-area/Factory_Automation. [Accessed July 31, 2013, no longer available].

[97] T. Shea, "EtherNet/IP Leads the Way in Industrial Automation Connectivity," December 28, 2011. [Online]. Available: http://blog.vdcresearch.com/industrial_automation/2011/12/ethernetip-leads-the-way-in-industrial-automation-connectivity.html. [Accessed May 7, 2020].

[98] M. Phelps, "AEA Report Illustrates Massive Size of Avionics Market," June 6, 2013. [Online]. Available: www.flyingmag.com/news/aea-report-illustrates-massive-size-avionics-market. [Accessed May 7, 2020].

[99] Airbus, "Global Market Forecast 2012–2031," 2012. [Online]. Available: www.airbusgroup.com/dam/assets/airbusgroup/int/en/investor-relations/documents/2012/presentations/2012-31-Global-Market-Forecast/Global%20Market%20Forecast%202012-2031.pdf. [Accessed July 31, 2016, no longer available].

[100] IMAP, "Automotive and Components Global Report 2010," 2010. [Online]. Available: www.proman.fi/sites/default/files/Automotive%20%26%20component%20global%20report%202010_0.pdf. [Accessed May 7, 2020].

[101] Semicast, "OE Automotive Semiconductor Market Grew Twelve Percent in 2012," March 11, 2013. [Online]. Available: http://semicast.net/wp-content/uploads/2013/07/110313.pdf. [Accessed May 7, 2020].

[102] J. Happich, "Mobile Video Streaming Drives Demand for Networking Semiconductors in Cars," April 3, 2013. [Online]. Available: www.eenewsautomotive.com/news/mobile-video-streaming-drives-demand-networking-semiconductors-cars. [Accessed May 7, 2020].

[103] A. Zankl, "Zur Geschichte der Automatisierungstechnik," October 6, 2009. [Online]. Available: https://docplayer.org/19309413-Zur-geschichte-der-automatisierungstechnik.html. [Accessed May 7, 2020].

[104] AIM GmbH, "What Is AFDX," 2016. [Online]. Available: www.afdx.com/. [Accessed May 6, 2020].

[105] Ethernet Alliance, "Ethernet Alliance Panel #2: Bandwidth Growth and The Next Speed of Ethernet," 2012. [Online]. Available: www.ethernetalliance.org/wp-content/uploads/2012/04/Ethernetnet-Alliance-ECOC-2012-Panel-2.pdf. [Accessed May 7, 2020].

[106] S. Carlson, T. Hogenmüller, K. Matheus, T. Streichert, D. Pannell, and A. Abaye, "Reduced Twisted Pair Gigabit Ethernet Call for Interest," March 2012. [Online]. Available: www.ieee802.org/3/RTPGE/public/mar12/CFI_01_0312.pdf. [Accessed May 6, 2020].

[107] IEEE 802.3, "Approved Minutes IEEE 802.3 Ethernet Working Group Plenary Vienna, Austria," July 15–18, 2019. [Online]. Available: www.ieee802.org/3/minutes/jul19/0719_minutes.pdf. [Accessed December 7, 2019].

[108] Wikipedia, "Point-to-Point (Telecommunications)," Wikipedia, February 15, 2020. [Online]. Available: https://en.wikipedia.org/wiki/Point-to-point_(telecommunications). [Accessed May 7, 2020].

[109] Wikipedia, "MAC Address," Wikipedia, May 3, 2020. [Online]. Available: http://en.wikipedia.org/wiki/Mac_address. [Accessed May 7, 2020].

[110] S. Simeonov, "Metcalfe's Law: More Misunderstood Than Wrong?" High Contrast, July 26, 2006. [Online]. Available: http://blog.simeonov.com/2006/07/26/metcalfes-law-more-misunderstood-than-wrong/. [Accessed May 7, 2020].

2 A Brief History of In-Vehicle Networking

2.1 The Role of In-Vehicle Networking

An explanation of the needs, the development, and some of the choices in in-vehicle networking starts with windows. When automobiles were invented they were simple machines on wheels without windows. It was only later that windows were added, first at the front, then at the sides and back. The windows were fixed or insertable as one piece. This obviously was not very comfortable, neither for the handling of the windows nor for the temperature regulation in the passenger cabin. Thus, in 1928, the first mechanical window winder, able to hold a window at any position desired, was presented to the public [1]. The first power windows were introduced in 1941 [2]. BMW was the first company to introduce power windows in Europe, and the first BMW with all electric power windows was a "Series 2 BMW 503," which had a start of production (SOP) at the end of 1957 [3]. This is where it gets interesting.

It is quite straightforward to imagine a switch in a vehicle door that actuates the electric motor for a window located in the same door. Everything is in one physical location and the wiring is short. The wiring gets longer when – in addition to the "local" control in every door – all movable windows must also be controllable by the driver. More wiring between almost exactly the same locations is needed if a central door lock with discrete wiring is added, even more with an additional electronic side-mirror adjustment. Figure 2.1(a) gives an idea that, with only these basic comfort functions, the size, weight, and number of wires will soon become prohibitive. In the case of discrete wiring, inventiveness quickly circles around the question of "how is it possible to fit another wire onto this inline connector or through this opening between, e.g., body and door?" instead of fully exploring the possibilities of creating a new feature. On top of this, large wiring bundles are not only heavy, costly, and hard to install but also error-prone and difficult to diagnose [4].

Figure 2.1(b) shows the same communication structure – the logic stays in every door – realized with a bus system. The amount of cabling and the associated weight are significantly reduced. While the diagram shows door-to-door communication as a stand-alone system, this is not necessarily the case. A bus might easily be connected to other functions in the car. Examples of new combinations of functions could be to activate the light when the car is being unlocked, to remotely open a window if the key has been locked inside, or to automatically close the windows if it rains. Figure 2.1(c) shows a different option when using in-vehicle networking. The intelligence is

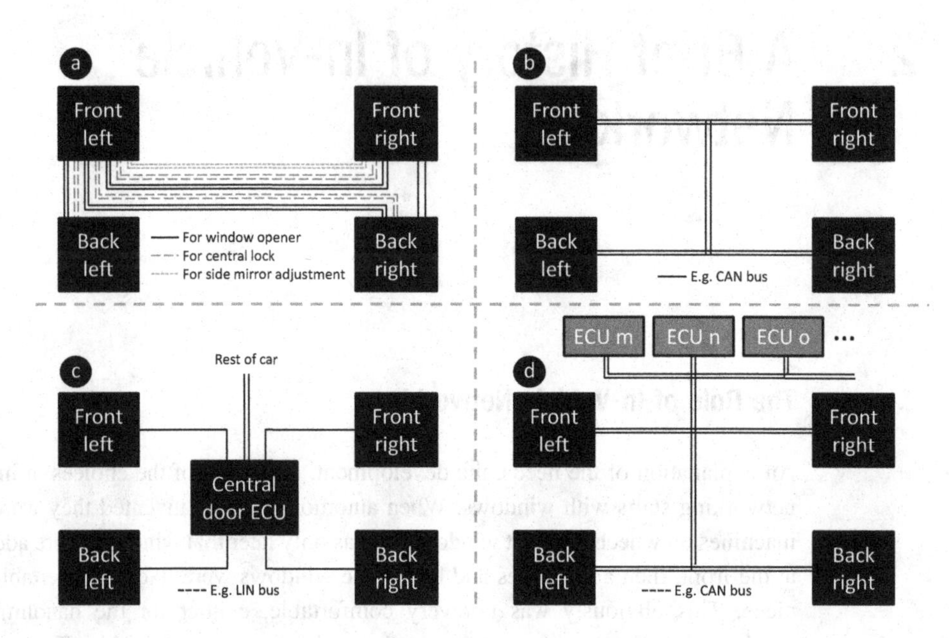

Figure 2.1 Wiring options for electric window control, central look, and side-mirror adjustments (see also [4])

concentrated in one electronic control unit (ECU) that controls all door functions. This potentially allows the complexity of the features provided to be increased. An efficient solution might also be found with a (not depicted) mixture of (b) and (c), in which a central ECU processes the information that is the same for all doors or car models, whereas smaller door ECUs handle door- or car-specific functions.

As a next step, the in-vehicle network is an enabler for an Electrics and Electronics (EE) architecture in which the software is distributed. It can be realized such that there is no distinct ECU for the door control but the processing power of idle ECUs is used instead (see Figure 2.1(d)). Considering the little amount of time that some functions are used in the car – those related to the door provide good examples – this potentially reduces costs and/or frees processing resources that can be used for other innovations.

The simple example just described gives an idea of the complexity involved. There are various other choices in relation to the architecture and various more criteria to consider. Depending on the car model envisioned and technologies available, a car manufacturer will enable and partition functions with respect to cost, weight, number of ECUs, installation effort, harness diameter, available communication bandwidth, sales prognosis, and more (see also [4] or Chapter 8 for the specific impact Automotive Ethernet has on the architectural choices). The important point is that innovation and technical possibilities form a virtuous circle. Because inventors want to realize a certain function, they push on the bounds of the technical resources. The availability of technical resources encourages innovators to make use of them until they reach their limits, and those limits must again be pushed out.

Another important aspect to consider is the type of data transmitted. In the window example described above, the information content of most of the messages is short: on/off, open/close, switch is activated / switch is no longer activated, window is open / window is closed, velocity of the window movement, etc. The use of networking technology can save weight, space, and costs. Nevertheless, the communication mechanisms behind it all can be kept simple.

This changes in the case of more complex electronics. For example, if the engine does not function as expected, it must be possible to receive differentiated messages from the engine control on the status of the system. The communication system needs protocols to allow it to distinguish between, e.g., special diagnostic messages, standard control messages, regular status messages, and software updates. Once all of this information is available on an in-vehicle networking system it can be reused in other units inside the car to, e.g., display OK or a warning to the driver. Then some type of message classification/middleware, more sophisticated addressing, and channel use concepts might be required. This in return needs yet more intelligence and refinement with the in-vehicle networking technology.

Figure 2.2 shows the development of the average number of networked nodes in cars for different regions. As can be seen, the number is continuously increasing; every car produced in 2019 contained on average 55 ECUs that needed to communicate! Ever more new functions are realized by electronics and ever more mechanical functions are being replaced by electronics. In 2019, all major car manufacturers were working feverishly on bringing the first fully autonomously driving car on the road [5]. In these cars, the driver function is taken over by even more electronics that again need to be connected very reliably to the in-vehicle network. Thus, not only the "window example" shows that automotive in-vehicle networking technologies are a

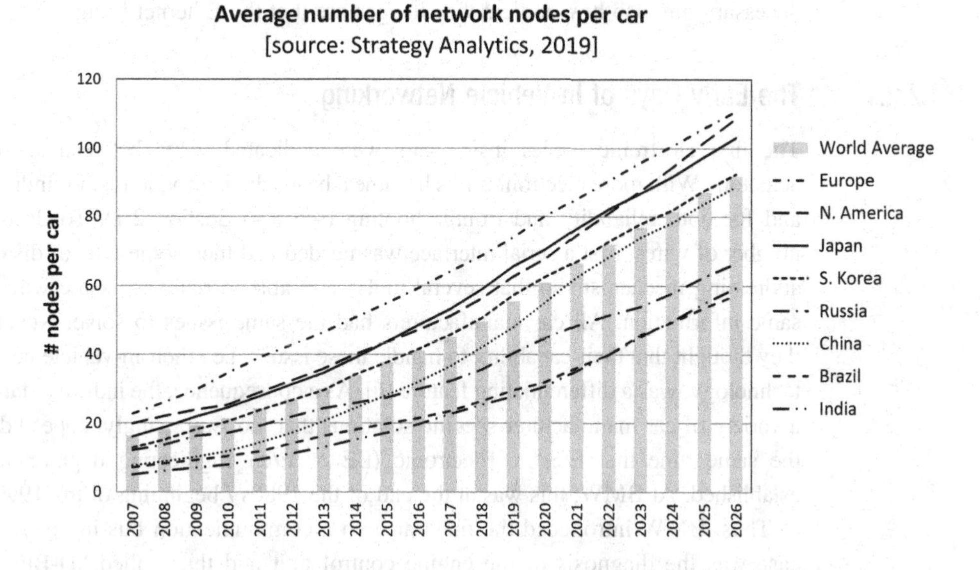

Figure 2.2 Average number of networked nodes per car, depending on region

fundamental technical resource. The more flexible and scalable the in-vehicle networking technology is, the better it provides a reliable resource for the innovations needed for ever more sophisticated customer functions.

2.2 Traditional In-Vehicle Networking

Section 2.1 illustrated the principal need for in-vehicle networking technologies. This section describes the actual technical developments in the automotive industry.

Each of the traditional in-vehicle networking technologies is different in respect to its characteristics. To those who have worked in the field for some time, they have become almost like old friends whose strengths and weaknesses are sometimes more and sometimes less enjoyable to deal with. None of the systems is ideal, but each is particularly suitable for a specific use case or physical location inside the car.

The following subsections discuss the early days of in-vehicle networking. They also describe "legacy" in-vehicle networking technologies that for one reason or another have an influence on how in-vehicle networks were traditionally seen in 2019 or impact(ed) the use of Automotive Ethernet: CAN, LIN, FlexRay, and MOST. Additionally, the use of pixel links/SerDes and some consumer links is described. The use cases, technological features, strengths, and limits of the technologies described represent only one side of the coin. As is the case for almost all technical developments, additional motives to use a certain technology depend on urgency, economics, and politics. Furthermore, choices depend on the way of thinking, capabilities, and preferences of the individuals working on a solution. The following explanations help the reader understand where automotive in-vehicle networking is coming from and where the industry is heading. They also explain how necessary but also how radical the changes are that the Ethernet brings.

2.2.1 The Early Days of In-Vehicle Networking

The first electronic cables inside cars were dedicated wires between sensors and actuators. With more electronics this became a headache in production, in finding space, and for both reliability and troubleshooting (see also Section 2.1). To decrease the number of wires, first a serial interface was needed and then some type of distribution/addressing mechanism so that several units were able to reuse the same wire and the same information. All car manufacturers had the same issues to solve. Nevertheless, they thought that their capability to handle these issues, i.e., their in-vehicle networking technology, was a differentiating feature [6]. As a consequence, the industry started with a variety of car manufacturer-specific solutions that, not surprisingly, appeared around the same time that Electric Electronic (EE or E/E) engineering departments were established. At BMW, this was at the end of the 1980s / beginning of the 1990s [7].

Thus, BMW introduced the first car with a communication bus in 1987. The use case was the diagnosis of the engine control unit and thus called "D-Bus" (D for *Diagnose*, English: Diagnosis). The communication method used was based on

Table 2.1 Principal choices to make when designing a networking technology

Channel access	Data rate	Robustness
• Arbitration (message priority based) • Carrier sensing/collision detection • Master–slave systems • Multiplex solutions – Time-multiplex – Frequency-multiplex – Code-multiplex • P2P/switched network • Token based	• Clock rates • Modulation and coding • Half-/full-duplex • Directed communication • Baseband communication	• Transmission media • Differential signaling • EMC immunity • EMC emissions • Signal integrity • Worst case channel – Impedance – Reflections • Crosstalk • Retransmissions, CRC

Note: A fundamental requirement in all cases is that the costs match the advantages of the solution.

"K-Line," a single-ended, i.e., 1-wire bus for asynchronous data up to 10.4 kbps. K-Line was later standardized as ISO 9141 and is similar to RS-232 [8, 9]. As engine control information was now available digitally, it was desired to reuse the data for information to the driver. This led in 1991 to the I-bus (*Instrumentierungsbus*, English: Instrument bus) and in 1993 to the K-bus (*Karosseriebus*, English: Body domain bus).

So, BMW used the I/K-bus, Daimler used CAN, Volkswagen used the A-bus [10], PSA and Renault used the Vehicle Area Network (VAN, standardized in ISO 11519-2), and the US car makers somewhat later introduced J1850 [6], standardized in ISO 11519-3. Yet another early in-vehicle bus system used in the industry is J1708 [11]. At some point, the car manufacturers realized that the various solutions had more disadvantages than advantages. The volumes were small for the semiconductor vendors and the suppliers had to support different automotive solutions without adding any distinct value to the products.[1]

The descriptions of CAN, LIN, MOST, FlexRay, plus pixel/SerDes, and consumer links in the next sections make no claim to be complete but focus on the aspects the authors consider relevant for understanding in-vehicle networking in the context of Ethernet.

One last general remark: In the end, all networking technologies have to solve the same basic issues. They start with a serial interface that needs to be shared by several users. For these users, the access to the medium needs to be organized, a certain data rate needs to be provided, and the transmission needs to be robust. Table 2.1 gives a rough outline of the topics that need to be addressed and some basic choices that need to be made when designing a communication technology.

2.2.2 Controller Area Network (CAN)

2.2.2.1 Background of CAN

The Controller Area Network (CAN) was one of the first in-vehicle networking technologies to have been developed, but, in contrast to other early in-vehicle

Table 2.2 Overview on the most important CAN ISO standard(s); the specifications released in 2015 or later include CAN-FD related updates

Identifier	Content	Year of latest release
ISO 11898-1	Data link layer and physical signaling, identical to BOSCH CAN 2.0 specification	2015
ISO 11898-2	High-speed medium access unit up to 1 Mbps	2016*
ISO 11898-3	Medium access unit for low-speed, fault-tolerant Media Dependent Interface (MDI) up to 125 kbps	2006
ISO 11898-4	Time-triggered communication	2004
ISO 11898-5	High-speed medium access unit with low-power mode	2007, withdrawn, content integrated in 11898-2
ISO 11898-6	High-speed medium access unit with selective wake-up functionality	2013, withdrawn, content integrated in 11898-2
ISO 16845-1	CAN conformance test plan: Data link layer and physical signaling	2016
ISO 16845-2	CAN conformance test plan: High-speed medium access unit	2018
ISO 15765-2	Diagnostic communication over CAN: Transport protocol and network layer services	2016
ISO 15765-5	Diagnostic communication over CAN: Specification for an in-vehicle network connected to the diagnostic link connector	Expected 2021

* Third release planned for 2021, to include ringing suppression for CAN-FD (H. Zeltwanger, email correspondence, January 3, 2020).

networking technologies, it continues to be used. Its development started at BOSCH in 1983, and in 1987 the first CAN controller was presented to the public. Daimler was the first car manufacturer to introduce CAN in 1992 [12]. In 1993, the first CAN ISO Standards were published: ISO 11898 on protocol and High-Speed PHY layer and ISO 11519-1 on protocol and Low-Speed PHY layer. Since then, the ISO standards have been continuously updated, and the structure of the most important latest developments is shown in Table 2.2. In 2019, the CAN bus was the most widely used in-vehicle networking technology, and – to a lesser extent – it was also used in other industries such as industrial and building automation, aerospace, medical engineering, etc. Nearly every car model is equipped with a CAN bus [10]. This might seem surprising, as only one company owns all the intellectual property and, at least to begin with, did not rely on any standards organization for its distribution. In the authors' opinion, the reasons for the success of CAN are as follows:

1. **BOSCH decided on an open licensing policy:** The technology was brought to a standards-setting organization relatively early, and the key elements for the licensing model are easily accessible to anyone interested [13].
2. **Early cooperation with leading semiconductor companies** ensured a good product portfolio for the automotive industry. Intel, NXP (then Philips

Semiconductors), and Freescale (then Motorola, now NXP) introduced their first CAN controller products in 1987–88 ([12], S. Singer, email correspondence, October 18, 2013).

3. **BOSCH is a customer for its own technology:** As one of the largest automotive suppliers [14], BOSCH is active in various roles in the automotive sector. The scope ranges from chip vendor to holistic system supplier of automotive components. BOSCH thus has a significant impact on the features provided in the chips of their own semiconductor division, as well as in the automotive semiconductor industry as such (see also Section 2.3 for the car manufacturer-to-supplier relationships). BOSCH also proved to have the necessary stamina and strategy in place to make such technology successful.
4. **Partners in the ecosystem** engaged in completing the usability of the technology by, e.g., providing test specifications and, thus, enhanced the acceptance of the technology in the community [15].
5. Last, but not least, **the technology proved to be robust** and usable in all areas of the difficult automotive environment (for details on the latter see also Chapter 4) with, at least at the beginning, one company taking responsibility for it.

2.2.2.2 CAN Technology

CAN is a bus system, i.e., all ECUs are attached to it and thus share the same wiring (see Figure 2.3). The key element of CAN is its method to decide which unit gets access to the medium/bandwidth. The method CAN uses is referred to as "arbitration," and functions on the principle that the message – and thus the ECU sending that message – with the highest priority / lowest value identifier can transmit. The method is called arbitration, because it is at the moment of transmission that the message with the highest priority wins over competing messages with lower priority.

The idea behind the CAN arbitration is based on pure electric principles. The arbitration method distinguishes between "dominant (0)" and "recessive (1)" bits in the message identifiers. The dominant bits, which electrically result in a low ohmic resistance on the channel, override the recessive bits, which result in a high ohmic resistance. So, if two ECUs start transmitting simultaneously, the ECU whose message starts with the larger amount of dominant "0" bits succeeds. In other words, the more contiguous "0s" a message identifier starts with, the higher its priority. As soon

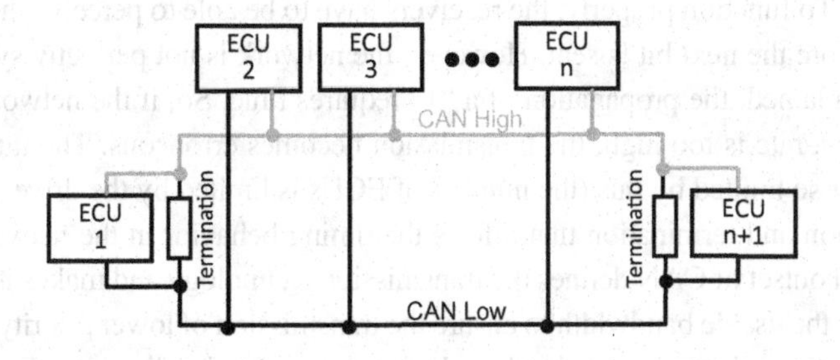

Figure 2.3 Typically used CAN topology ("linear" topology)[2]

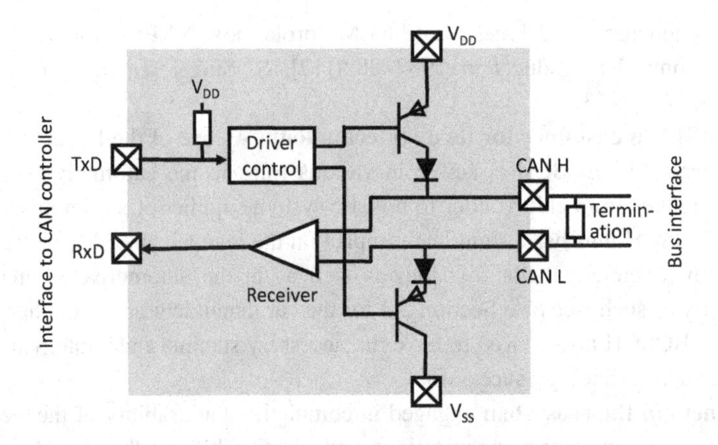

Figure 2.4 Simplified circuit diagram of a HS-CAN transceiver

as a unit perceives that the message on the bus is no longer the message it is sending, it stops its own transmission, waits for the actual transmission to terminate, then waits for the expiration of the interframe gap and tries to send its message again. This bears the risk that a message with a lower priority never has the chance to succeed if the network is very busy. A good rule of thumb is to design the CAN bus for a maximum load of 50% [4]. This means that every CAN bus automatically loses 50% of its theoretical data rate to its arbitration mechanism.

Figure 2.4 shows a simplified circuit diagram of a CAN transceiver (here for the High-Speed CAN, HS CAN version). It visualizes the electrical principles behind the arbitration. CAN uses a "Non-Return-to-Zero" coding, meaning that its symbols are represented by constant voltage levels on the channel. In the case of CAN there are two levels. In the case of a dominant "0," CAN_H is actively pulled onto the positive voltage (VDD) and CAN_L is pulled onto the negative voltage (VSS), which results in a somewhat lower voltage level of VDD – VSS. In the case of a recessive "1," the transistors are inactive and the voltage on the bus will stabilize around (VDD – VSS)/2. The key is that the system is not symmetric in its behavior. The change from "1" to "0" is active and immediate. The change from "0" to "1" happens as a discharge in the complete network. This results in a timing behavior that depends on the network, especially the lengths and number of stubs as well as the terminations and their locations in the network. To function properly, the receivers have to be able to perceive the "1" on the channel before the next bit is sent. However, the network is not perfectly synchronized and, as explained, the propagation of a "1" requires time. So, if the network is too large, or the data rate is too high, the transmission becomes erroneous. The number of ECUs is not per se limited by this (the number of ECUs is limited by the driver output). It is their location and termination that affects the timing behavior in the network.

The physical outset of CAN defines the transmission technology and makes it unique. Next to halving the usable bandwidth to ensure the transmission of lower priority packets within a certain time, the unique technology has various other implications. First of all, the transmission rate during the arbitration phase is generally limited to 500 kbps, independent of the data rate selected for the payload. Second, there is an interrelation between the possible topology and the usable payload data rate. A CAN topology used at 500 kbps can be set up with star or hybrid topologies, i.e., with stubs of varying lengths. At higher payload data rates this is not as simple as ringing can occur, which corrupts the transmission. To counteract this, a ringing suppression mechanism has been developed [16, 17]. Either new, more complex transceivers supporting this mechanism have to be used, or the topologies need to be limited to, e.g., passive linear topologies. Then, it is not possible to also transmit power over the CAN line. The need to be able to "ground" all transmission during the arbitration prohibits this. Last, but not least, there is no scrambling and CAN is thus DC-coupled (and not DC-free). This means more vulnerability of the hardware in case an error occurs in the system. For example, ground loops can damage the components more easily.

Because of its focus on messages, CAN is sometimes referred to as a "message-based" system. When a vehicle is developed, all possible messages on the CAN bus and their priorities have to be defined up front. The priorities get encoded into "identifiers," and the developers can choose either a system that supports 2^{11} different identifiers or a system that supports 2^{29}.

At the time of writing, there were three CAN versions deployed in cars: The HS CAN, LS CAN, and CAN FD. The HS CAN is specified for gross data rates up to 1 Mbps, but for the reasons given here is generally used at 500 kbps. The LS CAN is used for data rates up to 125 kbps.[3] CAN FD was specified for 2 and 5 Mbps, while not precluding the use of higher data rates [18]. However, at the time of writing, the majority of the implementations was at 2 Mbps for robustness reasons [19]. It is expected that the 5 Mbps version will be deployed more widely once the signal improvement (ringing suppression) is more common [20]. CAN FD allows for payloads of up to 64 bytes. LS and HS CAN payloads can have 8 bytes only. Together with the higher transmission rate for the payload – owing to the physics behind the arbitration as described, the data rate during arbitration remains at HS CAN level – this noticeably increases the throughput. To ensure the robustness of CAN FD, the payload is protected by a more powerful channel coding mechanism than is used for the HS CAN [21]. Figure 2.5 depicts the difference in the packet structure and data rate between a HS CAN and a CAN FD packet. Table 2.3 shows the elements that

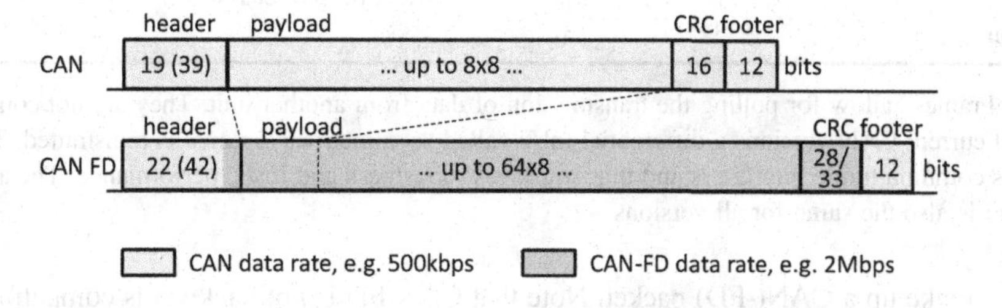

Figure 2.5 CAN-FD packet structure in comparison to a HS CAN packet; header length increases by 20 bits in cases where a 29 bit identifier is used, the CRC length for CAN-FD varies depending on the length of the payload

Table 2.3 Structure of CAN packets (also called "CAN messages") [22]

Arbitration part Field name	Length (bits)		Details
	11 bit id.	29 bit id.	
Start of frame	1	1	Denotes the start of frame transmission (always 0).
Identifier (A)	11	11	First part of the unique identifier that includes the message priority.
Remote Transmission Request (RTR) or Substitute Remote Request (SRR)	1	1	RTR normally 0 ("dominant"), 1 ("recessive") in case of "Remote Frames"*; 1 ("recessive") if SRR and 28-bit identifier.
Identifier extension bit	1	1	0 ("dominant") for 11-, 1 ("recessive") for 28-bit identifier. In the case of 1, the next two fields are added.
Identifier B	n/a	18	Second part of the unique identifier for the data including the message priority.
Remote Transmission Request (RTR)	n/a	1	Normally 0 ("dominant"), 1 in case of "Remote Frames."*
Remaining packet **Field name**	**Length (bits)**		**Details**
	CAN	CAN FD	
Flexible Data rate Format (FDF)	n/a	1	Recessive 1 plus next bit dominant 0 initiates the FD.
Reserved bits (r0, r1)	1 or 2	1 or 2	Reserved bits (1 in case of 11-, 2 in case of 29-bit identifier).
Bit Rate Switch (BRS)	n/a	1	Recessive 1.
Error State Indicator (ESI)	n/a	1	Recessive 1 (if error passive).
Data length code	4	4	Number of bytes of data (0–8/64 bytes).
Data field	8 × 8	8 × 64	Data to be transmitted (length as specified before).
CRC	15	27 or 32	Cyclic redundancy check, for CAN-FD the 32 bit CRC is used in case the payload >16 × 8 bit.
CRC delimiter	1	1	Must be 1 ("recessive").
ACK slot	1	1	Transmitter sends 1 ("recessive"). All receivers send a 0 ("dominant") ACK, if they have been able to receive the packet in this same slot.
ACK delimiter	1	1	Must be 1 ("recessive").
End of frame	7	7	Must be 1 ("recessive").
Intermission	3	3	

* "Remote Frames" allow for polling the transmission of data from another unit. They are not commonly used.In all current CAN versions a differential (also called "symmetrical") signal is transmitted. This suppresses common mode interferers and thus improves robustness and EMC performance. The arbitration mechanism is also the same for all versions.

make up a CAN(-FD) packet. Note that CAN FD is not backwards compatible to HS (or LS) CAN. They cannot be used on the same bus.

CAN does not prevent two different ECUs from using the same identifiers. It is therefore the implementer who has to make sure that every ECU receives its own

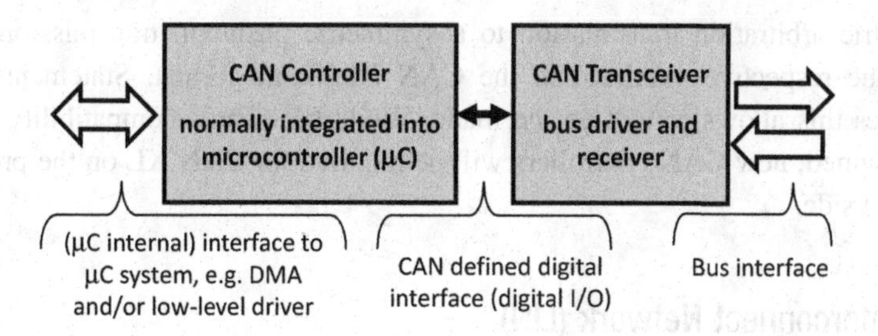

Figure 2.6 Elements and interfaces of a CAN node

unique set of identifiers. A transport protocol standardized in ISO 15765-2 enables messages longer than 8 or 64 bytes to be spread over several packets. CAN messages do not contain address values in the manner that one expects with Ethernet. Instead, a transmitter transmits its message on the bus, and all connected ECUs can potentially receive it. It is defined in the receiver whether a message identifier triggers the receiving ECU to store and process the offered data or not.

All participants on the bus acknowledge the reception of every error-free CAN frame received with a dominant bit in the same acknowledge space in the packet, independent of whether they actually use the data. The transmitter, in return, will recognize the acknowledgment. The uncertainty in the CAN system is that one acknowledgment is sufficient for the transmitter to perceive a correct transmission. The transmitter cannot discern which unit(s) have sent the acknowledgment and if the intended receiver unit was among them [23].

Figure 2.6 shows the elements needed for a CAN communication. Generally, the transceiver is a separate semiconductor, while the controller is integrated into the micro-controller. The clock rate, i.e. the transmission rate, is determined by each controller; there is no synchronization in the network other than what can be evaluated from observing the traffic on the channel. An important advantage of CAN is its robustness. It allows ECUs to be connected in almost all areas of a car.[4] For wiring, Unshielded Twisted Pair (UTP) cables and multi-pin connectors can be used.

For some (see e.g. [19]) the development of CAN-FD was already a reaction to Automotive Ethernet. While this may have been a possibility – despite the large data rate difference between 2/5 and 100 Mbps – the most recent development in CAN, called CAN-XL, is explicitly a competition to the 10 Mbps automotive Ethernet PHY 10BASE-T1S [20]. The technical definition of CAN XL was not completed at the time of writing. Therefore, only an outline of the published and likely stable parameters can be given here.

CAN XL follows the same principles as CAN FD in that it maintains the arbitration, but increases the data rate for the payload and extends it even further. The target data rate between arbitration (still running at 500 kbps) and payload phase is 10 Mbps. The maximum payload size is 2048 bytes; explicitly large enough to be able to tunnel an Ethernet packet [20]. Because of the physics of CAN, the high data rate during the transmission of the CAN XL payload can only be achieved by changing from the asymmetric arbitration transmission to a symmetric push/pull transmission and by adding the respective overhead in the CAN XL frame format. Statements on the topologies this allows cannot yet be made. While backwards compatibility to CAN FD is planned, new CAN controllers will be required for CAN XL on the processing (e.g., μC) side.

2.2.3 Local Interconnect Network (LIN)

2.2.3.1 Background of LIN

There are many applications inside a car, which require only simple sensor–actuator communication and the robustness of a 1-wire communication system. The use cases comprise comfort functions like power windows, central locks, electronic mirror adjustments (see also Section 2.1), electronic seat adjustments, rain sensors, light sensors, electric sunroofs, control of air conditioning, etc. These application areas are very cost sensitive and have few requirements. CAN over-performs for these applications and is thus deemed too expensive.

With the discussions that arose around the first in-vehicle networking systems in the industry, car manufacturers noticed that many had the same requirements for a simple networking technology. Thus, in 1998, Audi, BMW, Daimler, Volkswagen, Volvo, NXP (originally Motorola, then Freescale), and Mentor Graphics (originally Volcano) founded the Local Interconnect Network (LIN) consortium to standardize a respective solution [24]. In the end, this was the turning point for the industry, away from local/individual solutions toward commonly deployed standards. The accepted version, LIN v1.3, was published in November 2002, and LIN 2.0 followed in September 2003 [25]. In the US, the SAE published J2602 in September 2005, a LIN 2.0 version with some minor deviations that are supposed to meet the cost targets even better [24]. In 2013, LIN was transferred to ISO 17987 [26].

2.2.3.2 LIN Technology

The key requirement of LIN was to be cost efficient, which naturally influenced the technology. One of the first choices was to base the physical layer on the K-Line ISO 9141 standard, which had been known in the industry from the early days of in-vehicle diagnostics (see also Section 2.2.1). LIN was designed as a single-ended, i.e., 1-wire unshielded, system with several consequences:

1. The effort to provide LIN hardware was small.
2. LIN is single-ended and not differential/symmetrical. Common-mode noise can affect the system, which limits the immunity and increases the emissions. To meet the EMC requirements nevertheless, the data rate is limited to 19.2 kbps. This in return means that a basic clock synchronization mechanism and simple drivers are sufficient.
3. Ground is used as a back channel. This means that the system needs to be designed such that a certain amount of ground shift can be supported. At the receiver the recessive "1" state is thus defined for when a voltage level above 0.6 V_{Batt} has been

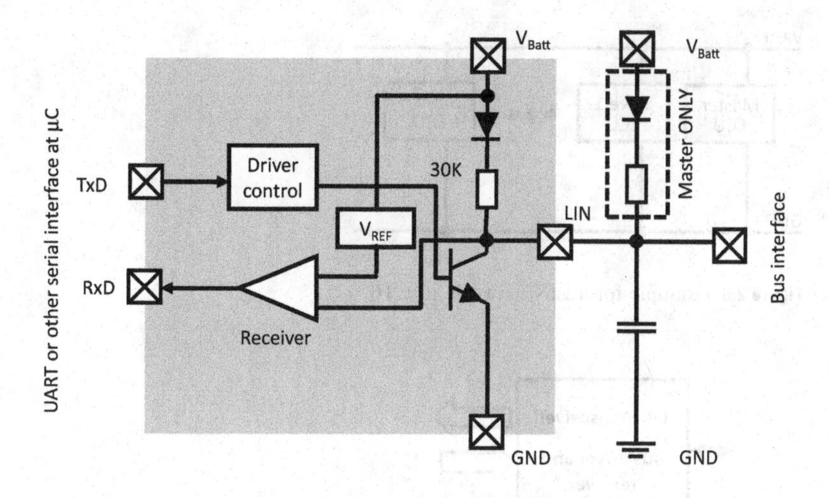

Figure 2.7 Simplified circuit diagram of a LIN transceiver

detected. The dominant "0" state is detected when the voltage level is below 0.4 V_{Batt} [25].

4. The behavior of the network also depends on the layout. Consequently, the physical expansion of the LIN bus inside the car needs to be restricted accordingly.

Figure 2.7 shows the circuit diagram of a LIN transceiver. The differentiation between a Master and a Slave node is achieved by external components. The Master terminates the complete network.

LIN is a good example of how a simple bus system can be derived from enhancing an existing serial interface with a communication protocol. LIN has been designed such that up to 16 ECUs can share the media the bus provides. The multi-user access protocol is a Master–Slave concept for the channel access, i.e., one unit on the bus is assigned Master (see also Figure 2.8) and Slaves can transmit only after having been polled by the Master with a respective header. As LIN is a bus system, master-initiated communication can also occur between two slaves – and all information on the bus can potentially be read by any unit attached to it.

Additionally, the scheduling of the communication on the LIN bus is predetermined in the design phase of a vehicle. The scheduling tables are transmitted to all LIN ECUs in so-called LIN Description Files. With LIN 2.0 a transport protocol was defined. This means that one message can be split over several packets – each packet has a maximum payload of 8 bytes – and in consequence LIN can also be used for diagnostics and software updates. It is not uncommon for a more complex ECU to be connected to several LIN busses and to serve as several LIN masters.

Figure 2.9 shows the elements and interfaces needed for a LIN node. As can be seen, LIN only requires a comparably simple transceiver. To connect to the micro-controller a UART interface is sufficient. A UART interface is so common that it is supported by even the smallest micro-controllers. Slaves can synchronize

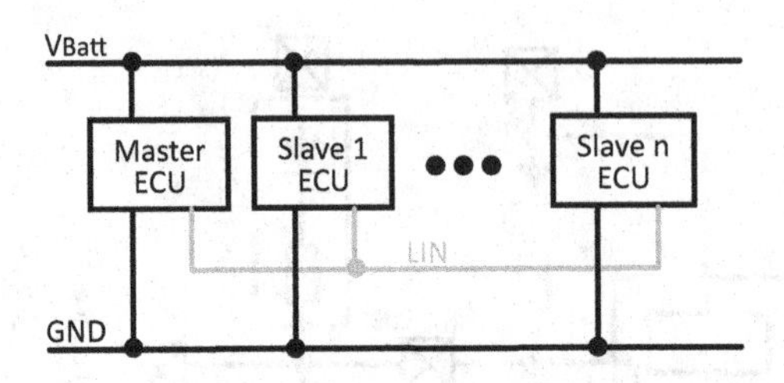

Figure 2.8 Example for a LIN network, n < 16

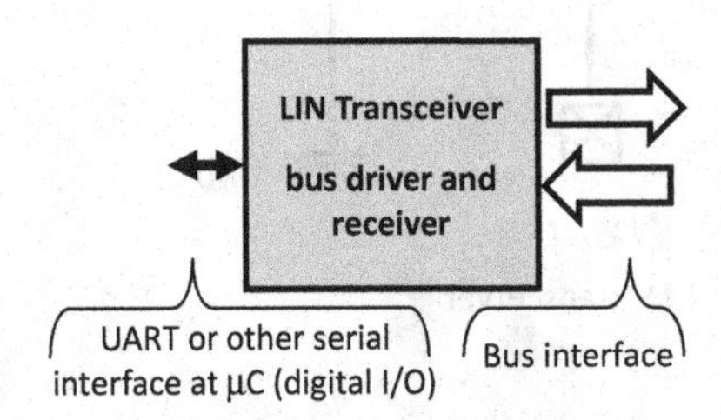

Figure 2.9 Elements and interfaces of a LIN node

independently, and do not need an external clock. Furthermore, LIN uses the battery voltage on the channel and thus does not require voltage regulators (see Figures 2.7 and 2.8).

Because LIN achieved its goals of cost efficiency and unifying the industry to this solution, LIN has been successfully established in the industry. The use cases that it addresses will not disappear anytime soon from vehicles and LIN can thus be expected to persevere in the in-vehicle networking landscape. LIN is an example that proves that there is likely never going to be one ideal in-vehicle networking solution, even if Automotive Ethernet can address many application areas and might reduce the number of technologies prevailing.

2.2.4 Media Oriented Systems Transport (MOST)

2.2.4.1 Background of MOST

At the end of the 1990s the need to support complex audio applications in cars became urgent. Not only were audio CDs irreversibly replacing analog music storage media but the customers were also getting more interested in navigation systems and mobile phones usage. Inside the car, the different audio streams these applications generated had to be coordinated with each other and with additional warning notices from driver assist functions, while at the same time providing an optimum sound experience.

This, first of all, required a significant increase in the data transmission rate, and it became obvious relatively quickly that only an optical system would solve the task. At that time, optical systems were the only systems promising the expected data rate with an EMC-compliant solution at a reasonable cost level. Daimler already had experience from using the Domestic Digital Bus (D^2B),[5] a technology later standardized as IEC 61030. The Daimler experience was particularly valuable as it proved the feasibility of the less costly and more robust Polymeric Optical Fiber (POF) as a transmission media in contrast to Glass Optical Fiber (GOF).

Nevertheless, the intention was not only to have a new PHY technology with a higher data rate available but to have a system that covered all networking aspects, from the physical layer to the application layer. The technology was to enable the complex control sequences the desired use cases needed in the already challenging automotive environment. This was unprecedented in the industry, and was seen as a challenge that required a new approach.

The industry decided to cooperate with a strong supplier, who in the role of a type of "general contractor" was to coordinate the fortunes of the new networking system, while the requirements were still provided by the car manufacturers. The partner chosen was OASIS (which later became part of SMSC, which today is part of Microchip). Some of the founding members of OASIS had previously worked for (Harman) Becker, and thus had significant experience with automotive infotainment systems and their requirements. Additionally, OASIS had a proposal for technology supporting optical transmission. So, in 1998, BMW, (Harman) Becker, Daimler, and OASIS founded the MOST Cooperation (MOST Co) to develop MOST as a networking technology for in-vehicle applications and to establish MOST as an industry standard. Other interested companies were welcome to join the MOST Co. The agreements include royalty-free licensing among the members for all developments except for the Data Link Layer (DLL)/PHY, for which the IPR belonged to OASIS and thus today to Microchip. To license the DLL/PHY technology a royalty-based license was/is possible but also necessary.

The MOST technology is often associated with a monopoly, as indeed there is only one supplier of the respective hardware (Microchip). With the selection of a supplier in the role of a "general contractor" the participants never intended to have a closed market and the MOST Co was set up in a way that allowed for competition. Competition was not the focus, however. The main goal was to minimize the risks associated with the development of such a complex system, and having a "general contractor" seemed to minimize those risks. One of the reasons given today for the missing competition is that the MOST Co did not provide a specification to which interoperability and compliance testing of the DLL could be performed, but that Microchip would have instead licensed an IP core. Potential other vendors thus had limited chances to differentiate their product and face(d) the risk of incompatibility with the technology of the dominant vendor, Microchip, if they tried. Because the MOST Co focused on doability, the important issue of interoperability was missed at the time and the fact is that no other company took the opportunity to enter the market.[6] In 2016, the ISO project 21806 was started in order to transfer the MOST specifications, including the DLL specification, to ISO.

MOST exists in four variants: MOST 25, with optical transmission first introduced by BMW in 2001; MOST 50, with electrical UTP cabling first introduced by Toyota in 2007 [27]; MOST 150, first introduced in the optical version by Audi in 2012 [28]; and MOST 150, using coaxial cabling, for which Daimler showed interest [29] and which was first introduced by Honda. As the following discussion is concerned with the basic principles only, it will use MOST 25 as an example.

The importance of the MOST technology in the car industry is diminishing. However, for understanding the developments that lead to the introduction of Automotive Ethernet, it is still of interest.

2.2.4.2 MOST Technology

The MOST technology defines the communication on all seven layers of the ISO/OSI layering model. The MOST protocol is thus more complex than for the previously described CAN or LIN. MOST is the first in-vehicle networking technology to support service-based methodologies, which means that functions and services can be requested during operation on demand. The interfaces to the available functions are described in detail in the MOST Function-Blocks (FBlocks). Also, for the first time, message sequences are provided. On the higher layers a MOST message is defined to consist of the following parts: DeviceID.FBlockID.InstID.FktId.OpType. Length. Table 2.4 lists how the elements of a control message form a 32 byte message.

A MOST message does not necessarily correspond to a MOST frame. Instead a message might have to be distributed over several frames. Frames represent the constantly repeated structure in which the traffic on the MOST bus is organized, with the bus topology generally being a (physical or virtual) ring. One frame is partitioned into 64 byte-sized slots. Two of these are administrative and two are for control information. This means that the 32 byte control message of one unit is spread over 16 frames and that the next unit has to wait before it can use the 2-byte control channel. MOST supports the two system frequencies: 48 kHz (from professional audio) and 44.1 kHz (from audio CDs) [30]. When frames are sent at 44.1 kHz the bandwidth for MOST 25 that can be shared on the control channel is 705.6 kbps. The reception of control messages must be acknowledged.

The remaining 60 bytes, i.e., 21.2 Mbps@44.1 kHz or 23 Mbps@48 kHz, can be divided into a synchronous and an asynchronous part. Note that once a MOST 25 system has been set up, the division between the two parts is not changeable during operation.

- **Synchronous data:** MOST has been optimized to transmit audio in continuous data streams; hence the frame frequencies of 48 kHz for general audio or 44.1 kHz for audio CDs. Anywhere from 24 to 60 (i.e., all) of the available bytes can be allocated to the conveyance of synchronous data. Multiple access for these bytes is managed by Time Division Multiplex, i.e., in every frame a certain unit transmits its data in certain byte(s)/slots. In MOST jargon the assigned slot in

Table 2.4 Structure of a MOST control message (from "FBlockID" to "Data" represents the data field of the control message)

Element	Length (bytes)	Content
Priority decision	4	The value of this field allows a unit to identify whether it can send a control message. Priority of the message, availability of the media, and transmission history are taken into account.
Target address	2	Address of the ECU which requested the function. As the data is available on the bus, potentially all units can listen to it.
Source address (DeviceID)	2	Address of the ECU which offers the function. The address is defined in the network layer and depends on the position of the unit in the network.
Message type	1	Identifies the type of control message.
FBlockID	1	ID of the function block.
InstID	1	Instance inside the FBlock. One FBlock can consist of several instances.
FktID	1½	Function to be called.
OpType	½	Defines the function type (e.g., could be an error message or a request).
Tel ID	½	Identification of parameters.
Tel Len	½	Length of parameters.
Data	12	Content of the parameters.
CRC	2	
ACK	2	Defines the function type (e.g., could be an error message or a request).
Reserved	2	
Overall	32	

the synchronous section is called a "channel." There are no retransmits for lost synchronous data packets.

- **Asynchronous data:** Zero to 36 bytes can be used for the transmission of application data such as map information from navigation systems or TCP/IP traffic. This corresponds to a maximum gross data rate of 12.7 Mbps@44.1 kHz and 13.8 Mbps@48 kHz. Access to the channel is granted by a token system. When a unit has the token, it can use all available bytes assigned to asynchronous data. A unit can transmit one message with a maximum of 1014 bytes of user data before it passes the token on. As for control messages, such a message is transmitted over several frames – up to 29 frames if the unit has been allocated the theoretical maximum bandwidth. There are no acknowledgements or retransmits for asynchronous data.

MOST generally uses a ring topology, virtual or actual (see Figure 2.10 for an example), which can handle up to 64 ECUs. Each ECU is addressed according to its location in the ring. One ECU functions as a "timing master" that continuously

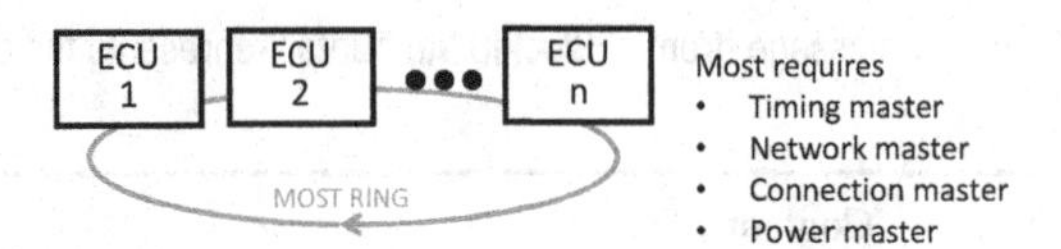

Figure 2.10 Example MOST ring, n < 65

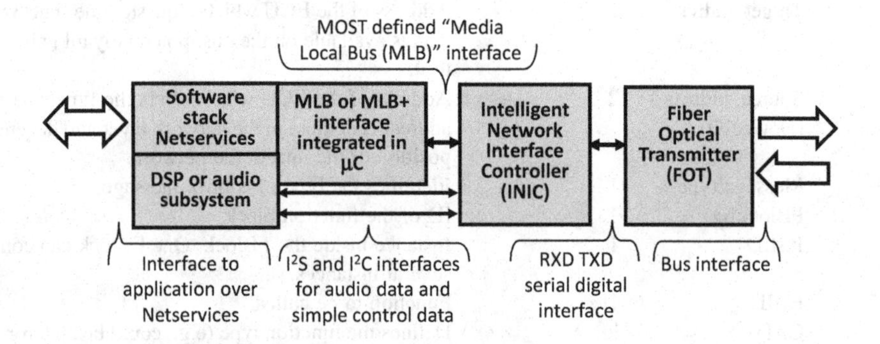

Figure 2.11 Elements and interfaces of a MOST node

sends the preamble that begins every frame onto the ring and that allows all ECUs to synchronize. The additional coordination functions of network, connection, and power master do not have to be handled by the same unit but generally are. The connection master administers the synchronous channels, the network master controls the system status, and the power master supervises start-up and shut-down of the MOST network.

Figure 2.11 shows the elements needed for a MOST communication node. As can be seen, the capabilities, but also the complexity have increased significantly. The communication functionality is provided by MOST network services, for which Microchip owns the trademark "Netservices." The Fiber Optic Transmitter (FOT) represents the PHY and the MOST Network Interface Controller (NIC) represents the DLL. The Intelligent Network Interface Controller (INIC) is controlled by Netservices, which are associated with the network, transport, and session layers of the ISO/OSI layering model [31]. The application sockets and FBlocks share layers six and seven. With the boundaries between the layers blurred, it is not practical to exchange protocols or even adjust the functionality on individual layers but to use the complete stack as is.

For efficient use of the INIC it is advisable to use the Media Local Bus (MLB). Since MLB was defined in tandem with MOST, it is not a very common interface at host processors in an ECU. Thus, in some cases the use of an additional companion chip might be required (see Figure 2.11). The audio data (and simple control functions) can use the Inter-IC Sound (I^2S) and Inter-Integrated Circuit (I^2C) interfaces. The FOT converts the electrical information into an optical signal. Communication on the fiber is unidirectional only.

Table 2.5 Overview on FlexRay standards

Identification	Content	Year
FlexRay Protocol V3.0.1	Data Link Layer Specification, designed to be fully backwards compatible. Most car manufacturers use Version V2.1 Rev A	2010 (V2.1, 2005)
FlexRay Electrical Physical Layer V3.0.1	Physical Layer Specification, designed to be fully backwards compatible. Most car OEMs use Version V2.1 Rev B	2010 (V2.1, 2005)
ISO 17458-1	General information and use case definition	2013
ISO 17458-2	Data Link Layer specification	2013
ISO 17458-3	Data Link Layer conformance test specification	2013
ISO 17458-4	Electrical Physical Layer specification	2013
ISO 17458-5	Electrical Physical Layer conformance test specification	2013

2.2.5 FlexRay

2.2.5.1 Background of FlexRay

FlexRay was developed at a time when the automotive industry became interested in "X-by-Wire" applications. The idea of X-by-Wire is to eliminate all mechanical fallback from the car and to have pure electric functions only.[7] Target applications were the steering, braking, and other safety critical systems. Security and timing are particularly important in this case. BMW had already gained some experience with time-triggered communication prior to the development of FlexRay with the proprietary development of a technology called Byteflight. This optical system was used in a few BMW models for airbag control and other safety related systems. Nevertheless, it did not persevere, as it proved to be too expensive.

Instead, in 2000 BMW and several other automotive companies agreed to develop a new technology in the FlexRay Consortium. The core partners were Freescale (formerly Motorola, now NXP), NXP (formerly Philips), BMW, Daimler, and somewhat later BOSCH, General Motors (Opel), and Volkswagen [32]. In 2009, after the finalization of FlexRay 3.0, the task was seen as completed. The FlexRay Consortium was disbanded [33] and the FlexRay standards were transferred to ISO (see Table 2.5 for an overview).

2.2.5.2 FlexRay Technology

As can be seen in Table 2.5, FlexRay defines the Physical (PHY) and Data Link Layers (DLL) only. Other layers are covered by other committees. For example, the AUTomotive Open System ARchitecture (AUTOSAR) standard explicitly addresses the higher layer software protocols needed for a FlexRay communication. The key requirement for FlexRay is reliability and consequently FlexRay provides a number of respective features such as determinism and redundancy.

FlexRay communication is based on timeslots and cycles, which are configurable by the developer. Every cycle consists of a static time segment and a network idle time, but can additionally comprise a dynamic time segment. The multi-user access is handled differently in the static and dynamic segments: In the static segment, the access is

Table 2.6 Elements of a FlexRay packet [32]

Field name	Length (bits)	Detail
Reserved bit	1	Reserved bit
Payload preamble indicator	1	Indicates whether packet has a payload
Null frame indicator	1	Indicates whether packet is a Null frame
Sync frame indicator	1	Indicates whether packet is a Sync frame
Start-up frame indicator	1	Indicates whether packet is a Start-up frame
Frame ID	11	Packet identifier
Payload length	7	Length of the packet's data payload as measured in 16-bit words.
Header CRC	11	Provides header integrity protection from Null frame indicator to Payload length
Cycle count	6	Indicates the actual cycle
Data field	0–254 bytes	
Payload CRC	24	Data integrity protection for payload
Overall	8 + (0–254) bytes	

defined in TDM, i.e., the units are assigned certain timeslots in every cycle up front. If a unit has nothing to transmit in its timeslot in the static segment, it transmits a "null" frame, so that the respective receiver always receives something as expected and knows that the communication is not unintentionally disrupted. The dynamic segment uses a so-called mini-slot method, wherein a preset order of FrameIDs combined with counters for multi-user access. Other than in the static segment, however, a unit whose turn it is to transmit does not do so unless it has data to send. If the unit has nothing to transmit, all units increase their counters and the one having the next FrameID can start the transmission in the next mini-slot instead of having to wait. To increase the throughput the mini-slot duration is shorter than the static slot and can be configured during the design of the system [32]. The mini-slot method is inherited from the Byteflight development and a similar method is specified under the name PHY Level Collision Avoidance (PLCA) for the bus use of 10BASE-T1S Ethernet (see also Section 5.4.3).

Table 2.6 shows the set-up of a FlexRay packet. Each packet consists of a header, a payload, and a trailer, which comprises the CRC for the payload. The gross data rate is 10 Mbps. A system's effective data rate depends on the configuration of the system and the ratio between the lengths of the dynamic and static sections. Unused assigned slots in the static segment are wasted. Additionally, the use of 8B10B Non-Return-to-Zero (NRZ) coding as well as a header and trailer reduce the net bit rate [34].

Like CAN, FlexRay also transmits a differential signal. In contrast to CAN though, the FlexRay transceiver has two separate push/pull entities (see also Figure 2.12). This means that current is actively driven for both the high and low voltage levels and that the behavior of the signaling is less influenced by the layout of the network for FlexRay than it is for CAN. FlexRay provides a 10-fold bit rate (or 20-fold when considering that CAN is normally used at 500 kbps) and requires higher investment into the network infrastructure and the terminations in order to meet EMC requirements. This is especially so as FlexRay uses, like LIN and CAN, unshielded cabling

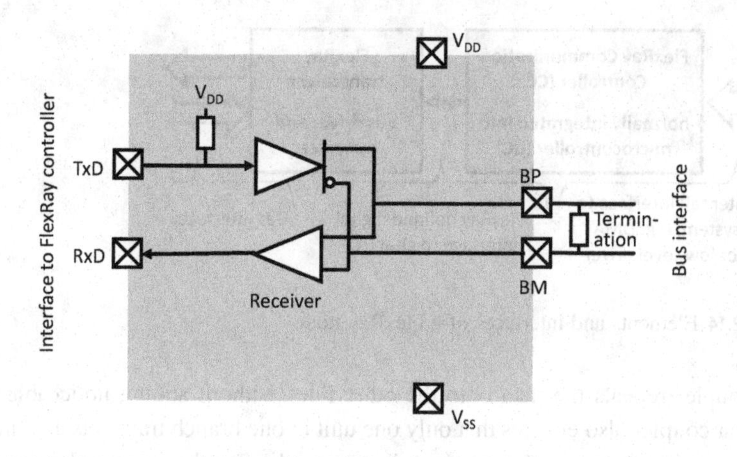

Figure 2.12 Simplified circuit diagram of a FlexRay transceiver, the transmitter circuit has not been included for complexity reasons

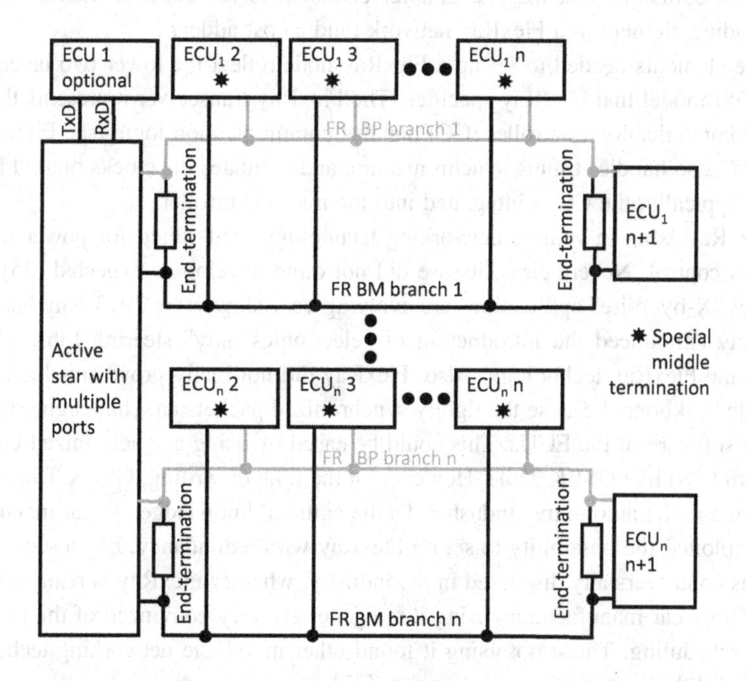

Figure 2.13 Large FlexRay network with active star

and multi-pin connectors. The cables need to be of higher quality, and fewer ECUs may be connected to one branch, as is explained here.

A small FlexRay system typically consists of four or five ECUs using a linear topology, but FlexRay also supports different topologies with more ECUS. With use of an "active star/star coupler" several linear topologies can be combined to one network (see Figure 2.13), though it is not possible to cascade the architecture. The

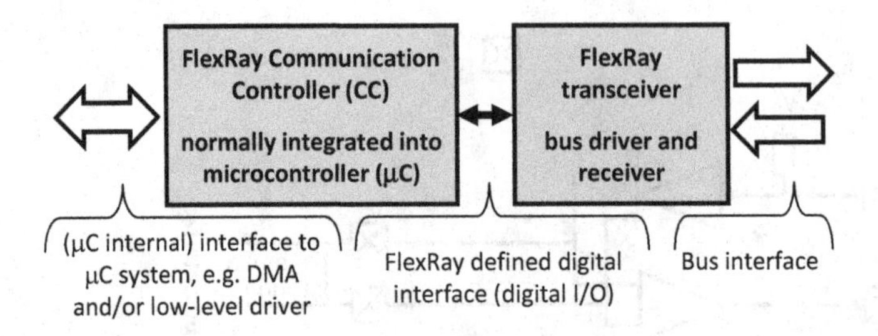

Figure 2.14 Elements and interfaces of a FlexRay node

star coupler repeats the data onto the other lines without adding noticeable latency. The star coupler also ensures that only one unit in one branch transmits at a time while all other units are in listening mode. It is noteworthy that the star coupler does not use scheduling but observes the voltage levels on the channel. The star coupler is thus also called a "moderator." The challenge is the speed of signal propagation, which can result in collisions that the star coupler cannot resolve. The star coupler is thus a demanding element in a FlexRay network (and a cost adder).

The elements needed to set up a FlexRay node reflect the lower two layers of the ISO/OSI model that FlexRay specifies. The FlexRay transceiver represents the PHY. The communication controller (CC) handles communication logic (see Figure 2.14). The CC also handles timing synchronization and regulates the clocks onto "FlexRay" time. Typically, the CC is integrated into the micro-controller.

FlexRay is an in-vehicle networking technology well suited for power train and chassis control. Nevertheless, its use did not quite develop as expected [35]. Safety critical "X-by-Wire" applications are evolving very slowly; by 2013 only Nissan had publicly announced the introduction of "electronics only" steering [36], while not using the FlexRay technology. Also, FlexRay did not really prove suitable as an in-vehicle backbone, because the tightly synchronized packets are challenging to handle in the software of the ECUs. This could be eased by using a synchronized Operating System (OS) like OSEK Time. However, at the time of writing, OSEK Time was not common in the automotive industry. To the authors' knowledge, no car manufacturer has exploited the possibility to set up FlexRay with redundancy, i.e., a second link.

It is controversially discussed in the industry, whether FlexRay is really needed or not. Those car manufacturers using it are generally very convinced of the need of its strict scheduling. Those not using it found other in-vehicle networking technologies that fulfill the same timing requirements [37].

2.2.6 SerDes Interfaces (Pixel Links)

The motivation to develop MOST 25 was to be able to handle sophisticated digital audio applications. Compared with the available in-vehicle networking technologies at the time of development, MOST 25 provided a very large increase in data rate. Nevertheless, the data rate of even an uncompressed audio stream is small when compared with some of the video or sensor data car manufacturers (want to) use in their vehicles today. In particular, cameras or displays with high quality resolutions ("high-definition (HD)," "4k," "8k") or sophisticated Light Detection and Ranging (LiDAR) sensors impress with data rates well beyond 1 Gbps.

Video data rates are derived by the following elements: The pixel resolution of the camera imager or display, the bit depth of the pixels (which encodes the colors), the rate at which the image frame is refreshed (frames per second, fps) and blanking.[8] For example, 4k video with a resolution of, e.g., 3840×2160 pixels, 60 fps, and a bit depth of 16, leads to a data rate of 8.76 Gbps (without blanking). In the context of Blu-ray, the resolution would be an "Ultra HD Blu-ray" format [38]. While important consumer standards influence the automotive infotainment features, the relevance of the HD sensor data is fundamental in the context of automated driving. Consequently, 8k+ resolutions are being discussed in the industry (see e.g. [39]). Most common at the time of writing were HD pixel resolutions of 1280×720 or 1920×1080 [40, 41], which, depending on the color depth and frame rate, results in data rates between 0.22 Gbps and ~3 Gbps (see Table 5.13 for more details).

The best communication technology to use for such data rates depends on the actual implementation and choices in the Electrics and Electronics (EE) architecture. Figure 2.15 shows example use cases, in which data rates above 1 Gbps occur. Recorded video, camera data, or a graphics processor might be the source. Figure 2.15 deliberately does not show which of these three blocks for each use case are in the same ECU, i.e. which units are connected on a circuit board, which units are connected by a communication link, and whether data is distributed further in the network or not.

For the example of Blu-ray (Figure 2.15(b)) the disk is read and the data is transmitted at 54 Mbps only [42].[9] The decoder needs to be integrated with the

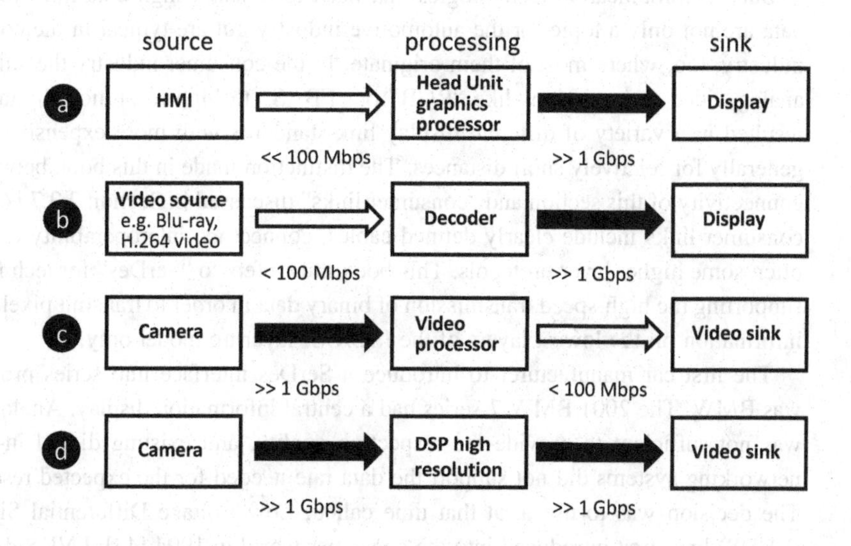

Figure 2.15 Example use cases for high speed link applications

display, which is likely to be at a different location. For video data with less content protection, it is also possible to directly decode the data in the same device and then to forward uncompressed data to the display. So, in one scenario the in-vehicle communication is between disk reader and decoder. In the other scenario, it is between decoder and display.

In the case depicted in Figure 2.15(c), the processor can be integrated with the camera, where it performs, e.g., image processing and compression, or with the video sink. This is a question of where to have the intelligence: Distributed in intelligent sensors or centralized in the processing unit (the video sink) with "dumb" sensors connected. Alternatively, the camera sends unprocessed data to a unit that then performs the processing (Figure 2.15(d)) before internally or externally passing unprocessed data on to the display. Often there is a choice, and whatever is being selected varies on a number of parameters, that include quality concerns (compression losses, latency), technical feasibility, costs and personal preferences. The conclusion, nevertheless, is that there are use cases in which (video) data at data rates significantly higher than 1 Gbps need to be transmitted from one ECU to another.

The next question is whether the communication is P2P only (in contrast to data needing to be distributed in the network). At the time of writing, only P2P SerDes[10] – also called LVDS, pixelink, or high speed video link – technologies supported the respective data rates. However, high-speed Automotive Ethernet PHY technologies were also being developed at the time (see also Section 5.4). Distributing the data in the automotive network is always easier with a networking technology like Ethernet. However, when only a P2P connection is needed (see Endnote 4 of Chapter 1 for the definition of P2P) it might be more efficient to use a SerDes technology rather than an Automotive Ethernet network when only a subset of features is needed. It will be interesting to see the choices the car industry will make eventually.

Such communication technologies that need to transmit high data rates for video data are not only a topic for the automotive industry but are typical in the consumer industry, too, where most of them originate. In the consumer industry the shift from analog video transmissions like "RGB" or "FBAS" to high-resolution digital video resulted in a variety of different display link standards with more expensive cables, generally for relatively short distances. The distinction made in this book between the connectivity of this section and "consumer links" discussed in Section 2.2.7 is that the consumer links include clearly defined cables, connectors, interoperability tests, and often some higher layer protocols. This book thus refers to "SerDes" for technologies supporting the high-speed transmission of binary data in order to transmit pixel precise information on the lowest layers of the ISO/OSI layering model only.

The first car manufacturer to introduce a SerDes interface into series production was BMW. The 2001 BMW 7-series had a central information display. Analog video was not sufficient to provide the respective quality, and existing digital in-vehicle networking systems did not support the data rate needed for the expected resolution. The decision was to use a, at that time called, Low-Voltage Differential Signaling (LVDS) link, first introduced into the consumer world in 1994 [43]. LVDS describes the physical principle with which digital data is transmitted as a differential signal over

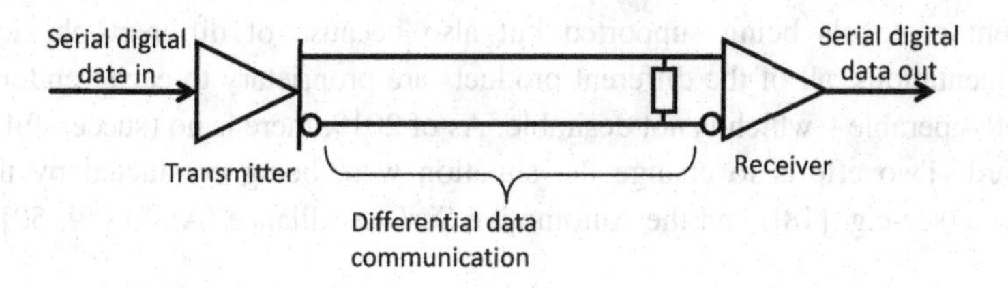

Figure 2.16 LVDS principle

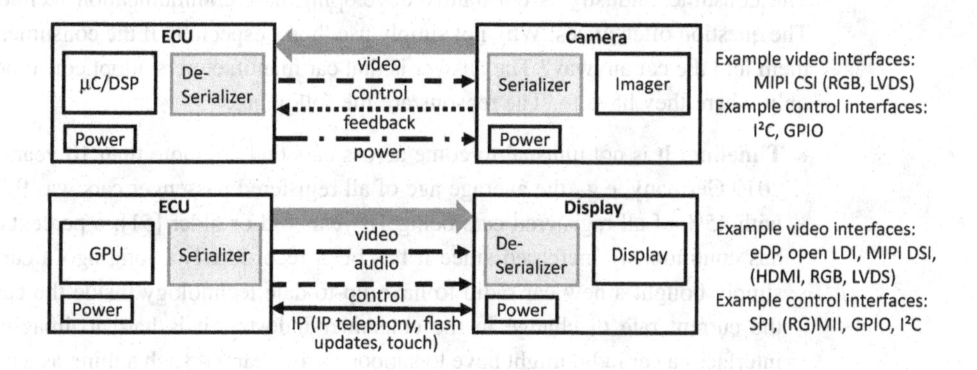

Figure 2.17 Differentiation potential for serializer and deserializer used for SerDes interfaces

a serial link (see also Figure 2.16). To support the high data rates, correct termination is important and shielded cables are used. Additionally, de-emphasis is used in the transmitter [44].

Some significant enhancements have been made to the technology since its introduction. Today, some SerDes interfaces are, e.g., current driven. This means that the information is not reflected in the voltage but in the current level. These systems are based on Current Mode Logic (CML) [44]. Also, the use of coaxial cables instead of shielded twisted pairs was introduced [45] and with it the possibility to simultaneously transmit power, i.e., Power-over-Coaxial (PoC). PoC allows for reducing cabling costs, production effort, and connector size.[11] This is an interesting option, especially for sensors at the edge of the car with limited power consumption.

Furthermore, early SerDes links transmitted pixels only. Control data had to use an additional communication technology. As a consequence SerDes link products started to appear that include a (bidirectional) control channel like I^2S or I^2C (e.g., [45, 46]), or an Ethernet channel (e.g., [47], see also Section 5.2.3). Additionally, products were and still are being differentiated by optimizing them for the different use cases (see also Figure 2.17). Cameras and displays have different requirements for the types of data that must be transmitted alongside the video data, and they support various physical interfaces.

The implementer can therefore choose from a variety of SerDes link solutions to optimize the implementation – which is desirable. However, not only because of different protocols being supported but also because of different physical layer implementations, all of the different products are proprietary to each vendor and are not interoperable – which is not desirable. As of 2019, there is no (successful) SerDes standard. Two efforts to change the situation were being conducted by the MIPI Alliance (see e.g. [48]) and the Automotive SerDes Alliance (ASA) [49, 50].

2.2.7 Consumer Links

The consumer industry is constantly developing new communication technologies. The question often arises: Why not simply use those, especially if the consumer brings them into the car anyway? The answer is that car manufacturers adopt consumer links only where they have to. The reasons are the following:

- **Timeline:** It is not unusual to come across cars that are more than 10 years old. In 2019 Germany, e.g., the average age of all registered passenger cars was 9.5 years, with 45% of all registered cars being 10 years old or older [51]; a percentage that has continuously increased since it has been recorded. Not long ago a car owner simply bought a new car radio to have up-to-date technology inside the car. With the current rate of change in the consumer industry, it is hard to imagine what interfaces a car radio might have to support in 10 years; if such a thing as a car radio still exists. Therefore, to use consumer technologies for in-vehicle networking would mean working on very unstable ground that the car industry has little control over.
- **Quality:** The quality requirements of the consumer industry are not nearly as stringent as those inside the car (see also Chapter 4). If technology of a suitable automotive quality can only be met with expensive cabling and expensive qualification programs, its attractiveness decreases drastically. Obviously, car manufacturers do rely on consumer interfaces when integrating consumer devices. This is one of the situations in which the use of a consumer link cannot be avoided. This leads to yet another quality issue: Generally, the perceived quality of the integrated functionalities is associated with the car despite its dependency on the Consumer Electronics (CE) device.
- **Networking functionalities:** A very popular consumer technology that most cars support in one way or other is the Universal Serial Bus (USB). USB is widely deployed and supported in many infotainment and communication related microcontrollers (μCs) and Digital Signal Processors (DSPs). Its use offers itself to the designers and since 2006 USB can be bought as an in-built interface to consumer devices inside the car [52, 53]. Additionally, it is sometimes used for ECU internal communication. Nevertheless, USB was designed to connect peripherals to a computer [54]. The topology it supports, and the communication schemes and networking functions are specific to that application and would incur significant costs or require significant workarounds if used for an extensive network inside a car. For example, USB is intended for a star topology with one master, the computer, controlling individually connected slaves, the peripherals. Such

EE-architecture would lead to extensive wiring inside a car. Additionally, automotive USB requires the use of expensive shielded cables (see also Section 3.1.2.2) with limited reach. It is a good example of a popular consumer link that is not really suitable for in-vehicle networking use, though, of course, not all communication use cases in the car require networking functionality.

- **New requirements:** The digitization of audio and especially video not only results in an increase in quality but also in an increase in complexity and in at least one function, which is not user friendly: Digital Rights Management (DRM). DRM requires data encryption and with that causes effort and costs in the components. The use of DRM is neither a choice of the consumer nor of the car manufacturers. Even if a car represents effectively a closed system, car manufacturers are required to provide DRM in their infotainment systems for the respective links, particularly if the communication technology used is one of those easily accessible to every consumer. One copyright protection method the automotive industry has to consider in this context is High-bandwidth Digital Content Protection (HDCP), which is needed for the High-Definition Multimedia Interface (HDMI) or its evolution, Mobile High-definition Link (MHL).
- **Dependency:** The car is a very complex system consisting of a large number of parts and technologies that all have to function together in an absolutely safe way. Any dependency on trends, developments, and manufacturers in other industries is counterproductive. Whatever the car industry integrates by choice needs to be controllable to a certain extent.

For these reasons, car manufacturers are careful when considering the use of consumer links inside the car. For the integration of consumer devices they often have to be supported, but in a clearly defined, limited, and isolated environment. So far, none of the wired consumer links has proved to be suitable as an in-vehicle networking technology.[12]

2.2.8 Trends and Consequences

The previous sections described important communication technologies prevailing in vehicles today. The first important message is that each of the technologies described was developed and/or is used with a specific application field in mind: CAN for robust ECU communication, LIN for low cost, MOST for high-end audio, FlexRay for X-by-Wire, SerDes interfaces for unprocessed video, and consumer links for consumer device integration. The car manufacturers actively drove some of the standardization work behind these technologies. Table 2.7 shows that this has led to very different technologies, not only with respect to the data rates supported but also with respect to the communication mechanisms and robustness methods used for the technologies.

Each new use case has thus led to new requirements, new standardization efforts, new communication principles, and new qualification processes. This is highly resource intensive in development and testing, especially as the technological complexities have increased significantly. Each new technology requires training and

Table 2.7 Comparison of discussed in-vehicle networking technologies, first overview, outlining the main, high level differences between Ethernet and the existing (in-vehicle) networking technologies

Technology	Multiple access scheme	Gross data rate	Robustness	Target use case
CAN (FD)	Priority based messages	Generally 500 kbps (2 Mbps) shared	Differential signal, comparably small data rate	Robust ECU control
LIN	Master–Slave and schedule tables	19.2 kbps shared	Small data rate	Low cost control
MOST	Priority based, TDMA, token	<25, 50, 150 Mbps shared	Optical for MOST 25/150	Complex, high-end audio
FlexRay	(Flexible) TDMA	10 Mbps shared	Differential signal	Real-time control, X-by-Wire
SerDes links	None, P2P communication between two partners only	Asymmetric, up to 12 Gbps uni-directional	Differential signal, shielded cables, short links	Links for unprocessed video
Consumer	None, P2P communication between two partners only	Up to 5 Gbps	Shielded cables, short links	Integration of consumer devices
Automotive Ethernet	Predominantly switched, for each link on the network, queuing	100/1000 Mbps, 2.5, 5, 10 Gbps per link and direction, 10 Mbps switched or shared	Differential signal, intelligent modulation and filtering	Use case independent packets

specialists who can solve inconsistencies and problems with the technology. Thus, the second important message is: While powerful in-vehicle networking is a fundamental requirement for functional innovations, the number of networking technologies needs to be as small as possible. After all it is the customer experience a customer buys with a car, not the in-vehicle networking technology enabling it. With respect to many of the early developments in in-vehicle networking (e.g., VAN, I/K-bus, K-Line, J1708, Byteflight, and D^{2}B – see the previous sections) a certain consolidation is already noticeable. In the authors' opinion, Ethernet/IP, while right now just another technology to be introduced into automotive networks, has the chance to drive this desired consolidation further,[13] even if it is highly unlikely that there will only ever be one in-vehicle networking technology.

In addition, car manufacturers face the challenge of rapidly changing customer expectations and product diversification in the form of new models, new derivatives, and potentially shorter life cycles. Modularization is a prominent way to handle this [55, 56]. In-vehicle networking technologies have to be flexible and have to support this. If with a derivative a higher speed grade is necessary for a communication link and if with this higher speed grade a completely different technology needs to be used, this is counterproductive. Message three thus is: The automotive industry requires a future-proof, in-vehicle networking technology that is flexible and scalable and can

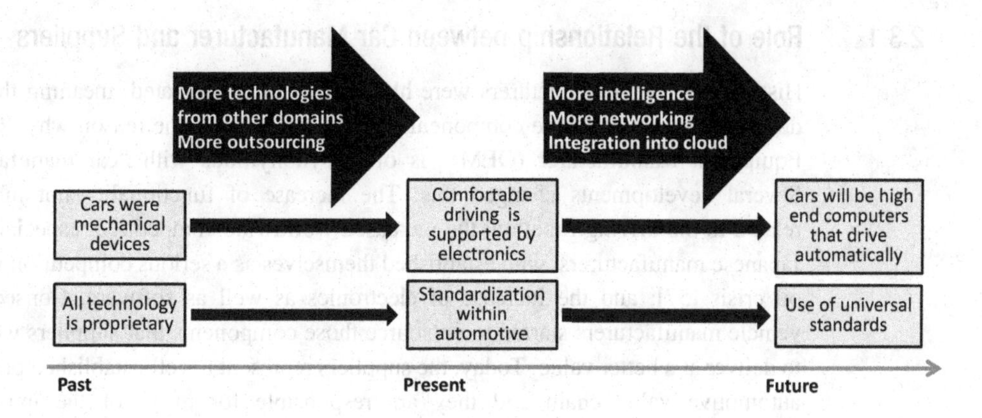

Figure 2.18 Long-term trends in automotive (networking)

grow with the requirements. More details in this context and Ethernet for in-vehicle use are described in Chapter 3.

One more aspect to consider is that, independent from what happens in the automotive industry, communication in general is irrevocably changing. There will always be specific physical environments that need to be addressed, such as short- or long-distance links, wired or wireless communication, EMC sensitive or insensitive environments, all of which require special treatment. No matter what the PHY looks like, the application data reduces to one type of data: Packets. This is one aspect of true digitization. Audio and video are compressed into packets. Circuit-switched telephone networks are being replaced by packet-switched networks (see also Section 1.2.4). The global Internet is packets anyway. As shown in Table 2.7, Ethernet is conceptually different from the traditional in-vehicle networking technologies: It is innately packet-based, high speed, and switched.

Figure 2.18 summarizes the phases the automotive industry is going through and indicates the direction that it is heading. When the automotive industry started, cars were purely mechanical devices. Over time electronics were added and are increasingly replacing traditionally mechanical functions as cars move toward being high-end computers that drive autonomously. In the beginning all the technologies in cars were proprietary. Then the automotive industry started to standardize especially non-differentiating functions like in-vehicle networking technologies. As always, the industry is in a constant search for larger economies of scale [57]. With all the electronics, cars are facing the same challenges that have been solved in other industries. The industry therefore moves toward the use of industry independent standards.

2.3 Responsibilities in In-Vehicle Networking

While the previous sections described the need for in-vehicle networking technologies as well as the technologies themselves, Sections 2.3.1 and 2.3.2 discuss the responsibilities for these technologies in the industry.

2.3.1 Role of the Relationship between Car Manufacturer and Suppliers

Historically, car manufacturers were highly vertically integrated, meaning they produced a great share of the components themselves. This is the reason why "Original Equipment Manufacturer (OEM)" is often synonymous with "car manufacturer." Several developments changed this: The increase of functionality not inherently related to the driving function; the increased worldwide competition, especially from Japanese manufacturers, who established themselves as a serious competition with the oil crisis [57]; and the increase of electronics as well as software. Consequently, vehicle manufacturers started to outsource those components that suppliers were able to deliver at a better value. Today, the suppliers represent a well-established part of the automotive value chain and they are responsible for many of the innovations happening in the automotive industry. The car manufacturers generally retain those parts in their own development that they have identified as their Unique Selling Point (USP).

Key to the unique customer experience is the composition, design, and overall functionality of the cars. To achieve this with the large number of parts from a large number of sources, these parts need to be precisely defined and assiduously integrated. A single day of delay in the Start of Production (SOP) of a car causes huge losses for the manufacturer. Every single component and its interaction with the rest of the car need to be faultless. Reliability is extremely important.

The V-cycle[14] (see also Figure 2.19) helps to structure the respective division of responsibilities between a car manufacturer and Tier 1s for electronic control units (ECUs). The car manufacturer defines the overall system requirements, distributes the individual functions needed to fulfill the system requirements to specific ECUs (i.e., defines the EE architecture of the system, see also Section 8.1), and then defines the interfaces to the rest of the car into which it will be integrated (i.e., the in-vehicle network). The Tier 1 supplier receives the respective component specification. The supplier is then responsible only for the ECU and its functionality according to the specification. The car manufacturer has to do the integration work up to the proof of functionality inside the car.

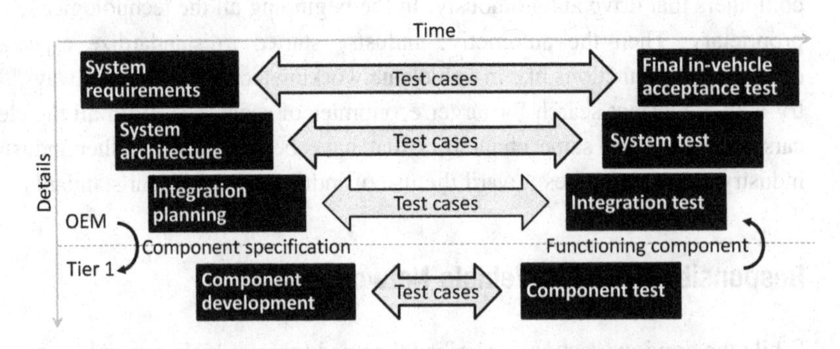

Figure 2.19 Division of responsibilities between car manufacturer and Tier 1 supplier along the V-cycle

In the OEM–Tier 1 relationship, the component definition is an interactive process. Car manufacturers often want what suppliers can provide (at a reasonable price point) and suppliers often build up that knowhow they expect to be reusing with various car manufacturers. This can lead to a chicken and egg problem, if both expect the other to propose new technologies and innovations. Close development partnerships between car manufacturers and Tier 1s are therefore common.

The V-cycle shows two things:

1. **The car manufacturer has a direct business relationship with Tier 1 suppliers only** and not, e.g., with a semiconductor vendor. The semiconductor vendor, who is a Tier 2 has, by definition, only dealings with the Tier 1s. The Tier 1 is the customer of the Tier 2 and the primary source for product requirements for the Tier 2, not the car manufacturer.
2. **The car manufacturer is responsible for the in-vehicle communication.** This comprises the correct design of the distributed functionalities, i.e., a precise definition of the communication interfaces,[15] and the provision of the data transmission functionalities, i.e., the right choice of the in-vehicle networking technology. This explains why the car manufacturers drive the development of in-vehicle networking technologies (see also Section 2.2 and Chapter 3). The Tier 1 needs to be able to handle the networking technology the OEM requires (meaning they generally have to be able to support more networking technology than any individual OEM [58]) but in the end it is the car manufacturer who is responsible for its choice. It requires Tier 2 semiconductor vendors to provide respective semiconductors, i.e., the basis for a new networking technology. This means that the Tier 2's customer and the responsible decision maker are not the same entity.

Semiconductor vendors play more of a role than just providing semiconductors for networking technologies. In general, many different types of semiconductors enable numerous innovations in automotive electronics. It is said that 90% of automotive innovations are driven by electronics and software [56] and that the value of semiconductors per vehicle is expected to increase from US$250–300 in 2011 to US$400–450 in 2020, not counting the value of semiconductors needed additionally in electrically powered vehicles [59]. Thus, not only the Tier 1s but also the Tier 2s play an important role for the innovations in the automotive industry. Like the OEM–Tier 1 relationship, the OEM–Tier 2 relationship can also present a chicken and egg problem. That is, the car manufacturers ask the Tier 2 what functions its semiconductors will enable, and the Tier 2 asks the car manufacturers what functions they want to have enabled, or it can also be a partnership in which innovations are driven together. What is different in the OEM–Tier 2 relationship is that there is no direct business relationship between the car manufacturer and semiconductor vendor. Furthermore, OEMs as well as Tier 1s would like to avoid a monopoly situation for any Tier 2. Therefore, car manufacturers do not like to require the use of a particular semiconductor in their specifications (which could offer business to Tier 2 suppliers indirectly). Car manufacturers require that certain functions are achieved, but not how it is done.

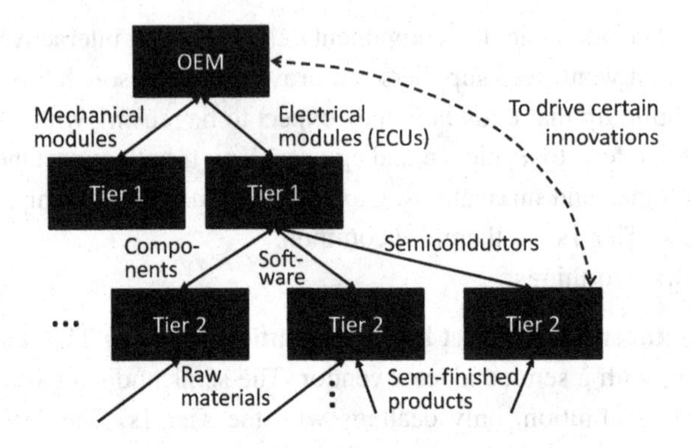

Figure 2.20 Car manufacturer (OEM) and supplier relationships

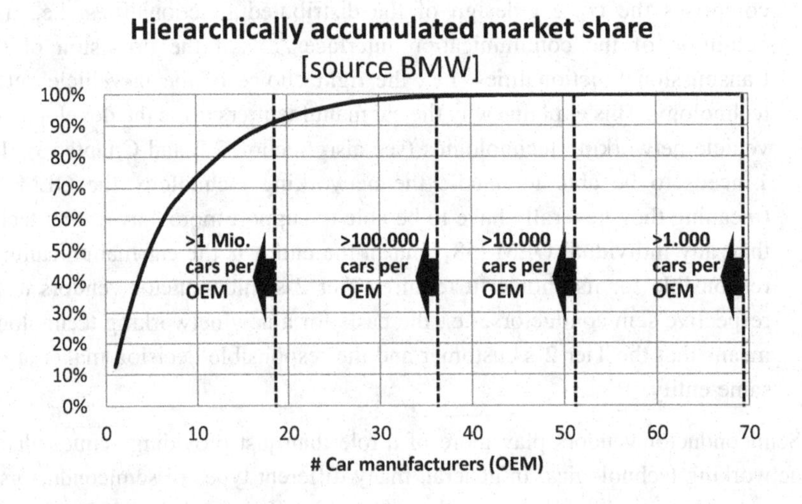

Figure 2.21 Accumulated market share in the automotive industry 2012

Hence, the OEM–Tier 2 relationship is not as clearly defined and the levers for successful ones are trickier to pin down. The role of personal relationships and long-term experiences should not be underestimated. Figure 2.20 visualizes the relationships between car manufacturers and the different supplier levels.

One more aspect to discuss is related to the specific structure of the automotive market. Figure 2.21 shows the accumulated market share in the automotive industry, in which the largest vendors are considered first. In 2012, for example, there were 69 car manufacturers who each had more than 1000 new cars registered[16] by customers. The cars of these 69 manufacturers were spread over 1310 different car models for 144 brands, and represented 99.99% of the market. Nevertheless, just five car manufacturers accounted for more than 50% of the new cars registered, and another 13 covered the next 40%. This means that the whole market consolidates to relatively few players and is therefore often referred to as an oligopoly (see e.g. [60]).

Each of the 18 largest car manufacturers is thus a powerful customer to the Tier 1 suppliers; not only because of sheer volume but also because of the type of product and the long product cycles. Traditionally, a car model generally runs for seven years,[17] so once a Tier 1 has passed the hurdle of having their technology designed into a car, it has a long-standing business for a customized product that is not easily replaceable, even in the case of an emergency. However, if the technology has not been designed into the car, that particular share of the market is lost for seven years. This results in a significant effort by the Tier 1 to make a good impression on the car manufacturers.

2.3.2 Role of the Relationships among Car Manufacturers

Car manufacturers create unique products out of a large number of components. It is not so much each individual component that makes the difference, but the right combination of thousands of parts integrated into a specific brand design. The competition, which is fierce also among car manufacturers, focuses on the final product and of course the design, but, with exceptions, not so much on the individual elements. This is particularly true for non-differentiating elements like in-vehicle networking. Section 2.2 described some of the communal efforts the car manufacturers made and some of the organizations that were founded in this context. The interactions between car manufacturers nevertheless go way beyond this. Cooperation is found at all levels, from purchasing, through development, to delivering parts to each other, to producing almost the same car. Even though this specific cooperation has come to an end, the VW–Crafter and Daimler–Sprinter production is a prominent, long-standing example of the latter [61]. A more recent example is the BW–Z4 and Toyota–Supra cooperation [62].

Figure 2.22 shows a snapshot of some of the 2019 interrelations between the 17 car manufacturers who had the most cars registered in 2018. Chinese car manufacturers have not been included. This is not because of their volumes – after all, most cars are produced in China [63], which includes cars from Chinese manufacturers – but because their interrelations, especially with non-Chinese car manufacturers, have a regulative, i.e., government enforced, edge to them. This blurs the assessment of the Chinese car manufacturers, their car market, as well as the relations to other car manufacturers somewhat [64, 65]. Most relationships shown in Figure 2.22 are for pure economic advantages and can change quickly depending on actual ownership changes, trends in the market, and other political considerations. Despite these changes (visible also when comparing the same figure among the different editions of this book), the diagram emphasizes that in an industry with so much interaction, cooperating in the development of in-vehicle networking or other non-differentiating functions is a matter of course.

Next to the direct bilateral relationships between car manufacturers, the automotive industry is divided into a multitude of organizations for all different kinds of topics. Depending on the topic and the gain from unification, these organizations are international, or – in a lot of cases – national. It is also not so unusual that a national unity is sought first, before an international unity is attempted. This is not only traditional

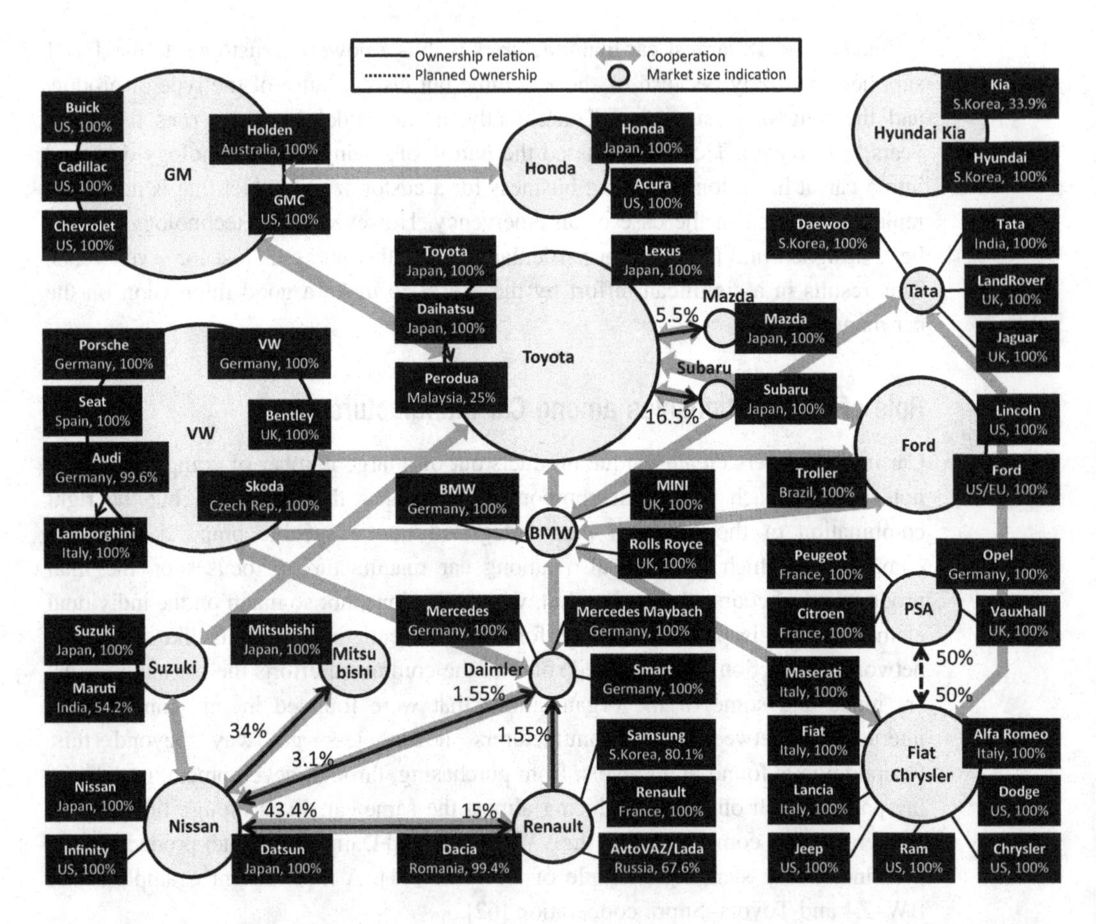

Figure 2.22 Example snapshot of some of the relationships between the 17 car manufacturers having sold most cars in 2018 (e.g., [66, 67])

but often very practical. After all, for the countries in which the car manufacturers are located, the automotive industry often plays a significant role in the national and regional economies in terms of direct and indirect employment, exports, Gross Domestic Product (GDP), etc. [68]. The industry is often nationally supported and the respective structures have existed for some time. A lack of language barriers and the reduced effort to meet in person add to the seeming nationalism. It is thus not surprising that early in-vehicle networking technologies originated and prevailed in different countries. With the globalization and the omnipresence of digital communication media this is changing. The developments around Automotive Ethernet (see Chapter 3) are a good example.

Another important aspect to understand when discussing relationships among car manufacturers, and why and how new networking technologies are introduced, is the driving forces behind innovations and their diffusion in the industry. Most innovative functions and features enter the market top down, meaning they are introduced in the high-end car segment first before they are sold in middle class or even small cars.

Table 2.8 The top 10 high-end (HE) car manufacturers and HE car brands in order of number of HE cars registered, example data from 2012 (source: BMW)

OEM	HE model ratio (%)	HE sales ratio (%)	Innov. ranking
Toyota Motor Corp.	26	16	5
General Motors Company	26	15	6
Nissan Motor Company	25	21	12
Volkswagen AG	23	8	1
Ford Motor Company	33	13	4
Hyundai Kia Automotive Group	24	10	6
BMW AG	47	32	2
Daimler AG	52	38	3
Chrysler Group LLC	30	26	15
Honda Motor Company	22	11	14
Brand			
Toyota	28	16	4
Nissan	20	20	9
Ford	27	11	3
BMW	50	39	1
Mercedes	52	41	2
Chevrolet	19	12	17
Hyundai	21	13	7
Audi	39	28	8
Honda	17	10	15
Dodge	27	31	20

HE model ratio = number of HE models/overall number of models, HE sales ratio = number of HE cars sold/overall number of cars sold, Innovation ranking of the same year according to [69].

This has several reasons. First, high-end car customers tend to be early adopters, willing to pay a premium for innovations. It is not unusual that a high-end car is bought fully equipped with all of the options. Second, and closely related to the first point, is that part of being high-end is that innovative features are offered in this segment first. Last, but not least, the relative cost of a new feature in a high-end car is significantly smaller than in a small car. It is therefore likely that the economies of scale necessary to allow that innovative feature to eventually be offered in smaller cars can only be achieved by starting at the high end.

It is the high-end cars and thus the high-end car manufacturers who drive automotive innovations and who bring new features into the industry. The innovation leaders need the support from a more powerful in-vehicle networking system first (see also Section 2.1) and are therefore likely to drive respective developments.

About 10% of the cars produced can be classified "High-End (HE)."[18] Table 2.8 lists the 10 car manufacturers and brands that sold the highest number of HE cars in 2012, with the manufacturers and brands listed top down according to the number of HE vehicles sold. What can be seen is that those car manufacturers who sell the most HE cars have neither the highest ratio for HE models to overall models ("HE model

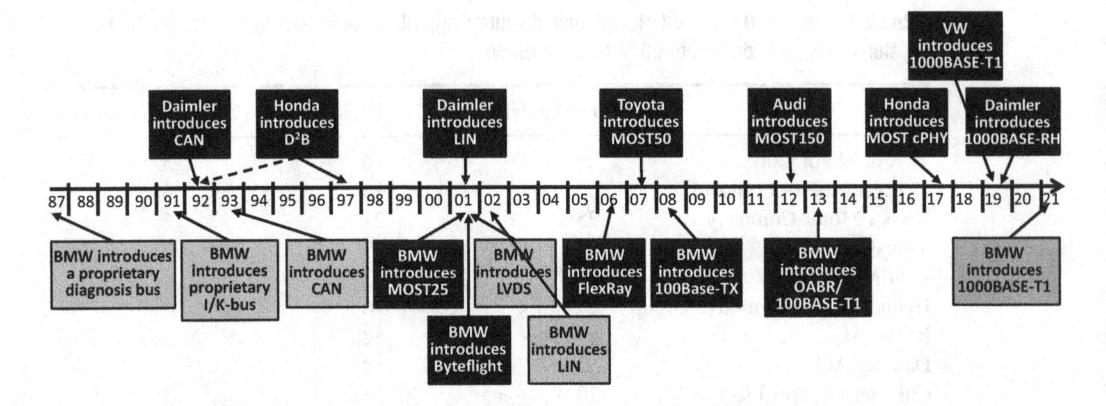

Figure 2.23 Timeline of the introduction of in-vehicle networking systems, at BMW and the industry as such. The event that first introduced a new technology is reflected by the dark boxes

ratio") nor for HE cars sold to cars sold ("HE sales ratio") nor the highest innovation ranking. Those who do have the highest HE model and HE sales ratio are those for whom HE cars are intrinsic for the company and its existence (BMW AG and Daimler AG). They also receive top marks when it comes to innovations. As a consequence, these companies are likely to experience the limits of existing in-vehicle networking technologies first, and they are also the ones accustomed to driving innovation. Figure 2.2 confirms, as well, that European cars have the highest number of networked nodes. It is thus not surprising to see these companies being particularly active when it comes to the development of new in-vehicle networking technologies, as has been described in the subsections of Section 2.2.

Last, but not least, Figure 2.23 shows the timeline in which in-vehicle networking systems have been introduced at BMW and in the car industry as such. It can be seen that BMW, as one of the innovation leaders, was often the first or one of the first to introduce a new in-vehicle networking technology.

Notes

1 Today, there are still a variety of networking technologies inside cars. A survey performed in 2015 showed that the major car manufacturers use, on average, eight different digital communication systems inside their cars today [58]. Additionally, despite the use of digital communication systems, there is still a lot of discrete wiring being used. In one car model BMW investigated, 50% of the weight of the harness consisted of power supply cables, 43% was discrete wiring, and only 7% was for the bus systems discussed in this book, even though they handle most of the communication. This emphasizes how impossible it would be today to design a car without in-vehicle networking systems. The vehicle would be choked with wiring.

2 With the help of passive coupling elements, it is also possible to realize more star-like topologies. The coupling elements function as a sort of ground that separates the different branches. Today, however, its deployment in the automotive industry is not very common.

3 The LS CAN transceiver does not only use the differential signal that HS CAN uses but additionally evaluates the absolute voltage to ground. In case one wire fails (e.g., breaks or gets disconnected), LS CAN will theoretically still function in an emergency mode. Nevertheless, this function has not been unambiguously defined for the receiving units and is therefore not really used [79].

4 Occasionally CAN is referred to as an "automotive fieldbus." As the terminology originated in industrial control (see also Section 1.2.3), this can only be attributed to its ability to being deployable in almost all areas, i.e., physically dispersed, locations in the car.

5 The information available on D^2B is not conclusive. What is likely is that the technology had been developed in the late 1970s [77], with a focus on home entertainment, and that it was transferred to an IEC standard in the (late) 1980s [75, 78]. Philips definitely played an important role [78], but also Matsuhita [77, 80], and Sony [75] seem to have been involved. For the first in-vehicle use, 1992 [75] and, which in the authors' opinion is more likely, 1997 [76] are quoted, with Honda as the respective car manufacturer [75]. Certainly, Daimler introduced D^2B in some of the Daimler models, and 1998 is realistic.

6 In the same year the MOST Co was founded, the car manufacturers Chrysler, Daimler, Ford, General Motors, Renault, and Toyota also founded the Automotive Multimedia Interface Corporation (AMIC) in order to define suitable hardware and software interfaces for automotive information, communication, and entertainment systems [81]. In this context, FireWire, i.e., IEEE 1394, was also discussed as a possible solution. Nevertheless, with the MOST Co gaining more traction and showing faster progress, the activity was disbanded in 2004 [71].

7 That all aspects of driving – including steering – can be controlled electronically is fundamental to enabling autonomous/automated driving. While autonomous driving is becoming a reality, electronic control of driving functions is so commonplace that the term "X by Wire" has become outmoded.

8 Blanking in digital image processing resulted from the need to retrace back from the end of a line to the beginning of the next or the end of a frame to the beginning of the next frame [72]. For the respective time, "blanking" data was sent. The overhead needed for this was reduced with progress, while parts of the blanking overhead was used to transmit other, system-related data. MD-Elektronik and Texas Instruments [39] considers 10% blanking overhead, independent of the actual resolution used. Real implementations have varying overheads depending on the actual parameters, with 10% presenting an upper limit.

9 Compressing the data supports content protection for Blu-ray if its data is transmitted over cables in a compressed and encrypted way. This is why Blu-Ray data is generally transmitted this way.

10 The term "SerDes" or "SerDes Interface" comes from the Serializer and Deserializer needed to realize the technology (see Figure 2.17).

11 SerDes technologies saw a significant cost reduction after Automotive Ethernet was introduced, with its first use case replacing SerDes-connected cameras (see also Section 3.4.2). All competition changes the market, and Automotive Ethernet is a strong competition for many in-vehicle networking technologies (see also the developments in CAN, Section 2.2.2).

12 The additional discussion of wireless (consumer) technologies such as Bluetooth or WiFi would open up a whole new set of topics to consider. While these are very interesting and are being discussed in the industry, we consciously decided not to address these in the context of this book (other than that WiFi would seamlessly integrate into the Ethernet network) in order to concentrate on the aspects important for understanding Automotive Ethernet.

13 In a survey performed within the car industry, the participants stated that, while every car manufacturer currently supports an average of eight in-vehicle networking technologies in their cars, the majority preferred between one and four technologies [58].

14 The V-cycle "V"isualizes the process how and in what order tasks must be performed in a project. It matches the idea, specification, and realization with the respective steps on the test and integration side. The model was first proposed at the end of the 1970s and has become especially popular in automotive (software) development [70, 74].

15 Naturally, in the case of malfunctions it is not always obvious whether the Tier 1 has implemented the communication requirements incorrectly or whether the car manufacturer has described them insufficiently. With the consequences that this can have on the SOP of a car, this is a very serious issue in the industry.

16 Counting "registered cars" results in somewhat different numbers than looking at the number of cars produced. Not every car is sold and/or registered in the same year it was built. The numbers are nevertheless similar enough to make absolutely no differences for the points made. This also applies to the year of ascertainment of the data. It can be expected that the market has consolidated somewhat more in 2020. However, the market principles will not have changed.

17 Seven years is a typical model cycle for German car manufacturers. Those seven years generally include a "facelift" in year four to ensure the cars stays up to date with the latest developments. It will be interesting to see whether the car industry will retain this in a world that seems to turn ever faster. There are early indications that model cycles are being accelerated [55]. The desire to offer new features independent of model cycles is also a common argument in the discussions around the use of Automotive Ethernet.

18 There are various different ways to classify cars into market segments (see [73] for an overview). In the investigation described here, "High-End" consists, in the classification of the European union, of the E-, F-, and the upper segment of the J-segments. In American/British English classifications this translates into: Full-size and mid-size luxury cars/executive cars, full-size luxury cars/luxury cars, grand tourers, supercars, and full-size SUVs/large 4×4s.

References

[1] Brose, "Gründung und Aufbau; 1908–1955," 2012. [Online]. Available: www.brose.com/de-de/unternehmen/historie/1908–1968/. [Accessed May 8, 2020].

[2] J. Donelly, "Needing a Lift, (Maybe) Finding It," August 8, 2008. [Online]. Available: www.hemmings.com/hcc/stories/2008/08/01/hmn_feature9.html. [Accessed May 7, 2020].

[3] Freundeskreis BMW 503, "Die zwei Serien im Vergleich," 2006. [Online]. Available: www.fb503.de/serien.htm. [Accessed May 8, 2020].

[4] T. Streichert and M. Traub, *Elektrik/Elektronik-Architekturen im Kraftfahrzeug*, Berlin, Heidelberg: Springer, Vieweg, 2012.

[5] Wikipedia, "History of Self-Driving Cars," December 2019. [Online]. Available: https://en.wikipedia.org/wiki/History_of_self-driving_cars. [Accessed December 22, 2019].

[6] A. Grzemba, *MOST, the Automotive Multimedia Network; from MOST 25 to MOST 150*, Poing: Franzis Verlag GmbH, 2011.

[7] P. Thoma, *Die EE-Geschichte*, München: BMW, 1997.

[8] Wikipedia, "On-board Diagnostics," April 28, 2020. [Online]. Available: http://en.wikipedia.org/wiki/On_Board_Diagnostics. [Accessed May 8, 2020].

[9] ARC Electronics, "RS232 Tutorial on Data Interface and Cables," unknown. [Online]. Available: www.arcelect.com/rs232.htm. [Accessed May 8, 2020].

[10] CiA, "CAN Newsletter Special Issue Automotive," October 2006. [Online]. Available (subscription only): www.can-cia.org.

[11] Wikipedia, "J1708," 24 October 2019. [Online]. Available: http://en.wikipedia.org/wiki/J1708/J1708. [Accessed May 8, 2020].

[12] CiA, "History of CAN Technology," 2019 (continuously updated). [Online]. Available: www.can-cia.de/can-knowledge/can/can-history/. [Accessed May 6, 2020].

[13] BOSCH, "License Conditions CAN Protocol and CAN FD Protocol," June 2013. [Online]. Available: www.bosch-semiconductors.de/media/automotive_electronics/pdf_2/ipmodules_3/can_protocol_license_1/Bosch_CAN_Protocol_License_Conditions.pdf. [Accessed February 23, 2014, no longer available].

[14] Automotive News, "Top Suppliers," 17 June 2013. [Online]. Available: www.autonews.com/assets/PDF/CA89220617.PDF. [Accessed May 8, 2020].

[15] C&S, "Munich Group Server," C&S, 2013. [Online]. Available: www2.cs-group.de/MunichGroupServer/public/testspec.php. [Accessed October 16, 2013, no longer available].

[16] T. Islinger and Y. Mori, "Ringing Suppression in CAN-FD Networks," (1) 2016. [Online]. Available: https://can-newsletter.org/media/raw/1f408a2b48db3f512e99754878396e70.pdf. [Accessed December 23, 2019].

[17] T. Adamson, "Migrating to CAN FD," March 27, 2017. [Online]. Available: www.can-cia.org/fileadmin/resources/documents/slides/icc_2017_slides_adamson.pdf. [Accessed December 23, 2019].

[18] ISO, *ISO/DIS 11898-2:2016 – Road vehicles – Controller Area Network (CAN) – Part 2: High-speed Medium Access Unit*, Geneva: ISO, 2016.

[19] M. Schreiner, L. Donat, and S. Köngter, "Introduction of CAN FD into the Next Generation of Vehicle E/E Architectures," 2017. [Online]. Available: www.can-cia.org/fileadmin/resources/documents/conferences/2017_schreiner.pdf. [Accessed December 23, 2019].

[20] A. Mutter, "CAN XL – an Introduction and Comparison to Available IVN-Technologies," in *Hanser Automotive Networks*, Munich, 2019.

[21] BOSCH, "CAN with Flexible Datarate; White Paper Version 1.1," 2011. [Online]. Available: www.bosch-semiconductors.de/media/pdf_1/canliteratur/can_fd.pdf. [Accessed January 15, 2013, no longer available].

[22] BOSCH, *CAN Specification Version 2.0*, Stuttgart: BOSCH, 1991.

[23] W. Lawrenz, *CAN Controller Area Network, Grundlagen und Praxis*, 2nd edition, Heidelberg: Hüthig, 1997.

[24] D. Marsh, "LIN Simplifies and Standardizes In-Vehicle Networks," *EDN*, pp. 29–40, April 28, 2005. Available: www.edn.com/lin-simplifies-and-standardizes-in-vehicle-networks/. [Accessed November 27, 2020].

[25] LIN Consortium, "LIN Specification Package, Revision 2.2A," LIN Consortium, 2010.

[26] Consortiuminfo, "Local Interconnect Network Consortium (LIN)," July 8, 2015. [Online]. Available: www.consortiuminfo.org/links/linksdetail2.php?ID=428. [Accessed May 6, 2020].

[27] H. Schoepp, "In-Vehicle Networking Today and Tomorrow," January 17, 2012. [Online]. Available: www.automotive-eetimes.com/en/in-vehicle-networking-today-and-tomorrow.html?cmp_id=71&news_id=222902008. [Accessed August 28, 2013, no longer available].

[28] MOST Cooperation, "MOST 150 Inauguration in Audi A3," *MOST Informative*, no. 8, p. 2, October 2012.

[29] Microchip, "Daimler Adopts Microchip's MOST150 Devices With Coaxial Physical Layer for Future Infotainment Network System," January 5, 2017. [Online]. Available: www.microchip.com/en/pressreleasepage/daimler-adopts-microchip-s-most150-devices-with-coaxial-physical-layer-for-future-infotainment-network-systems. [Accessed December 23, 2019].

[30] H. Kaltheuner, "Das Universalnetz, Ethernet-AVB: Echtzeitfähig und Streaming-tauglich," *c't*, no. 13, pp. 176–181, June 2013.

[31] A. Grzemba, *MOST – Das Multimedia-Bussystem für den Einsatz im Automobil*, Poing: Franzis Verlag GmbH, 2007.

[32] FlexRay Consortium, "FlexRay Communications System Protocol Specification Version 3.0.1," FlexRay Consortium, 2010.

[33] K. R. Avinash, P. Nagaraju, S. Surendra, and S. Shivaprasad, "FlexRay Protocol Based an Automotive Application," International Journal of Emerging Technology and Advanced Engineering, pp. 50–55, May 2012.

[34] K. Leiß, "Kommunikationssysteme: FlexRay," 26 January 2010. [Online]. Available: www6.in.tum.de/pub/Main/TeachingWs2009Echtzeitsysteme/20100126_FlexRay_K_Leiss.pdf. [Accessed October 29, 2013, no longer available].

[35] C. Hammerschmidt, "FlexRay not Dead, Chipvendors Claim," October 25, 2012. [Online]. Available: www.automotive-eetimes.com/en/flexray-not-dead-chip-vendors-claim.html?cmp_id=7&news_id=222902569. [Accessed October 30, 2013, no longer available].

[36] heiseAUTOS, "Schon in einem Jahr kommt ein Infiniti mit elektronischer Lenkung auf den Markt," October 17, 2012. [Online]. Available: www.heise.de/autos/artikel/Nissan-bringt-Steer-by-Wire-in-Serie-1732435.html. [Accessed May 8, 2020].

[37] OPEN Alliance, All Member F2F Meeting, *Results from a Structured Discussion on Capabilities of IVN Technologies*, Detroit: OPEN Alliance, 2019.

[38] Wikipedia, "Blu-ray," 12 December 2019. [Online]. Available: https://en.wikipedia.org/wiki/Blu-ray/Blu-ray. [Accessed December 25, 2019].

[39] MD-Elektronik and Texas Instruments, *Driving Higher Resolutions In Automotive Display and Camera Applications*, Dallas, TX: Texas Instruments, 2016.

[40] Wikipedia, "Display Resolution," April 30, 2020. [Online]. Available: https://en.wikipedia.org/wiki/Display_resolution. [Accessed May 8, 2020].

[41] ISO, *ISO/DIS 17215:2013 – Road Vehicles – Video Communication Interface for Cameras (VCIC) Parts 1–4*, Geneva: ISO, 2013.

[42] Blu-ray.com, "Blu-ray FAQ, How Fast Can You Read/Write Data on a Blu-ray Disc?," unknown. [Online]. Available: www.blu-ray.com/faq/#bluray. [Accessed May 8, 2020].

[43] Wikipedia, "Low-voltage Differential Signaling," September 6, 2019. [Online]. Available: http://en.wikipedia.org/wiki/LVDS. [Accessed May 8, 2020].

[44] Texas Instruments, "LVDS Owner's Manual," 2008. [Online]. Available: www.ti.com/lit/ml/snla187/snla187.pdf. [Accessed August 29, 2013, no longer available].

[45] Wikipedia, "FPD-Link," January 28, 2020. [Online]. Available: http://en.wikipedia.org/wiki/FPD-Link/FPD-Link. [Accessed May 8, 2020].

[46] Maxim, "Gigabit Multimedia Serial Link with Spread Spectrum and Full-Duplex Control Channel," January 2011. [Online]. Available: http://datasheets.maximintegrated.com/en/ds/MAX9259-MAX9260.pdf. [Accessed May 8, 2020].

[47] J. Schyma, "Big Data Rates in the Car: Ethernet & More Over One Single Link System," in 5th Ethernet & IP @ Automotive Technology Day, Yokohama, 2015.

[48] Green Car Congress, "MIPI Alliance A-PHY v.1.0 to Support Lidar, Radar and Camera Integration for Autonomous Driving at 12–24 Gbps and Beyond," August 4, 2018. [Online]. Available: www.greencarcongress.com/2018/08/20180804-mipi.html. [Accessed December 25, 2019].

[49] I. Kuss, "Automotive SerDes Alliance, A New Standard for the Automotive Industry," Elektroniknet.de, 25 July 2019. [Online]. Available: www.elektroniknet.de/international/a-new-standard-for-the-automotive-industry-167729.html. [Accessed December 25, 2019].

[50] Automotive SerDes Alliance, "Automotive SerDes Alliance (ASA) Hompage," 2019 (continuously updated). [Online]. Available: https://auto-serdes.org. [Accessed December 25, 2019].

[51] Kraftfahrbundesamt, "Bestand an Kraftfahrzeugen und Kraftfahrzeuganhängern nach Fahrzeugalter 1. Januar 2019," January 1, 2019. [Online]. Available: www.kba.de/SharedDocs/Publikationen/DE/Statistik/Fahrzeuge/FZ/2019/fz15_2019_pdf.pdf?__blob=publicationFile&v=7. [Accessed December 26, 2019].

[52] bookofjoe, "Microsoft Windows Automotive – World's First Production Car with a Factory-installed USB Port," March 12, 2006. [Online]. Available: www.bookofjoe.com/2006/03/microsoft_windo.html. [Accessed May 6, 2020].

[53] just-auto, "USA: Cars to Get USB Ports Soon – CD Players to Become Obsolete?," 31 October 2005. [Online]. Available: www.just-auto.com/news/cars-to-get-usb-ports-soon-cd-players-to-become-obsolete_id84660.aspx. [Accessed May 6, 2020].

[54] R. Sheldon, "USB Communications Introduction," 2008. [Online]. Available: www.controlanything.com/Relay/Device/A0003. [Accessed September 2, 2013].

[55] T. Grünweg, "Modellzyklen der Automobilhersteller: Eine Industrie kommt auf Speed," February 10, 2013. [Online]. Available: www.spiegel.de/auto/aktuell/warum-lange-entwicklungszyklen-fuer-autohersteller-zum-problem-werden-a-881990.html. [Accessed May 8, 2020].

[56] Automobil Konstruktion, "Marktforderung: modularer Mehrfachnutzen," May 22, 2009. [Online]. Available: https://automobilkonstruktion.industrie.de/top-news/topbeitraege/marktforderung-modularer-mehrfachnutzen/. [Accessed May 8, 2020].

[57] G. Volpato, "The OEM–FTS Relationship in Automotive Industry," *International Journal of Automotive Technology and Management*, vol. 4, no. 2/3, pp. 166–197, 2004.

[58] K. Matheus, "Die Zukunft von Vernetzungstechnologien im Fahrzeug – Ergebnisse einer Umfrage," February 2016. [Online]. Available: www.hanser-automotive.de/zeitschrift/archiv/artikel/die-zukunft-von-vernetzungstechnologien-im-fahrzeug-ergebnisse-einer-umfrage-1299263.html. [Accessed May 15, 2020].

[59] K. Sievers, "Semiconductors – Enablers of Future Mobility Concepts," February 22, 2012. [Online]. Available: www.nxp.com/wcm_documents/news/download-media/presentations/nxp-sievers-zvei-elektromobilia-final.pdf. [Accessed May 8, 2020].

[60] J. Bouman, "Section 3: Characteristics of an Oligopoly Industry," Inflate your mind, 2011. [Online]. Available: https://inflateyourmind.com/microeconomics/unit-8-microeconomics/section-3-characteristics-of-an-oligopoly-industry/. [Accessed August 17, 2013].

[61] Auto.de, "Mercedes Sprinter/VW Crafter: Die Kooperation wackelt," October 20, 2012. [Online]. Available: www.auto.de/magazin/mercedes-sprinter-vw-crafter-die-kooperation-wackelt/. [Accessed May 8, 2020].

[62] G. Kacher, "Neue Angstgegner für Boxster, Cayman und TT," August 27, 2017. [Online]. Available: www.sueddeutsche.de/auto/bmw-und-toyota-neue-angstgegner-fuer-boxster-cayman-und-tt-1.3638062?print=true. [Accessed December 26, 2019].

[63] I. Wagner, "Worldwide Automobile Production through 2018," September 11, 2019. [Online]. Available: www.statista.com/statistics/262747/worldwide-automobile-production-since-2000/. [Accessed December 27, 2019].

[64] M. Grzanna, "China's Love–Hate Relationship with the German Automotive," 22 April 2013. [Online]. Available: https://worldcrunch.com/business-finance/china039s-love-hate-relationship-with-the-german-automobile. [Accessed May 8, 2020].

[65] E. Nefzger, "Die Partei entscheidet immer mit, Interview mit Jochen Siebert," November 27, 2019. [Online]. Available: www.piegel.de/auto/aktuell/uiguren-in-china-warum-colkswagen-ein-werk-in-xinjiang-gebaut-hat-a-1298154.html. [Accessed December 26, 2019].

[66] K. Laing, "Ford, GM and Toyota Collaborate for Self-Driving Safety Rules," April 3, 2019. [Online]. Available: https://eu.detroitnews.com/story/business/autos/2019/04/03/ford-gm-toyota-announce-robot-car-partnership/3350539002/. [Accessed December 27, 2019].

[67] T. Harloff and G. Stegmaier, "Wer mit wem in der Autoindustrie," March 21, 2019. [Online]. Available: www.auto-motor-und-sport.de/tech-zukunft/alternative-antriebe/welche-autohersteller-miteinander-kooperieren-ueberblick/. [Accessed May 25, 2020].

[68] OECD, "Chapter 2: The Automobile Industry in and beyond the Crisis," 2009. [Online]. Available: www.oecd.org/eco/outlook/44089863.pdf. [Accessed May 8, 2020].

[69] Center of Automotive Management, "Automotive Innovations Award 2012," May 2, 2013. [Online]. Available: www.auto-institut.de. [Accessed May 9, 2013, no longer available].

[70] J. Schäuffele and T. Zurawka, *Automotive Software Engineering – Grundlagen, Prozesse, Methoden und Werkzeuge*, Wiesbaden: Vieweg + Teubner Verlag, 2005.

[71] The Hansen Report, *AMI-C nearly spent*, Portsmouth: Hansen, 2004.

[72] G. Eitzmann, "Video Format Compiler Programmer's Guide; Chapter 3. Building Blocks of a Video Format," 1996. [Online]. Available: http://csweb.cs.wfu.edu/~torgerse/Kokua/Irix_6.5.21_doc_cd/usr/share/Insight/library/SGI_bookshelves/SGI_Developer/books/VFC_PG/sgi_html/ch03.html. [Accessed December 25, 2019].

[73] Wikipedia, "Car Classification," March 31, 2020. [Online]. Available: http://en.wikipedia.org/wiki/Car_classification. [Accessed May 8, 2020].

[74] Die Beauftragte der Bundesregierung für Informationstechnik, "Häufg gestellte Fragen zum V-Model XT," Die Beauftragte der Bundesregierung für Informationstechnik, 2012. [Online]. Available: www.cio.bund.de/Web/DE/Architekturen-und-Standards/V-Modell-XT/Haeufig-gestellte-Fragen/haeufig_gestellte_fragen_node.html#doc2157266bodyText1. [Accessed May 8, 2020].

[75] M. Pöllhuber, "Plastic Optical Fiber," 2001. [Online]. Available: http://tkhf.adaxas.net/cd1/Optik.pdf. [Accessed May 8, 2020].

[76] Tyco Electronics (TE Connectivity), "High Speed Data Networking for the Automotive Market," 2007. [Online]. Available: http://docplayer.net/19160927-High-speed-data-networking-for-the-automotive-market.html. [Accessed May 8, 2020].

[77] Wikipedia, "D²B," November 23, 2018. [Online]. Available: http://de.wikipedia.org/wiki/D2B. [Accessed May 8, 2020].

[78] Wikipedia, "IEC 61030," January 19, 2020. [Online]. Available: http://en.wikipedia.org/wiki/IEC_61030. [Accessed May 8, 2020].

[79] S. Meier, "CAN, Controller Area Network," unknown. [Online]. Available: www.siegfriedmeier.de/Download/X/CAN.pdf. [Accessed May 8, 2020].

[80] IT Wissen, "D2B (Domestic Data Bus)," unknown. [Online]. Available: www.itwissen.info/definition/lexikon/domestic-data-bus-D2B-D2B-Bus.html. [Accessed September 17, 2013, no longer available].

[81] PR Newswire, "Automotive Multimedia Interface Collaboration Announces Formalization of Objectives and Operating Procedures," 27 April 1998. [Online]. Available: www.prnewswire.com/news-releases/automotive-multimedia-interface-collaboration-announces-formalization-of-objectives-and-operating-procedures-74245312.html. [Accessed September 17, 2013, no longer available].

3 A Brief History of Automotive Ethernet

When a new technology is being developed and adopted in an industry, there are various factors that impact the success of that technology. In the authors' opinion the most important ones are its benefits, its costs, and the framework that allows an industry to develop around it. This chapter will discuss these factors with respect to Automotive Ethernet. However, as is frequently pointed out (see e.g. [1–3]), it is not only technical facts but also individuals who act as the driving force behind a new technology. With respect to Automotive Ethernet, both authors feel that they have had a role in the events described herein. As a consequence, the descriptions in this chapter will sometimes reflect personal viewpoints.

3.1 The First Use Case: Programming and Software Updates

3.1.1 Architectural Challenges

In 2004, BMW decided to introduce a central gateway Electronic Control Unit (ECU) in its cars starting from 2008 Start of Production (SOP) onwards. This central gateway was to perform two functions: (1) to route data between the different CAN, FlexRay, and MOST (Media Oriented Systems Transport) buses inside the cars and (2) to function as the diagnostic and programming interface with the outside world. For the latter, BMW has always used a centralized approach. This means that software can only be flashed with an external tester device that is connected via the OnBoard Diagnostics (OBD) connector [4] and, when flashing is performed, all of the flashable ECUs inside the car, without exception, are updated with their newest software versions. This approach assures that the customer always has the latest software in the car and that there are no sudden software inconsistencies in functional domains, in which some units have been updated and others have not. This is an architectural choice that, as it happens, was decisive for the introduction of Automotive Ethernet. Nevertheless, there are car manufacturers that use decentralized approaches for software updates and handle the version management differently. They might update, for example, the multimedia/infotainment unit only and use USB or DVD for it, while using the OBD connection to update other, individual ECUs.

In 2004, BMW used a High-Speed CAN (HS CAN) interface with the OBD connector for connecting the tester to the in-vehicle network. The physical limit of

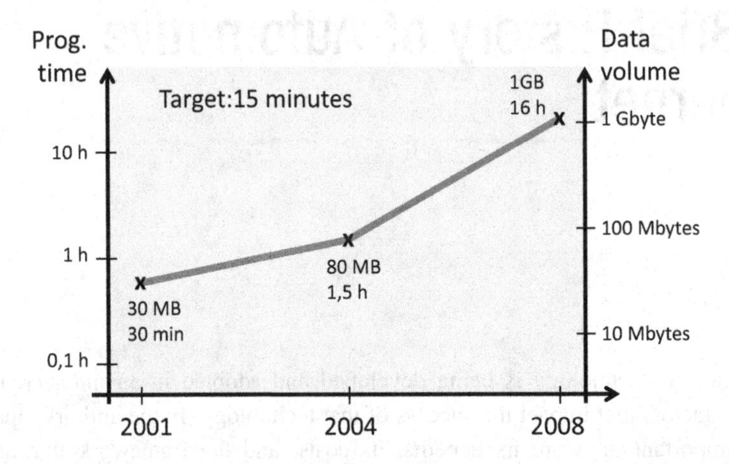

Figure 3.1 Flash data volume and programming time predicted in 2004 for 2008 at BMW [5]

the HS CAN was/is 500 kbps (see also Section 2.2.2); the additional overhead due to the protocols needed for this application reduced the net data rate to about 200 kbps. The prognosis for the accumulated amount of flash data in 2008 exceeded 1 Gbyte; some of the multimedia devices had several hundreds of Mbytes of software (including map data), and also the new FlexRay connected devices in the chassis domain needed a significant amount of software. Taking everything into account, the complete software update of a well-equipped, high-end car would have exceeded 16 hours. At a dealer, this would have meant sending customers home with a replacement car and making them come back the next day, even though all they potentially required was an update of the map data. Also, in the factory a flash time of 16 hours was unthinkable. The target duration for the software update was set to 15 minutes (see also Figure 3.1). Obviously, an interface technology other than HS CAN was needed.

3.1.2 Potential Car Interface Technologies

The new technology needed to fulfil several basic requirements:

1. The data rate had to be sufficiently high. The intention was to flash all units/busses connected to the central gateway in parallel. The flash memory in the Head Unit (HU) at the time allowed for write access at 15 Mbps. Additionally, a FlexRay bus with 4 Mbps, 2 HS CAN buses, and 1 LS CAN had to be serviced. This led to a required net data rate of about 20 Mbps from the interface.
2. The flash process is only performed very few times during the lifetime of a car. It was therefore not acceptable to require additional processing resources in the central gateway for the flash process only. It was thus important that the selected interface technology would not overstress the available resources.
3. It was intended to have the flash process as part of a (worldwide) networking function. Within the local garage, the network allows more than one external tester

to be connected to the car at a time, or to have one tester connected to multiple cars. On a worldwide scale, with the latest software on a central server, good integration into the network would allow this software to be flashed directly into a car at any dealership in the world.

4. The solution needed to be cost-efficient, both inside the car and in the test equipment used at the dealer and factory. BMW intended to introduce a new flash concept. Backwards compatibility with existing systems was not required.

3.1.2.1 Evaluation of MOST

In principle, MOST provided a higher data rate than was needed and it had been introduced in BMW serial cars three years earlier, in 2001. Nevertheless, MOST 25 was also considered unsuitable. The primary use case for MOST 25 was synchronous audio communication. In respect to the required high-speed data communication, this had the following disadvantages:

- **Insufficient data rate:** The maximum net bandwidth on the MOST 25 asynchronous data channel is only about 7 Mbps (the gross data rate of 12.7 Mbps mentioned in Section 2.2.4.2 is significantly reduced due to protocol overheads).
- **High resource demand:** To achieve the maximum net bandwidth of 7 Mbps, it is necessary to use data packets of 1014 bytes, i.e. ~1 kbyte. Additionally, it requires using a block acknowledgement for 64 packets that is part of the so-called MOST-high protocol. For the software update use case this would have meant completing the reception of 64 MOST packets before being able to send them on. In consequence, 64 kbyte RAM would have been needed for this procedure only. Additionally, the block acknowledge would have affected the routing between MOST 25 and the other bus systems, which alone was estimated to take up the computation power of a complete Central Processing Unit (CPU).
- **Wrong topology:** It is a fundamental quality of a tester that it is attached to the car only temporarily. Because MOST requires a ring topology, this would have meant either adding a second MOST ring between tester and gateway during testing, or extending the ring when the tester was attached. Both concepts were unattractive for complexity reasons.
- **No IP support:** MOST 25 did/does not speak IP, i.e., does not provide for routers, switches, or even hubs (i.e., repeaters). To integrate the system into the diagnostics network at a BMW dealer or BMW, as such, would have required significant effort and workarounds. IP support was only introduced later with MOST 150. MOST 25 relied on the MOST-high protocol only.
- **New interface:** MOST would have been a completely new interface for the external testers that are developed and used with different technical background and focus. Adding MOST to the testers would have required adding the respective hardware and software interfaces to the testers along with the introduction of the communication paradigms of MOST to the diagnostic application.
- **High costs:** The interface is comparatively costly.

In 2004, the development of a next generation MOST that would support IP and data communication better was being discussed. Some of the disadvantages thus might have been easier to overcome. However, the development timeline was too long and the concepts were too immature to base a decision on. In consequence, BMW decided against using MOST as the diagnostic car interface; and rightly so. MOST 150 eventually saw market introduction in 2012 (see also Figure 2.23).

FlexRay's gross data rate of 10 Mbps was obviously too small. Also, in 2004, FlexRay had not yet been introduced, and as its SOP was planned for 2006 this raised concerns about the maturity of the technology.

3.1.2.2 Evaluation of USB

The next interface investigated was USB 2.0. USB was well known as a consumer interface and was on the roadmap of many car manufacturers to be introduced as an interface for consumer devices (see also Section 2.2.7). It was very common in the PC environment and thus suitable for the external testers as well. Also, with a data rate of 480 Mbps, the bandwidth of USB 2.0 was more than sufficient. Nevertheless, when investigating USB in detail, the following disadvantages led to the decision not to use USB as the diagnostic interface:

- **Insufficient robustness/immunity:**[1] To achieve sufficient signal integrity, expensive cables and connectors would have been needed.
- **Insufficient cable length:** USB allows only for a cable length of about four meters, which is a disadvantage in a large garage.
- **No network support:** As said, the idea was to have more than one external tester connected to the car at a time, or to have one tester connected to more than one car. With USB this would have led to a collision of multiple USB controllers, or to very complex, non-standard compliant workarounds.
- **New protocol:** The automotive protocol stack and driver would have to be developed for the use case.

As a consequence, USB was also not the right solution. LVDS/pixel links/SerDes interfaces were never investigated, because they could not support networking (see also Section 2.2.6). FireWire (IEEE 1394) had disadvantages similar to those of USB. Additionally, the physical interface was unclear, and the automotive industry had not yet accumulated any experience with the technology.

3.1.3 The Solution: 100BASE-TX Ethernet

The next technology to investigate was Ethernet. It provided a sufficient data rate, was readily available in computers and laptops, and was a networking technology, i.e., it promised to fulfill the idea of handling the car as a node in the worldwide network and in a larger (diagnostic) network at the dealer. In this network, multiple cars are connected to one tester or multiple testers are connected to one car, while all being connected to the backend at BMW. In 2004, the idea of using Ethernet in the automotive industry was unheard of. But, Ethernet was/is a well-documented technology and provides a good infrastructure, so it was reasonable for BMW to assess its suitability.

3.1.3.1 The Physical Layer

The anticipated, most critical element for Ethernet deployment was the physical layer. The expectation was that, as for USB, the automotive robustness requirements would result in (too) expensive cabling and connectors inside the car. This turned out to be wrong. The first experiments in the ElectroMagnetic Compatibility (EMC) lab were run with two PCs connected via two pairs of simple Unshielded Twisted Pair (UTP) CAN cables that BMW had used in serial production for some years. These cables did not at all comply with the standard CAT 5 cable defined for 100BASE-TX Ethernet. Yet, the very first measurement results for the immunity showed that the set-up met the immunity requirements for in-vehicle communication, without any modifications being necessary!

The situation was different for the EMC emissions.[2] The emissions were way beyond the limit lines and would have caused audible reception distortions in the FM radio, if used at runtime.[3] Nevertheless, in the case of software updates at the dealer or in the factory the car is stationary and not in use, e.g., by a customer driving and listening to the radio. To ensure that the 100BASE-TX UTP Ethernet connection could not cause distortions during runtime, BMW added an "activation line" in a later implementation. This activation line ensured that the 100BASE-TX UTP Ethernet connection inside the car would be active only when the external tester was connected.

To start with, it was expected that RJ-45 connectors had to be reused, and all the first investigations used these. In the end, this turned out to be unnecessary. It was possible to add the two wire pairs necessary for the 100BASE-TX connection to the OBD connector (see Figure 3.2). The well-established and standardized connector offers four vehicle manufacturer specific pins, and BMW decided to use those for the 100BASE-TX connection. Measurements in the EMC lab proved that the immunity still met the requirements. So, after the initial evaluation, in 2005, 100BASE-TX was considered a promising technology and a decision was taken to seriously investigate its use.

3.1.3.2 Protocol Stack and Software

With the CAN interface, BMW used the Unified Diagnostic Services (UDS) protocol. UDS describes the handling of diagnostic information in the automotive industry and is specified in ISO 14229-1 [7]. When moving to a new networking technology, BMW wanted to avoid defining a new protocol and new sequences for the software update, even though BMW did not require backwards compatibility to the existing implementation. At the same time, Ethernet was an "IT-Technology" with a pool of available protocols and technologies. Thus, the next step after establishing the principle suitability of the PHY was to investigate the reuse and adaptability of standard IT protocols for Ethernet-based diagnostic communication in the automotive industry.

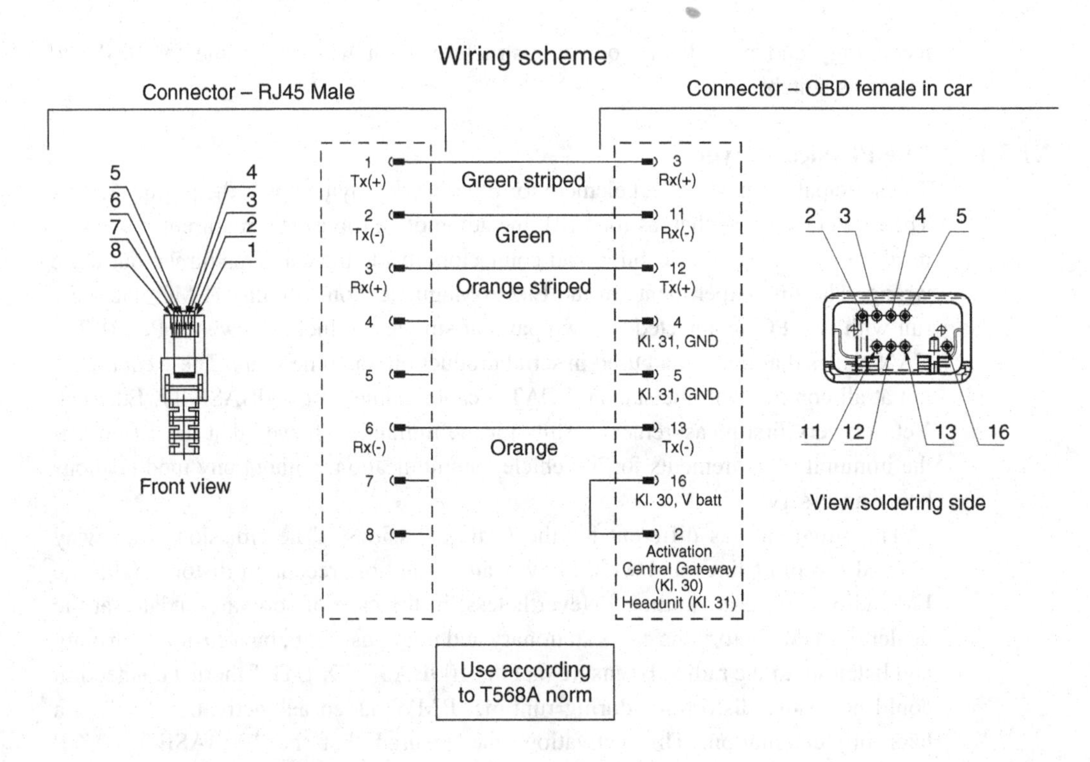

Figure 3.2 RJ-45 100BASE-TX Ethernet connector in relation to the vehicle OBD connector [6]

Table 3.1 shows the result. It is unique because it showed for the first time how IT and automotive standard protocols can be matched. It was the first time at BMW that no new protocol had to be developed from scratch for automotive use – instead the focus was on reuse and synergies. With only a small addition, called "High-Speed Fahrzeug Zugang (HSFZ)," which was needed in order to enable the parallel flash process and to map the UDS onto TCP, it proved to be perfectly possible.

At the same time as collecting the protocols and solutions, their portability from Linux[4] to automotive operating systems needed to be investigated. For multimedia ECUs this would not have been so much of an issue, as they are normally based on modern operating systems like Linux or QNX.[5] The gateway, however, was a typical automotive ECU using an OSEK[6] Operating System (OS) with much lower memory resources. The question was whether it would be possible to have a suitable software stack on such a typical automotive ECU without overstressing the available resources. It was, indeed, possible and all needed functionality was implemented in the central gateway within the given resource bounds.

As said, implementing an Ethernet-based communication system had not been done before at BMW, nor in the automotive industry as such. Hence, there was a significant amount of skepticism and anxiousness about the feasibility. Yet, the results were impressive. The gateway project also included the implementation of CAN, FlexRay, and MOST 25 buses onto which the data being flashed into the car via

Table 3.1 Comparison of OBD-Stacks – migrating from HS CAN to Ethernet

OBD over CAN	OBD over Ethernet	The interface
UDS	UDS	The same diagnostic devices and the same protocol.
n/a	HSFZ	Maps UDS onto TCP and organizes parallel flashing.
CAN transport protocol	TCP/IPv4	IPv4 and TCP are used instead of the CAN transport protocol.
CAN controller	Ethernet MAC	Use of the Ethernet MAC instead of the CAN controller.
CAN transceiver	100BASE-TX PHY	The CAN transceiver is changed for a 100BASE-TX Ethernet PHY.
Two pins at OBD interface	Four pins at OBD interface	Ethernet 100BASE-TX uses four pins instead of the two used for CAN.

Ethernet had to be distributed. The Ethernet implementation, despite being new, caused the fewest error tickets during the qualification of the ECU compared with the implementations of other supported networking technologies. Furthermore, with Ethernet, it was for the first time possible to use freeware software stacks in the development process; one example being the "lightweight IP stack" [8]. This helped tremendously to prove that Ethernet was at a time when no one would have dared to invest heavily in the solution. Available test specifications and programs as well as existing test infrastructure for interoperability tests were an additional bonus.

3.1.3.3 The Car as a Node in the Network

With using Ethernet as the car interface, i.e., with the respective SOP in 2008, it was possible to treat the car as a network node connected to the external world, i.e., the dealer's or BMW's network. Figure 3.3 visualizes this. In the example in Figure 3.3, *n* cars are connected to various utilities such as testers, programming devices or a server using an external, standalone switch. One car can be connected to various utilities and one utility can be connected to various cars at the same time. The Ethernet interface thus provides more than just a high-speed link to the car.

In the depicted use case, IPv4 addresses are assigned to the cars by a Dynamic Host Configuration Protocol (DHCP) server. With the help of the unique Vehicle Identification Number (VIN) each car is unambiguously identified and in combination with the temporary, but also unique IP address the car can be located in any workshop around the world. The diagnostic application in the external equipment communicates via the UDS protocol to the car's internal devices. The car's internal switched architecture – see the example of car *n* in Figure 3.3 – provides for this. With Ethernet, the test software can be installed on normal PCs instead of needing proprietary hardware, which is another benefit.

The described BMW efforts are a BMW-specific solution. Nevertheless, in 1998 the United Nations initiated the World Wide Harmonized OnBoard Diagnostics (WWH-OBD) effort, with the goal of a harmonized standard for emissions control [9]. In this context, IP was selected as the communication protocol

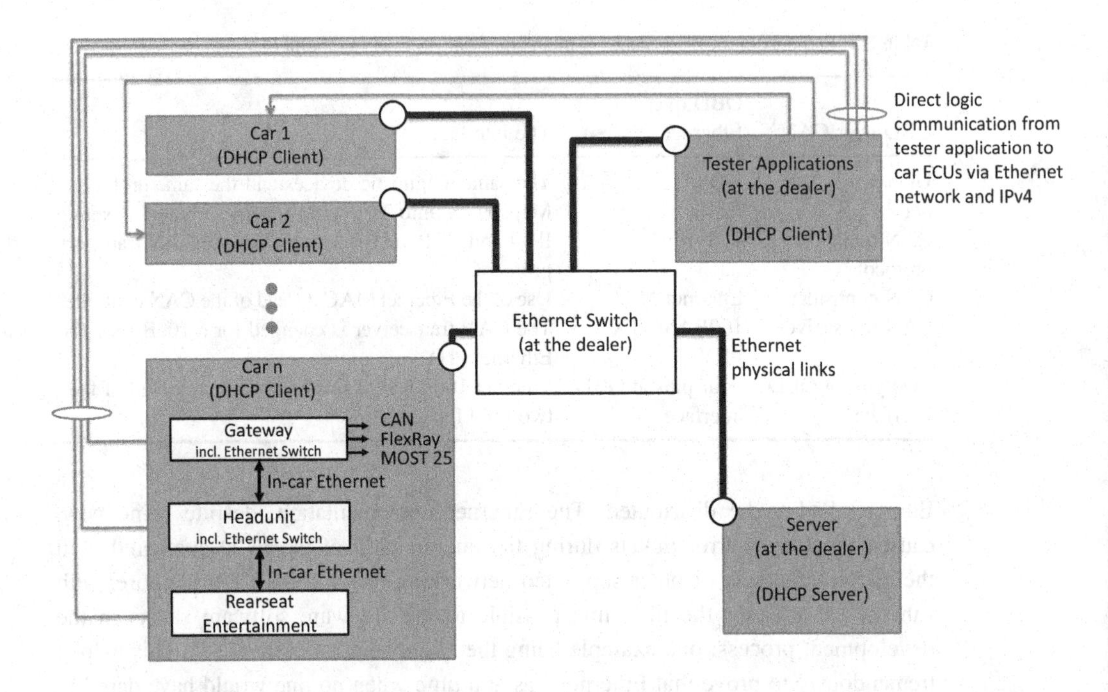

Figure 3.3 The car as a network node

between on-board and off-board diagnostic applications [10]. The resulting Diagnosis-over-IP (DoIP) ISO 13400 standard is (and this is not accidental) based on the same principles as the BMW solution: It enables the UDS applications via TCP/IP and a 100BASE-TX Ethernet interface [11, 12].

Even if it was for diagnostic and flash purposes only, car *n* in Figure 3.3 indicates that the central gateway as well as the Head Unit (HU) use an internal Ethernet switch. BMW started with a small network and limited use cases, but it provided an excellent learning base and allowed the derivation of guidelines for future in-vehicle high-speed networks. Note, this was developed in 2005/2006. In other words, it took about 10 years from the first assessments to rolling out Ethernet as an extensive system bus in BMW cars in 2015 [13].

3.1.3.4 Automotive Semiconductors for 100BASE-TX

One last aspect needed to be solved: The availability of automotive suitable Ethernet chipsets. Various vendors sold and still sell 100BASE-TX PHYs and switches, but the automotive industry has severe requirements that need to be fulfilled in order for semiconductors to be used inside vehicles. Broadly speaking, these are [14]:

- cost-effectiveness
- fast start-up
- reliability
- long-term maintainability

- scalability and flexibility (in case of extras/options)
- suitability for critical environmental conditions (temperature, vibration, humidity)
- high EMC fitness
- low weight
- small size
- low power consumption.

Some of the above requirements are technology dependent (e.g. scalability), some depend on the willingness of semiconductor suppliers (e.g. long-term maintainability), and some depend on their capabilities (all quality related aspects as well as size, weight, power consumption, etc.). In 2005, when BMW started looking for parts to use in production, Ethernet in the automotive industry was a completely new idea. In general, this meant that the traditional automotive semiconductor suppliers did not sell Ethernet chips and that the traditional Ethernet suppliers did not consider the automotive market to be particularly promising in order to justify investing in automotive qualification.

In the end, it turned out that Micrel (now Microchip), who was selling a similar portfolio into industrial automation, was interested. In a joint effort between Micrel and BMW a qualification plan was devised. BMW benefited in two ways from this approach:

1. BMW had direct knowledge of potential risks and weaknesses and was able to set up appropriate actions in parallel with the qualification of the semiconductors in order to ensure the SOP in 2008.
2. BMW was able to gather experience with handling semiconductors that had no base in the automotive eco-system. This allowed BMW to amend the qualification program accordingly and to learn for the future growth of Ethernet in the automotive industry.

The results of the qualification program were good. Only the package of the chips had to be changed in order for the chips to pass the tests for ElectroStatic Discharge (ESD).[7] For the diagnostic application under discussion, electromagnetic emissions were not an issue, so it was not necessary to investigate this aspect. With the diagnostic interface having been introduced consecutively in all BMWs since the SOP in 2008, Micrel (and now Microchip) has made a good choice and can now – at a time when the success of Ethernet in the automotive industry is no longer questioned – rightfully claim to have been the first company with AEC-Q100[8] qualified Ethernet products [15]. For BMW, the target of needing only 15 minutes for flash updates was met closely enough to call the project a full success.

This episode shows an, often overlooked, advantage of a multi-vendor environment. It is not all about costs and prices. It is also about product differentiation. When there are enough vendors, it is more likely to find one for whom it fits into the portfolio to explore edges and niches (which the automotive use case represented at the time). When there are very few or even only one, they are more likely to focus on core customers and core markets. Automotive Ethernet would probably not have happened without the well-established IT Ethernet market.

3.2 The Second Use Case: A "Private" Application Link

In parallel with the programming and diagnostic use case described in Section 3.1, a special use case was planned for the 2008 high-end Rear Seat Entertainment (RSE). This application was to reuse the navigation data stored in the Head Unit (HU) in the RSE and required that the navigation data be transmitted from the HU at about 20 Mbps. MOST 150 was not available at the time and MOST 25 did not accommodate 20 Mbps for data communication, as discussed. So, together with Harman Becker (now Harman International), the supplier of the respective HU and RSE at the time, it was decided to use 100BASE-TX Ethernet for the communication between the units. As the link was a private link between two units of the same supplier, the development was the responsibility of the supplier, who chose a QNX-based implementation.

Unlike the programming use case described in Section 3.1, this link was to be used during the runtime of the car and ElectroMagnetic Emissions (EME) did make a difference. Consequently, the cabling for the HU–RSE link required shielding, which made it heavier, more expensive, and less attractive.

With the harness in the car being the third heaviest and the third most expensive component in the car [16], the weight and costs of any connection inside the car are important. Also, while economies of scale and cost reductions are expected for semiconductors, cabling does not comply with the same market mechanism. The price for copper is very volatile [17]. This means that it is unrealistic to expect a cost reduction in cabling in the same way as for semiconductors.

So in 2007,[9] BMW was at a turning point. Ethernet looked promising, but was not quite there yet. As visualized in Figure 3.4, using 100BASE-TX as a PHY technology either meant a restriction of use cases or not being competitive cost-wise. It required the advent of an automotive-capable Unshielded Twisted Single Pair (UTSP) Ethernet (also called BroadR-Reach, OPEN Alliance BroadR-Reach (OABR) or now 100BASE-T1, IEEE 802.3bw)[10] to make Ethernet attractive for the automotive

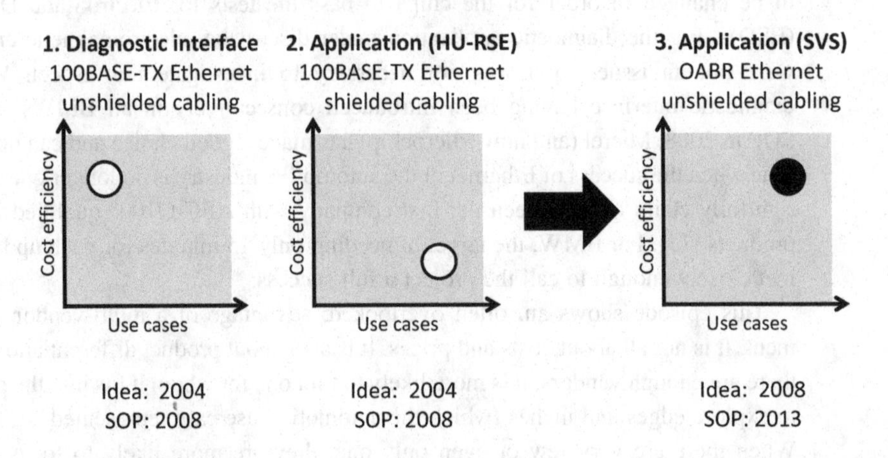

Figure 3.4 Limitations of 100BASE-TX Ethernet at BMW in 2008 and target achieved with OABR/100BASE-T1 in the first application, the Surround View System (SVS) [5, 25]

industry (as will be described in Sections 3.3–3.6). In 2007, before the discovery of 100BASE-T1, the situation was totally different. BMW had been one of the founding members of the MOST Corporation and the first carmaker to introduce that technology. Using Ethernet with shielded cabling had no cost advantage over MOST. Also, MOST had been developed for streaming audio, and seemed much more suitable for high-quality customer experiences than best-effort Ethernet.

As a consequence, some engineers thought it would be more useful to enhance MOST with better data and IP capabilities than to invest in Ethernet; the results of which can be found in MOST 150. Nevertheless, BMW also started several research programs on the use of Ethernet and IP in the automotive industry [18]. Their early focus was on Quality-of-Service (QoS) and timing behavior. This coincided with the efforts at IEEE, where the Audio Video Bridging standardization projects had been started [19] and interesting material was available (see also Section 7.1). The BMW activities yielded good results (see e.g. [20–24]) and led to the start of a project funded by the German government called SEIS (Security in Embedded IP-based Systems) in 2009.[11] Among other outcomes, setting up and pursuing this project served to create a community in the German automotive industry for Automotive Ethernet. However, the PHY remained the roadblock.

3.3 The Breakthrough: UTSP Ethernet for the Automotive Industry

BMW decided to synchronize all relevant knowledge with the future requirements on IP-based communication systems. A key learning was that a PHY useable with unshielded cabling was decisive for the future of Ethernet in the automotive industry. Another was that the EMC properties, at least the immunity, had been surprisingly good the first time round. In consequence, BMW decided to look more closely at the possibilities to reduce the emissions of 100BASE-TX Ethernet when using unshielded cabling.

Together with Lear, who had supplied the central gateway, BMW performed measurements. The starting point was the existing gateway. The EMC performance of the ECU itself, when Ethernet transmissions were idle, was very good. The hope was thus to be able to isolate the source(s) of the strong emissions. And indeed, the Ethernet output driver stage was identified as the root cause. The gateway hardware allowed the output driver of the Ethernet PHY to be deactivated, while leaving the internal MII and all other interfaces live. In this case, the unit was well under the emission limit lines. Unfortunately, irrespective of the means taken – filters, ferrite beads, etc. – it was not possible to get below the emission limit lines when the output drive stage was switched on.

Thus, in summer 2007, BMW approached four well-known vendors of Ethernet PHYs and asked for their opinions and solutions. Colleagues at the automotive semiconductor supplier Freescale (now NXP) had suggested that this might be worthwhile. Of the companies addressed, only Broadcom responded positively and in September 2007, the first meeting was held in Munich. During this meeting, the

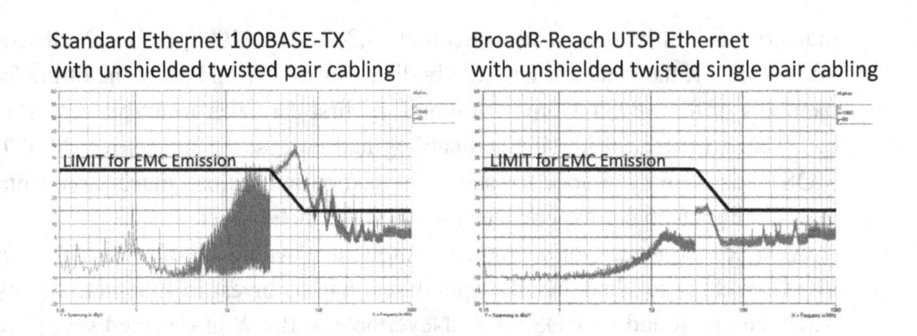

Figure 3.5 The world's first automotive measurements of the ElectroMagnetic Emissions (EME) of an Ethernet BroadR-Reach link performed in January 2008. The results even exceeded the performance of many traditional networking technologies

results of the gateway measurements were discussed and the automotive industry requirements were aligned with the performance value of a solution Broadcom had originally developed for Ethernet in the First Mile (EFM). In January 2008, the Broadcom technology called BroadR-Reach went into the EMC labs at BMW. Figure 3.5 shows the results of the first emission measurements performed.

Of the other three companies, one did not reply and the other two thought the BMW request impossible. Some years later, after BMW and Broadcom had proved that transmitting Ethernet packets at 100 Mbps over unshielded cabling was possible in the automotive environment and were promoting BroadR-Reach in order to attract other customers and suppliers, every one of the three other companies originally approached developed other, incompatible solutions. One solution was even based on 100BASE-TX (see Section 4.3.1.3); something BMW would have greatly appreciated a few years earlier.

On the one hand these solutions created confidence in the industry that transmitting 100 Mbps Ethernet packets over unshielded cabling in the automotive environment is really feasible. On the other hand, for those not having yet decided on the use of BroadR-Reach, it caused additional validation and decision effort and some uncertainty for all. In a fragmented market, no one wants to go with the technology with the smaller and potentially decreasing market share.

In hindsight we know that BroadR-Reach/100BASE-T1 succeeded, but at the time the situation was not always that clear. All car manufacturers have a long lead time to introduce new in-vehicle networking technologies. The decision has to be taken at least three years ahead of SOP, meaning that another year before investigations on the technology have to have started. For BMW all other proposals were simply too late to consider. This meant that for BMW the other solutions were mainly a source of discomfort as they posed an economical risk.

With BroadR-Reach, the door opener to Automotive Ethernet was found. BroadR-Reach promised to transmit Ethernet packets at 100 Mbps at vehicle runtime over a single pair (100BASE-TX requires two pairs) of unshielded cabling (Unshielded Twisted Single Pair, UTSP), i.e., the same cabling the industry used for CAN or

FlexRay networks. This would be the most cost-efficient high-speed network in the automotive industry, providing a higher data rate than MOST at a lower price level than MOST, any SerDes interface, or consumer technology. Nevertheless, this was still only the beginning. In 2008 all that existed was a good technical prototype some engineers at BMW had had the chance to investigate. The technical and economic feasibility had yet to be proven over all levels of decision-making within BMW. Also, the automotive industry as such had yet to be convinced that Automotive Ethernet was the right way forward. Before Ethernet, all communications technologies used in the automotive industry were initiated or developed by car manufacturers or automotive tier 1 suppliers. If only BMW was to use the new technology, it would have been of limited advantage (see also Section 2.4).

3.4 BMW Internal Acceptance of UTSP Ethernet

3.4.1 Yet Another In-Vehicle Networking Technology

BMW was one of the first car manufacturers to introduce in-vehicle networking as such and one of the first to introduce CAN and LIN. BMW was a founding member of the LIN, FlexRay, and MOST consortia and the first car manufacturer to introduce MOST 25, FlexRay, and 100BASE-TX Ethernet in serial production cars (see also Section 2.2). The company had especially invested in the MOST technology, and built up know-how and experts. Additionally, MOST 150 was going to offer a higher data rate than MOST 25 as well as better data/IP support. So, why adopt yet another networking technology?

It is true that BMW has invested a lot in in-vehicle networking technologies in the past. BMW is one of the innovation leaders in the industry (see also Section 2.4) and therefore always one of the first car manufacturers to need new in-vehicle networking technologies with different properties. After all, the in-vehicle networking provides an essential infrastructure for distributed applications. At the same time, having worked with all the networking systems means to have accumulated significant networking know-how, to have observed the increase in complexity in the systems, and to realize that to constantly completely change technologies is not sustainable in the long run. A more future-proof system was needed that is flexible, that scales, and that allows for reuse. The bandwidth requirement in cars is expected to continue to increase and it is no longer acceptable to constantly change the technology because of it.

Ethernet- and IP-based in-vehicle networking provides all of this (see Figure 3.6). As it builds on the ISO/OSI layering model, changing to a higher data rate requires only changing the PHYsical layer (PHY) technology, while from the Data Link Layer (DLL) upwards the software can potentially be reused. It is also possible to use a different medium, e.g., wireless or optical, without many changes. If a new protocol needs to be added on the application layers, this can be added without touching the layers below. Ethernet will eventually allow a reduction in the number of networking

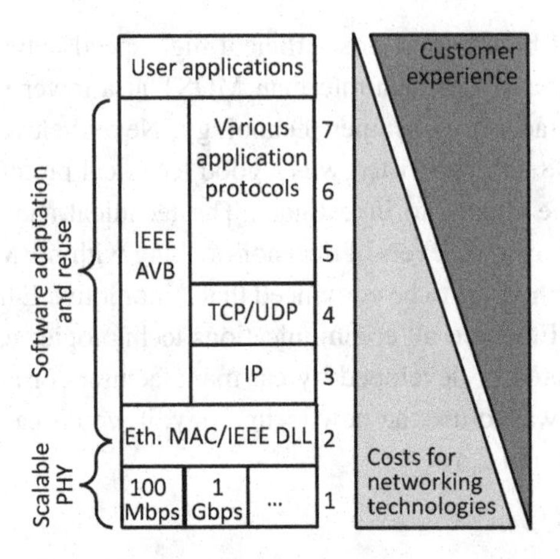

Figure 3.6 Flexibility, scalability, and reuse in Automotive Ethernet

technologies as well as the resources bound to them. Instead, those development resources will be able to focus on innovations with direct customer use.

At a higher level, Ethernet-based communication also addresses a general challenge that the automotive industry faces: The ever increasing product differentiation combined with the trend toward shorter model and innovation cycles [26]. Car manufacturers handle this by modularization and developing building block systems that allow designers to compose certain domains of a new car from sets of building blocks. The in-vehicle network must support this model. Ethernet-based communication provides for scalability in respect to data rates and transmission media. Additionally, a switched Ethernet network adds new possibilities and flexibility to the network design [27]. A switched network can have all kinds of topologies and is not restricted to a ring, line, or star. Increasing or decreasing the number of ECUs is significantly simplified (see also Section 8.3). Furthermore, Ethernet offers the possibility to separate networks virtually with the help of Virtual LANs (VLANs), even if they use the same physical network (see Section 7.2).

So, in principle, it was understood that Ethernet-based in-vehicle networking was the right way forward. The question was: How to introduce Ethernet on a larger scale into the vehicle? The application area that presented itself for the introduction was the infotainment domain. The first calculations that compared MOST 25 with shielded 100BASE-TX Ethernet, however, did not yield any obvious cost advantage. In the end, how do you quantify "future-proof"? BroadR-Reach Ethernet was too new. The first measurement results were promising, but many voices also within BMW doubted that UTSP cabling would really work. Ultimately, the infotainment domain was seen as one of the keys for the customer experience of a car. The existing MOST solution had a well-established, automotive experienced supplier base. Ethernet, at that time, did not. Clearly, a different pilot application was needed in order to prove the feasibility, strength, and maturity of Automotive Ethernet and the BroadR-Reach technology.

3.4.2 A Suitable Pilot Application

BMW chose to use the Surround View System (SVS) as a pilot application for BroadR-Reach/100BASE-T1 Ethernet. The purpose of a SVS is to show the surroundings of a car when it is being parked and in the following it is explained why the SVS was particularly suitable. The existing SVS system was already using digital LVDS/SerDes links to transport the uncompressed data streams of each individual camera to the ECU for generating the surround view picture. This surround view picture was sent "ready to be displayed" to the Head Unit (HU) via an analog FBAS connection (see also Figure 3.7), while the HU sent its control data to the SVS via CAN. The SVS controlled the cameras via LIN. For risk minimization reasons in the pilot application, only the SerDes links and the LIN control links were replaced by UTSP Ethernet, while the connection between SVS ECU and HU remained unchanged. As the Ethernet links provide with 100 Mbps a much smaller data rate than the SerDes interfaces even in 2009 – and a smaller data rate than the new High-Definition (HD) imager would generate – the video streams from the cameras needed to be compressed.

From the application point of view this raised two concerns: First, whether the loss of information caused by the compression would impair the performance of the image processing algorithms, and second, whether the latency introduced by compression and decompression would be acceptable. An early prototype that included the use of an Ethernet link with compression and decompression had been set up by the research department. The results were encouraging and the investigations were subsequently sufficiently refined to remove any concerns on the feasibility (see e.g. [28–30]). Concerning the latency H.264 and Motion JPEG (MJPEG) were investigated. Not all modes of H.264 were suitable; those suitable were not available in hardware at the time of investigation, though they yielded good results in simulation. In the end, it was

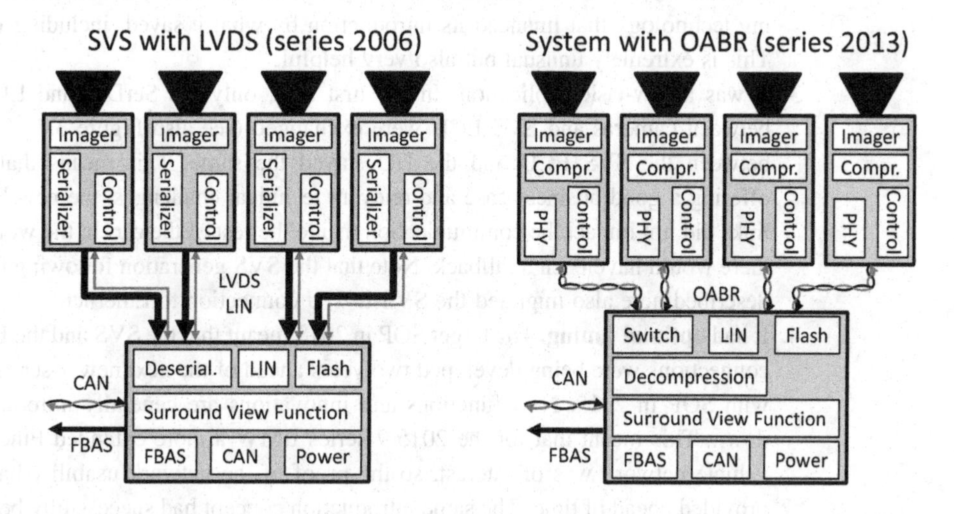

Figure 3.7 The pilot surround-view system application [32]

the joint effort with the μC supplier Freescale (now NXP), which resulted in a product allowing for a low latency implementation using MJPEG compression, which sufficiently addressed the original concerns.

Therefore, the SVS was selected as the pilot application. It turned out to serve as an optimal pilot use case for several reasons:

1. It provided the **right technical challenges**. The main focus was on proving that the **EMC** requirements could be met using **UTSP cabling** in a real-life application. This included the selection of **standard cables and connectors**, the choice and development of a μC with low power dissipation and low EMC emissions [31], the decision on the transformers/common-mode chokes (**CMC**)/filtering to use (or not to use), and the investigation of the influence of **temperature** changes (for more details see Chapter 4). As the spatial constraints of a camera are particularly tight, a camera can be seen as a worst-case scenario with respect to thermal influences and operating temperature. The small size of the camera was also a challenge in terms of software, which needed to reuse as much of the available IT technology (see also Chapter 7) while also being portable onto the **small embedded controller** available [31]. Last, but not least, the **automotive qualification** of all previously non-automotive parts, such as the BroadR-Reach semiconductors, had to be achieved.
2. It had an **excellent business case.** The cameras providing the respective images in an SVS need to be located in the extremities of the car. As a consequence, the cables leading there are long, some pass through several inline connections, and some end in wet areas, i.e., cables and connectors need to be water-resistant. Shielded Twisted Pair (STP) cabling as well as the respective shielded and partially waterproof connectors in small spaces result in significant costs. The Ethernet system required some more effort in the cameras due to the compression, but the savings in the harness more than outweighed these extra costs [31]. In fact, the OABR/100BASE-T1 Ethernet technology was the first high-performance networking technology that financed its introduction by what it saved, including interest. This is extremely unusual but also very helpful.[12]
3. It was a **low-risk** application. In the first step, only the SerDes and LIN links between cameras and SVS ECU were exchanged (see also Figure 3.7). The link between the SVS ECU and the HU stayed the same. This means that, while offering a good business case and relevant technical challenges, the new Ethernet links did not impact the communication inside the rest of the car. In the worst case, there would have been a fallback. Note that the SVS generation following the pilot described here also migrated the SVS-to-HU connection to Ethernet.
4. It had **optimal timing.** The target SOP in 2013 meant that the SVS and the Ethernet connections were being developed two years ahead of the next new 7-series BMW with SOP in 2015. New functions and innovations are generally introduced top down. This meant that for the 2015 7-series BMW a more extended Ethernet in-vehicle network was of interest, so the proof of the network usability had to be provided ahead of time. The same introduction concept had successfully been used with FlexRay, so it was seen as the right way to proceed with Ethernet, too.

5. It **proved the commitment**. Some additional risk was seen in working with suppliers inexperienced in the automotive industry. After all, the automotive industry has a long return on investment period. It often takes four to five years after semiconductors have been developed before the first cent comes rolling back. Especially for companies who are focused on the consumer industry, this is completely unheard of. Additionally, each car model is produced for about seven years and might need replacement parts for another 20 years. So, the car manufacturer has to trust not only the technical solution but also the long-term commitment of the semiconductor supplier. The supplier can prove the commitment with a local support network, product roadmaps, etc. The pilot project offered a comfortable time window in which new suppliers were able to familiarize themselves as well as comply with the necessities of the automotive industry.

3.4.3 The Future of Automotive Ethernet at BMW

The fulfillment of technical requirements is generally not sufficient for deciding on a technology. For example, the technology also needs to be affordable, for which a promising business case always provides a strong argument. In an environment with limited resources such as the engineering workforce of a company, these have to be used wisely. Even if there is money to be saved in one case, maybe it is (even) better for a company to use the same resources in another project. Thus, the long-term implications of the decision must also be taken into account. In the case of Automotive Ethernet it provides technical solutions for an otherwise unsustainable situation. Also, the suppliers showed commitment to the automotive market by heavily investing in the automotive qualification. Despite all this, it additionally needed to be possible that a market can develop around the technology. The related aspects – multi-sourcing, future developments, Tier 1 suppliers and other car manufacturers, etc. – were essential for the BMW internal decision, too. As those aspects will be discussed in Sections 3.5 and 3.6, this section concentrates on the elements relevant for BMW.

The first EMC measurements with the BroadR-Reach technology were performed at BMW at the beginning of 2008, the decision to use the technology for the pilot application was taken in March 2010, and the SOP of the respective Surround View System (SVS) was in September 2013 [33]. This means that BMW decided to investigate the technology thoroughly during the world economic crisis of 2008 and 2009, in order to be able to decide on series production in March 2010. At a time when many predevelopment projects in the industry were stalled for lack of funding, BMW allocated money and engineering power to Automotive Ethernet, which was thought technically infeasible at the time, even by many players in the Ethernet industry.

The obvious explanation is that Automotive Ethernet had a strong case, technically as well as financially. In the authors' opinion this is not sufficient though. In the authors' opinion the spirit which makes BMW one of the most innovative car manufacturers [34] has its role; and not only because a powerful in-vehicle networking system is an enabler for innovations. It is an essential part of being an innovation leader to dare to go into unknown terrain while at the same time being able to assess the risk correctly and being able to handle the challenges. No innovation leader would be an innovation leader without this being part of the company's culture. It implies motivated engineers and capable management, too.

During the preparation of the pilot project decision and in the first year after, many important technical questions and challenges were addressed (for the results see also Chapters 4–8). From the nucleus of the project group, the knowledge of the achievements was passed onto larger groups within the company (and to the outside world, see Section 3.6). Personal networks and selected partners in, e.g., qualification or research, who generally have a good exposure to management, helped spreading the knowledge to a critical mass. Decisions as consequential as using Automotive Ethernet as the basis for a large-scale network inside the car, require broad acceptance; over all involved departments and hierarchy levels. When making decisions of this scale, not everything can be expected to run smoothly. When problems arise, the engineers must (want to) overcome the hurdles and management must back them up. After all, there is a social component in all technical developments.

Ultimately, these efforts were successful. In March 2011, BMW made the decision to migrate the infotainment domain from MOST 25 to 100BASE-T1 Ethernet instead of to MOST 150, with target SOP in 2015. It was a goal of BMW to digitize all video streams inside the car. The existing MOST 25 system did not provide sufficient bandwidth for this, so a migration to a new system was necessary. BroadR-Reach/100BASE-T1 Ethernet was the more cost-efficient solution. In October 2011, the decision followed to migrate part of the driver assist domain, also starting in 2015. In this case more bandwidth was needed for new innovations and the integration of Ethernet seemed more future-proof than to add yet another CAN or FlexRay, or even two of each.

3.5 The Industry Framework for a New Technology

"The discussions regarding standard adoption are technical, but it is people and firms that must agree to the standards. Not surprisingly, this means that there is a social component to this process" [2]. Adoption of a new concept, standard, or technology is multi-faceted. Not only do technical and economic questions need to be answered, a framework needs to be in place that serves as a breeding ground in which the new technology can thrive. The greater the number of suppliers that expect to profit from a new standard, the more likely it will succeed [35]. Additionally, there are individual preferences, animosities, and paradigms. These influence the decision processes, but they themselves are influenced by the availability of a structure that allows an industry to develop as well.

3.5.1 From a Proprietary Solution to an Open Standard

The BroadR-Reach technology was developed by Broadcom, who also own the respective Intellectual Property Rights (IPR) on the technology and its trademark.

Table 3.2 Options on how to open a technology and their consequences [5]

	More suppliers	Acceptance	Influence on technology	Timing
Give the standard to an SSO	Interest depends on market prospect. A promising market finds interested suppliers	Good, well-established and transparent	A not yet established technology is likely to change	Complete process >3 years
Create a SIG that publishes the standard		Transparent, but not established at the beginning	Technology does not need to change, control shifts to SIG	Faster, depends on founders
Leave it to the IPR holder		Not transparent, very dependent on the IPR holder	Only IPR holder defines technology	Fast

This means that to start with, the technology was proprietary,[13] i.e., closed to competitors. Closed technologies lead to monopolies and these are undesirable. Not only can it be expected that the prices the customers pay in a monopoly situation are unfavorable, also the customer depends on one supplier for reliability, availability, future developments, and innovations. Products in a competitive situation are simply better for the customer and generally the industry as such.

Fundamental for not getting into a monopoly situation and for allowing intra-standard competition is that the IPR holder embraces this. There are numerous examples, of which Ethernet itself is actually one, that show how even an allegedly inferior technology can win over a superior technology. This is simply because the IPR holder of the inferior technology pursues a truly open licensing policy, while the technology owner of the superior technology does not or does so only halfheartedly. According to Burg and Kenny [2], IBM, in the hope of a market advantage or unawareness, lost the whole Token Ring LAN market to Ethernet because of IBM's inconsequential technology opening, even though they did offer it for standardization in IEEE 802.5.

In the end, there are three possible ways to make a technology an open standard (see also Table 3.2 and note 11): (1) the standard is offered to a Standard Setting Organization (SSO) for publication; (2) the standard is published via a Special Interest Group (SIG)/industry consortia; and (3) the standard is simply published directly by the IPR holder. Provided the IPR holder executes a Reasonable and Non-Discriminatory (RAND) licensing policy, all three ways are viable and have been chosen successfully in the past [35, 36]. While the first is probably the most accepted, it bears the risk of delays and the risk of changes to the original technology, unless the original technology has already been successfully established as a de-facto standard. The third option is the least transparent and the success relies very much on the IPR holder. If it is mainly the customer that requests the opening of the technology, this is not a good start – full support of the IPR holder is a fundamental requirement for the success – and can lead to misunderstandings.

In the case of Automotive Ethernet, companies from two industries with different cultures had to rely on each other, while at the same time timing was crucial. As a consequence, the second option was chosen. In November 2011, NXP, Broadcom, and BMW started the One Pair EtherNet (OPEN) Alliance.[14] The companies that joined within November 2011 were C&S, Freescale (now NXP), Harman, Hyundai, Jaguar Landrover, and University of New Hampshire Interoperability Lab (UNH-IOL) [35].[15] As it happened, the OPEN Alliance became one of the fastest growing automotive consortia and had more than 400 members in July 2019 [37].

The overall goal of the OPEN Alliance was to help establish Ethernet-based communication as an in-vehicle networking technology. An early focus was on the 100 Mbps BroadR-Reach/100BASE-T1 technology. In order to support other semiconductor vendors in developing competitive BroadR-Reach products, the specification was reviewed, clarified, and enhanced. Functional as well as EMC compliance tests were defined and interoperability tests were developed. After all, the OPEN Alliance had the right members, with UNH-IOL the preeminent entity in compliance and interoperability tests in the Ethernet world, C&S a known entity in the automotive certification world for traditional in-vehicle networking systems, and FTZ with expertise on EMC testing (see also Section 3.6). Other test houses followed. For usability by, e.g., the Tier 1s and car manufacturers, OPEN specified the components (cable, connectors, harness manufacturing) and identified suitable tools.

Multiple sources are essential for a market to prosper, and the prognosis of a market to prosper is essential for multiple sources to be offered. The mentioned activities of the OPEN Alliance aimed at achieving more planning security for semiconductor and other vendors wanting to enter and invest in the market. Not only were technical risks reduced by OPEN, e.g., avoiding ending up with a non-interoperable solution, the members of OPEN also represent the interest of the market. Additionally, OPEN was set up to address any issue that hampers the adoption of Automotive Ethernet, either by finding/defining a solution within OPEN or by cooperating with other organizations better suited to take up the task at hand. In consequence, the technical work has grown. The OPEN Alliance started with five technical committees and, at the time of writing, it had 15; the latest being "2.5, 5, 10GBASE-T1 Interoperability and Compliance Tests" [38].

One last aspect to discuss in the context of successful technology development and deployment is the difference between intra- and inter-standard competition. The worst thing that can happen to a customer is to be faced with a monopoly that leaves no alternatives. This is nevertheless very rare. If there is a promising market and one company has a proprietary solution for this market but is not going to license it, generally other, technically different solutions will be created by competitors seeing an opportunity in the same market [39]. This leads to a market with inter-standard competition. If the product is a standalone product, i.e., it requires no complementary products, no minimum distribution, or no interoperability of any kind, the customers have healthy competition despite the fact that the solutions differ. In the case of communication technologies, however, interoperability, standardization, and network effects[16] are key. The more manufacturers that produce products with/for/of the same technology the better for the customer. Technologies with inter-standard competition can work to some extent – Automotive Ethernet itself has spurred the competition of other technologies – but generally speaking inter-standard competition slows the market development for communication technologies.

Inter-standard competition leads to insecurity among the customers [40]. No one wants to invest in a technology that does not succeed and that might be discontinued. Some customers in such situations even delay their decision to adopt a technology. This in return leads to smaller volumes and reduced economies of scales, which again makes the whole market less attractive. If there is no intra-standard competition, i.e., a number of competitors offering products that are interoperable, inter-standard competition is better than no competition at all. If there is a chance for intra-standard competition, inter-standard competition generally harms the development of the industry.

In the authors' opinion the OPEN Alliance played a strong role (next to the IEEE adoption of the BroadR-Reach standard discussed in Section 3.5.2) in aligning the market to a single solution and ensuring that – despite other solutions being proposed – the market became an intra-standard competition market.

3.5.2 Shaping the Future at IEEE

Ethernet-based communication is attractive because it potentially scales, i.e., the MAC and software layers can stay the same, while the PHY is being replaced with one that supports higher data rates. The IEEE has a Gbps PHY technology for copper wiring available, but 1000BASE-T requires four pairs of twisted cables and was expected to additionally need shielding if used in the automotive environment. This meant that, while in principle Ethernet provided the possibility to scale to a higher data rate in the automotive environment, in practice a suitable technology had yet to be developed. So, not only the present, but also the future of Automotive Ethernet had to be initiated.

At BMW, it was estimated that Gbps Ethernet would be needed for serial production starting from 2018. In order to achieve this, the final decision would have to be made by 2015, with a preceding opportunity to evaluate the technology. So, after the first critical milestones for 100 Mbps had been met, efforts started in the middle of 2011 to standardize an automotive-suitable Gbps Ethernet at the IEEE in order to meet this timeline. In March 2012 the Call for Interest (CFI) for the "Reduced Twisted Pair Gigabit Ethernet (RTPGE)" passed [16] and a respective IEEE study group was established, which was successfully turned into the task force IEEE 802.3bp by the end of 2012 [41]. In January 2014 the task force agreed on renaming RTPGE to 1000BASE-T1 and IEEE concluded the standard in June 2016 (see Section 5.2.2 for technical details).

Unfortunately, this process took longer than expected and was too late for 2018 SOP. But, independent from the details of the standardization process, an important structural step had been achieved with starting 1000BASE-T1 at IEEE. IEEE 802.3 represents the home of Ethernet, with the respective experts and interested industry representatives present. With 1000BASE-T1, the automotive industry was established as a new application field in IEEE 802.3 and with that opened a path for the future. Table 3.3 shows the other standardization efforts useful for the automotive industry

Table 3.3 Overview on Automotive Ethernet related projects at IEEE 802.3 in March 2020 [43]

Name	Abbreviation	CFI-date	First TF meeting	Publication date	See Section
1000BASE-T1	802.3 bp	Mar. 2012	Jan. 2013	Aug. 31, 2018	5.3.1
1000BASE-RH	802.3 bv	Mar. 2014	Jan. 2015	Mar. 14, 2017	5.3.2
100BASE-T1	802.3 bw	Mar. 2014	Sep. 2014	Mar. 7, 2016	5.2.1
10BASE-T1S	802.3 cg	Jul. 2016	Jan. 2017	Nov. 19, 2019	5.4
2.5, 5, and 10GBASE-T1	802.3 ch	Nov. 2016	May 2017	Jun. 30, 2020	5.5
>10 G electrical	802.3 cy	Mar. 2019	Jun 2020	n/a	5.6
≥10 G optical	802.3 cz	Jul. 2019	Jul 2020	n/a	5.6
PoDL	802.3 bu	Jul. 2013	Jan. 2014	Feb. 17, 2017	6.2.1
IET	802.3 bv	Nov. 2012	Jan. 2014	Mar. 14, 2017	7.1.4.4
Multidrop Enhancements	802.3 da	Jul. 2019	Jul 2020	n/a	5.4.3

that had followed. In July 2013 the CFI for 1 Pair Power over Data Line (1PPoDL) passed [42] and the respective task force IEEE 802.3bu was established by the end of the same year [41]. PoDL refers to a concept in which power is transmitted over the cables that are also and originally used for data transfer. In the Automotive Ethernet context, this is particularly attractive for sensors, such as cameras, that are located in the extremities of the cars. PoDL therefore allows a reduction in the number of cables and hence a reduction in the harness weight and volume. The IEEE had standardized Power-over-Ethernet (PoE) first in 2003 for the two pair 100BASE-TX, which, of course, is not suitable in the case of a one pair Ethernet variant (see Section 6.3 for more details).[17]

Table 3.3 shows that yet another 1 Gbps solution was standardized for automotive use (plus home and other markets): 1000BASE-RH. Last, but not least, BroadR-Reach was rubberstamped as 100BASE-T1. Despite this being the latest standard to have been initiated in this first round it needed the shortest time, with only one meeting cycle as a study group and one as a task force before the document moved into the ballot phase. The Interspersed Express Traffic (IET)/IEEE 802.3br standard is somewhat of an outsider. First, it was initiated by Industrial Automation (albeit potentially useful for the automotive industry) and, second, it serves to increase the efficiency of the channel while reducing some latencies; something that is discussed with Time Sensitive Networking (TSN) in Section 7.1.4. At the time of writing the 10 Mbps Ethernet, 10BASE-T1S, had just been completed and the automotive multi Gbps Ethernet (2.5, 5, 10GBASE-T1) was being completed, while developments beyond 10 Gbps were being investigated (see also Chapter 5).

3.5.3 Supporting Structures and Organizations

Ethernet-based communication in the automotive industry, or Automotive Ethernet, is not only about the physical layer. It covers all layers of the ISO/OSI layering model (see also Section 1.2.5). One of the prime attractions of using Ethernet-based communication is the opportunity for reuse. This applies to technical solutions as well as to organizations developing the solutions. However, the reuse and adaptation from the IT industry are just one side of the coin, with the integration of Ethernet-based communication into various existing automotive industry efforts being the other. The following list gives a brief overview of the main organizations and activities BMW engaged with other than IEEE and OPEN in order to establish Ethernet-based communication in the automotive industry: AUTOSAR, Avnu, GENIVI, and ISO.

- In 2003, as the amount and complexity of software in the automotive industry was continuously increasing, key players of the industry launched the AUTomotive Open System ARchitecture (**AUTOSAR**) development partnership [44]. The main goal was to enable the exchange and update of software and hardware over the service life of a vehicle. For this, AUTOSAR developed a software architecture standard that also covers communication interfaces. The AUTOSAR Operating System (OS) was designed to be suitable for many types of applications and today is an integral part of many ECUs. It was therefore fundamental for the introduction of Ethernet-based communication in the automotive industry that AUTOSAR support the respective protocols. To start with, AUTOSAR 4.0, which was published at the end of 2009, provided means to support Diagnosis-over-IP (DoIP), i.e., Ethernet communication-based diagnosis and software flashing via IP and UDP [45]. Since then the Ethernet capabilities have continuously increased with the consecutive AUTOSAR versions: Version 4.1 (2014) added, e.g., TCP, Service Discovery (SD), and the connection to the MAC and PHY layers [46]; Version 4.2 (also 2014) optimizes the resources [47]; and Version 4.3 details the support of Ethernet switches. The latest development is Adaptive AUTOSAR that was developed for functions requiring more computing/compute power as needed for, e.g., autonomous driving that inherently uses the Automotive Ethernet suitable Scalable service-Oriented MiddlewarE over IP (SOME/IP, see also Sections 7.4 and 8.2).
- The **Avnu Alliance** was founded in August 2009 [48] with the goal of promoting the emerging IEEE 802.1 Audio Video Bridging (AVB) and the related IEEE 1722 and 1733 standards (see Sections 7.1.2 and 7.1.3 for the technical description). These IEEE standards enhance Ethernet-based communication systems with QoS functions. However, the standards offer a wide variety of choices. Avnu set out to overcome potential ambiguities with profiles, certification, and plug fests. Avnu focused originally on the three application areas of professional audio, mobile devices, and the automotive industry [49] (industrial was added as a fourth application field later [50]). In order to support and guide the ongoing standardization activities around TSN with harmonized automotive requirements (see also Section 7.1.4), Avnu, e.g., established the so called "Avnu sponsored Automotive AVB gen 2 Council (AAA2C)" [51, 52]. From day one, Avnu promoted the use of Ethernet in the automotive industry and the availability of Audio/Video QoS for respective applications. With the lack of QoS being seen as a major flaw in Ethernet in comparison with, e.g., MOST, Avnu thus provided an important contribution to

the cause. Avnu published the first automotive profile [53] in order to simplify the qualification process for Ethernet-based communication systems in the automotive industry. With IEEE 802.1 having taken up the task to develop the automotive profile for the newer TSN standards in IEEE P802.1DG [54], Avnu is set up to provide the certification and plug fests.

- When a car manufacturer, e.g., buys a Surround View System (SVS) from a supplier today, it buys the cameras and the control ECU from the same supplier. This is not always the optimal solution as a supplier who is good at image processing is not necessarily good at building cameras. In order to allow for buying cameras separately from the ECUs, the automotive industry initiated in 2009 a standardization activity at ISO: **ISO 17215, Road vehicles – Video Communication Interface for Cameras** (VCIC). At that time, BMW was at the beginning of the surround view project. Various technologies were being discussed for the networking technology to use. In the end, Ethernet-based communication succeeded and BroadR-Reach was recommended for the physical layer technology [55].

Note that in addition to the early ISO efforts of VCIC, in 2016 ISO started the project **ISO 21111, Road vehicles – In-vehicle Ethernet**. This project was initiated by Japanese industry players in order to support the deployment of optical Gigabit Ethernet (see also Section 5.2.2.2) in the vehicle [56] and the project was originally named "Road vehicles – In-vehicle Gigabit Ethernet" [57, 58]. However, ISO agreed later in 2016 to rename and restructure the project such that it can comprise all specifications needed to enable Automotive Ethernet, independent of speed grade and medium (see Figure 3.8 for an overview). At the time of writing the standardization activity was completing the first standards, incorporating also a multitude of the specifications the OPEN Alliance had published. For the OPEN Alliance EMC specifications, a transfer to IEC 62228-5 was ongoing [59].

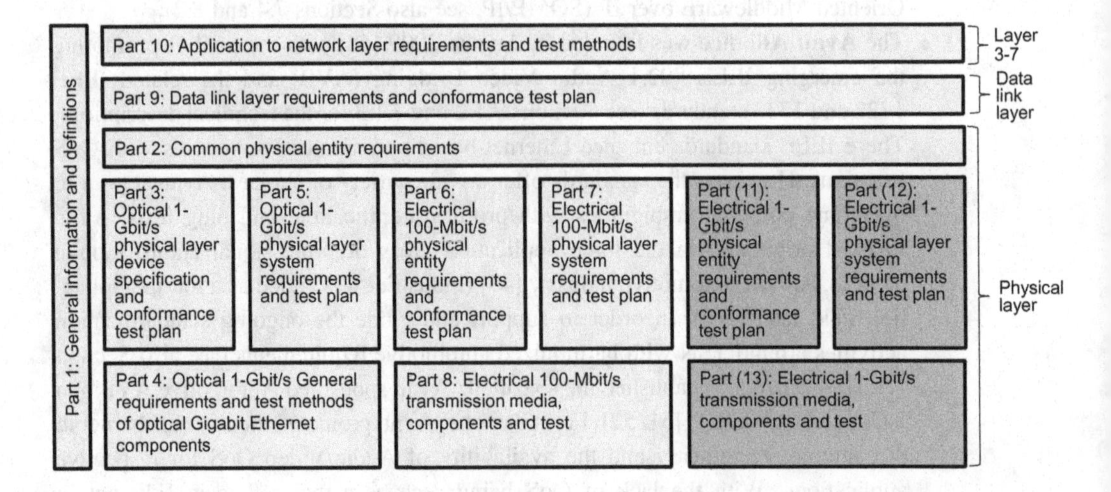

Figure 3.8 Overview of the planned ISO 21111 Automotive Ethernet Specifications in 2019 [60]

- The **GENIVI** Alliance was also founded in 2009 with the goal of driving the broad adoption of an in-vehicle open source development platform [61]. The idea was to spur software development and to achieve shorter product life-cycles by collaborating on a common, Linux-based reference platform and by fostering an open source development community. The GENIVI platform includes Linux-based services, middleware, and open application layer interfaces. As a consequence, the principles intended to be used for the communication middleware of Automotive Ethernet had to be integrated into GENIVI. SOME/IP (see also Section 7.4) was thus made available as a GENIVI library. GENIVI is a good example of how diverse all the activities that need to be taken into account are, when introducing a new networking technology in the automotive industry.

3.6 Industry-Wide Acceptance of Ethernet

The historic development of in-vehicle networking technologies had taught the industry that such non-differentiating functionalities are more beneficial if broadly accepted and widely deployed in the industry (see also Chapter 2). A high probability of industry wide acceptance was therefore also required inside BMW to move ahead. Both the adoption by Tier 1 suppliers as well as the adoption by other car manufacturers was and is relevant. Tier 1 suppliers function as multipliers of new technologies. The car manufacturers are their customers who might request those technologies or who adopt the new technologies because the Tier 1 offers them at good value.

To assure that a Tier 1 supplier is interested in a new technology, a car manufacturer can simply ask for it. But for a Tier 1 to really embrace that technology the Tier 1 needs to experience or at least expect many car manufacturers to be interested. For the success of Automotive Ethernet, it was therefore important to convince other car manufacturers, i.e. the competitors, of the advantages of Ethernet-based communication. How do you convince someone you do not have a business relationship with? In the end, every car manufacturer has to evaluate the benefits and economic impacts internally, in line with their key market segments and other economic considerations.

Nevertheless, this can be supported. Other car manufacturers had similar EMC requirements and could be expected to pay similar prices for components as BMW did. Thus, expecting interest, BMW pursued a proactive information policy and actively approached competitors, as well as Tier 1 and Tier 2 suppliers. Also, every interested company was welcome to discuss Ethernet-based communication on their own accord. BMW encouraged competitors to perform their own EMC measurements, in order to reduce skepticism of the feasibility. After all, seeing is believing. These efforts had two important results:

1. The **inclusion of independent organizations**. The University of Applied Science in Zwickau (FTZ) got involved. As an independent entity FTZ had performed EMC tests on in-vehicle networking technologies for the automotive industry in the past. With the development of test methodologies for Automotive Ethernet, FTZ played an important role in substantiating the feasibility of Automotive Ethernet.

As they did so as an independent entity, this added to the credibility of the concept. Many of their results later became part of various OPEN Alliance specifications (e.g. [62–64]). The UNH-IOL had been included, too, at a very early stage for assessing the feasibility of product development on the basis of the early BroadR-Reach specification. At a later stage UNH-IOL added credibility to the testing of BroadR-Reach/100BASE-T1 components [65].

2. When companies discussed Automotive Ethernet with BMW, they were not only interested in what BMW was doing but also in what everybody else thought. BMW thus perceived a significant market interest and hosted the first **Ethernet & IP @ Automotive Technology Day** in November 2011; an event which sold out completely. In alignment with the first Ethernet & IP @ Automotive Technology Day, the OPEN Alliance was started [66], NXP announced their development of a BroadR-Reach compliant PHY [67], and BMW completed the internal decisions on the wide introduction of Automotive Ethernet (see Section 3.4.3). In the authors' view, November 2011 represents the turning point. Automotive Ethernet stopped being just an idea of some engineers at BMW and became the future of in-vehicle networking. The Ethernet & IP @ Automotive Technology Day allowed for a non-committal information exchange, with all players present. In 2020, the event had its tenth anniversary. In the meantime the event had been hosted by Continental (with support from Harman) in Regensburg, by BOSCH near Stuttgart, by GM in Detroit, by Jaspar in Yokohama, by Renault in Paris, by US Car in San Jose, by JLR in London, and by Ford in Detroit [68].

Thus, the foundation for Automotive Ethernet was laid. Integrating other organizations involved and necessary for Automotive Ethernet (examples are AUTOSAR, Avnu, GENIVI, ISO 17215); setting up future developments at IEEE (especially for higher data rates) and supporting the creation of assurance in the market (open information policy, starting technology days) set additional, reinforcing impulses. In the end, in a growing market a virtuous cycle is achieved between customers, suppliers, and supporting/complementary organizations. For Automotive Ethernet the same cycle is happening. That this is independent of the exact form of the end result, i.e. what PHY technologies, speed grades, or protocols the industry will use is one of the strengths of Automotive Ethernet.

The industry acceptance is good. In 2016, three car manufacturers (BMW, JLR, and VW with various brands) publicly stated that they had Automotive Ethernet in series production cars on the road (see e.g. [69]). At the various events (e.g., the Ethernet & IP @ Automotive Technology Days, the Automotive Ethernet Congress, the Hanser Automotive Networks event, now renamed the Automotive Systems Architecting Conference, the Nikkei Automotive Ethernet seminar), Daimler, GM, Hyundai, PSA Peugeot Citroën, Renault, Toyota, and Volvo Cars have also spoken publicly about their use of Automotive Ethernet. At the time of writing the OPEN Alliance had been chaired by BMW, GM, Hyundai, Renault, and Toyota. And, last but not least, alongside Broadcom, NXP, Realtek, Marvell, and Microchip, the OPEN Alliance had publicly announced 10, 100, and/or 1000BASE-T1(S) products [70–72].

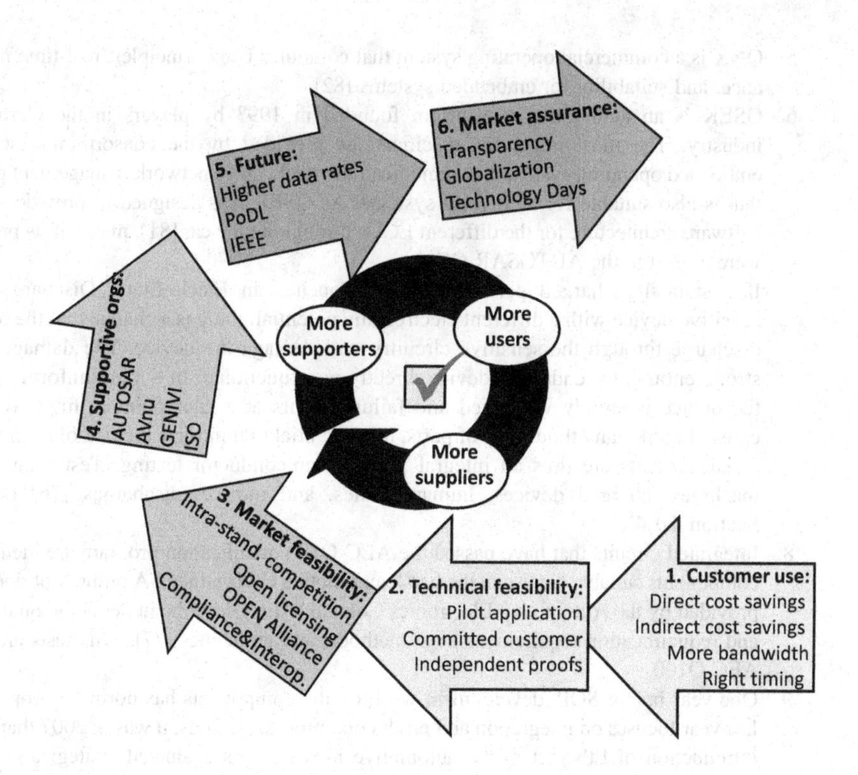

Figure 3.9 Path to Automotive Ethernet

Figure 3.9 summarizes the interrelations between the different aspects relevant for the market success of Automotive Ethernet. There is, of course, one element that cannot be structurally captured: The element of chance. The right people with the right skills and ideas have to come across the right potential technical solution in the right innovative environment at the right time. This is what starts technical revolutions.

Notes

1 EMC immunity, sometimes also referred to as Electro Magnetic Susceptibility (EMS), is the ability of a system to function despite external interference. The EMC immunity tests answer the question of whether a system is stable enough to function correctly in a very bad EMC noise environment (see also Section 4.1).

2 EMC emissions (EME) is the electromagnetic noise generated by the system that, via air or cabling, can impact the performance of other systems (see also Section 4.1).

3 "At runtime" means that the car is being used for its primary purpose of driving, in contrast to service mode at a garage.

4 Linux is the name of one of the most widely used operating systems among software developers. It originated during the time that AT&T was engaged in an IP battle with the University of California, Berkeley over the use of Unix. It was developed as freeware to be POSIX compatible and Unix-like [79].

5 QNX is a commercial operating system that combines Unix principles, real-time performance, and suitability for embedded systems [82].

6 OSEK is an automotive consortium founded in 1993 by players in the German car industry. The most important specifications provided by the consortium describe an embedded operating system, a communications stack, and a network management protocol that is also suitable for embedded systems. As OSEK was designed to provide standard software architecture for the different ECUs throughout the car [81], many of its principles were reused in the AUTOSAR OS.

7 If a statically charged person or object touches an ElectroStatic Discharge (ESD) sensitive device with a different electrostatic potential, there is a chance that the resultant discharge through the sensitive circuitry will damage the device. The damage can be strong enough to render the device directly non-functional. In a more unfortunate case, the device is simply weakened and failure occurs at a later point in time. With cars consisting of many thousands of parts, it is essential to minimize the risk of such failures, and ESD tests are thus an integral part of semiconductor testing. Tests can emulate machines, charged devices, human bodies, and indirect discharges [76] (see also Section 4.1.4).

8 Integrated circuits that have passed the AEC-Q100 qualification program are identified as components suitable for use in the harsh automotive environment. A number of documents provided by the Automotive Electronics Council (AEC) describe in detail the qualification and requalification requirements, test methods, and guidelines [77]. ESD tests are part of AEC-Q100.

9 One year before SOP, development work on the components has normally stopped. The last year focuses on integration and production processes. Thus, it was in 2007 that the first introduction of Ethernet in the automotive industry was evaluated strategically and the next steps were being discussed.

10 The same technology has several names. First, it is called "BroadR-Reach" as this is the name Broadcom, the inventor of the technology, gave the technology and has a trademark on. First, "BroadR-Read" is the trademarked name given to it by its inventor, Broadcom. With BroadR-Reach being facilitated by the OPEN Alliance (see Section 3.5.1), it is also called OPEN Alliance BroadR-Reach, or OABR, for short. The main characteristic of OABR is that it can be used with Unshielded Twisted Single Pair (UTSP) cabling. At the time of its invention using an Ethernet technology with a single pair worked for just this technology, which is why it is also called "UTSP Ethernet." We will use this name at the one instance where this is of major importance. However, since the development of the automotive-suitable Gbps Ethernet (see Section 5.2.2), there are other PHYs using UTSP Ethernet (with more in 2019) and "UTSP Ethernet" was no longer unambiguous. At IEEE, the UTSP versions have received the suffix "T1," with the "1" representing one pair. Therefore, 100BASE-T1 is the name BroadR-Reach received as an IEEE standard. As this is the latest name. We will use it whenever possible.

11 Other than what the title of the project suggests, the project actually has a strong focus on all protocol layers needed for Automotive Ethernet, while security represented only one of the six work packages defined (see e.g. [75]). However, this project had a stronger focus on the protocol layers than on the PHY layers. More information on the project and its results can be found in [78].

12 The business case was calculated in 2009/2010. At that time, SerDes interfaces were available for shielded twisted pair cables only and required a separate control channel. Possibly because of the competition from Ethernet, SerDes interfaces have become significantly cheaper. Not only have the semiconductor prices as such dropped, but newer SerDes developments eliminated the need for a separate control channel, enabled the use of coaxial cabling (which is less expensive than STP), and allowed for power transmission

within the coaxial cable (see also Section 2.2.6). This is a good example of how customers can (occasionally) profit also from inter-technology competition.

13 The terms "proprietary," "open," and "public" standards are used in this book in accordance with the definitions in [35]. "Proprietary" is therefore used only for technologies whose IPR is owned by one company that does not license the technology to others under Reasonable and Non-Discriminatory (RAND) terms, but which either licenses very selectively or not at all. In contrast, there can be "open" or "public" technologies. "Public" refers to technologies described in standards developed by Standard Setting Organizations (SSOs) such as IEEE or ISO. SSOs follow established rules for IPR; normally requiring that owners of essential patents declare prior to the publication of the standard that they will license their essential patents to interested parties under RAND conditions. "Open" technologies mean that they are at least RAND licensed, regardless of whether the IPR is owned by one or by many companies, or whether this is organized in an SSO, a Special Interest Group (SIG), or other consortium, or whether the patent holders simply agree to it. A "public" standard can therefore also be described as "open." It nevertheless helps for the distinction to refer to "public" standards if it is a standard published by an SSO.

BroadR-Reach started as a proprietary solution. The OPEN Alliance ensured that it became an open technology. With its standardization as IEEE 100BASE-T1, it has become a public standard; this being independent from how many companies own the IP. CAN is another example of a technology that is perceived and accepted as an open and public industry standard, despite the facts that the technology was defined before being made public and that all IPR is owned solely by BOSCH, who licenses it under RAND conditions. Both BroadR-Reach and CAN are not "open" following [74]. Here, an open standard requires equal contributions from multiple companies without the dominance of one company. The authors accept this as a different way of looking at it and agree that it is generally (though not always) more motivating for multiple companies to participate in the market if this is the case. However, for the purposes of this book, the key point is that a technology or standard is not necessarily proprietary just because one company owns the IPR. It might be open(ed).

14 Like probably all selections of names, the naming of the OPEN Alliance took some time and effort. In the end, One Pair EtherNet (OPEN) reflected best that the BroadR-Reach technology was going to be licensed "openly," i.e., under RAND terms. To the authors, who were involved in the naming, the "One Pair" part of the name was less relevant. It did name a major feature of the BroadR-Reach technology, but in the end OPEN's purpose was and is facilitating Ethernet-based communication in the automotive industry, independent of whether one or multiple pairs or even other media are used.

15 The public announcements on this event are not 100% correct. The reader must trust that the authors, as founders and first chair of the OPEN Alliance, know better.

16 Network effects refer to products whose use increases with distribution [73, 74]. Communication technologies per definition require communication partners; the more there are, the higher the use of the technology for all. The so-called Metcalfe's law put this into a formula (see also Endnote 1 of Chapter 1). Inside vehicles, it is a little different because, after all, the car manufacturer decides on (almost) all communication interfaces inside the car. Nevertheless, a large distribution among other car manufacturers has more advantages than just better economies of scale. It leads to a better educated workforce, better tooling, better infrastructure in terms of independent test houses, more reliability from the Tier 1s, etc. The network effects are thus indirect.

17 The name "Power over Ethernet (PoE)" is tied to the IEEE 802.3af standards and their successors. It is also referred to as "clause 33" into which it was incorporated in the standards revision IEEE 802.3–2005. Clause 33 implies a specific method requiring two twisted pairs of cabling. Therefore, a standard that discusses the transmission of power

over just one pair needed a different name. Instead of "1 Pair Power over Ethernet" the activity is thus called "1 Pair Power over DataLine," in short PoDL (pronounced poodle). In contrast, the IEEE 802.3 activity that investigates the transmission of power over data lines with four pairs is called "4 Pair Power over Ethernet" [80].

References

[1] A. L. Russell, "OSI: The Internet That Wasn't," July 30, 2013. [Online]. Available: http://spectrum.ieee.org/computing/networks/osi-the-internet-that-wasnt. [Accessed May 6, 2020].

[2] U. v. Burg and M. Kenny, "Sponsors, Communities, and Standards: Ethernet vs. Token Ring in the Local Area Networking Business," *Industry and Innovation*, vol. 10, no. 4, pp. 351–375, December 2003.

[3] R. M. Metcalfe, "The History of Ethernet," December 14, 2006. [Online]. Available: www.youtube.com/watch?v=g5MezxMcRmk. [Accessed May 6, 2020].

[4] Wikipedia, "On-board diagnostics," April 28, 2020. [Online]. Available: http://en.wikipedia.org/wiki/On_Board_Diagnostics. [Accessed May 8, 2020].

[5] K. Matheus, "OPEN Alliance – Stepping Stone to Standardized Automotive Ethernet," in 2nd Ethernet & IP @ Automotive Technology Day, Regensburg, 2012.

[6] BMW, "Ethernet Diagnose Stecker," April 2007. [Online]. Available: http://bmwtools.info/uploads/enet_doku.pdf. [Accessed May 13, 2020].

[7] ISO, *ISO 14229-1:2013-03 (E) Road Vehicles – Unified Diagnostic Services (UDS) – Part 1: Specification and Requirements*, Geneva: ISO, 2013.

[8] Free Software Foundation, "lwIP – a Lightweight IP Stack – Zusammenfassung," unknown. [Online]. Available: http://savannah.nongnu.org/projects/lwip/. [Accessed May 13, 2020].

[9] UNECE, "Test Procedure for Compression–Ignition Engines and Positive-Ignition Engines Fuelled with Natural Gas or Liquefied Pertoleum Gas with Regards to the Emission of Pollutants," June 25, 1998. [Online]. Available: www.unece.org/fileadmin/DAM/trans/main/wp29/wp29wgs/wp29gen/wp29registry/ECE-TRANS-2005-124-04e.pdf. [Accessed May 13, 2020].

[10] M. Johanson, P. Dahle, and A. Söderberg, "Remote Vehicle Diagnostics over the Internet using the DoIP Protocol," *Proceedings of the Sixth International Conference on Systems and Networks Communications*, pp. 226–231, October 2011.

[11] ISO, *ISO 13400-1:2011 – Road Vehicles – Diagnostic Communication over Internet Protocol (DoIP) – Part 1: General Information and Use Case Definition*, Geneva: ISO, 2011.

[12] ISO, *ISO 13400-2:2012 – Road Vehicles – Diagnostic Communication over Internet Protocol (DoIP) – Part 2: Transport Protocol and Network Layer Services*, Geneva: ISO, 2012.

[13] G. Mascolino, "Ethernet – der Standard. Jetzt auch im Automobil," in Automotive Kongress, Ludwigsburg, 2012.

[14] R. Bruckmeier, "Ethernet for Automotive Applications," in Freescale Technology Forum, Orlando, 2010.

[15] M. Jones, "Ethernet im Automobil," Automobil Elektronik, pp. 40–42, January 2011.

[16] S. Carlson, T. Hogenmüller, K. Matheus, T. Streichert, D. Pannell, and A. Abaye, "Reduced Twisted Pair Gigabit Ethernet Call for Interest," March 2012. [Online].

Available: www.ieee802.org/3/RTPGE/public/mar12/CFI_01_0312.pdf. [Accessed May 6, 2020].
[17] InvestmentMine, "Historical Copper Prices and Price Chart," September 26, 2013. [Online]. Available: www.infomine.com/investment/metal-prices/copper/all/. [Accessed May 13, 2020].
[18] Heise Online, "BMW erforscht Bordnetz mit Internet-Protokoll," October 27, 2007. [Online]. Available: www.heise.de/autos/artikel/BMW-erforscht-Bordnetz-mit-Internet-Protokoll-466769.html. [Accessed May 13, 2020].
[19] R. Brand, S. Carlson, J. Gildred, S. Lim, D. Cavendish, and O. Haran, "Residential Ethernet, IEEE 802.3 Call for Interest," July 2004. [Online]. Available: http://grouper.ieee.org/groups/802/3/re_study/public/200407/cfi_0704_1.pdf. [Accessed May 6, 2020].
[20] J. Hillebrand, M. Rahmani, R. Bogenberger, and E. Steinbach, "Coexistence of Time-Triggered and Event-Triggered Traffic in Switched Full-Duplex Ethernet Networks," in Industrial Embedded Systems, 2007. SIES '07, International Symposium on Digital Object Identifier, Lisbon, 2007.
[21] M. Rahmani, J. Hillebrand, W. Hintermaier, R. Bogenberger, and E. Steinbach, "A Novel Network Architecture for In-Vehicle Audio and Video Communication," in 2nd IEEE/IFIP International Workshop on Broadband Convergence Networks, 2007. *BcN '07*, Munich, 2007.
[22] M. Rahmani, R. Steffen, K. Tappayuthpijarn, E. Steinbach, and G. Giordano, "Performance Analysis of Different Network Topologies for In-Vehicle Audio and Video Communication," in 4th International Telecommunication Networking Workshop on QoS in Multiservice IP Networks, 2008, Venice, 2008.
[23] M. Rahmani, A. Pettiti, E. Biersack, E. Steinbach, and J. Hillebrand, "A Comparative Study of Network Transport Protocols for In-Vehicle Media Streaming," in *IEEE International Conference on Multimedia and Expo, 2008*, Hanover, pp. 441–444, 2008.
[24] M. Rahmani, K. Tappayuthpijarn, B. Krebs, E. Steinbach, and R. Bogenberger, "Traffic Shaping for Resource-Efficient In-Vehicle Communication," *IEEE Transactions on Industrial Informatics*, pp. 414–428, May 15, 2009.
[25] C. Salzmann, Modified from non-public document, Munich, 2009.
[26] T. Grünweg, "Modellzyklen der Automobilhersteller: Eine Industrie kommt auf Speed," February 10, 2013. [Online]. Available: www.spiegel.de/auto/aktuell/warum-lange-entwicklungszyklen-fuer-autohersteller-zum-problem-werden-a-881990.html. [Accessed May 8, 2020].
[27] K. Matheus, "Ethernet-basierte Kommunikation: Der skalierbare Vernetzungsbaukasten für die Fahrzeugentwicklung," *Elektonik Automotive*, January 2013.
[28] M. Rahmani, E. Steinbach, W. Hintermaier, A. Laika, and H. Endt, "A Novel Network Design for Future IP-based Driver assistance camera systems," in *International Conference on Networking, Sensing and Control, 2009. ICNSC '09*, Okayama, 2009.
[29] W. Hintermaier and E. Steinbach, "A System Architecture for IP-Camera Based Driver Assistance Applications," in *2010 IEEE Intelligent Vehicles Symposium (IV)*, San Diego, 2010.
[30] T. Hase, W. Hintermaier, A. Frey, T. Strobel, U. Baumgarten, and E. Steinbach, "Influence of Image/Video Compression on Night Vision Based Pedestrian Detection in an Automotive Application," in *2011 IEEE 73rd Vehicular Technology Conference (VTC Spring)*, Yokohama, 2011.
[31] T. Königseder and S. Singer, "Development Partnership BMW-Freescale Enables Ethernet for Automotive Applications," in Freescale Technology Forum, Austin, 2011.

[32] K. Balszuweit, "Einführung Ethernet&IP als applikativer Fahrzeugbus," in Steinbeis Symposium, Fellbach, 2013.

[33] K. Matheus, "Structural Support for Developing Automotive Ethernet," in *3rd Ethernet & IP @Automotive Technology Day*, Leinfeld-Echterdingen, 2013.

[34] Center of Automotive Management, "Automotive Innovations Award 2012," May 2, 2013. [Online]. Available: www.auto-institut.de. [Accessed May 9, 2013, no longer available].

[35] F. Borowitz and E. Scherm, "Standardisierungsstrategien, eine erweiterte Betrachtung des Wettbewerbs auf Netzeffektärkten," Fernuniversität Hagen, Hagen, 1999.

[36] H. L. Gabel, *Produktstandardisierung als Wettbewerbsstrategie*, London: McGraw-Hill, 1993.

[37] D. Martini, "Meeting Minutes of the OPEN Alliance Steering Committee," OPEN Alliance, July 2019.

[38] OPEN Alliance, "OPEN Alliance Homepage," 2020 (continuously updated). [Online]. Available: www.opensig.org/. [Accessed December 28, 2019].

[39] C. Hill, "Establishing a Standard: Competitive Strategy and Technological Standards in Winner-Take-All Industries," *Academy of Management Executive*, vol. 11, no. 2, pp. 7–25, 1997.

[40] B. D. Abramson, "The Patent Ambush: Misuse or Caveat Emptor?," January 5, 2011. [Online]. Available: www.ipmall.info/sites/default/files/hosted_resources/IDEA/idea-vol51-no1-abramson.pdf. [Accessed May 13, 2020].

[41] IEEE 802.3, "Approved Minutes IEEE 802.3 Ethernet Working Group Plenary Grand Hyatt, San Antonio, TX USA," November 12–15, 2012. [Online]. Available: www.ieee802.org/3/minutes/nov12/minutes_1112.pdf. [Accessed May 6, 2020].

[42] D. Dwelley, "Power over Data Lines Call for Interest," July 16, 2013. [Online]. Available: www.ieee802.org/3/cfi/0713_1/CFI_01_0713.pdf. [Accessed May 6, 2020].

[43] IEEE 802.3, "IEEE 802.3 Ethernet Working Group Homepage," IEEE 802.3, 2020 (continuously updated). [Online]. Available: www.ieee802.org/3/. [Accessed May 13, 2020].

[44] F. Leitner, "AUTOSAR AUTomotive Open System ARchitecture," 2007. [Online]. Available: www.inf.uni-konstanz.de/soft/teaching/ws07/autose/leitner-autosar.pdf. [Accessed October 9, 2013, no longer available].

[45] AUTOSAR, "Requirements on Ethernet Support in AUTOSAR, Release 4.0, Revision 1," December 7, 2009. [Online]. Available: www.autosar.org/fileadmin/user_upload/standards/classic/4-0/AUTOSAR_SRS_Ethernet.pdf. [Accessed May 13, 2020].

[46] AUTOSAR, "Specification of Ethernet Interface, Release 4.1., Revision 3," March 31, 2014. [Online]. Available: www.autosar.org/fileadmin/user_upload/standards/classic/4-1/AUTOSAR_SWS_EthernetInterface.pdf. [Accessed May 13, 2020].

[47] AUTOSAR, "Specification of Ethernet Interface, Release 4.2.2," 2015. [Online]. Available: www.autosar.org/fileadmin/user_upload/standards/classic/4-2/AUTOSAR_SWS_EthernetInterface.pdf. [Accessed May 13, 2020].

[48] Business Wire, "AVnu Alliance Launches to Advance Quality of Experience for Networked Audio and Video," August 25, 2009. [Online]. Available: www.businesswire.com/news/home/20090825005929/en/AVnu-Alliance-Launches-Advance-Quality-Experience-Networked. [Accessed May 6, 2020].

[49] Avnu Alliance, "Homepage of the Avnu Alliance," Avnu Alliance, 2020 (continuously updated). [Online]. Available: http://avnu.org/. [Accessed May 13, 2020].

[50] Avnu Alliance, "Industrial Control & Monitoring," 2020 (continuously updated). [Online]. Available: http://avnu.org/industrial/. [Accessed May 13, 2020].

[51] M. Jochim and J. Specht, "AAA2C – Automotive Requirements for a Flexible Control Traffic Class," July 14, 2014. [Online]. Available: www.ieee802.org/1/files/public/docs2013/new-tsn-jochim-aaa2c-requirements-for-control-traffic-0713-v01.pdf. [Accessed May 13, 2020].

[52] Avnu Alliance, "Avnu Knowledge Base, Avnu Automotive Advisory Council (AAA2C) Materials," Avnu Alliance, 2016. [Online]. Available: http://avnu.org/whitepapers/. [Accessed July 7, 2016, no longer available].

[53] G. Bechtel, B. Gale, M. Kicherer (Turner), and D. Olsen, *Automotive Ethernet AVB Functional and Interoperability Specification Revision 1.4*, Beaverton: Avnu, 2015.

[54] IEEE 802, "P802.1DG – TSN Profile for Automotive In-Vehicle Ethernet Communications," 2019 (continuously updated). [Online]. Available: https://1.ieee802.org/tsn/802-1dg/. [Accessed December 7, 2019].

[55] ISO, *ISO/DIS 17215:2013 – Road Vehicles – Video Communication Interface for Cameras (VCIC) Parts 1–4*, Geneva: ISO, 2013.

[56] O. Sugihara and S. Takahashi, "Standardization Activities on Gigabit Ethernet Operation Over Plastic Optical Fiber," in *5th Ethernet & IP @ Automotive Technology Day*, Yokohama, 2015.

[57] F. Horikoshi, *NWIP on Road Vehicles – In-vehicle Gigabit Ethernet System – Part 1: General Requirements of Gigabit Ethernet System and Physical and Data-link Layer Requirements of Optical Gigabit Ethernet System*, Geneva: ISO, 2015.

[58] F. Horikoshi, *NWIP on Road Vehicles – In-vehicle Gigabit Ethernet System – Part 3: General Requirements and Test Methods of Optical Gigabit Ethernet Components*, Geneva: ISO, 2015.

[59] IEC, *IEC 62228-5 ED1, Integrated Circuits – EC Evaluation of Transceivers, Part 5: Ethernet Transceivers*, Berlin: VDE, 2019.

[60] ISO, "Online Browsing Platform; ISO/DIS 21111-4(en)," 2018. [Online]. Available: www.iso.org/obp/ui#iso:21111:-4:dis:ed-1:v1:en. [Accessed December 28, 2019].

[61] B. Wuelfing, "CeBIT 2009: BMW and Partners Found GENIVI Open Source Platform," March 3, 2009. [Online]. Available: www.linuxpromagazine.com/Online/News/CeBIT-2009-BMW-and-Partners-Found-GENIVI-Open-Source-Platform. [Accessed May 13, 2020].

[62] B. Körber, *EMC Test Specification for BroadR-Reach™ Transceivers, v1.1*, Irvine, CA: OPEN Alliance, 2013.

[63] B. Körber, *EMC Test Specification for BroadR-Reach™ Common Mode Chokes, v1.1*, OPEN Alliance, 2013.

[64] B. Körber, S. Buntz, M. Kaindl, D. Hartmann, and J. Wülfing, *BroadR-Reach® Physical Layer Definitions for Communication Channel, v1.0*, Irvine, CA: OPEN Alliance, 2013.

[65] Business Wire, "OPEN Alliance and UNH-IOL Rev up Evolution of Connected Car," August 20, 2012. [Online]. Available: www.businesswire.com/news/home/20120820005361/en/OPEN-Alliance-UNH-IOL-Rev-Evolution-Connected-Car. [Accessed May 13, 2020].

[66] PR Newswire, "Broadcom, NXP, Freescale, and Harman Form OPEN Alliance Special Interest Group," November 9, 2011. [Online]. Available: www.prnewswire.com/news-releases/broadcom-nxp-freescale-and-harman-form-open-alliance-special-interest-group-133514928.html. [Accessed May 6, 2020].

[67] PHYS ORG, "NXP Develops Automotive Ethernet Transceivers for In-Vehicle Networks," November 9, 2011. [Online]. Available: https://phys.org/news/2011-11-nxp-automotive-ethernet-transceivers-in-vehicle.html. [Accessed May 6, 2020].

[68] IEEE-SA, "IEEE Standards Association (IEEE-SA) Ethernet & IP @ Automotive Technology Day," IEEE-SA, Continuously updated. [Online]. Available: https://standards.ieee.org/events/automotive/. [Accessed December 28, 2019].

[69] OPEN Alliance, "Automotive Ethernet Hits the Road in Wide Range of New Vehicles," October 14, 2015. [Online]. Available: http://opensig.org/news/press-releases/. [Accessed May 6, 2020].

[70] Marvell, "Marvell Unveils Industry's First 1000BASE-T1 Automotive Ethernet PHY Transceier," October 19, 2015. [Online]. Available: www.marvell.com/company/newsroom/marvell-unveils-industrys-first-1000base-t1-automotive-ethernet-phy-transceiver.html. [Accessed May 13, 2020].

[71] OPEN Alliance, "Automotive OPEN Alliance Demonstrates Multi-Vendor Interoperability," October 25, 2015. [Online]. Available: http://opensig.org/news/press-releases/. [Accessed May 13, 2020].

[72] A. Vollmer, "Microchip ist ein Konsolidator in der Halbleiterindustrie," February 14, 2019. [Online]. Available: www.all-electronics.de/microchip-ist-ein-konsolidator-in-der-halbleiterindustrie/4/. [Accessed December 28, 2019].

[73] J. Farrel and G. Saloner, "Standardization, Compatibility, and Innovation," *RAND Journal of Economics*, vol. 16, no. 1, pp. 70–83, 1985.

[74] M. Katz and C. Shapiro, "System Competition and Network Effects," *Journal of Economic Perspectives*, vol. 8, no. 2, pp. 93–115, 1994.

[75] M. Glaß, D. Herrscher, H. Meier, M. Piastowski, and P. Schoo, "SEIS – Sicherheit in Eingebetteten IP-Basierten Systemen," *ATZelektronik*, vol. 5, no. 1, pp. 50–55, 2010.

[76] ON Semiconductors, "In Vehicle Networking ESD Performance," [Online]. Available: www.onsemi.com/pub_link/Collateral/TND391-D.PDF. [Accessed September 29, 2013, no longer available].

[77] Automotive Electronics Council, "AEC Documents," July 17, 2012. [Online]. Available: www.aecouncil.com/AECDocuments.html. [Accessed May 13, 2020].

[78] Universität Erlangen, Lehrtstuhl für Informatik 12, "SEIS – Sicherheit in Eingebetteten IP-Basierten Systemen," June 6, 2013. [Online]. Available: www12.informatik.uni-erlangen.de/research/seis/. [Accessed May 13, 2020].

[79] Wikipedia, "Geschichte von Linux," April 10, 2020. [Online]. Available: http://de.wikipedia.org/wiki/Geschichte_von_Linux. [Accessed May 13, 2020].

[80] Business Wire, "IEEE Forms 4-Pair Power Over Ethernet (PoE) Study Group," April 1, 2013. [Online]. Available: www.businesswire.com/news/home/20130331005002/en/IEEE-Forms-4-Pair-Power-Ethernet-PoE-Study. [Accessed May 13, 2020].

[81] Wikipedia, "OSEK," April 22, 2020. [Online]. Available: http://en.wikipedia.org/wiki/OSEK/OSEK. [Accessed May 25, 2020].

[82] Wikipedia, "QNX," March 1, 2020. [Online]. Available: http://en.wikipedia.org/wiki/QNX/QNX. [Accessed May 25, 2020].

4 The Automotive Environment

The intention when designing communications systems is to neither over- nor under-design the technology. Overdesign means the technology is more expensive than it needs to be and has less chance to be competitive. Underdesign means it does not meet the performance requirements and is therefore not usable the way it was intended. Underdesign results either in wasted investments or, more often, in very costly measures to improve the technology in hindsight. It is therefore crucial to well understand the environment in which the technology is supposed to function before defining the technology.

The design of a communication system thus generally starts with the definition of the transmission channel. Key input parameters are the target data rate, the desired transmission medium, and the transmission environment. One of the main challenges in the automotive environment is meeting the stringent ElectroMagnetic Compatibility (EMC) requirements. This chapter therefore starts in Section 4.1 with explaining EMC in the automotive context before explaining in Section 4.2 the (partially resulting) parameters that define the automotive transmission channel in general. Section 4.3 addresses another distinctive feature in automotive: The quality requirements that active and passive components must meet in order to be used in cars. After all, simply using consumer grade parts is not an option and it needs to be understood where the different requirements come from. This impacts the component design, but goes beyond and/or is independent of the actual PHY specification. Figure 4.1 gives an overview of how the different aspects interrelate. The Ethernet PHY technology specific channels are discussed with the PHY speeds in Chapter 5. Chapter 6 discusses Ethernet in relation to the power supply.

4.1 ElectroMagnetic Compatibility (EMC)

If a device is electromagnetically compatible this means that it functions in its intended surroundings without being impaired by electromagnetic emissions of other devices in the same physical location while not disturbing the performance of other devices by its own emissions [1]. Both the ElectroMagnetic Immunity (EMI)[1] against interference from others as well as a device's own ElectroMagnetic Emissions (EME) are integral parts of the EMC performance of a device [2].

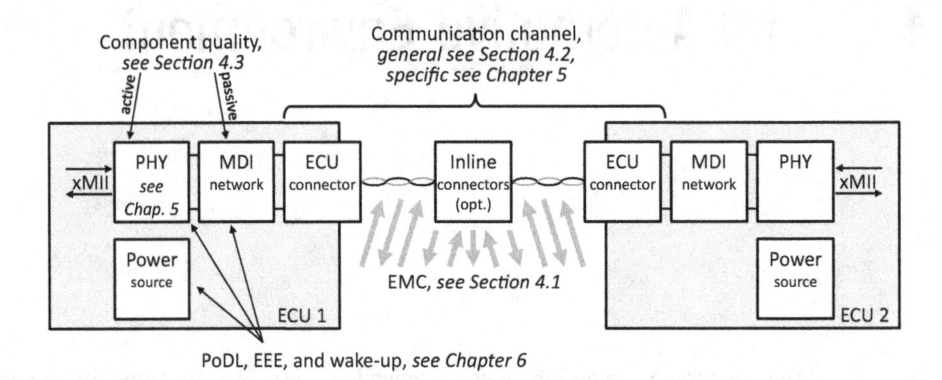

Figure 4.1 Interrelation between automotive environment and physical transmission

EMC has a long history and the automobile has actually accelerated the respective legislation. The first ever law on the topic was passed in 1892 in Germany in the context of the upcoming telegraph and telephone business [2]. It had become evident at an early stage that physically close cables can interfere with each other's transmissions. This interference was especially painful in the case of telegraph and telephone lines. The law thus dealt with the impact such interference had on respective devices and installations and how to handle it.

However, the EMC topic received a push in Germany on December 22, 1920, with a live radio transmission of a Christmas concert southeast of Berlin. The German chancellor at the time was invited to a nearby location in order to be charmed by the latest technical achievements, but instead was angered by the crackling that every passing car induced in the speakers. Countermeasures had to be taken and – what was only later called – EMC was by 1927 the reason for the first German law on the use and installation of high frequency radio transmitters. The law included limit lines and a clearance process, which were, with adaptations, valid in Germany until 1995 [2].

The international community saw similar developments. In 1933 the Comité International Spécial des Perturbations Radioélectriques (CISPR) was founded in Paris in order to develop guidelines on a European level. In the US, the American National Standards Institute (ANSI) and Federal Communications Commission (FCC) also produced respective rules for the US [3]. However, the need to regulate EMC on an even much broader scale arose with the invention and spread of the transistor. In 1973 the International Electrotechnical Commission (IEC) created a special technical committee with the purpose of handling EMC topics [2]. Today, with electronics having penetrated into every part of everyday life, EMC is more important than ever.

This book discusses four perspectives on EMC:

1. The coupling mechanisms, i.e., how the electric and electronic activity of one unit can actually affect the performance of another (see Section 4.1.1).
2. The standards addressing EMC (see Section 4.1.2).
3. The test methods to evaluate EMC behavior (see Section 4.1.3).
4. The sources of interference (see Section 4.1.4).

All four approaches are briefly described in order to generate the necessary understanding of the requirements for Automotive Ethernet in general. Next to emissions and immunity, the ElectroStatic Discharge (ESD) is also important for the quality and life expectancy of an electronic device. Even though ESD is not strictly part of EMC, it is part of the qualification tests that need to be performed and are thus included in this subsection (see Section 4.1.5).

4.1.1 Coupling Mechanisms of Electromagnetic Interference

In principle, every electronic device can, at the same time, be the cause as well as the victim of electromagnetic interference. It thus makes sense to select one device as the victim while identifying possible sources and coupling mechanisms that cause the interference (see Figure 4.2 or, e.g., Learn EMC [4]). The coupling mechanisms can be grouped into conducted coupling and coupling caused by a field. The field coupling can be far-field or near-field. In the latter case source and sink are less than a sixth of a wave length apart [5] and the coupling can be inductive from a magnetic field and capacitive from an electric field [6]. All four coupling paths can coexist and disturb a device at the same time. In order to counteract the interference, the correct identification of coupling paths and sources is important.

- In the case of **conducted coupling**, unintended signal energy leaves a unit via its cables. An example is High Frequency (HF) energy coupling into and leaving a device via the power supply cable, where it is not meant to be and from where it can cause interference to other devices directly that share a path in the power supply [6]. This type of interference is not inhibited by inline connectors. Often insufficient or defect ground connections, causing so-called ground loops, enhance this interference. If two units theoretically use the same ground but one has a significantly longer distance to it, or one ground connector is simply not functioning well, an interference that would otherwise simply be led to ground might find another path with lower impedance and cause disruptions on that path. A common mode signal might be coupled as well as a differential signal. A common mode signal coupled onto a wire pair causes currents flowing in the same direction on both wires. A differential signal causes opposing effects on both wires. Filtering and proper ground measures, which need to take the complete car design into account, combat this type of coupling.

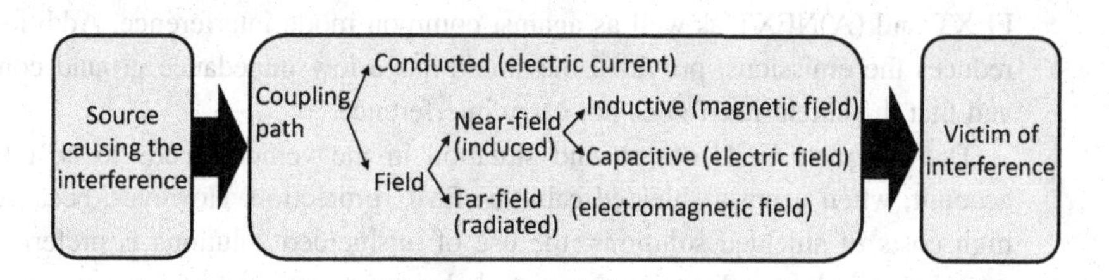

Figure 4.2 EMI model with source, coupling (types), and victim (sink)

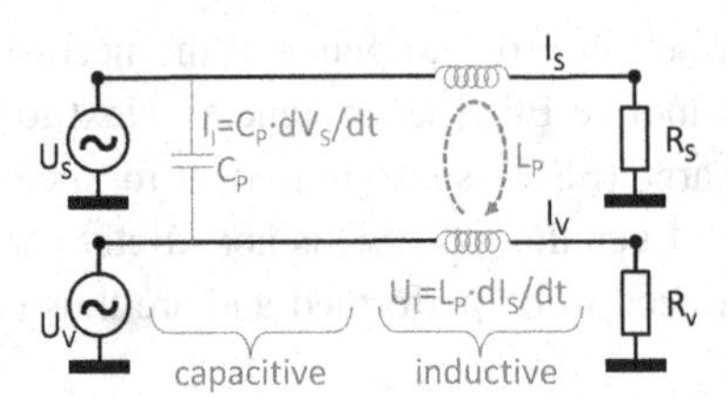

Figure 4.3 Near-field coupling via a parasitic capacitor or inductor [6]

- In the case of **near-field coupling**, interference is induced into a victim by a changing electric or magnetic field that is at a closer distance than a sixth of a signal's wavelength. The interference consequently increases with faster changes in the field (i.e., dv/dt), higher frequencies, and shorter distances [6]. Figure 4.3 shows the principle functioning of capacitive and inductive coupling in one schematic. For **capacitive coupling**, i.e., coupling from an electric field, the voltage of the interference source, V_S, causes an electric field across the gap between its own wire and the wire of an adjacent victim (V) system. The induced/interfering current, I_I, depends on the change of the voltage, U_I, and the parasitic capacitor, C_P, shared by the units. Typical sources for electric, i.e., capacitive coupling, are high-voltage power lines, ignition systems, or communication transceivers [6]. They represent very different technologies, but all have a high impedance in common. For **inductive coupling**, i.e., coupling from a magnetic field, the current in the wire from the source (S) system induces a magnetic field and thus a voltage into the victim system that depends on the parasitic inductor, L_P. Typical examples for inductive interferers are highway control transmitters, wireless stations, and radio frequency transmitters. As said, it is circuits with high impedance that are more likely to couple capacitively. Circuits with low impedance are more likely to cause interference from inductive coupling [6].

In communication systems, including Ethernet, one type of near-field electromagnetic interference is referred to as crosstalk (XTALK). In the case of crosstalk, a differential signal couples into another differential signal. Near End Cross Talk (NEXT) and Far End Cross Talk (FEXT) cause interference induced by an electric or magnetic field from wires of the same system (see Figure 5.12 for an example), while Alien NEXT (ANEXT) and Alien FEXT (AFEXT) are from neighboring wires of another system. Using shielded cables increases the immunity of cables against (A) FEXT and (A)NEXT as well as against common mode interference. Additionally, it reduces the emissions, provided the shield has a low impedance ground connection and that the shield itself does not carry interference.

The complete ECU design and situation in the vehicle needs to be taken into account, when using a shielded cable as EMC protection. However, because of the high costs of shielded solutions, the use of unshielded solutions is preferred in the automotive industry. Twisting the wires helps to improve the performance in cases of differential transmissions, as the two wires of the twisted pair are subject to the same

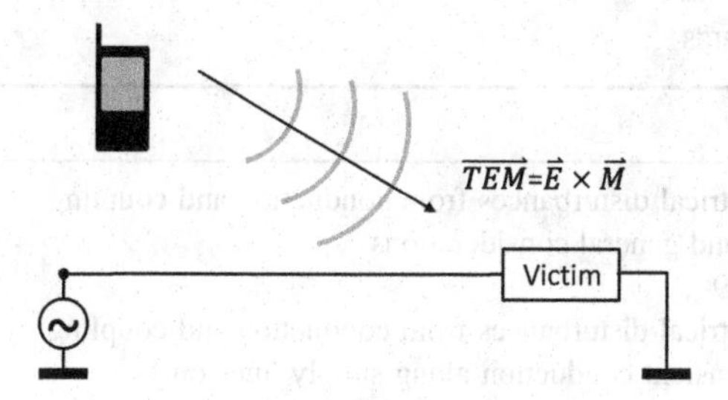

Figure 4.4 Radiated far-field coupling (also known as Radio Frequency Interference (RFI)) [7]

electromagnetic field. The coupling, therefore, is the same on both wires, which means that the differential signal is not affected as much, because the common mode interference is eliminated when the differential signal is combined at the receiver. The more symmetric, i.e., the more balanced a twisted pair is, the better. Furthermore, the EMC crosstalk performance can be improved if the distance between two potentially conflicting cables or wires is increased. In Automotive Ethernet using a jacket makes a critical difference (see e.g. Section 5.1.2 for the 1000BASE-T1 channel). For more information about EMC interference such as crosstalk or transient interferers see also Section 4.1.4.

- When the distance between the interference source and the interfered sink increases, only radiated **far-field coupling** can be the cause of electromagnetic interference. Mobile devices like mobile phones that use a transmitter are per se a potential source of such interference as it is their purpose to transmit electromagnetic energy through space in order to communicate. The phones radiate so-called Transversal ElectroMagnetic (TEM) waves that consist of an electric and a magnetic component (see also Figure 4.4). Any circuit that contains antenna-like elements for the right frequency will receive some of the energy transmitted and thus experience interference [6]. Cell phones used in a car may produce noise on both signal and power lines. However, the coupled energy is normally significantly smaller than the required automotive limits that are tested, e.g., with the BCI test (see also Section 4.1.3).

4.1.2 Standards for EMC

The topic of EMC is complex and requires a significant amount of experience and references. The existing standards are thus often based on prior versions and reuse or reference the experience that has been collected over decades. In addition to the relevant ISO standards that are listed in Table 4.1, various earlier and national norms, or norms that are used with national preference, exist, e.g., from the VDE, CISPR, or IEC (see also Section 4.1.3). Selecting which EMC specification to use and how to use them generally varies among car manufacturers who generate additional norms (see e.g. [8]). However, with the increasing globalization, the international applicability

Table 4.1 Overview on ISO EMC and ESD standards

Standard	Content
ISO 7637-1 (2015)	Road vehicles: Electrical disturbances from conduction and coupling Part 1: Definitions and general considerations Replaces DIN 40839
ISO 7637-2 (2011)	Road vehicles: Electrical disturbances from conduction and coupling Part 2: Electrical transient conduction along supply lines only
ISO 7637-3 (2016)	Road vehicles: Electrical disturbances from conduction and coupling Part 3: Electrical transient transmission by capacitive and inductive coupling via lines other than supply lines
ISO 10605 (2008)	Road vehicles: Test methods for electrical disturbances from electrostatic discharge
ISO 11451-1 (2015)	HYPERLINK "www.iso.org/contents/data/standard/06/24/62477.html" \o "ISO 11451-1:2015 Road vehicles – Vehicle test methods for electrical disturbances from narrowband radiated electromagnetic energy – Part 1: General principles and terminology" Road vehicles: Vehicle test methods for electrical disturbances from narrowband radiated electromagnetic energy Part 1: General principles and terminology
ISO 11451-2 (2015)	Road vehicles: Vehicle test methods for electrical disturbances from narrowband radiated electromagnetic energy Part 2: Off-vehicle radiation sources
ISO 11452-1 (2015)	Road vehicles: Component test methods for electrical disturbances from narrowband radiated electromagnetic energy Part1: General principle and terminology
ISO 11452-2 (2019)	Road Vehicles: Component test methods for electrical disturbances from narrowband radiated electromagnetic energy Part 2: Absorber-lined shielded enclosure
ISO 11452-3 (2016)	Road Vehicles: Component test methods for electrical disturbances from narrowband radiated electromagnetic energy Part 3: TEM-Cell
ISO 11452-4 (2011)	Road Vehicles: Component test methods for electrical disturbances from narrowband radiated electromagnetic energy Part 4: Harness excitation methods (Bulk Current Injection (BCI))
ISO 11452-5 (2002)	Road Vehicles: Component test methods for electrical disturbances from narrowband radiated electromagnetic energy Part 5: Stripline
ISO 11452-8 (2015)	Road Vehicles: Component test methods for electrical disturbances from narrowband radiated electromagnetic energy Part 8: Immunity to magnetic fields

and up-to-datedness makes the ISO standards the most comprehensive documents. Table 4.1 serves as an introduction and orientation.

4.1.3 Measuring EMC

Cars are the skillful combination of thousands of parts from different sources. In order to provide optimal quality, tests, and validations are performed at all levels of car

Table 4.2 Example hierarchy of EMC measurement methods in the automotive industry

	Semiconductor		ECU		Vehicle
Immunity	• Direct Power Injection (DPI), IEC 62132-4	➔	• Bulk Current Injection (BCI), ISO 11452-4 • Transversal ElectroMagnetic (TEM) cell, ISO 11452-3 • Antenna measurements in absorber lined chambers, ISO 11451-2	➔	• Antenna measurements in absorber lined chambers, ISO 11451-2 (orig. CISPR25) • Stripline, ISO 11452-5 with OEM adaptations for large cars
Emissions	• 150 Ohm method, IEC 61967-4	➔	• Stripline, ISO 11452-5 • Antenna measurements in absorber lined chambers, ISO 11451-2	➔	• Measurements with vehicle on-board antennas, ISO 11451-2 (orig. CISPR12, EN55025) • Antenna measurements in absorber lined chambers ISO 11451-2 (orig. CISPR25)

Note: The precise EMC test requirements generally vary from car manufacturer to car manufacturer

development and production. Proving EMC performance is an integral part of this. Respective tests are performed at the semiconductor, ECU, and vehicle level (see also [9] and Table 4.2).

Performing tests at the semiconductor level is relatively new for car manufacturers, but is especially crucial when introducing a new in-vehicle networking technology such as Automotive Ethernet. After all, it is the Tier 1 suppliers that are responsible for the correct functioning of the ECUs they deliver, but it is the car manufacturer who decides on the in-vehicle network and who is responsible for a functioning in-vehicle network (see also Section 2.3.2). The earlier in the development process potential error sources are detected, the less likely that malfunctions will occur at a later point in time. To solve potential EMC issues at the semiconductor level is about as early as it is possible to find errors in a system.

The most conclusive results on semiconductor immunity are achieved with Direct Power Injection (DPI) [10]. The DPI test is defined in the IEC 62132-4 standard. When deploying the DPI test with Automotive Ethernet, the developer must pay attention to the fact that the DPI test impacts the return loss performance. In the case of in-vehicle networking technologies with a shared medium – e.g., CAN, LIN, or FlexRay – this is of no consequence as the link is always used for one transmission direction only. However, in the case of full-duplex Automotive Ethernet, the test set-up of the DPI test can make the link perform worse than it would be without the test set-up. In order to avoid any test artifacts, the coupling termination in the DPI test set-up must be matched carefully in order to ensure that the impedance of the transmission link does not change (M. Tazebay, email correspondence, December 21, 2013).

An important (conducted) emissions test at the semiconductor level is the "150 Ohm direct coupling method," also described in IEC 62132-4 or, apparently, in ISO 61967-4 [11]. The typical impedance of an in-vehicle wiring harness is 150 Ohms. For the tests, a decoupling network with 150 Ohms impedance is used to measure the frequency spectrum of the output voltage at certain IC pins or IC pin groups [12]. With test boards similar to the ones used for the DPI tests and special measurement receivers, the emissions can be assessed. If a semiconductor passes the DPI and 150 Ohms tests, this is a first assurance that its design will show a good EMC performance when integrated into an ECU and later a car. In consequence, car manufacturers might request the positive test results before recommending the semiconductor for use by their Tier 1s.[2]

The tests at the ECU level endeavor to emulate the automotive environment and consist of measurements performed on the ECU in a laboratory. For the tests, the communication interfaces and their communication partners are modeled in order to achieve useful results. In the laboratory, EME as well as EMI are measured extensively. The ISO Bulk Current Injection (BCI) tests and stripline measurements are common. Also, TEM-cell as well as antenna tests in an anechoic chamber are typical. Current injection during the BCI test simulates an external EMC noise being coupled into the wiring harness. For the stripline and TEM-cell measurements, special antennas are used to simulate specific interference with repeatable conditions.

Finally, the technology is integrated and measured inside the car. This is extremely important, as the conditions in the laboratory are not identical to those inside the car. Not 100% of the in-vehicle effects can be modeled; sometimes simply because they are not known in advance. Examples are the effects of the ground connection and the electric fields that depend on the exact body form. Simulations and laboratory tests are continuously being improved, but because of the complexity of the coupling effects in the real product, simulation results only give an indication. Before a technology has been qualified with in-vehicle measurements, it is not ready for production. Table 4.2 gives an overview of the hierarchy of important EMC measurement methods in the automotive industry; without making claims to be complete.

4.1.4 Sources of (EMC) Interference

Any communication system must be designed such that it can handle the anticipated EMC interference. A well-founded estimation of the types of interference and their characteristics is thus key to ensuring that a system is neither over nor underdeveloped. Naturally, the exact sources and levels of interference must not be known but need to be modeled realistically. As this leaves room for interpretation and preferences (conservative or less conservative), the authors have witnessed serious debates in standardization efforts just on this point. It does not help those debates that individual OEM norms sometimes define different requirements for the same effect (see e.g. [8]).

A first approach to distinguishing the noise sources is separating between noise that the system generates itself and noise that comes from external/alien sources [13], i.e.,

Table 4.3 Example for alien EMC noise sources [13, 14, 16, 17]

Category	Source	Type of error
AWGN	Various (accumulated) noise sources	Random
NBI/CW interference	Radio phones, ham radio, AM/FM/TV broadcasting, safety and security Radios (police, ambulances), etc.	Continuous (has a defined frequency range, <1 kHz according to CISPR)
Pulsed interference	E.g., (airport) radar	Pulsed (tested for frequencies >800 MHz)
Impulse/burst/ transient noise	Ignition system, Electric motors' commutator, switched electronics, power supplies, etc.	Bursts (shorter than 120 ns for e.g., 1000BASE-T1, broadband by nature, but often relatively narrow band damped sign wave)
(Alien) XTALK	Other data communication wires close by	Random

noise = self-noise + alien noise. One type of self-noise is NEXT or FEXT. This only occurs when systems use multiple cable pairs like 1000BASE-T, see also Section 5.2.1.2. It is therefore not relevant for the 1-pair Automotive Ethernet systems. Another type of self-noise is echo/InterSymbol Interference (ISI). As self-noise can be known, it is possible to implement measures to combat self-noise, (often) leading to a relevant trade-off question in the system design: How much effort/costs versus how much residual self-noise is sustainable.

This is different for alien noise. It cannot be cancelled, and the system therefore has to be able to cope with it. Sources of alien noise can be alien XTALK (AFEXT and ANEXT) from neighboring communication wires, Narrow Band Interference (NBI, of which Continuous Wave (CW) interference represents a specific case [14]), e.g., from other radios, impulse/burst/transient noise from ignition, etc. Sometimes pulsed modulation interference is discussed, as may be caused by some radars [15]. Additive White Gaussian Noise (AWGN) is a method for modeling the accumulation of a number of more or less specified noise sources for calculation or simulation purposes. Table 4.3 provides an overview of the possible interference types.

Which noise type is included in the considerations depends predominantly on the frequency range in which the system is going to function and whether the noise source is expected to be in-band or out of band. Out of band noise can be filtered and is less critical [13]. With Automotive Ethernet PHYs now covering a range from 10 Mbps to 10 Gbps (and beyond), the operating frequencies vary accordingly and thus also the noise/channel models (see Section 5.1 for details for the different speed grades).

The OPEN Alliance EMC test specifications include tests for immunity to transient interference for 100 and 1000BASE-T1 transceivers [18, 19]. This is a requirement of the automotive industry. However, because the transmit spectrum of Automotive Ethernet is relatively high, the low frequency transients are out of band. For some of the legacy in-vehicle networking systems discussed in Section 2.2 that operate in a lower frequency range the interference from these transient disturbances is a more serious concern.

4.1.5 ElectroStatic Discharge (ESD)

In the case of ESD, a short, high voltage impulse is caused by a spark at an electronic device owing to a large electric difference between the electronic device and the touching entity. Under disadvantageous circumstances the high voltage discharge can damage the device. Field-effect transistors are especially susceptible to such damage [20]. The cause of the electric potential difference is generally a triboelectric effect or electric induction. A well-known example is walking on a carpet, which can charge a human up to 30,000 V.

In the automotive industry, ESD has many facets and needs to be considered in the whole chain, from ESD during assembly, to ESD caused by service staff, to ESD caused by a car's occupants. To test the robustness of electronic devices to ESD, four ESD-simulation models have been introduced with ISO 10605, which also describes the respective test methods for road vehicles. A good summary of the models can be found in [21]:

- The **Human Body Model (HBM)** models the discharge of an electrostatically charged person when touching an electronic hardware element. The induced current is assumed to pass between different pins of the touched hardware in question.
- The **Machine Model (MM)** is similar to the HBM, but instead of a person an electrostatically charged machine discharges in contact with the electronic hardware element. Like in the case of HBM, the induced current is assumed to pass between different pins of the touched element.
- The **Charged Device Model (CDM)** is fundamentally different from the HBM and the MM. In the case of the CDM the whole electronic device is assumed to have been electronically charged and suddenly discharged against a low resistance electrode. In this case, no current passes through the discharging device.
- The **Field induced Charge Device Model (FCDM)** also assumes that an electronic device is suddenly discharged. The difference is that the charging happened in an electric field or via electric load shift.

For enabling Automotive Ethernet this means that all transceiver semiconductors need to be tested accordingly; individually and in the respective application. In principle, ESD tests are a standard procedure for all automotive semiconductors, which is part of the AEC-Q100 qualification and recommended design process. A detailed description of ESD tests for 100BASE-T1 and 1000BASE-T1 transceivers is given in the OPEN Alliance specifications [18, 19], which reference IEC 61000-4-2. Körber [18, 19] makes a distinction between unpowered and powered ESD, which refers to the transceiver actually tested. The scenario of unpowered discharge is more likely in the production process, whereas powered discharge can happen during (other) tests or even in the field, when passengers enter or exit a running car and touch the covering near the respective ECU.

In the case of a part that is integrated into an ECU, the situation is not quite as straightforward. The key question is how much of the 8 kV discharged at the outside connector contacts of an ECU (see ISO 10605) actually reaches the transceiver chip.

Figure 4.5 Example of a test set-up that measures the voltage at the 100BASE-T1/OABR PHY pins in case of ESD at the entrance of the test board
[photo by Tim Puls, Semtech]

This is of particular interest, because transceiver chips are per definition connected to the outside of an ECU and because the continuous miniaturization in the semiconductor industry makes ICs ever more sensitive to ESD. It can be expected that those portions of the Printed Circuit Board (PCB) that lie between the connector and the transceiver chip somewhat reduce the voltage; in case of 100BASE-T1/OABR these are a CMC, a coupling capacitor, and potentially other filter elements. The question is, and this also depends on the particular PCB design, how much does the PCB reduce the discharge voltage? Figure 4.5 shows the example of a respective test set-up.

In the case of the BMW test boards, about 500 V of the 8 kV reached the transceiver pins. If the ICs used cannot handle this voltage, additional ESD protection elements must be added. These elements in return have to be dimensioned such that their parasitic capacitance does not influence the transmission channel (too much). A generic answer or recommendation on how to handle ESD protection for Automotive Ethernet is not possible. In the best case, the PCB design and Ethernet transceiver can handle the residual ESD voltage. In the worst case, additional ESD protection must be added.

4.2 The Automotive Communication Channel in General

The (automotive) communication channel generally constitutes of cables and connectors used in the wiring harness. Figure 4.6 gives an impression of how complex the wiring harness can be in the automotive industry. Not for nothing is the harness the third heaviest and third most expensive part of a car (after the engine and chassis, see e.g. [22]). Ground connections throughout the car are not necessarily less complex: Because of the use of new compound materials, the body is no longer one huge,

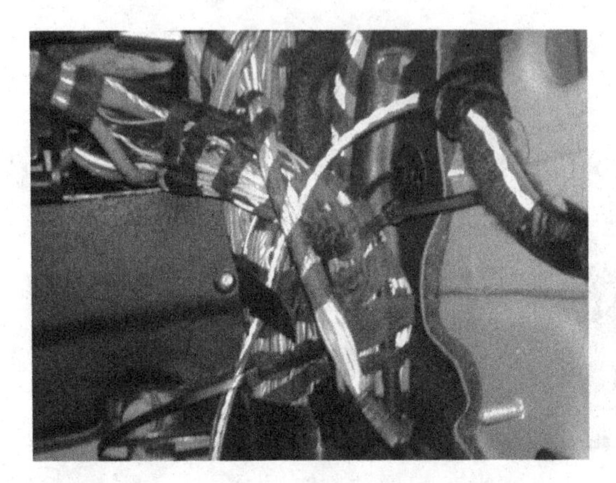

Figure 4.6 An impression of the complexity of a car's harness

conducting sheet of metal. Furthermore, the harness must pass through separations in order to reach doors, the trunk, or the engine compartment.

Harnesses are manufactured in a large variety: For every car model and depending on the options the customer selected, a different harness is built. In every harness, a large number of cables with different uses and functions are in closest proximity to one another, which means that the possibility of crosstalk and asymmetries cannot be neglected. Also, a harness uses components from various suppliers. A harness may consist of many different types of cables and connectors: UTP cables are often twisted and connected as part of the harness manufacturing. Shielded, coax, and optical cables are delivered pre-manufactured with connectors. All these aspects need to be taken into consideration when defining the channel. Sections 4.2.1 and 4.2.2 explain the framework for the Automotive Ethernet channel and the parameters used.

4.2.1 Channel Framework

Figure 4.7 shows the different elements needed for two ECUs to be able to communicate at the PHY level. Before discussing the channel parameters, it is essential to define the elements the channel is comprised of and to identify the elements that are part of the PHY but not part of the channel. Figure 4.7 shows that parts of the ECU – the PHY transceiver IC, the Media Dependent Interface (MDI) network, and the connector parts attached to the ECU – are not part of the channel. Instead, the channel consists of the cable, up to four inline connectors,[3] and the end connector parts attached to the cables. The standard maximum channel length for automotive is 15 m.[4]

As an example, Figure 4.8 shows the elements of the MDI network for 100BASE-T1/OABR and the location of the MDI. The MDI is where the media changes from PCB to wire. The MDI network is in principle independent from the function of the channel. The channel and the MDI network are designed to meet certain requirements and limit lines and require that the corresponding other parts meet theirs as well. If the goal is to further improve, e.g., the MDI performance, this simply improves the overall

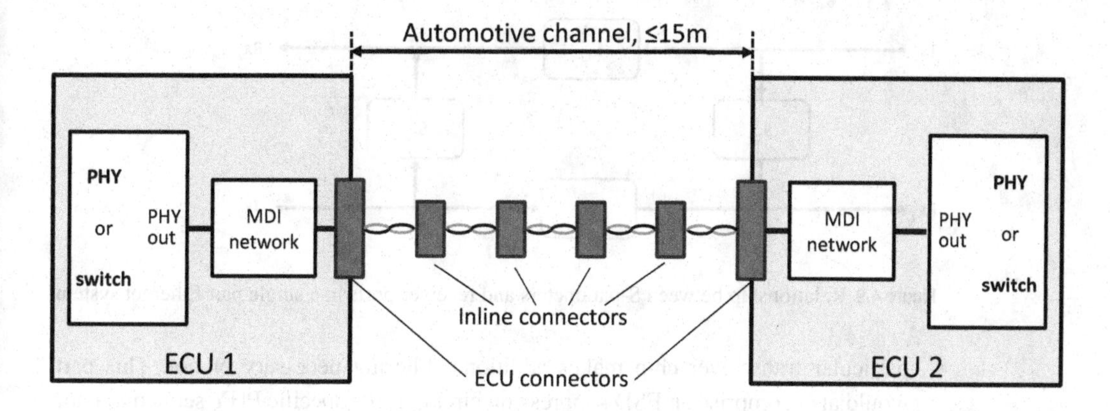

Figure 4.7 Ethernet transmission elements on PHY level identifying the communication channel [23]

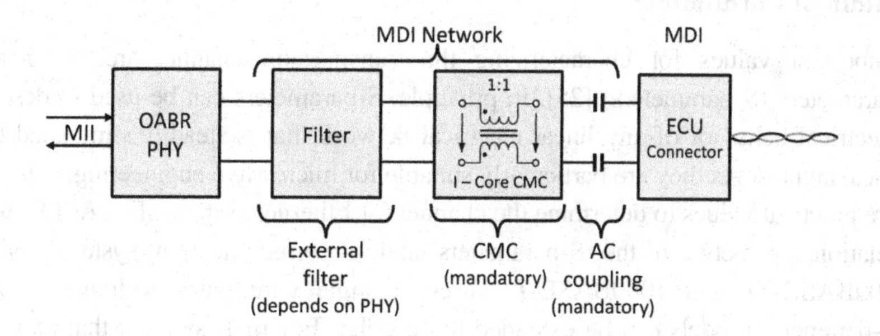

Figure 4.8 Elements of the MDI network for 100BASE-T1/OABR Ethernet [23]

performance without having any (negative or positive) impact on the channel performance. This leaves room for manufacturers to optimize their parts.

The MDI network consists of three major parts:

- **AC coupling (DC blocking):** Some coupling mechanism is mandatory in order to suppress DC from the transmission. Section 4.3.2 explains in more detail why it is possible to use capacitors instead of transformers for 100BASE-T1/OABR and 1000BASE-T1. Capacitors are technically sufficient and more cost efficient than transformers.
- **CMC:** The CMC is one of the most important components in the system. Its function, common mode suppression, is vital for the EMI. Section 4.3.2 details why an I-core CMC can be used with 100BASE-T1. This is desirable, because I-core CMCs allow for fully automated production. As the bandwidth for 1000BASE-T1 is about 10-fold the bandwidth of 100BASE-T1, the 1000BASE-T1 CMC needs to perform wide band suppression at 10 dB better performance. For details of the 1000BASE-T1 CMC specification see Körber [24].
- **Filter:** The additional filter in the MDI network performs spectral shaping in order to improve the EMC performance. The use of the filter depends on whether a

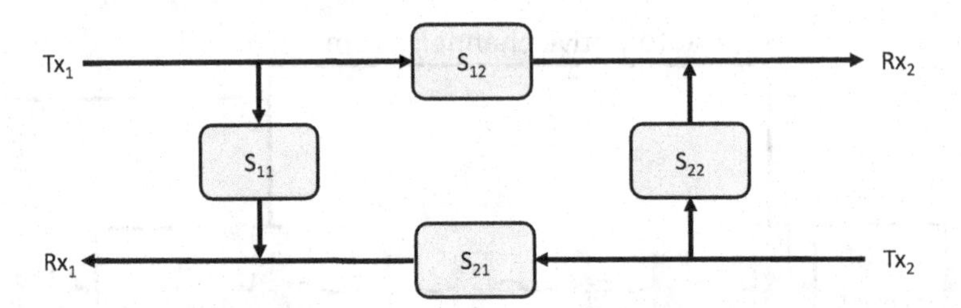

Figure 4.9 Relationship between S-parameters and receiver ports in a single pair Ethernet system

particular transceiver chip makes additional filtering necessary or not. This part would also comprise an ESD suppression circuit, if the specific PHY semiconductor requires its use (see also Section 4.1.4).

4.2.2 Channel Parameters

Important values for characterizing the transmission channel are the Scattering-parameters (S-parameters) [25]. In principle, S-parameters can be used to describe the electrical behavior of any linear electrical network that is steadily stimulated by electrical signals. As they are particularly suitable for microwave engineering, S-parameters are practical values to determine the channel for Ethernet systems. Figure 4.9 shows the relationship between the S-parameters and a communication system that, as in 100BASE-T1 and 1000BASE-T1, uses a single wire pair for transmission only. S-parameter models can be extended to describe (Ethernet) systems that require more cable pairs. As none of the Automotive Ethernet PHYs need more pairs as of now, it is not discussed here. The S-parameters for the one pair system are S_{11}, S_{12}, S_{21}, and S_{22}. Their actual characteristics for a given channel can be measured with a network analyzer, whose output results can then be compared with the allowed limit lines.

For robustness reasons, even communication systems with data rates significantly below 100 Mbps generally use differential signaling (see also Section 2.2) because the communication channel not only carries the differential signal, but also interference with common mode characteristics. The whole idea of the differential signal is that the common mode interference is suppressed at the receiver, when both parts of the differential signal are recombined to one signal, i.e., converted back from a differential to a common receive signal. This works completely when the interference is symmetric and it is thus a goal for Automotive Ethernet to have the channel set up as symmetrically as possible. Unfortunately, even very carefully designed networks can have some asymmetries, be it slightly different lengths of the two wires within the UTSP cable or an unavoidable neighboring power supply in a multi-pin connector. The EMC performance of a system is significantly influenced by these asymmetries. In order to be able to define limits for this as well, the S-parameters thus distinguish between values for the differential transmission performance "dd" and values for the common mode to differential mode "cd" and differential mode to common mode "dc" conversion performance.

The following parameters are (commonly) used to describe the channel:

- **Insertion Loss (IL):** The IL describes the power loss that happens between transmitter and receiver on a communication line. The main causes for this attenuation of signal strength are: Impedance mismatch between link and load, losses in the dielectric material and power dissipation due to the conducting surfaces [26]. The IL is directly linked to the differential performance, with $IL = f(S_{dd,12}, f) \approx f(S_{dd,21} f)$. The IL increases with the frequency, meaning the higher the transmit frequency (the symbol rate and Nyquist Frequency) the larger the attenuation and the lower the signal strength at the receiver.
- **Return Loss (RL):** The RL is the echo strength a received signal is impaired by from reflections at discontinuities of its own transmission. It also directly relates to the differential performance and is defined as $RL_1 = f(S_{dd,11}, f)$ and $RL_2 = f(S_{dd,22}, f)$. In Section 4.1.4 the echoes are identified as one source of self-noise. The interference strength of the echoes also increases with the frequency.[5] Note that the respective scattering parameters are not the same as either the IL or the RL. However, the S-parameters can be measured (relatively easily) and provide a close enough approximation in order to evaluate whether a particular channel (consisting of cable, connectors, inline-connectors) meets the limit lines. Sometimes, also the relation between IL and RL is of interest. If the signal strength at the receiver is small (e.g., when the channel is very long), the strength of the echoes will also be small. If the signal strength is very large, the echo strength will also be large (this might happen for very short channels, or short distances to the first inline-connector).
- **Mode Conversion Loss (MCL):** It is important to have parameters that describe the symmetry (also called balance) between two wires of one cable pair that is so important for EMC. Originally, Körber et al. [27] used the Transverse Conversion Loss (TCL) and the Equal Level Transverse Conversion Transfer Loss (ELTCTL) for this purpose. The TCL measures the echo of the common mode signal as a function of $S_{cd,11}/S_{cd,22}$. The ELTCTL measures the common mode to differential mode conversion in relation to the intended signal as a function of $S_{cd,12} - S_{dd,12}$ and $S_{cd,21} - S_{dd,21}$. Newer publications (e.g., [28–30]), use the Longitudinal Conversion Loss (LCL, $S_{dc,11}/S_{dc,22}$) and the Longitudinal Conversion Transmission Loss (LCTL, $S_{cd,12}/S_{cd,21}$) instead as parameters for the mode conversion loss. LCL and TCL as well as LCTL and TCTL provide the same technical information [31]. TCTL replaced ELTCTL because the "Equal Level" turned out to be hard to maintain in real systems that always experience some kind of attenuation. Especially in the case of channels with high attenuation, this can lead to results that do not represent the actual situation (B. Körber and T. Wunderlich, email correspondence, March 17, 2016).
- **Power Sum for Alien Near-End Cross Talk (PSANEXT):** Crosstalk describes the interference caused by surrounding cable pairs. With the Automotive Ethernet technologies from 10 Mbps to 10 Gbps being single pair, all crosstalk comes from sources unknown and thus "alien" (see also Section 4.1.4). The limit lines for the ANEXT) are described by the PSANEXT.

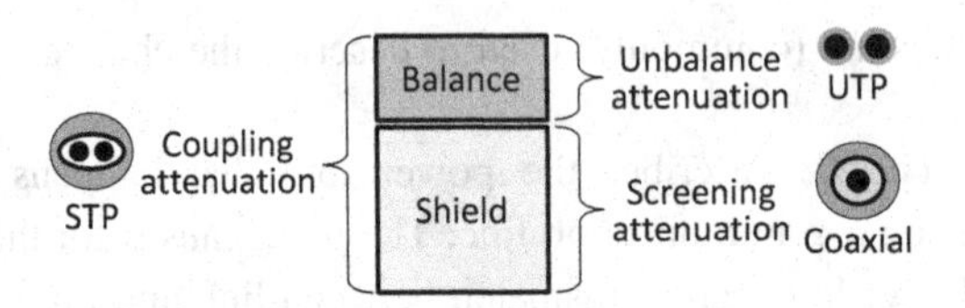

Figure 4.10 Simplified depiction of the relationship between screening, coupling, and unbalance attenuation [35]

- **Power Sum for Alien Attenuation to Cross talk Ratio at Far-end (PSAACRF):** The PSAACRF describes the AFEXT impact on the far-end.
- **Coupling Attenuation (CA):** When unshielded links are deployed, the symmetry of the used components is the essential parameter for EMC behavior. The developments around 100BASE-T1 have been an immense contribution to the knowledge in the industry to that end. However, with increasing data rates and frequencies, the attenuation increases and the resulting signal strength may no longer be sufficient to combat the also-increased EMC noise. To use a shielded channel instead is one way to obtain an additional EMC-margin.[6,7] Coupling attenuation is thus used as an additional parameter for shielded links to describe the overall effectiveness of EMC behavior of shielded links.
- **Screening Attenuation (SA):** The coupling attenuation takes both the unbalance attenuation and the screening attenuation into account (see Figure 4.10). The unbalance attenuation reflects the quality of the symmetry (which might actually be negatively impacted by the screen) that accounts for the amount of power that is coupled from the common mode to the differential mode and the other way around. For frequencies up to 1 … 2 GHz the unbalance attenuation is very similar to the LCTL (B. Bergner, email correspondence, January 21, 2020). The screening attenuation reflects the effectiveness of the screen for shielded cables and connectors [32] to prevent coupling. It is listed as an additional channel parameter in, e.g., [33, 34], despite being part of the coupling attenuation. Figure 4.10 depicts that, for single ended shielded cables like coaxial cables, the coupling attenuation consists only of the screening attenuation, while UTP cables only exploit the symmetry. STP cables can make use of both and, therefore, in principle, combat EMC the best of the three. Generally, the variable distance of the shield to the wire pair is a potential source for asymmetry in STP cables.
- **Impedance:** Last but not least, the data transmission links have a characteristic impedance that describes their property for electromagnetic wave propagation.

Having limit lines for the channel available is crucial for developing the PHY transceiver solutions. From the channel limit lines, it is possible to deduce requirements that allow manufacturing a harness and its components in a way that meets the automotive quality requirements and to identify optimization potential. This is done by reverse engineering, i.e., measuring the S-parameters of certain samples against the limit lines.

4.3 The Quality Strain

Addressing quality in cars means considering the complete production chain. It starts with the IC silicon design, includes PCB design on component level, ECU design, the manufacturing process at Tier 1 suppliers, and the manufacturing process of the car at the OEM. The quality has to last for a minimum of ten years of customer use and for use in a very challenging physical environment that faces extreme temperature variations, very wet as well as very dusty environments, vibration, dirt, and electrical stress to name but a few. In order to handle this, clear processes are needed for every aspect, which altogether define "Automotive Quality." As this book is about Automotive Ethernet, this section will focus on the quality of the active semiconductors and passive components needed for deploying networking technologies.

4.3.1 Automotive Semiconductor Quality Standards

A common quality value well suited to visualize the difference between automotive and other industries is the statistic defect rate for one Million parts, indicated in Parts Per Million (PPM). Naturally, zero defects is the ideal target value. But, the lower the PPM value the more expensive the part will be, which means that every industry and every company must find its own compromise between the price a company is willing to pay and the target quality it wants to achieve.

Imagine a car manufacturer who builds one million cars a year. Every car has 100 ECUs and every ECU consists of 400 electronic parts. This means 40,000 electronic parts per car, and 40 Billion parts the car manufacturer builds into cars per year. Note that 400 electronic parts per ECU does not represent the average of today, but it represents a simplified future toward which the car industry is heading. Figure 4.11 shows the probability of having a defect ECU and the probability of having delivered a car with a defective ECU to a customer, both depending on the PPM on part level.[8] The additional parameter is the percentage of defective parts that are not or cannot be detected in an ECU at the time of production.

At 30 PPM on electronic part level – a very good quality value in the consumer industry [36] – a car manufacturer would deliver about 70% of cars with at least one defective ECU were none of these faulty ECUs were detected in time in the production chain. The vertical lines in Figure 4.11 indicate the respective PPM values at the ECU level. It can be seen that 100 PPM at the ECU level means about 0.25 (!) PPM on part level. This would still result in every 100th car containing a defective ECU if all defects were missed and every 1000th car if 10% of the defective ECUs were missed; quality values completely unacceptable from a customer's perspective. The solution, of course, would be to simply find all defective ECUs before they are built into the car. Unfortunately, even with very good final qualification methods, this is not possible as the malfunctions in defective semiconductors often only materialize in one of the many borderline situations of extreme temperature, humidity, or mechanical stress. The solution is thus to improve the production process of semiconductors further and to standardize the qualification of the semiconductors such that not only faulty

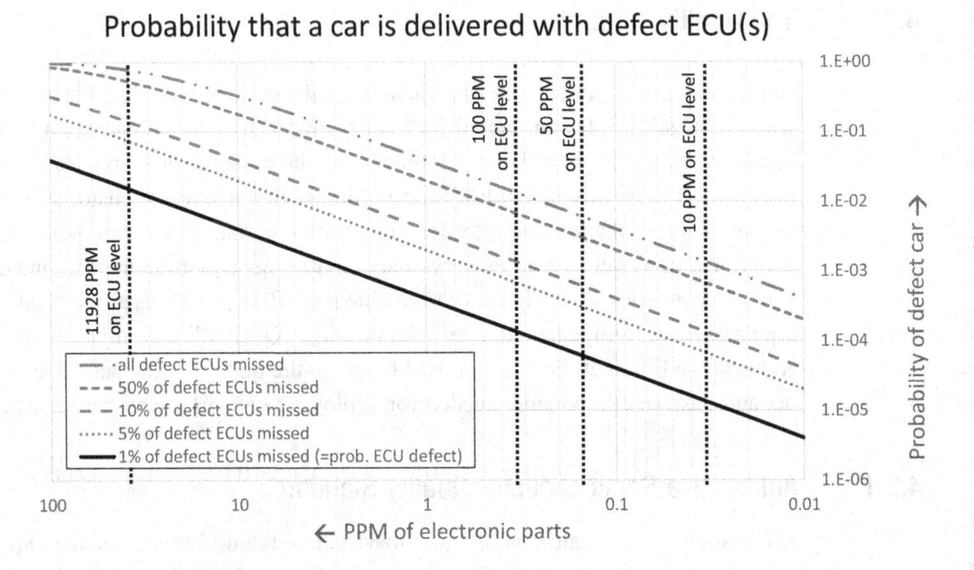

Figure 4.11 Relationship between PPM and the percentage of defective cars, depending on the percentage of defective ECUs not detected ("missed") during the qualification process; the assumption is that every car consists of 100 ECUs and that every ECU consists of 400 parts. Today, not many cars have this complexity, but it shows the more complex a car is and the more electronics a car contains, the more critical the quality of every component is

semiconductors but also those more likely to be faulty at a later point in time are not used in products in the first place.

In the beginning, when electronics started to become more prevalent in the automotive industry, every car manufacturer had its own requirements and every supplier its own qualification plan that, in turn, needed to be reviewed by the car manufacturer. So, in 1992, US car makers came up with the idea that led to the Automotive Electronics Council (AEC), which then standardized quality in the yet small but increasing market of automotive semiconductors [37]. The initial standard developed by the AEC was the AEC-Q100 for Integrated Circuits (ICs). After having been reviewed by primary IC suppliers, it was available in June 1994. All three US car makers, Chrysler, Ford, and GM, accepted only AEC-Q100 qualified semiconductors and thus achieved a major milestone in standardized automotive quality [37].

Today, the AEC-Q100 is a minimum standard for automobile manufacturers and suppliers worldwide. Later specifications for two more categories of semiconductors were added: AEC-Q101 for discrete semiconductor devices and AEC-Q200 for passive components. To this very day, the AEC still organizes annual reliability workshops [38].

To give an overview and an idea about the complexity of the quality requirements, Tables 4.4, 4.5, and 4.6 list the respective specifications.

The AEC quality specifications are important for Automotive Quality, but are not comprehensive. Car manufacturers often have additional requirements and work in

Table 4.4 Overview on AEC specifications on quality test methods for integrated circuits AEC-Q100

Standard	Content
AEC - Q100	Stress qualification for integrated circuits (base document only with no test methods)
AEC - Q100-001	Wire bond shear test
AEC - Q100-002	Human Body Model (HBM) electrostatic discharge test
AEC - Q100-003	Machine Model (MM) electrostatic discharge test
AEC - Q100-004	IC latch-up test
AEC - Q100-005	Non-volatile memory program, erase endurance, data retention, and operational life test
AEC - Q100-006	Electro-thermally induced parasitic gate leakage test
AEC - Q100-007	Fault simulation and test grading
AEC - Q100-008	Early Life Failure Rate (ELFR)
AEC - Q100-009	Electrical distribution assessment
AEC - Q100-010	Solder ball shear test
AEC - Q100-011	Charged Device Model (CDM) electrostatic discharge test
AEC - Q100-012	Short circuit reliability characterization of smart power devices for 12 V systems

Table 4.5 Overview on AEC specifications on quality test methods for discrete semiconductors AEC-Q101

Standard	Content
AEC - Q101	Stress test qualification for discrete semiconductors (base document only, no test methods)
AEC - Q101-001	Human Body Model (HBM) electrostatic discharge test
AEC - Q101-002	Machine Model (MM) electrostatic discharge test
AEC - Q101-003	Wire bond shear test
AEC - Q101-004	Miscellaneous test methods
AEC - Q101-005	Charged Device Model (CDM) electrostatic discharge test
AEC - Q101-006	Short circuit reliability characterization of smart power devices for 12V systems

Table 4.6 Overview on AEC specifications on quality test methods for passive components AEC-Q200

Standard	Content
AEC - Q200	Stress test qualification for passive components (complete document with test methods)
AEC - Q200-001	Flame retardance test
AEC - Q200-002	Human Body Model (HBM) electrostatic discharge test
AEC - Q200-003	Beam load (break strength) test
AEC - Q200-004	Measurement procedures for resettable fuses
AEC - Q200-005	Board flex / terminal bond strength test
AEC - Q200-006	Terminal strength / shear stress test
AEC - Q200-007	Voltage surge test

close relation with key automotive semiconductor manufacturers with respect to quality management, design rules, test coverage, test strategy, and process technologies for the different stages of development, ramp up, and serial production. At ECU level, additional requirements exist and they might vary depending, e.g., on the use of an ECU in a so-called wet area or dry area, or if additional temperature requirements must be met. These additional requirements often vary among car manufacturers, which, in the end, also reflects the quality customers can expect from their cars.

4.3.2 The CMC (Quality) for Automotive Ethernet

While Section 4.3.1 gave an idea of the complexity of (semiconductor) quality in the automotive industry in general, this section gives an example of the impact the automotive quality requirements can have on the actual use of a technology. In the case of Automotive Ethernet, the quality of the transceiver (PHY) semiconductor was less critical. If a semiconductor company cannot afford to set up production according to the automotive requirements in house, foundries such as Taiwan Semiconductor Manufacturing Company (TSMC) can perform the automotive qualified production on their behalf. Basic automotive quality can thus be achieved with reasonable effort.[9]

However, Ethernet-based communication does not only require a suitable PHY chip but also some form of coupling to the network. For this, transformers are used. Transformers provide DC blocking in the form of a galvanic separation between PCB and transmission line and they perform a common mode rejection/suppression in order to optimize the design and to comply with EMC requirements. Figure 4.12 shows the schematic and x-ray of a standard 100BASE-TX transformer. The inductor cores are generally wound and soldered by hand before being cast into the housing. Figure 4.12 shows the irregular winding and how the cores are placed into the housing without any additional fixing. In principle, the wires are isolated. However, when the housing is soldered onto the PCB, such irregular winding is more problematic, as wires are more likely to run unintentionally close to pins other than the ones they are intended to be connected to. In the soldering process these wires are then easily affixed to those other pins as well by mistake.

The variation in quality of such parts tends to be high and is thus not acceptable for the automotive industry. Any handmade component is critical. This means that in

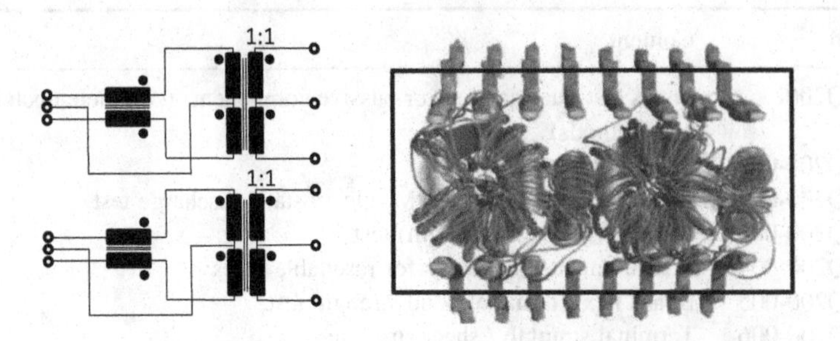

Figure 4.12 Schematic and x-ray of a typical 100BASE-TX transformer

order to be able to use Ethernet in the automotive industry, another solution was needed. At the time that BMW was preparing the introduction of Automotive Ethernet and was looking for a suitable solution, "planar transformers" were being discussed as a solution whose production is automated. However, the path was not pursued further as planar transformers were not competitive in terms of costs and size.[10] The next step therefore was to look at whether transformers were needed at all. In cars, the high insulation voltage standard Ethernet transformers have to deal with is not needed. Cars use 12 V only and have a common ground. This allows realizing the DC isolation with capacitors instead of using the transformer part of the schematic shown in Figure 4.12. However, the EMI is improved with common mode suppression. This function does require an inductance in form of a Common Mode Choke (CMC).

For 100BASE-T1/OABR, BMW first intended to use a ring core CMC. This reduced the complexity in comparison with the transformer shown in Figure 4.12. The challenge, however, was to fully automate the production of the CMC. It turned out not to be possible. In order to meet the automotive quality requirements nevertheless, this would have meant to test every single produced part individually. As a result, the CMC would have been more expensive than the PHY IC. This was not acceptable, but is a good example of how a quality requirement can potentially prohibit the use of a technology.

A different solution was needed. After some intensive research and development work an I-core variant was selected for the CMC. This variant allows fully automated production and the expected small variance in quality, which can be achieved with the CMCs used for other in-vehicle networking technologies like CAN and FlexRay. In the end, only the discovery of this CMC variant made Ethernet-based communication cost competitive in automotive use.

When a final, working solution is available, it is very difficult to recall the effort required to achieve it. Critics of Automotive Ethernet always suspected the complementary hardware to be the cost driver of Ethernet that would, in the end, make it too expensive. Without the usability of an I-core CMC, they might have been right. This example shows how important it is to always look at the complete system. It also explains why the automotive industry needs significant lead time to introduce a new technology.

Notes

1 EMI is sometimes also used as an abbreviation for ElectroMagnetic Interference, which might be used as a general term for EMC behavior. We find that using EME and EMI as defined in this book is less ambiguous. EME, emissions, is the disturbance to others. EMI, immunity, is the effect of the disturbance from others.

2 For 100BASE-T1 and 1000BASE-T1 a detailed description of the emission and immunity tests for transceiver components without power over functionality can be found in the respective OPEN Alliance specifications [18, 19]. Furthermore, at the time of writing, the IEC worked on an update of the IEC 62228-5 specification ("Integrated circuits – EMC evaluation of transceivers Part 5: Ethernet transceivers"), with the goal to incorporate the content of the published OPEN Alliance EMC specifications.

3 The requirement to support four inline connectors came out of a survey among the car manufacturers for the 1000BASE-T1 standardization effort [42]. As the authors recall, it was the maximum number requested by the Japanese car industry. Like the link length of 15 meters, four inline connectors are negotiated when the transmission environment becomes more challenging (see e.g. the MultiGBASE-T1 development, Section 5.5).

4 Fifteen meters has been identified as the maximum length in cars, vans, and light trucks. The original BroadR-Reach specification foresaw a link length of only 10 meters, which for passenger cars is more than sufficient [41]. By contrast, buses or large trucks require a link length of 40 meters [40].

5 Additionally, some of the specifications define MDI RL. The MDI RL is not related to the transmission channel but is measured from the MDI connector to the pins of the transceiver. It is a value useful for PHY design.

6 The IEEE 802.3ch specifications foresee a shielded link for 2.5, 5, and 10 Gbps data rates. 100BASE-T1 is defined for an unshielded link. The IEEE 802.3bp specification defines an unshielded but jacketed cable for a 1 Gbps data rate (see also Table 5.9). The jacket serves as a physical means to increase the distance to neighboring wires and thus decreasing the impact of alien crosstalk. This shows that transmitting 1 Gbps over unshielded cables is approaching a feasibility limit. It is thus not surprising that the OPEN Alliance specified an additional shielded link for 1000BASE-T1 [33].

7 Naturally, the shield also reduces the emissions!

8 Assuming that the probability of a defect is the same for all electronic parts used, the probability to have selected X defect parts in an ECU is

$$ProbXDefectsInECU = \frac{NoPartsECU!}{X!(NoPartsECU - X)!}\, ProbDefectPart^{X} \, (1 - ProbDefectPart)^{NoPartsECU-X}$$

which means that the probability of having at least one defect in an ECU is

$$\begin{aligned} ProbECUdefect = ProbMin1DefectInECU &= 1 - Prob0DefectsInECU \\ &= 1 - (1 - ProbDefectPart)^{NoPartsECU} \end{aligned}$$

Assuming that $y\%$ of defective ECUs are not detected before the ECU is built into the car, the probability of having an undetected defect and at a later point in time malfunctioning ECU in the car can be calculated similarly

$$\begin{aligned} ProbCarWithDefectECU = ProbMin1ECUdefect &= 1 - Prob0ECUdefect \\ &= 1 - (1 - yProbECUdefect)^{NoECUs} \end{aligned}$$

with $ProbDefectPart = \frac{PPM}{1{,}000{,}000}$. In the example $NoPartsECU = 400$ and $NoECUs = 100$.

9 Basic automotive semiconductor quality requirements address production and testing. Enhanced automotive quality starts with including specific quality considerations in the design and development of an automotive part.

10 Planar transformers can sometimes be found in the consumer and IT industries, especially in the case of high frequency applications. As their production is fully automated, they achieve better quality values than standard transformers. Planar transformers are directly integrated as part of the PCB and encased with ferrite material [20]. Apart from automated production, planar magnetics have performance advantages in low profile structure, low leakage current, reduced high frequency winding loss, and better thermal management. On the downside, they have a low window utilization factor and an increased parasitic capacitance [39].

References

[1] IEC, "The Electromagnetic Environment," not known. [Online]. Available: www.iec.ch/emc/explained/environment.htm. [Accessed May 14, 2020].

[2] D. E. Möhr, "Was ist eigentlich EMV? – Eine Definition," not known. [Online]. Available: www.emtest.de/de/what_is/emv-emc-basics.php. [Accessed May 6, 2020].

[3] Wikipedia, "CISPR," December 3, 2019. [Online]. Available: http://de.wikipedia.org/wiki/CISPR/CISPR. [Accessed May 14, 2020].

[4] Learn EMC, "Introduction to EMC," unknown. [Online]. Available: www.learnemc.com/emc-tutorials. [Accessed May 14, 2020].

[5] Williamson Labs, "EMC," 1999–2011. [Online]. Available: www.williamson-labs.com/ltoc/glencoe-emc-11.htm. [Accessed December 26, 2013, no longer available].

[6] R. Schmitt, *Electromagnetics Explained*, Amsterdam: Newnes, 2002.

[7] K. Budweiser, *A Theoretical Analysis of EMC Influence on Advanced Ethernet*, Munich: Technische Universität München, 2013.

[8] Texas Instruments, "ISO 7637 Standards & Solutions Series," June 1, 2016. [Online]. Available: https://training.ti.com/iso7637-and-related-standards-overview?cu=1127713. [Accessed January 1, 2020].

[9] S. Buntz, "EMI Adhoc – First Steps Generating a Simulation Model," November 2012. [Online]. Available: http://grouper.ieee.org/groups/802/3/RTPGE/public/nov12/buntz_01_1112_rtpge.pdf. [Accessed January 1, 2020].

[10] B. Körber, D. Sperling, and T. Form, "EMV Bewertung von CAN Transceivern," 2001. [Online]. Available: www.fh-zwickau.de/fileadmin/ugroups/ftz/Publikationen/Prof_Sperling/2001_EMV-Beurteilung.pdf. [Accessed December 17, 2013, no longer available].

[11] Toshiba Electronic Devices and Storage Corporation, "EMC Testing Laboratory Receives ISO/IEC 17025 Accreditation," 2018. [Online]. Available: https://toshiba.semicon-storage.com/content/dam/toshiba-ss/shared/docs/company/technical-review/technical-review-11_e.pdf. [Accessed January 1, 2020].

[12] B. Deutschmann, R. Jungreithmair, and G. Winkler, "Messmethode zur Messung der leitungsgeführten Störemissionen von ICs," 2003. [Online]. Available: www.eue24.net/pi/media/pdf/eue/emck_2003/ek3c0101.pdf. [Accessed 14 February 2014, no longer available].

[13] G. Parnaby, "Noise Considerations for RTPGE Objectives," September 25, 2012. [Online]. Available: http://grouper.ieee.org/groups/802/3/RTPGE/public/sept12/parnaby_01_0912.pdf. [Accessed January 2, 2020].

[14] W. Schäfer, "Narrowband and Broadband Discrimination with a Spectrum Analyzer or EMI Receiver," December 1, 2010. [Online]. Available: https://incompliancemag.com/article/narrowband-and-broadband-discrimination-with-a-spectrum-analyzer-or-emi-receiver/. [Accessed January 2, 2020].

[15] M. A. Richards, J. A. Scheer, and W. A. Holm, *Principles of Modern Radar; Vol.1 Basic Principles*, Raleigh, NY: SciTech Publishing, 2010.

[16] M. Tazebay, "Evolution of Ethernet for Next Generation Automotive Applications," *Nikkei Automotive Seminar*, Osaka, June 2019.

[17] Wikipedia, "Electromagnetic Compatibility," December 3, 2019. [Online]. Available: https://en.wikipedia.org/wiki/Electromagnetic_compatibility. [Accessed January 2, 2020].

[18] B. Körber, *IEEE 100BASE-T1 EMC Test Specification for Transceivers, v1.0*, Irvine, CA: OPEN Alliance, 2017.

[19] B. Körber, *IEEE 1000BASE-T1 EMC Test Specification for Transceivers, v.1.0*, Irvine, CA: OPEN Alliance, 2017.

[20] P. Leroux and M. Steyaert, *LNA-ESD Co-Design for Fully Integrated CMOS Wireless Receivers*, Cham: Springer (Kluwer), 2005.

[21] ESD Association, "Fundamentals of Electrostatic Discharge," 2010. [Online]. Available: www.esda.org/esd-overview/esd-fundamentals/part-5-device-sensitivity-and-testing/. [Accessed May 14, 2020].

[22] S. Carlson, T. Hogenmüller, K. Matheus, T. Streichert, D. Pannell, and A. Abaye, "Reduced Twisted Pair Gigabit Ethernet Call for Interest," March 15, 2012. [Online]. Available: www.ieee802.org/3/RTPGE/public/mar12/CFI_01_0312.pdf. [Accessed May 6, 2020].

[23] C. Arndt, "Determination of Relevant Parameters for the Diagnosis of BroadR-Reach. Prototypic Realization for the Riagnosis of the Communication Channel," Hochschule Deggendorf, Deggendorf, 2013.

[24] B. Körber, *IEEE 1000BASE-T1 EMC Test Specification for Common Mode Chokes, v1.0*, Irvine, CA: OPEN Alliance, 2017.

[25] Wikipedia, "Scattering Parameters," May 14, 2020. [Online]. Available: http://en.wikipedia.org/wiki/Scattering_parameters. [Accessed May 15, 2020].

[26] P. A. Rizzi, *Microwave Engineering: Passive Circuits*, Upper Saddle River: Prentice Hall, 1988.

[27] B. Körber, S. Buntz, M. Kaindl, D. Hartmann, and J. Wülfing, *BroadR-Reach® Physical Layer Definitions for Communication Channel, v1.0*, Irvine, CA: OPEN Alliance, 2013.

[28] IEEE Computer Society, *802.3bw-2015 – IEEE Standard for Ethernet Amendment 1 Physical Layer Specifications and Management Parameters for 100 Mb/s Operation over a Single Balanced Twisted Pair Cable (100BASE-T1)*, New York: IEEE-SA, 2015.

[29] IEEE Computer Society, *802.3bp-2016 – IEEE Standard for Ethernet Amendment 4: Physical Layer Specifications and Management Parameters for 1 Gb/s Operation over a Single Twisted Pair Copper Cable*, New York: IEEE-SA, 2016.

[30] B. Körber, S. Buntz, R. Pöhmeier, T. Müller, and J. Wülfing, *BroadR-Reach® Definitions for Communication Channel, v2.0*, Irvine, CA: OPEN Alliance, 2014.

[31] C. DiMinico, "802.3bp Cabling Parameters to S-parameter Naming," September 2014. [Online]. Available: www.ieee802.org/3/RTPGE/email/pdfDOLXj6Te3J.pdf. [Accessed May 14, 2020].

[32] B. Mund and C. Pfeiler, "Balunless Measurement of Coupling Attenuation," October 6–8, 2015. [Online]. Available: www.bedea.com/images/IWCS_2015_Balunless_measurement_of_coupling_attenuation.pdf. [Accessed January 8, 2020].

[33] B. Körber, B. Bergner, D. Marinac, D. Dorner, F. Bauer, H. Patel, J. Razafiarivelo, M. Dörndl, M. Kaindl, M. Rucks, M. Nikfal, P. Gowravajhala, T. Müller, T. Wunderlich, V. Raman, W. Mir, and Y. Bouri, *Channel and Component Requirements for 1000BASE-T1 Link Segment Type A (STP), v1.0*, Irvine, CA: OPEN Alliance, 2019.

[34] IEEE Computer Society, *802.3ch-2020 – IEEE Standard for Ethernet Amendment: Physical Layer Specifications and Management Parameters for 2.5 Gb/s, 5 Gb/s, and 10 Gb/s Automotive Electrical Ethernet*, New York: IEEE-SA, 2020.

[35] E. DiBiaso and B. Bergner, "Media Considerations – Insertion Loss and EMC," May 2017. [Online]. Available: www.ieee802.org/3/ch/public/may17/DiBiaso_3NGAUTO_01_0517.pdf. [Accessed January 15, 2020].

[36] C. Kymal and P. Patiyasevi, "Semiconductor Quality Initiatives," April 2006. [Online]. Available: www.qualitydigest.com/april06/articles/05_article.shtml. [Accessed May 14, 2020].

[37] Automotive Electronics Council, "AEC History," not known. [Online]. Available: www.aecouncil.com/AECHistory.html. [Accessed May 6, 2020].

[38] Automotive Electronics Council, "Annual Reliability Workshop," 2020 (continuously updated). [Online]. Available: www.aecouncil.com/AECWorkshop.html. [Accessed May 14, 2020].

[39] S. Wang, "Modelling and Design of Planar Integrated Magnetic Components," July 31, 2013. [Online]. Available: https://vtechworks.lib.vt.edu/handle/10919/34400. [Accessed May 14, 2020].

[40] S. Buntz, "IEEE RTPGE Information: Update on Required Cable Length," July 17, 2012. [Online]. Available: http://grouper.ieee.org/groups/802/3/RTPGE/public/july12/buntz_02_0712.pdf. [Accessed May 14, 2020].

[41] Broadcom, "OPEN Alliance BroadR-Reach (OABR) Physical Layer Transceiver Specification for Automotive Applications v3.2," 24 June 2014. [Online]. Available: www.ieee802.org/3/1TPCESG/public/BroadR_Reach_Automotive_Spec_V3.2.pdf. [Accessed 14 May 2020].

[42] H. Zinner, K. Matheus, S. Buntz, and T. Hogenmüller, "Requirements Update for IEEE802.3 RTPGE," July 18, 2012. [Online]. Available: http://grouper.ieee.org/groups/802/3/RTPGE/public/july12/zinner_02_0712.pdf. [Accessed January 2, 2020].

5 Automotive Physical Layer Technologies

Figure 5.1 depicts the different functional layers that make up a PHY system following the IEEE 802.3 nomenclature. The lowest layer is represented by the medium over which the data is to be transmitted. The medium, i.e., channel is normally one of the first decisions taken in every communication system development as the channel parameters impact many design choices in the physical layer directly. This chapter will therefore start, in Section 5.1, with a description of the channels for the Automotive Ethernet PHYs. The IEEE 802.3 PHY specifications themselves start/end at the Media Dependent Interface (MDI), that is the interface between the channel and the physical layer technology.

The two main sublayers of the PHY are the Physical Coding Sublayer (PCS) and the Physical Medium Attachment (PMA), which are normally implemented in one ASIC. The PCS receives digital data from the many Media Independent Interfaces (xMII) and encodes the data into symbols for consecutive processing by the PMA. It also decodes the received signal into a bit stream ready to be passed to higher layers via the xMII. The PMA has the task to physically prepare a signal for transmission and to prepare the receive signal such that the coded information can be extracted from it by the PCS.

Depending on the technology, more sublayers are added to the PHY. The Physical Medium Dependent (PMD) sublayer is, e.g., needed when, for the same transmission standard different media are used that require (each) additional and different handling of the signal. For example, for an optical transmission standard such as 1000BASE-RH (see Section 5.3.2), the PMD defines the translation between the electrical and optical analog signals.

A generally optional feature is described by the auto-negotiation layer. Auto-negotiation is a method to establish and select the best communication capabilities with respect to data rate, duplex mode, and flow control. This is crucial, when the same unit may have communication partners with different capabilities connected, as happens frequently in the ad hoc plug & play environments of the Consumer Electronics (CE) and Information Technology (IT) industries. In the predefined in-vehicle networks, the situation is slightly different and the scenarios where plug & play is needed are the exception and not the rule. Because they can occur nevertheless, auto-negotiation has been included as an optional feature in the 1000BASE-T1, 2.5, 5, 10GBASE-T, and 10BASE-T1 specifications (see Sections 5.3.1, 5.4, and 5.5).

Usually, the PHY specifications end with the respective Media Independent Interface (xMII) that connects the PHY with the MAC layer. The reconciliation layer

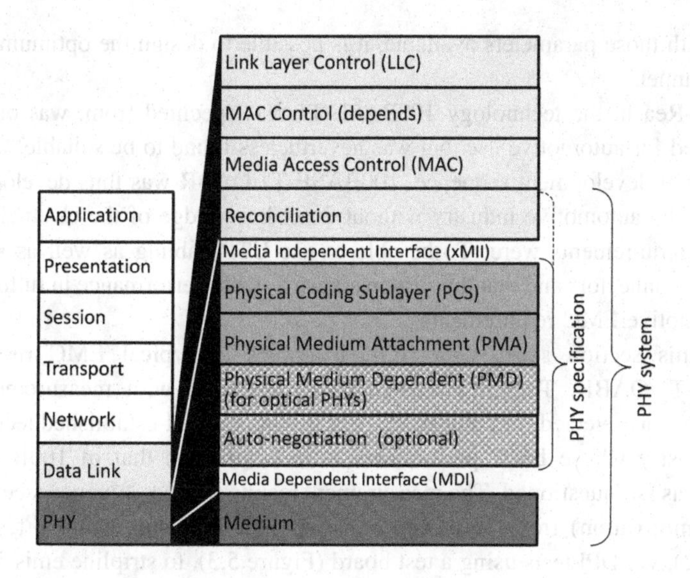

Figure 5.1 Overview on PHY sublayers in respect to OSI levels as referenced for various IEEE802.3 specifications

terminates this function by mapping the signals of the MII to the Physical Signaling Sublayer (PLS) service primitives used in the MAC. In most (Automotive Ethernet) PHY specifications the reconciliation layer is not changed and therefore not addressed. However, parts of the 10BASE-T1S specification touch the reconciliation layer[1] (see Section 5.4.3), which is why the IEEE 802.3cg specification includes the reconciliation layer and why Figure 5.1 extends the PHY specification bracket to it with a dashed line.

In this chapter, the PHY speeds are described in order of their development. That is, the descriptions start with 100 Mbps (Section 5.2), then move to 1 Gbps (Section 5.3), then to 10 Mbps (Section 5.4), then to multi Gbps (Section 5.5), and the chapter is closed with the latest initiatives (Section 5.6). This order is necessary as many decisions for later projects are based on learnings from the earlier ones. If a numerical order was chosen, some later decisions would be hard to understand and impossible to describe.

Describing the channels first can lead to a situation where the reader learns about the motivation for developing a particular speed grade only after having read the details on the channel of that speed grade. However, keeping the channels together allows for a better direct comparison of the channels.

5.1 The Automotive Ethernet Channels

5.1.1 The 100BASE-T1 Channel

When developing a communication technology, the technical properties of the channel the technology is meant to run on are generally the first parameters that need to be known. With those parameters available, it is possible to design the optimum system for this channel.

BroadR-Reach, the technology 100BASE-T1 was specified from, was originally not designed for automotive use, but was nevertheless found to be suitable. Contrary to the normal development sequence, 100BASE-T1/OABR was thus developed and adopted in the automotive industry without exact knowledge of the channel parameters. The requirements were: To be able to use UTP cabling as well as standard automotive connectors and that the system shows a stable performance in fulfilling the strict automotive EMC requirements.

Thus, this section starts with some examples of typical EMC results for 100BASE-T1/OABR. The measurements are compared with measurements for CAN. CAN was selected for comparison, because it is a well-established technology in the industry whose EMC performance – in contrast to that of 100BASE-T1/OABR – was not questioned. The measurement results show a sequence (see Section 4.1.3 for motivation) from 150 Ohm emission measurements using a test board (Figure 5.2), via DPI tests using a test board (Figure 5.3), to stripline emission tests with a reference ECU (Figure 5.4). For both CAN and 100BASE-T1/OABR, better or worse results can be achieved depending on the actual implementations. The examples shown provide an orientation and present the limit lines that need to be met.

EME for a typical CAN interface.
The "OEM Requirement" line is the limit.
"CM choke" has to be under that limit.

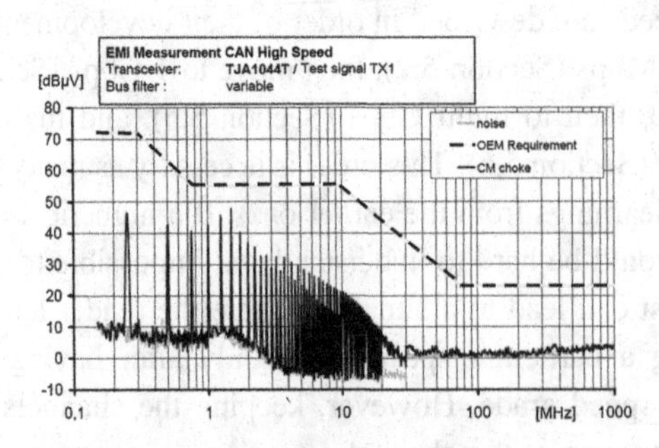

EME for a typical 100BASE-T1/OABR interface.
The "OEM requirement" line is the limit.
"symmetric" has to be under that limit.
"+2,5%" and "-2,5% unbalanced" show how OABR performs in case of disturbed symmetries.

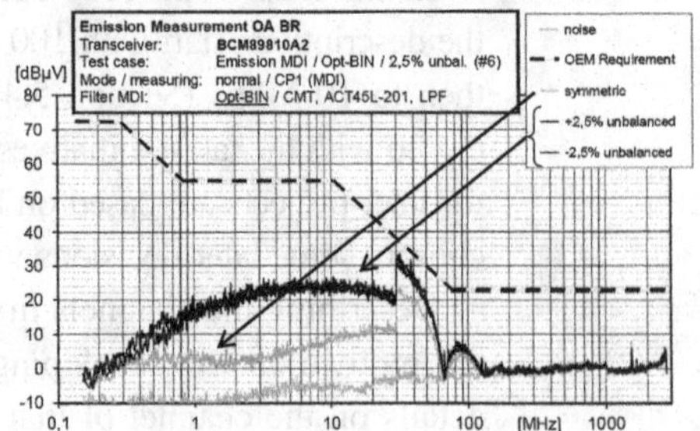

Figure 5.2 Example results for 150 Ohm EME tests according to IEC 61967-4. Conclusion: The 100BASE-T1/OABR Ethernet tests on board level meet the automotive requirements. They are also good when comparing the results with CAN. Of course, the frequency behavior of the EME is different as a different communication method is used. While CAN works in a range of 200 kHz to a few MHz, 100BASE-T1/OABR Ethernet is using a bandwidth up to 33⅓ MHz. The peak that can be seen at 30 MHz in the OABR curves is an artifact of the measuring equipment, which changes the resolution at this frequency

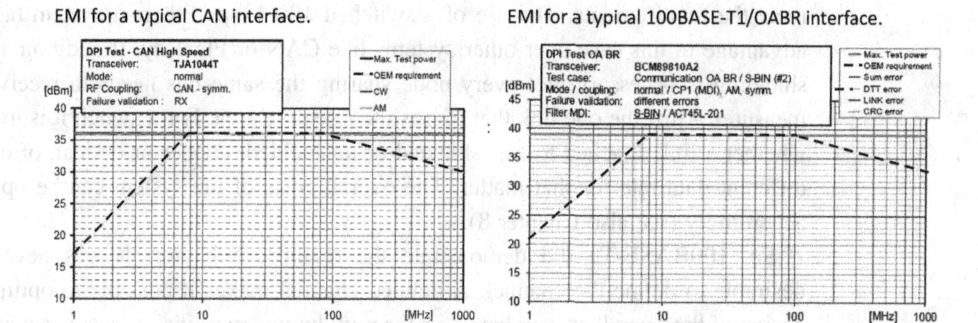

Figure 5.3 Example results for DPI tests according to IEC 61967-4; the level of noise stimulating the system has to be over the "OEM Requirement." If the communication in the system is working under such noisy condition, the test is passed. Note that all measured curves in both pictures are superposed; the limit line in the OABR picture is somewhat stricter because it represents a recently harmonized limit line. Conclusion: 100BASE-T1/OABR passes the immunity requirements of many car manufacturers, just like CAN does

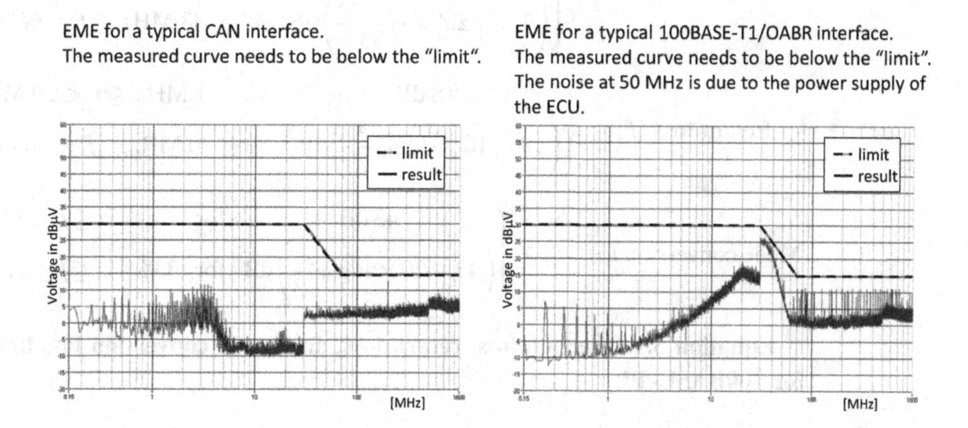

Figure 5.4 Example results for stripline measurements according to ISO 11452-5. Conclusion: 100BASE-T1/OABR Ethernet works under real automotive conditions and the limits of the EME can be met.[2] The results of the stripline emission test with the reference ECU fits very well to the results achieved with the test board using the 150 Ohm method (see Figure 5.2). As it is therefore seen as redundant, this book does not show additional immunity measurements for the reference ECU

The example results show that 100BASE-T1/OABR fulfills the EMC requirements of many[3] car manufacturers. Nevertheless, there are specific applications and harness routes inside a car, in which the shown limit lines are not sufficient. This can be the case if the Ethernet (or CAN or FlexRay) cable needs to pass a sensitive antenna system in close proximity. Even the use of shielded cables can be critical in these environments, and sometimes ferrites are added in addition to the shield in order to meet the requirements. The use of a switched 100 Mbps Ethernet system has some advantage in this case over other systems like CAN or FlexRay. If additional, costly shielding is necessary, not every node sharing the same bus needs to receive extra measures – just the one link that passes the sensitive area. For a switch it is irrelevant whether a link attached to it is shielded or unshielded, copper or optical, or carries a different data rate for that matter. The overall costs of the system can be optimized accordingly (see also Chapter 8).

So, 100BASE-T1 did demonstrate the required behavior. It was nevertheless desirable to define the channel. A defined channel makes it possible to optimize the design of the overall system based on the now-known margins. As a consequence, the 100BASE-T1 channel limit lines presented in Equations 5.1 and 5.2 and Figure 5.9 (see also [1]) were defined in hindsight.

$$\text{Insertion loss}(f) < \begin{cases} \left(1.0 + 1.6 \times \dfrac{f-1}{9}\right)\text{dB} & \text{for} \quad 1\text{ MHz} \le f < 10\text{ MHz} \\ \left(2.6 + 2.3 \times \dfrac{f-10}{23}\right)\text{dB} & \text{for} \quad 10\text{ MHz} \le f < 33\text{ MHz} \\ \left(4.9 + 2.3 \times \dfrac{f-33}{33}\right)\text{dB} & \text{for} \quad 33\text{ MHz} \le f \le 66\text{ MHz} \end{cases}$$

$$\text{Return loss}(f) \ge \begin{cases} 18\text{ dB} & \text{for} \quad 1\text{ MHz} \le f < 20\text{ MHz} \\ \left(18 - 10 \times \log_{10}\dfrac{f}{20}\right)\text{dB} & \text{for} \quad 20\text{ MHz} \le f \le 66\text{ MHz} \end{cases}$$

$$\text{Mode conversion loss}(f) \ge \begin{cases} 43\text{ dB} & \text{for} \quad 1\text{ MHz} \le f < 33\text{ MHz} \\ \left(43 - 20 \times \log_{10}\dfrac{f}{33}\right)\text{dB} & \text{for} \quad 33\text{ MHz} \le f \le 200\text{ MHz} \end{cases}$$

Equation 5.1 Insertion loss, return loss, and mode conversion loss limit lines for 100BASE-T1

$$\text{PSANEXT}(f) \ge \left(31.5 - 10 \times \log_{10}\frac{f}{100}\right)\text{dB} \quad \text{for} \quad 1\text{ MHz} \le f \le 100\text{ MHz}$$

$$\text{PSAACRF}(f) \ge \left(16.5 - 20 \times \log_{10}\frac{f}{100}\right)\text{dB} \quad \text{for} \quad 1\text{ MHz} \le f \le 100\text{ MHz}$$

Equation 5.2 Crosstalk limit lines for 100BASE-T1

Figure 5.5 shows example measurement results of channel parameters for a 100BASE-T1/OABR respective channel [2]. The impedance required for a 100BASE-T1 channel is 100 Ohm ± 10%.

Various aspects can impact the performance of an actual wiring harness: Ambient temperature, cable (length, material, and size), insulation (dielectric, color, and thickness), number of inline connectors, wire twist rate, consistency of lay length, untwist length at connectors, loops in the cable, whether the cable is jacketed or not [3], as well as pinning in the case of multi-pin connectors. Actually deploying 100BASE-T1 Ethernet, thus, requires careful design rules.

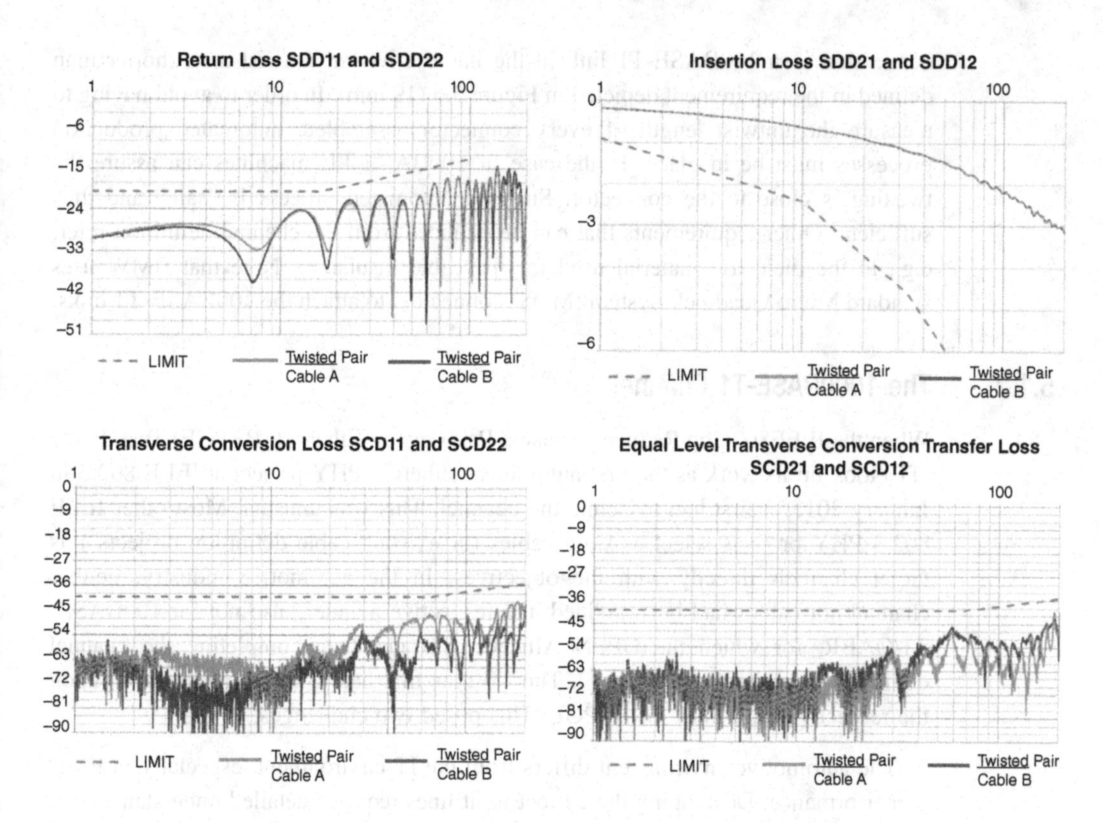

Figure 5.5 Example S-parameter measurements for a 100BASE-T1/OABR channel with two cables; for each measurement one cable was measured in both directions with a four-port network analyzer. Both cables measured are automotive qualified

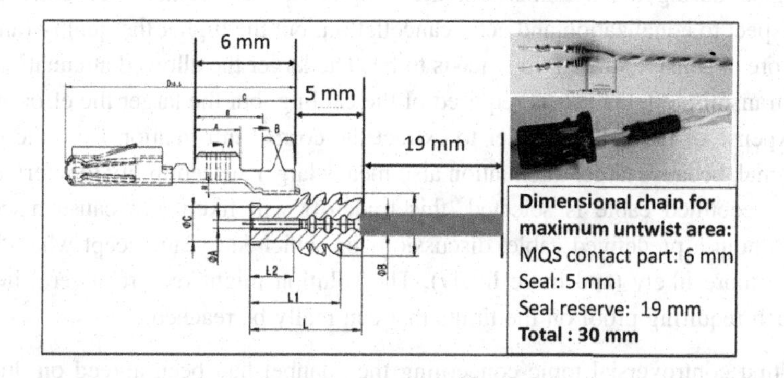

Figure 5.6 Maximum untwist area for 100BASE-T1 in the case of connector attachment [D. Gräber, Adopted from non-public document, 2013]

Figure 5.6 visualizes an example of one design criterion derived from evaluating actual cable measurements against the limit lines: The maximum untwist area 100BASE-T1 cables allow when being connected is 30 mm. This untwist length consists of two parts that are inside the connector, in which twisting is generally not possible, and the area just before the connector. Automotive quality requires that every connection in a 100BASE-T1 link in the harness has an untwist area shorter than defined in the requirement depicted in Figure 5.6 (19 mm). In order to avoid having to measure the untwist length of every connector assembled, automated production processes must be in place. In the case of 100BASE-T1, machines can assure the twisting is close to the connector. Such an automated process is "safe" and thus sufficient. Other requirements that can be deduced from the channel definition refer, e.g., to the dielectric material used for the cable insulation. Note that BMW uses standard Micro Quadlock System (MQS) connectors to attach the 100BASE-T1 links.

5.1.2 The 1000BASE-T1 Channel

When the IEEE802.3bp Reduced Twisted Pair Gigabit Ethernet (RTPGE) Task Force (TF) took up its work as the first automotive Ethernet PHY project at IEEE 802.3 in January 2013, it first had to define the channel. This was unusual. Most other IEEE 802.3 PHY projects select existing cables (or existing cable definition projects[4]) as target channels directly with the objectives. In the automotive industry, neither channel nor cables had been defined in a respective manner; also not for 100BASE-T1/OABR, for which the OPEN Alliance was just then completing the channel definition (see also Section 5.1.1). The TF thus first had to develop and agree upon the respective limit lines for RTPGE. This posed two challenges:

1. The automotive environment differs from the IT environment, especially in EMC performance. Developing the correct limit lines requires detailed understanding in the TF of the requirements and test set-ups relevant for automotive.
2. There is a trade-off between the capabilities the PHY needs to provide and the quality of the cables and connectors. For example, the smaller the attenuation of a signal during the transmission, the simpler the requirements of the PHY with respect to equalization and echo cancellation, but the higher the quality (and likely more expensive) the cabling needs to be. The larger the allowed attenuation during a transmission, the less is required of the cabling, but the larger the effort and thus expense of the PHY in order to extract the correct information from the received signal, because larger attenuation also means larger susceptibility to interference. If a predefined cable is selected, this trade-off will likely not cause discussions. Without a predefined cable, discussions on which side can accept what "burden" are more likely (and more lively). The solution might require several iterations, each requiring proof on the limits that can really be reached.

The first controversial topic concerning the channel had been agreed on during the preceding Study Group (SG) phase: The link length. For an Ethernet link in passenger cars, 3.5 m length is a good average value (see e.g. [4]). A 10 m link can easily connect a sensor in the right corner of the front bumper with an Electronic Control Unit (ECU) that sits in the left part of the trunk, even in a long passenger car. When including vans and light trucks and connecting cameras at exposed ends, 15 m is still sufficient, while long haul busses and large trucks can need up to 30 m [5]. For the industrial automation industry, 100 m is a standard link length requirement [6]. So, while it is attractive to address many applications, the volume, however, is in the shorter automotive links. A cost efficient successful technology for a volume market will more likely be adopted (potentially with adaptations) in a more challenging market than the other way around.

As a result, RTPGE was to be designed for a 15 m link segment with up to four inline connectors. Once this was achieved the same PHY was to be used to investigate how much longer the link segment can be – the objectives proposed an optional 40 m link segment – with better cabling [7].

The second controversial topic was the number of twisted pairs to use for the cables. RTPGE meant reducing the number of pairs to fewer than four – the number of pairs used for 1000BASE-T (see also Section 5.2.1.2). For cost efficiency reasons, Unshielded Twisted Single Pair (UTSP) was the preferred target cabling of the car manufacturers. If the environment becomes more challenging, this, in principle, allows car manufacturers to move first to the next more costly option of coax cables and then to shielded cables. Coax was not officially addressed in the IEEE project. However, the use of one pair of twisted cables in principle allows for this step, while the selection of two pairs would have prohibited it. People had been skeptical that the 100 Mbps PHY, 100BASE-T1, would be usable with UTSP. Now, it was desired to transmit a 10-fold data rate over similar cables.

The EMC requirements therefore became central to the development process. The project had to start with an EMC-based feasibility study with respect to the usable cabling types. This had not been done before, and therefore the methodology had yet to be proven and established. The TF decision in parallel was to work with single-pair cables. If the methodology failed (to prove a one pair solution would work), two pairs were to be considered as an alternative [8].

The basic process of the feasibility study is depicted in Figure 5.7. It consisted of three main steps:

1. Prove that cables and connectors (can) exist that meet the EMC emission criteria while allowing for enough power to be transmitted (to ensure EMC immunity).
2. Use those cables and connectors to derive the (EMC) noise the system has to cope with.
3. Define the PHY parameters such that with the power, noise, and resulting bandwidth they leave enough margin for the (always imperfect) implementation (see Section 5.3.1 for details).

The main task of step 1 was to find a suitable transmit power spectrum mask TX_{mask} that in return would allow setting the power of the transmit signal and its PSD. The procedure to obtain this mask (see [9]) is based on the direct correlation between the emissions and the transmit power spectrum TX_{power}, as well as the emissions transfer function (see Equation 5.3). This correlation was not known upfront, but had to be proven first [9].

$$\begin{aligned}
&Emissions\ [\mathrm{dB\mu V}] = TX_{power}\ [\mathrm{dB\mu V}] + EmissionsTransferFunction\ [\mathrm{dB}]\\
&\Rightarrow TX_{power}\ [\mathrm{dB\mu V}] = Emissions\ [\mathrm{dB\mu V}] - EmissionsTransferFunction\ [\mathrm{dB}]\\
&\Rightarrow TX_{mask}\ [\mathrm{dB\mu V}] = Emissions_{\lim it}\ [\mathrm{dB\mu V}] - EmissionsTransferFunction_{mask}\ [\mathrm{dB}]
\end{aligned}$$

Equation 5.3 Deriving the transmit power limit

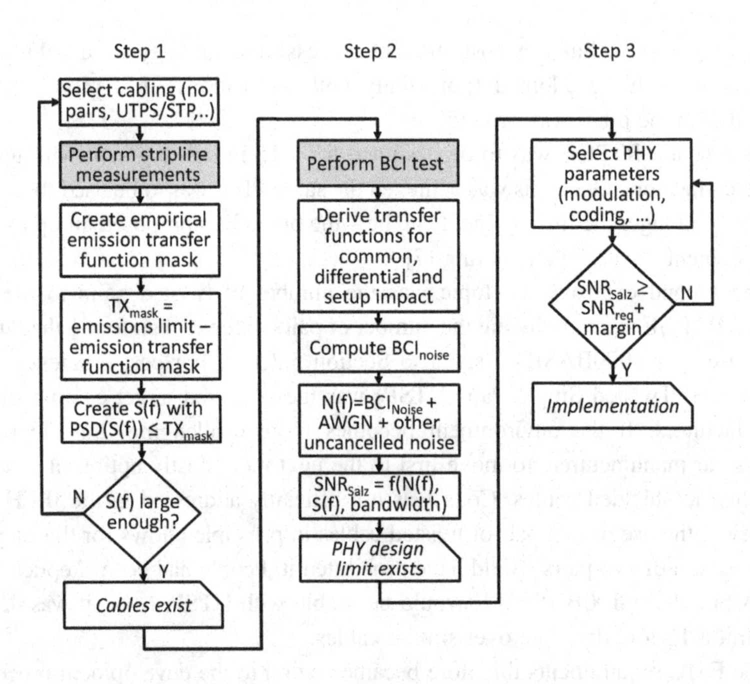

Figure 5.7 PHY specification development process based on EMC measurements and limit lines [9, 14, 15]

So, first stripline measurements were performed with various cables and, for those with promising emissions behavior, the transfer function was derived. Tazebay [9] showed example stripline measurements of various such cables. From these results an emission transfer function mask was derived empirically such that the measured emissions stayed below the mask. The difference between the BMW stripline limit of 15 dBμV and the emission transfer function mask was used to obtain the transmission power spectrum mask (TX_{mask}) of the differential signal. Tazebay [9] gives an example of the PSD of a signal *S(f)* with 1 V peak to peak (Vpp) that is well below the TX_{mask} limit. This proved that UTSP cabling was feasible for 1000BASE-T1. Note that the upper transmit power mask defined in the 1000BASE-T1 specification [10] actually allows for somewhat more power, up to approximately 1.4 Vpp.

The second step addresses the noise the channel is susceptible to, which in the automotive automotive needs to include the noise caused by EMC. In differential systems like Automotive Ethernet, the noise is the result of non-ideality and imbalances in the channel. Because of these imperfections, the interference has a different impact on each of the two wires in one pair. Therefore, an interference can no longer be completely cancelled out and results in common mode as well as differential mode noise. Tazebay [11] proposes to use the Bulk Current Injection (BCI) immunity test set-up in order to quantify and measure the common mode and differential mode transfer functions H_{CM} and H_{DM} of the cables found suitable with the emission tests. Key to computing the common mode and differential mode noise V_{CM} and V_{DM} for a given BCI test profile I_{BCI} (see Equation 5.4), is the correct mitigation of the effects the test set-up (test heads, clamp, termination [12], etc.) has on the results (reflected in H_{CIP}).

$$V_{CM}(f)\,[mV] = I_{BCI}(f)\ [mA]\ Z_{CM}(f)\ [\Omega]\quad \text{with}\quad Z_{CM}(f)\,[\Omega] = \frac{50\Omega}{\sqrt{2}} \left|\frac{H_{CM}(f)}{H_{CIP}(f)}\right|$$

$$V_{DM}(f)\,[mV] = I_{BCI}(f)\ [mA]\ Z_{DM}(f)\ [\Omega]\quad \text{with}\quad Z_{DM}(f)\,[\Omega] = 50\Omega\ \sqrt{2} \left|\frac{H_{DM}(f)}{H_{CIP}(f)}\right|$$

Equation 5.4 Computation of the common mode and differential mode noise V_{CM} and V_{DM} [11]

The so-obtained EMC noise for the existing (prototype) cables can then be added to the Additive White Gaussian Noise (AWGN) and other uncancelled noise sources to obtain the overall noise *N(f)*. With these inputs it is possible to calculate the Salz SNR as a function of the bandwidth *W* (see e.g. [13] or Equation 5.5). The Salz SNR represents the SNR value that is theoretically achievable with an infinite length equalizer and can be set as a maximum bound. The actual PHY parameters can be based on this value with the consideration of respective implementation margins [14] and the effects it has on the bandwidth *W*. For details of the actual PHY parameters selected, see Section 5.3.1.

$$SNR_{Salz}(W) = 10\log_{10} e^{\frac{1}{W}\int_0^W \log_e\left(1+\frac{S(f)}{N(f)}\right)df} \quad [\text{dB}]$$

Equation 5.5 True Salz SNR optimum

Next to the impedance, IL, RL, PSANEXT, PSAACRF, and the common mode to differential mode conversion loss (see [16] and [17] for details), the TF investigated wire gage and temperature impact (see e.g. [16, 18]) as well. Finally, cable and connector companies provided measurements as proof that these requirements can be met. In the end, 1000BASE-T1 was designed for UTSP (with jacketed cables, the reasons for this are explained later in this section) and the naming 1000BASE-T1 for the IEEE 802.3bp/RTPGE technology reflects the outcome of the single pair decision.[5]

Recall that the industry had serious concerns about transmitting 100 Mbps data over unshielded cabling. With 1000BASE-T1, 10-times the amount of data was being proposed to go across unshielded wiring. Convincing the industry to deploy 1000BASE-T1 with unshielded (even if jacketed) cables turned out to be (too) ambitious. For the purpose of facilitating timely introduction – especially in critical locations near antennas, etc. – the OPEN Alliance thus defined a shielded channel [19] as well. For this definition, the STP insertion loss, return loss, and crosstalk[6] values remained unchanged from UTP (see Equation 5.6), while the mode conversion limit line was relaxed for STP channels and values for coupling and screening attenuation were added (see Equation 5.7, also Figure 5.8, [10, 19]). Körber et al. [19] thereby define two environment classes. Class 1 reflects a partially shielded system, while Class 2 reflects a fully shielded system. The idea was to allow a system with partial shielding, provided this can be compensated for by better symmetry.

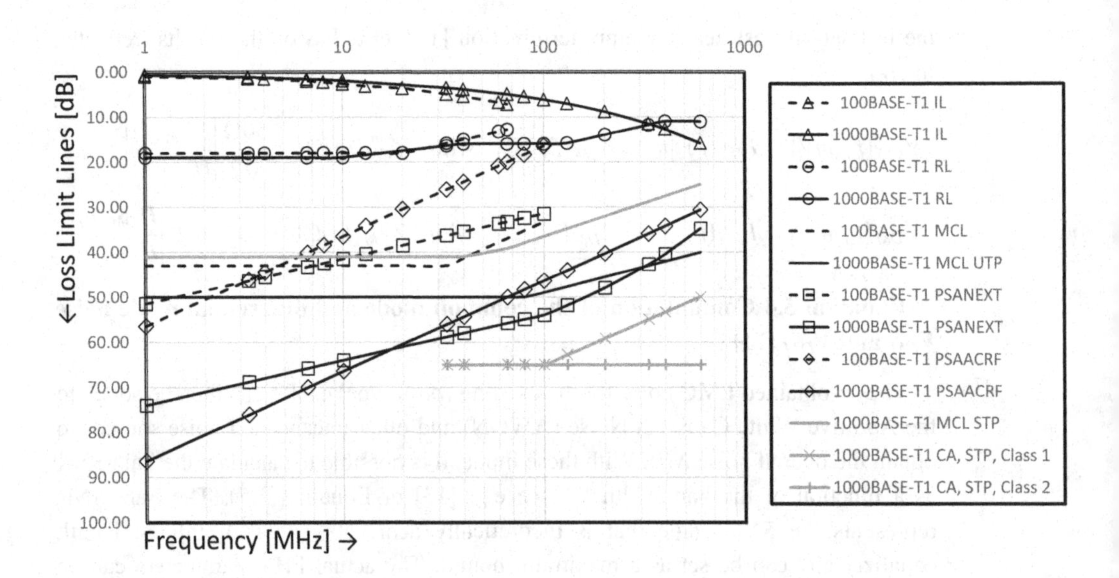

Figure 5.8 The 1000BASE-T1 channel in comparison

$$\text{Insertion loss}_{\text{UTP, STP}}(f) \leq \left(0.0023 \times f + 0.5907 \times \sqrt{f} + \frac{0.0639}{\sqrt{f}}\right) \text{dB} \quad \text{for} \quad 1\,\text{MHz} \leq f \leq 600\,\text{MHz}$$

$$\text{Return loss}_{\text{UTP, STP}}(f) \geq \begin{cases} 19\,\text{dB} & \text{for} \quad 1\,\text{MHz} \leq f < 10\,\text{MHz} \\ (24 - 5 \times \log_{10} f)\,\text{dB} & \text{for} \quad 10\,\text{MHz} \leq f < 40\,\text{MHz} \\ 16\,\text{dB} & \text{for} \quad 40\,\text{MHz} \leq f < 130\,\text{MHz} \\ (37 - 10 \times \log_{10} f)\,\text{dB} & \text{for} \quad 130\,\text{MHz} \leq f < 400\,\text{MHz} \\ 11\,\text{dB} & \text{for} \quad 400\,\text{MHz} \leq f \leq 600\,\text{MHz} \end{cases}$$

$$\text{PSANEXT}_{\text{UTP, STP}}(f) \geq \begin{cases} \left(54 - 10 \times \log_{10} \frac{f}{100}\right) \text{dB} & \text{for} \quad 1\,\text{MHz} \leq f < 100\,\text{MHz} \\ \left(54 - 15 \times \log_{10} \frac{f}{100} - 6\left(\frac{f - 100}{400}\right)\right) \text{dB} & \text{for} \quad 100\,\text{MHz} \leq f \leq 600\,\text{MHz} \end{cases}$$

$$\text{PSAACRF}_{\text{UTP, STP}}(f) \geq \left(46.33 - 20 \times \log_{10} \frac{f}{100}\right) \text{dB} \quad \text{for} \quad 1\,\text{MHz} \leq f \leq 100\,\text{MHz}$$

Equation 5.6 Insertion loss, return loss, and crosstalk limit lines for 1000BASE-T1 UTP and STP

$$\text{Mode conversion loss}_{\text{UTP}}(f) \geq \begin{cases} 50\,\text{dB} & \text{for} \quad 10\,\text{MHz} \leq f < 80\,\text{MHz} \\ (72 - 11.51 \times \log_{10} f)\,\text{dB} & \text{for} \quad 80\,\text{MHz} \leq f \leq 600\,\text{MHz} \end{cases}$$

$$\text{Mode conversion loss}_{\text{STP}}(f) \geq \begin{cases} 41\,\text{dB} & \text{for} \quad 10\,\text{MHz} \leq f < 50\,\text{MHz} \\ (66.2 - 14.83 \times \log_{10} f) & \text{for} \quad 50\,\text{MHz} \leq f \leq 600\,\text{MHz} \end{cases}$$

$$\text{Coupling attenuation}_{STP}(f) \geq \begin{cases} 65 \text{ dB} & \text{for } 30 \text{ MHz} \leq f \leq \begin{matrix} 100 \text{ MHz for Class 1} \\ 600 \text{ MHz for Class 2} \end{matrix} \\ \left(65 - 19.3 \times \log_{10}\frac{f}{100}\right) \text{dB} & \text{for } 100 \text{ MHz} < f \leq 600 \text{ MHz for Class 1} \end{cases}$$

$$\text{Screening attenuation}_{STP}(f) \geq \begin{cases} 25 \text{ dB for } 30 \text{ MHz} \leq f \leq 600 \text{ MHz for Class 1} \\ 40 \text{ dB for } 30 \text{ MHz} \leq f \leq 600 \text{ MHz for Class 2} \end{cases}$$

Equation 5.7 Mode conversion loss, coupling, and screening attenuation values for 1000BASE-T1 UTP and STP

Figure 5.8 shows the resulting channel limit lines for 1000BASE-T1 (UTP and STP) in comparison with the channel limit lines for 100BASE-T1.

The most obvious observation is that the 1000BASE-T1 channel has to deal with 10-fold the bandwidth. However, in respect to IL and RL this results in more stringent performance requirements at higher frequencies "only," while the requirements in the lower frequencies are very similar. This is different for the mode conversion and crosstalk values. The 10-fold bandwidth increase means that the signal is significantly more susceptible to interference. In order to maintain the same Signal-to-Noise-Ratio (SNR) without increasing the signal, the noise must be smaller and the channel thus needs to suppress the interference better. For the mode conversion this means about 7 dB difference. As can be seen, when using a shield with 1000BASE-T1 this is more than compensated for. For the ANEXT (PSANEXT) 1000BASE-T1 requires about 22.5 dB and for the AFEXT (PSACRF) about 30 dB better performance, for both also at the lower frequencies. In order to achieve these values, while still deploying UTSP cables, it has been proposed to use jacketed cables. The jacket ensures a certain physical distance to the next neighboring cables, while at the same time giving better stability to the twist and with that improving the balance, even in areas where the cable is bent (regularly). This additionally means that the connectors provide better performance, with shorter untwist areas and better physical separation to the neighboring pins than the (n)MQS-connector provides. For details on the harness components and their requirements see [20].

5.1.3 The 10BASE-T1(S) Channel

First of all, the IEEE 802.3cg specification defines two different PHYs. This book solely discusses the short reach PHY 10BASE-T1S ("S" for "short"). The long reach PHY 10BASE-T1L – "L" indicating "long" – is designed for reaches of up to 1000 m, as needed in industrial applications (see also Section 1.2.3). It is thus over-dimensioned for automotive use. Next, the 10BASE-T1S PHY has three different operating modes. The mandatary default mode is half-duplex P2P operation. Additionally, either an optional half-duplex "bus" mode is supported – called "multidrop" in the IEEE 802.3cg specification – or an optional P2P full-duplex mode can be implemented. The focus of the explanations of 10BASE-T1S is on the half-duplex multidrop implementation. However, the channel definition starts with the P2P channel, which is valid for both half- and full-duplex operation.

To be cost-efficient was one of the main design criteria for the short reach 10 Mbps Ethernet PHY (see also Section 5.4). From a BMW perspective, however, there was not much further cost saving potential in the wiring and connectors compared with 100BASE-T1 (whose wiring is also used for CAN FD). In the interest of cost efficiency, 10BASE-T1S used the 15 m with four inline connector 100BASE-T1 channels as a starting point. Naturally, the frequency range is somewhat shifted. For 10BASE-T1S it starts with 0.3 or 1 MHz and ends at 40 MHz, while for 100BASE-T1 it ranges from 1 MHz to 66 MHz. Otherwise the insertion loss and crosstalk limit lines are absolutely identical (see also Figure 5.9 and Equations 5.1 and 5.2).

Nevertheless, it was possible to markedly lower the return loss and to somewhat relax the mode conversion limits. This was possible because of the lower operating frequency – 10BASE-T1S transmits 25 instead of 66.6 MBaud, i.e., the Nyquist frequency is 12.5 instead of 33.3 MHz – and the larger signal margin owing to using a two level Differential Manchester Encoding (DME) instead of the three level PAM modulation of 100BASE-T1 (see also Sections 5.2.1 and 5.4). While it had not been a primary goal of BMW, this does allow the use of inferior cabling and connectors than those used for 100BASE-T1. Equation 5.8 lists the specific 10BASE-T1S P2P limit lines (see also [21]).

$$\text{Return loss } (f) \geq \begin{cases} 14 \text{ dB} & \text{for} \quad 0.3 \text{ MHz} \leq f < 10 \text{ MHz} \\ \left(14 - 10 \times \log_{10} \dfrac{f}{10}\right) \text{dB} & \text{for} \quad 10 \text{ MHz} \leq f \leq 40 \text{ MHz} \end{cases}$$

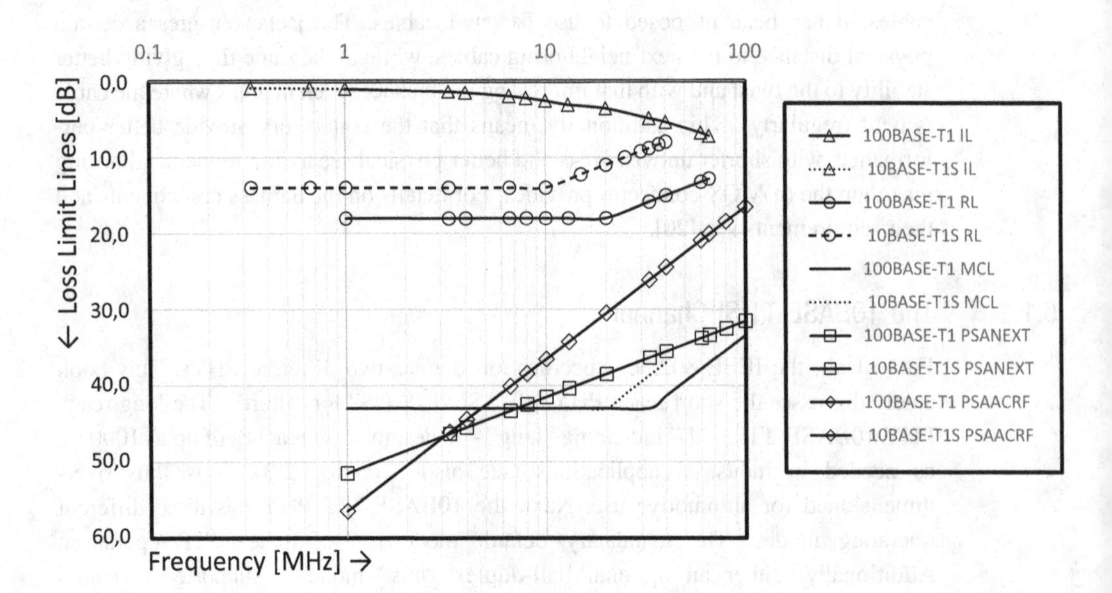

Figure 5.9 The 10BASE-T1 channel in comparison to the 100BASE-T1 channel

$$\text{Mode conversion loss}\,(f) \geq \begin{cases} 43\text{ dB} & \text{for } 1\text{ MHz} \leq f < 20\text{ MHz} \\ \left(43 - 20 \times \log_{10}\dfrac{f}{20}\right) & \text{for } 20\text{ MHz} \leq f \leq 200\text{ MHz} \end{cases}$$

Equation 5.8 10BASE-T1S limit lines different from 100BASE-T1S limit lines

The main cost savings for 10BASE-T1S are realized by a simplified PHY design (discussed in Section 5.4) and the possibility to use a multidrop topology. Figure 5.10 visualizes the main principle. A traditional switched Ethernet system requires two PHYs for every link (see scenario A in Figure 5.10), while a multidrop/bus topology (scenario B) requires one PHY per end node ECU. Just this brings a saving in the number of PHYs of $(N\text{-}2)/(2N\text{-}2)$, with N representing the number of nodes on the multidrop segment. As can be seen in the depiction of Figure 5.10, multidrop additionally saves having to use switches in the end nodes when compared with the switched daisy chain scenario.

While the cost saving concept is straightforward, bringing multidrop into the 10BASE-T1S specification was not. From the channel perspective, various additional parameters needed to be defined, key of which concerned the topology that could be supported. From a design, production, and maintenance perspective, it is easiest for a car manufacturer if a passive star topology can be supported as a bus. However, a scenario with many long stubs is also the most challenging scenario, as the propagation delays and echoes are strongest and the termination of the link is not obvious (see e.g. [22]). FlexRay introduced an active star coupler to allow for such a scenario at 10 Mbps (see Section 2.2.5.2), while for CAN FD the additional concept of ringing suppression was needed to provide for more flexibility even at a 2 Mbps data rate (see Section 2.2.2.2).

In the end, the 10BASE-T1S multidrop was deemed feasible for a passive linear topology, as shown in Figure 5.10 (see e.g. [23]).[7] It was targeted with the IEEE

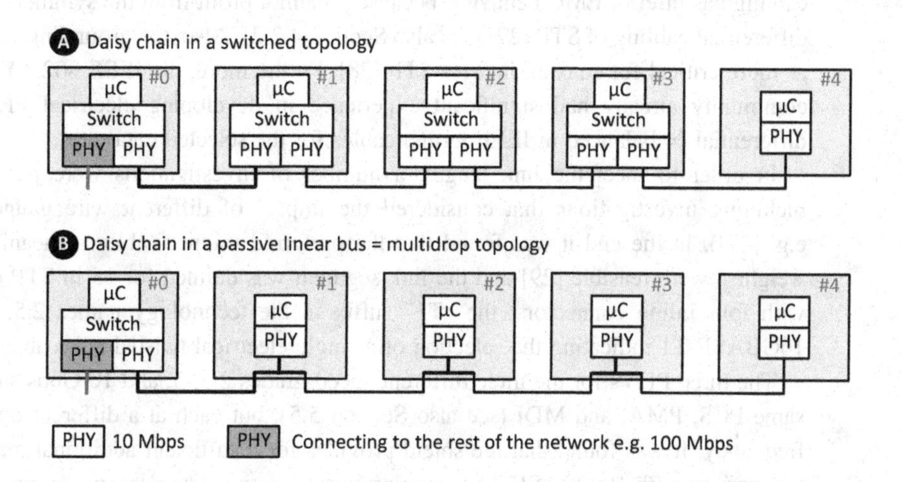

Figure 5.10 Daisy chain as a switched topology (A) and as a passive linear topology (B)

802.3cg specification that the IEEE 802.3cg PHY supports a passive linear topology of 25 m length, with 10 cm long stubs (likely to be realized on the PCB as indicated in Figure 5.10), and minimum eight nodes.[8] The multidrop mode is specified to function over links that meet the requirements as defined for the P2P link [21] with the difference that the reference impedance is 50 Ohms. As was also shown by Wechsler et al. [22], it is absolutely important that a multidrop link segment is terminated properly, with 100 Ohm termination resistors at exactly two nodes that are preferably at the physical ends of the link.

At the time of writing the technology had not yet been introduced into vehicles. However, first tests performed by various semiconductor vendors and interested users indicated that, from the channel and signaling side, significantly more nodes than eight can be supported [OPEN Alliance Members, email conversation on the OPEN Alliance Reflector, January 8, 2020].

5.1.4 The MultiGBASE-T1 Channel(s) for 2.5, 5, and 10 Gbps

With increasing data rates, data transmission is more challenging and the characteristics of the channel ever more critical. When the respective "NGauto" study group took up its work at IEEE 802.3, the target was therefore vague: To enable a multi-Gigabit data rate for automotive use [24]. A significant amount of study group effort was then applied to the feasibility and the possible link segment (see [25] for details), with the conclusion to limit the objectives for the IEEE 802.3ch task force to the development of full-duplex PHYs with 2.5, 5, and 10 Gbps data rates. Other than that the channel was to be electrical and of 15 m length, the channel was not specified in any more detail. The considered possibilities included single as well as multi-pair in a number of different cable varieties [26].

Concerning the cable type, the TF converged and eliminated UTP quickly and coaxial cabling relatively fast. While having better insertion loss than STP, coaxial cabling has inferior EMC behavior, because it cannot profit from the symmetry of the differential cabling of STP [27] (see also Section 4.2.2). Also, the grounding was seen as more critical for coaxial than for STP [28]. Furthermore, the IEEE 802.3 Ethernet community already had significant experience in developing electrical PHYs on differential (= balanced in IEEE terms) cables for the selected data rates.

In order to meet the link length, a number of investigations were performed, including investigations that considered the impact of different wire gauges (see e.g. [27]). In the end it was found that the preferred thinner cables – meaning less weight – were feasible [29] and the link segment was defined for 15 m STP cabling with four inline connectors; the "T1" suffix in the technology names 2.5, 5, and 10GBASE-T1 indicating the selection of a single electrical twisted pair cable [30].

The three PHYs for the three different speed grades 2.5, 5, and 10 Gbps share the same PCS, PMA, and MDI (see also Section 5.5), but each at a different operating frequency. It was found that the shield provides for a sufficient additional margin in comparison with 1000BASE-T1 to support the respective data rates, while limiting the EME at the same time.

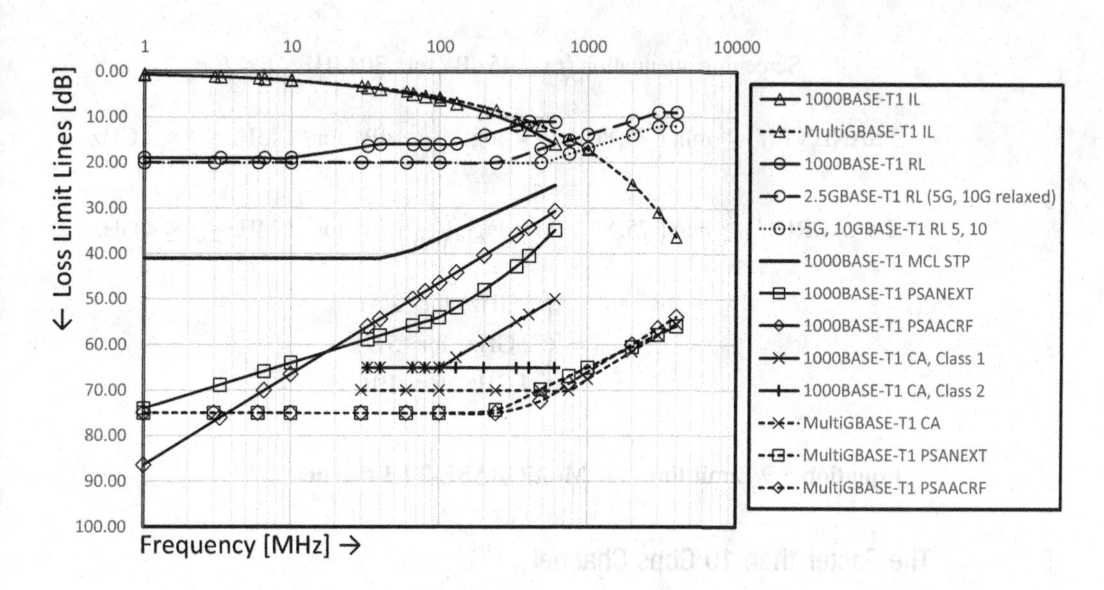

Figure 5.11 The (multi-)Gbps STP channels in comparison

Equation 5.9 shows the limit lines for the MultiGBASE-T1 channel. With the exception of the return loss, all limit lines are the same, independent of the speed grade, with some slight differences in the maximum frequency F_{max} they are defined for. The return loss limit lines are more stringent for 5 and 10GBASE-T1 than for 2.5GBASE-T1. However, should a channel with better insertion loss be used, it is possible to relax the required return loss to the level of the 2.5GBASE-T1 channel. This, in particular, was added when there was some uncertainty during the development whether the stringent return loss values could be met (which we now know that it can). Figure 5.11 compares the MultiGBASE-T1 limit lines with those of 1000BASE-T1 for STP. As can be seen, except for the lower frequencies, all limit lines present more stringent requirements for MultiGBASE-T1 than for 1000BASE-T1.

$$\text{Insertion loss } (f) \leq \left(0.002 \times f + 0.68 \times f^{0.45}\right) \quad \text{for} \quad 1\text{ MHz} \leq f \leq F_{max}$$

$$\text{Return loss } (f) \geq \begin{cases} 20\,\text{dB} \;\; \text{for} \;\; 1\,\text{MHz} \leq f < \dfrac{480}{2^N} \\ \left(20 - 10 \times \log_{10}\dfrac{2^N \times f}{480}\right)\text{dB for } \dfrac{480}{2^N}\text{MHz} \leq f \leq \min\left(F_{max}, 3\,\text{GHz}\right) \\ (12 - 3 \times N) \;\; \text{for} \;\; 3\,\text{GHz} \leq f \leq 4\,\text{GHz}, \text{if } F_{max} > 3\,\text{GHz} \end{cases}$$

$$\text{with } N = \begin{cases} 0 \text{ for } 15\text{ dB} < IL_{5G}@1.5\text{ GHz or } IL_{10G}@3\text{ GHz} \\ 1 \text{ for } 2.5\,G, \text{ if } IL_{5G}@1.5\text{ GHz or } IL_{10G}@3\text{ GHz} \leq 15\text{ dB} \end{cases}$$

$$\text{Coupling attenuation } (f) \geq \begin{cases} 70\,\text{dB} \;\; \text{for} \;\; 30\,\text{MHz} \leq f < 750\,\text{GHz} \\ \left(50 - 20 \times \log_{10}\dfrac{f}{7500}\right)\text{dB} \;\; \text{for} \;\; 750\,\text{MHz} \leq f \leq 4\,\text{GHz} \end{cases}$$

$$\text{Screening attenuation }(f) \geq 45\,\text{dB} \quad \text{for} \quad 30\,\text{MHz} \leq f \leq F_{max}$$

$$\text{PSANEXT }(f) \geq \min\left(75, 80 - 15 \times \log_{10}\frac{f}{100}\right)\text{dB} \quad \text{for} \quad 1\,\text{MHz} \leq f \leq 4\,\text{GHz}$$

$$\text{PSAACRF }(f) \geq min\left(75, 86 - 20 \times \log_{10}\frac{f}{100}\right)\text{dB} \quad \text{for} \quad 1\,\text{MHz} \leq f \leq 4\,\text{GHz}$$

$$F_{max} = \begin{cases} 1\,\text{GHz} & \text{for } 2.5G \\ 2\,\text{GHz} & \text{for } 5G \\ 4\,\text{GHz} & \text{for } 10G \end{cases}$$

Equation 5.9 Limit lines for MultiGBASE-T1 Ethernet

5.1.5 The Faster than 10 Gbps Channel

At the time of writing, the channel for the next higher data rate Automotive Ethernet project had not yet been determined. However, a decision on the channel yet again provides the basis for all subsequent decisions and must be made first. As all previous Automotive Ethernet projects used single pair channels, next to length, type and number of inline connectors, a relevant question is to what capacity a single pair channel can be used and improved, or whether it is better to use multiple pairs. First investigation results presented in Bergner and DiBiaso [31] show that, next to improvement potential in the harness components and manufacturing, simplex transmission on a single pair significantly increases the channel capacity. With the additional expectation that in any case higher data rates will mainly be used asymmetrically [32], a channel that supports asymmetric transmission on a single pair and symmetric transmission on two pairs would be a sensible solution.

5.2 PHY Technologies for 100 Mbps Ethernet

5.2.1 100BASE-T1

5.2.1.1 Background to 100BASE-T1

The basis for 100BASE-T1 can be found in the IEEE 802.3 1000BASE-T standard.[9] Broadcom engineers reused some of the basic principles of 1000BASE-T to propose a solution for Ethernet in the First Mile (EFM) standard development: Instead of four pairs of wiring, one pair was used and the channel coding was made more robust, so that it was possible to transmit 100 Mbps data over a worse, i.e., longer channel. IEEE selected a different solution for EFM, and Broadcom proposed their technology for EFM in China [33]. When BMW was looking for an Ethernet solution suitable for the automotive industry in 2007, another interesting use case was found for the Broadcom technology. This technology was named BroadR-Reach at the time, then published by the OPEN Alliance and called OPEN Alliance BroadR-Reach (OABR) in 2011, before finally being ratified as the IEEE 802.3bw standard in 2015, naming the technology 100BASE-T1. To create an understanding for 100BASE-T1, Section 5.2.1.2 addresses first some fundamentals of 1000BASE-T, before the 100BASE-T1 technology is explained in Section 5.2.1.3.

After the proof had been provided that it is possible to transmit Ethernet packets at 100 Mbps over UTP cabling in the automotive environment with BroadR-Reach, other companies started to provide different, i.e., incompatible solutions to the same end. For example, another semiconductor vendor presented a solution to BMW as early as 2009 that also used a single pair UTP (Unshielded Twisted Single Pair, UTSP) cable. Even though the vendor decided not to pursue its technology, it was an important milestone for BMW. In 2009, the market was not yet ready for Automotive Ethernet. Among most of the decision makers in the automotive industry, skepticism prevailed on the technical feasibility and on the need for such a technology. Therefore, there was no obvious market prospect for the decision makers in the semiconductor industry.

The authors therefore doubted at the time and still doubt that starting a public standardization of BroadR-Reach – which requires faith in the technical feasibility and a market prospect – in 2009 or earlier, would have been successful. Additionally, a solution was needed fast. The targeted SOP was 2013. If that had been missed, the complete market development of Automotive Ethernet might have been missed as well. The other solutions that were being proposed in the meantime were proof that multiple semiconductor vendors would be able to handle automotive UTSP Ethernet (see Section 3.1.3.4). This, in the end, motivated BMW to go ahead with enabling and qualifying BroadR-Reach (prior to pursuing a public standard) as well as establishing a multisourcing strategy for the technology (see also Sections 3.3–3.5).

Section 5.2.2 addresses one alternative development, which uses UTP cabling with 100BASE-TX for automotive use. It can provide some advantages, e.g., in the context of Diagnosis-over-IP (DoIP), which has been standardized to use 100BASE-TX in ISO 13400. Section 5.2.3 shows the flexibility that can be achieved with an Ethernet network, because of its strict layering approach. It explains how the MII interface properties can be exploited to transmit 100 Mbps Ethernet packets quite differently.

5.2.1.2 The Reference for 100BASE-T1/OABR: IEEE 802.3 1000BASE-T

One of the goals during the development of 1000BASE-T was to meet the same Federal Communications Commission (FCC) class A ElectroMagnetic Compatibility (EMC) requirements as has been done for 100BASE-TX [34]. As a result, 1000BASE-T uses more or less the same spectrum as 100BASE-TX (see also Figure 5.17 and Table 5.7). With the technologies available at the time, this was not possible to achieve with one or two pairs of wires only. Instead a cable with four pairs, CAT 5e, was selected. In order to keep the bandwidth required per cable pair low, it was necessary to implement a "true" full-duplex mode, i.e. to transmit and receive on the same wire. This, in return, led to the use of echo cancellation and hybrids. 100BASE-TX, in contrast, uses one of its two cable pairs for transmission and one for reception.[10] For 1000BASE-T, auto-negotiation was made mandatory. However, there are three reasons why auto-negotiation did not make it into 100BASE-T1. (A) It was not needed for the original use case (EFM in China). (B) When reused for the automotive industry, 100 Mbps Ethernet was the first speed grade used. Data rate and duplex mode were unambiguous for 100BASE-T1. (C) With the preset network inside a car, it was not seen as an important feature at the time, although now, with the number of speed grades available, there are some use cases that benefit from auto-negotiation. This is quite different from the plug & play in the IT and CE industries.

In the following, 1000BASE-T elements important for the understanding of 100BASE-T1 are being described:

- **Hybrid:** A hybrid is used in case data is transmitted and received simultaneously over the same wire pair ("true" full-duplex operation). Its function is to cancel the transmitted signal from the signal at its pins at t=0 in order to obtain the received signal such that the dynamic range of the receiver can be relaxed. In the original Carrier Sense Multiple Access with Collision Detection (CSMA/CD) Ethernet, the transceiver only either transmitted or received data, as the media was shared among all participants. A hybrid would have thus been useless. With 1000BASE-T transmitting and receiving on the same wire pair, the intention is to use 1000BASE-T in a switched Ethernet network with P2P links only, even if the original 1000BASE-T standard still supported CSMA/CD operation [35].

  Figure 5.12 visualizes a hybrid circuit and its use in 1000BASE-T and 100BASE-T1. The differential transmit signal A is led via two paths to ground; one of which is comprised of the two resistors R1 and R3 and on the other which is comprised of the two resistors R2 and R4. The output of the receiver sees only the transmit signal A that is passed via resistor R2. On the receive side, the differential receive signal B is connected to the middle of this bridge. This means that the

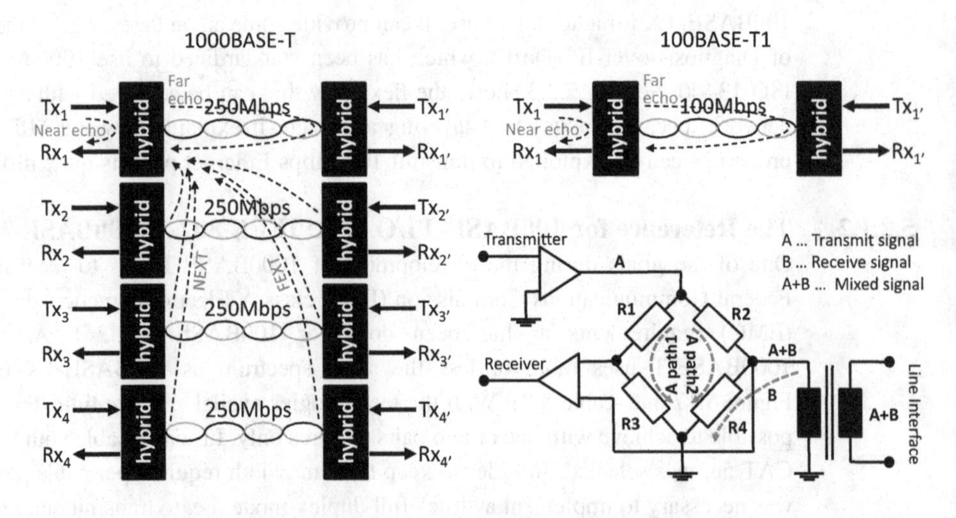

Figure 5.12 IEEE 802.3 1000BASE-T and 100BASE-T1 hybrid use and inner system interference shown for the receive signal Rx_1 as an example [34–36]

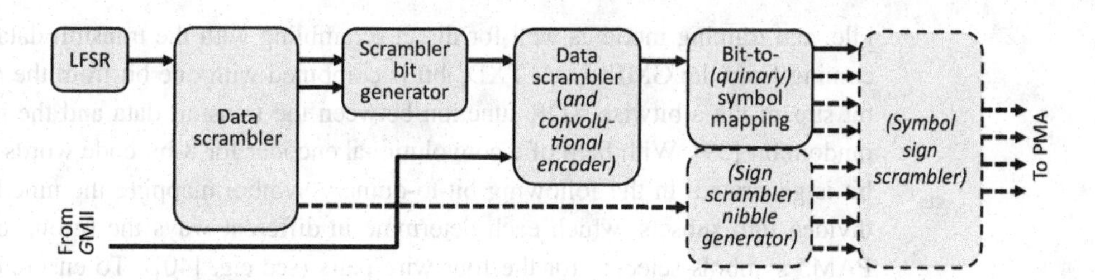

Figure 5.13 1000BASE-T PCS elements and their relation (dashed lines and italics indicate the differences to 100BASE-T1) [37, 38]

receive signal is terminated via R4, but also that the receive signal, at least theoretically, does not see the transmit signal. The receive path is completely decoupled from the transmit path. In a real system, however, the resistors are never perfectly matched and a perfect balance cannot be attained for the hybrid circuit. Therefore, non-ideal hybrid cancellation will cause a small leakage signal to occur at the receiver input. The line interface of the system sees the mixed signal A+B.

- **PCS:** The 1000BASE-T1 PCS processes the data received from the Gigabit MII (GMII) such that it passes "optimal" Four-Dimensional Five-Level Pulse-Amplitude Modulation (4D-PAM5) coded symbols on to the PMA. At the same time it receives coded symbols from which it extracts the information to send via the GMII to higher layers. The different elements of the 1000BASE-T PCS are shown in Figure 5.13.

  A main feature of the 1000BASE-T PCS is the use of a pseudo random bit "side stream." The goal of this side stream is that the transmitted symbols are as different from each other as possible and that the signal energy is evenly spread over the available frequency band. This improves the EMC behavior. EME on the wire are reduced and the transmission is less susceptible to EMI. As different pseudo random bit streams are used for the transmit and receive side – one will be master, the other slave – this also ensures that the receive and transmit signals are different enough to not continuously cancel each other out. Furthermore, it prevents the same symbol being transmitted too many times in a row, which might otherwise create a common signal (DC) on the channel.

  To generate the side stream, first a Linear Feedback Shift Register (LFSR) is used with the master and slave polynomials shown in Equation 5.10.

$$
\begin{aligned}
g_M(x) &= 1 \oplus x^{13} \oplus x^{33} \\
g_S(x) &= 1 \oplus x^{20} \oplus x^{33}
\end{aligned}
$$

Equation 5.10 100BASE-T1 and 1000BASE-T master and slave polynomials ($\oplus$ denotes the XOR function)

  A following data scrambler generates three bit streams, of which two are processed further with the scrambler bit generator into one bit stream that includes an odd and even timing distinction. This resulting bit stream is used for control, idle, and training mode as well for direct scrambling with the transmit data TXD_n coming from the GMII. Each TXD_n bit is combined with one bit from the random bit stream via a bitwise XOR function between the transmit data and the pseudo-random bit [39]. With help of a convolutional encoder for 8-bit code words a ninth bit is generated. In the following bit-to-quinary symbol mapping the nine bits are divided into subsets, which each determine in different ways the (group of) 4D-PAM5 symbols selected for the four wire pairs (see e.g. [40]). To ensure enough randomization each set of 4D-PAM5 symbols is scrambled with data derived from the first data scrambler following the LFSR.

The symbol mapping used is quite sophisticated. The 4D-PAM5 uses five voltage levels (–1 V, –0.5 V, 0 V, 0.5 V, 1 V) on four wire pairs. This theoretically allows for 5^4 = 625 different symbols, while the 8-bit code words require 2^8 = 256 different symbols only. 1000BASE-T1 exploits this for transmitting additional control information but also for making the transmission more robust by selecting the symbols only from a specifically reduced set of symbols [37]. The transmission signal experiences attenuation, InterSymbol Interference (ISI), ANEXT, AFEXT, and other interference while being on the channel. With a voltage difference of only 0.5 V between symbols, the signal is more susceptible to noise, and thus being decoded incorrectly, than if the difference was larger. In order to increase the voltage difference between subsequent symbols (see Figure 5.14 for a depiction of the basic principle) while at the same time profiting from five voltage levels, the symbol mapping distinguishes between four sets of symbols, two each derived of the voltage levels (–1 V, 0 V, 1 V) and (–0.5 V, 0.5 V). From these sets the 4D-PAM5 code words are selected. The methodology applied – of convolutional coding with an extra bit per 8-bit code word and signal mapping by set partitioning – is called Trellis Coded Modulation (TCM) [38]. At the receiver, the PCS blocks are applied in reverse order.

100BASE-T1 reuses many of the methodologies defined for 1000BASE-T (for details see Section 5.2.1.3). It uses the same polynomials and reuses parts of the first data scrambler, of the scrambler bit generator, and the second scrambler that XORs the random data with the transmission data. As 100BASE-T1 is a single pair and only using PAM3 (see Figure 5.14), it has no use for the 1000BASE-T elements implemented to cater for the four channels and the "maximum distanced"

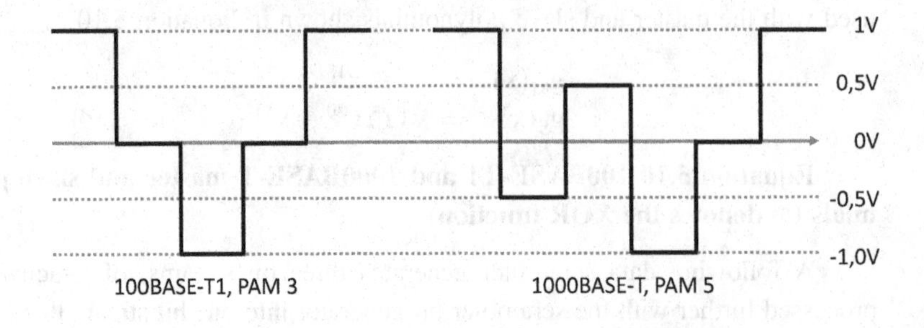

Figure 5.14 Comparison of 100BASE-T1, PAM3 and 1000BASE-T, PAM5 voltage successions

4D-PAM5 – i.e., the third data random data stream, the sign scrambler nibble generator, the symbol sign scrambler, and the TCM. One other important difference not immediately visible is that for 100BASE-T1 the bit stream used for control, idle, and training mode is not part of the data processing in the side stream, as is the case for 1000BASE-T, but inserted after the symbol mapping (see also Figure 5.19).

- **PMA:** Figure 5.15 shows an example PMA structure, which has the task to prepare 1000BASE-T data for transmission and received data for decoding by the PCS. This entails tasks such as clock recovery and reset. As there are four wire pairs the PMA must be instantiated four times for the transmit and four times for the receive path.

  The incoming transmit data first passes a partial response pulse shape filter with the transmission function $0.75+0.25z^{-1}$. This filter adds a multiple of 0.25 of the previous symbol to a multiple of 0.75 of the current symbol in order to reduce the power spectrum of the transmitter and meet the EMC requirements by reducing the EME. The symbols are then converted into analog signals before being overlaid onto the existing signals on the channel by the hybrid.

  The received analog signals are provided by the hybrid. A High Pass Filter (HPF) can follow in order to reduce the required dynamic range of the analog to digital (A/D) conversion, but especially to cut the ISI and, with that, reduce the number of taps needed later in the equalizer. Feedback from previously receive data is used to optimize the power of the receive signal with the help of Adaptive Gain Control (AGC) and an amplifier before the signal is actually converted. The now

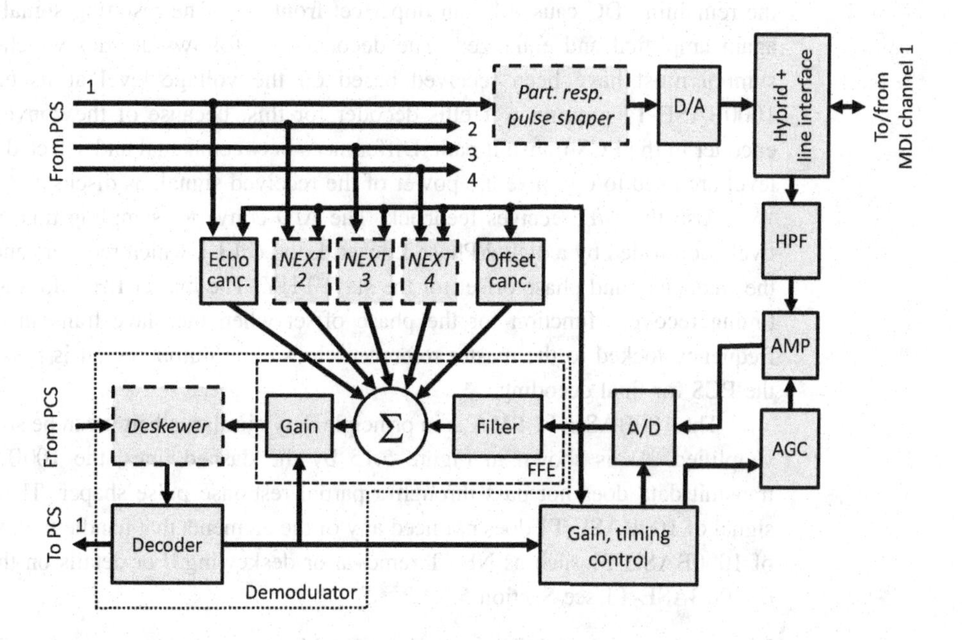

Figure 5.15 Example of a PMA structure for a 1000BASE-T PHY (see e.g. [34, 37, 41]), dashed lines and Italics indicate the blocks not needed for 100BASE-T1

digital signal is then processed in the demodulator, which, in the example given, contains a Feed Forward Equalizer (FFE), a deskewer, and a (Trellis) decoder. The deskewer aligns the delay differences between the four different pairs and is therefore not needed for 100BASE-T1/OABR. As the cables may have slightly different physical lengths, they can have different propagation times with respect to each other. Feedback from the PCS is required to achieve the right lock in the deskewer.

Inside the FFE, the signal first passes a pulse shaper with the transmission function $g+z^{-1}$. The value of g depends on the cable length, which is deduced from the receive signal strength. Then the inverse partial response is performed in order to reverse the artificial signal spread performed in the transmitter and to ease equalization. The transfer function $1+Kz^{-1}$ with K $[0,1] \in \mathbb{R}$ is constantly dynamically adapted in order to mitigate disturbances caused by the pulse shaping of the transmitter. It is entirely up to the implementer of the technology to decide on these blocks of the receiver. For example, the example does not contain a Decision Feedback Equalizer (DFE), which might well be seen as a suitable addition.

The signal is then freed from voltage portions resulting from echoes caused on the channel by the PHY's own transmit signal and from the non-alien Near End CrossTalk (NEXT), caused by the transmit signals on the other respective three wire pairs of the 1000BASE-T system. The interference caused by Far End Cross Talk (FEXT) cannot be known exactly, but could be estimated and then also subtracted. However, the level of FEXT is so low for 1000BASE-T that this is typically not done. Last but not least there is an offset subtraction, which removes the remaining DC caused by an imperfect front-end. The resulting signal is then again amplified and equalized. The decoder that follows decides which PAM5 symbol must have been received based on the voltage level at its entrance. 1000BASE-T1 requires a Trellis decoder for this, because of the convolutional encoder in the PCS transmit path. Differences between actual and selected voltage level are used to optimize the power of the received signal, as discussed.

Also the A/D receives feedback. The A/D converter sampling time is adaptively controlled by a digital Phase-Locked Loop (PLL), which recovers and tracks the frequency and phase offset for the slave PHY. The master PHY does a similar timing recovery function for the phase offset when the slave transmit clock is frequency locked to the master reference clock. The output signal is passed onto the PCS for final decoding.

The 100BASE-T1 PMA is in principle very similar, albeit it can be somewhat simplified. As is shown in Figure 5.15 by the dashed lines, the 100BASE-T1 transmit data does not pass through a partial response pulse shaper. The receive signal of 100BASE-T1 does not need any of the elements that handle the four pairs of 1000BASE-T1 such as NEXT removal or deskewing. For details on the PMA of 100BASE-T1 see Section 5.2.1.3.

- **Master-slave principle/Loop timing:** Establishing a link between two Ethernet PHYs or link partners requires several steps, which take into account whether a

PHY functions as slave or as master. Master and slave PHY definition is used for two reasons: First, in the case of true full-duplex operation, i.e. when data is transmitted and received simultaneously on the same wire pair, master and slave have to be assigned different, predefined scrambling polynomials (see PCS description just above). This ensures uncorrelated data and idle streams on the wire.

Second, master and slave distinction is used for the loop-timing concept, which synchronizes the clocks. The master PHY originates a reference transmit clock, which is recovered by the slave PHY receiver for determining its own frequency and phase offsets with respect to the master reference. Then, the slave PHY uses this receive clock to generate its own transmit clock in order to transmit back to the master PHY. At this stage, the master PHY only needs to recover the phase offset in its receiver from the slave signal, as the slave is already using the same frequency as the master. This action completes the loop for exact timing synchronization between master and slave PHYs, which is needed to ensure the correct reception of data on a full-duplex link. The wide deployment of 1000BASE-T [42] can be seen as an empirical proof for the robustness of this mechanism.

After power-up, master and slave PHYs go through a handshake process for start-up, sometimes also called a "PHY link acquisition process". This process uses three different signals:

- SEND_Z describes the transmission of zero-code (inactivity or "transmit silent").
- SEND_I describes the transmission of PAM3 idle symbols, which can have the voltage levels (–1 V, 0 V, 1 V).
- SEND_N describes the transmission of PAM5 data or idle symbols which can have the voltage levels (–1 V, –0.5 V, 0 V, 0.5 V, 1 V).

After the link is enabled the start-up process begins with only the master sending SEND_I idle symbols and the slave staying quiet, i.e., sending SEND_Z. During this time the master trains its echo-canceller and the slave synchronizes onto the master clock, adjusts its timing recovery, its equalizer, and locks its scramblers. In the second step, the slave sends idle symbols and trains its echo canceller while the master uses the information to adjust its timing, equalizer, and scrambler. In the third step, the transmitted idle frames of both master and slave are used to further improve the previously performed trainings. Afterwards the "scr_status," the "loc_rcvr_status," and the "rem_rcvr_status" are validated. If all are positive, the link set-up has been successful and both PHYs will then go into data mode SEND_N. If any of the status values is negative, the process restarts.

Table 5.1 gives an overview on the start-up sequence that has been defined for 1000BASE-T. 100BASE-T1 uses the same start-up procedure, except that all transmitted symbols can only have the values (–1 V, 0, 1 V) (see also Figure 5.25). Note that, during start-up, the system runs a timer that defines the maximum time the system can remain in slave silent and training (idle) state. If expired before the loc_rcvr_status is OK, data transmission will not be enabled and the system will not change into SEND_N.

Table 5.1 Training phase during start-up sequence in order to converge to minimum errors [43]

	Master	Slave
1	Transmit "idle" (SEND_I) Adapt echo canceller	Transmit "silent" (SEND_Z) Adapt AGC Clock recovery Adapt FFE Lock scrambler
2	Transmit "idle" (SEND_I) Adapt AGC Phase recovery Adapt FFE Lock scrambler	Transmit "idle" (SEND_I) Adapt echo canceller
3	Transmit "idle" (SEND_I) Refine adaptation	Transmit "idle" (SEND_I) Refine adaptation
4	Transmit data (SEND_N)	Transmit data (SEND_N)

5.2.1.3 Technical Description of 100BASE-T1/OABR

In autumn 2007, when BMW approached Broadcom and other Ethernet PHY semiconductor vendors in search of a 100 Mbps Ethernet technology suitable for use over unshielded cables in the harsh automotive environments, 1000BASE-T was long established and the 10 Gbps technology 10GBASE-T was just being introduced into the market. For optical systems, the standardization efforts for the 40 and 100 Gbps PHY standards were on the way [44]. With this in mind, 100 Mbps seems like a very low data rate. But, it was the experience and building blocks that had been built for the ever higher data rate PHYs that helped find a solution for the automotive requirements at 100 Mbps.

Figure 5.16 shows the dependency of the attenuation/Insertion Loss (IL) as well as NEXT and FEXT interference on the frequency for the example 50 m Cat 6 channel. As can be seen, the attenuation as well as the interference are smallest below 40 MHz. The bandwidth (Nyquist frequency) of both 100BASE-TX and 1000BASE-T was 62.5 MHz (see also Table 5.7). One of the key design points for BroadR-Reach was to reduce this bandwidth further, which helped with both the longer channel it was intended for and the more stringent EMC requirements of the automotive industry. The development of 1000BASE-T had shown that a system with a particular bandwidth (namely, 100BASE-TX) can be used as the basis to fulfill the requirements of a system with 10-times the data rate. With that perspective, halving the bandwidth for a system with the same data rate as 100BASE-TX, as has been done for BroadR-Reach, seemed realistic.

Figure 5.17 shows the power spectral density for 100BASE-TX, 1000BASE-T, and BroadR-Reach/100BASE-T1. As can be seen, 1000BASE-T and 100BASE-TX both use about 125 MHz and therefore have a Nyquist frequency of 62.5 MHz. At a baud rate of 66⅔ MHz, BroadR-Reach achieves a Nyquist frequency of 33⅓ MHz. The following technical description explains in detail the functioning of the

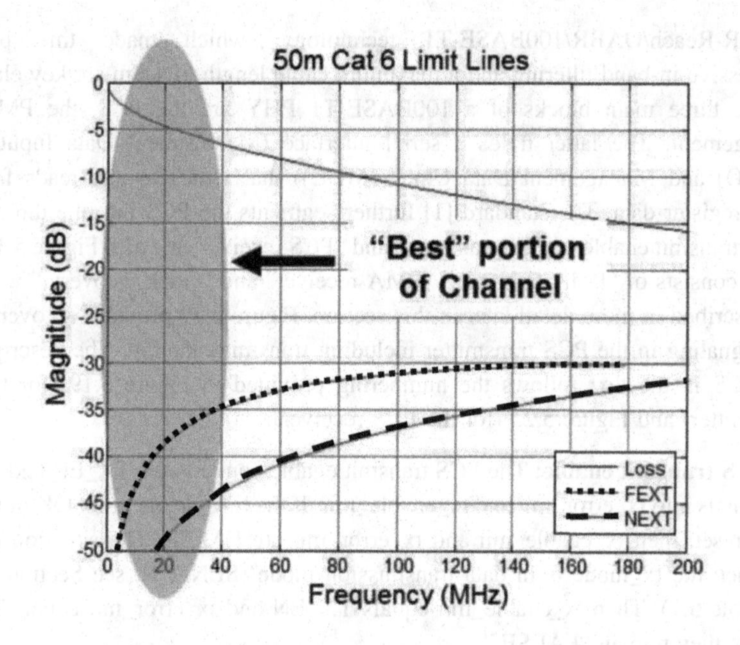

Figure 5.16 IL (attenuation), NEXT, and FEXT as a function of the frequency [45]. Figure reprinted with permission from Broadcom Inc., © 2012 Broadcom Corporation

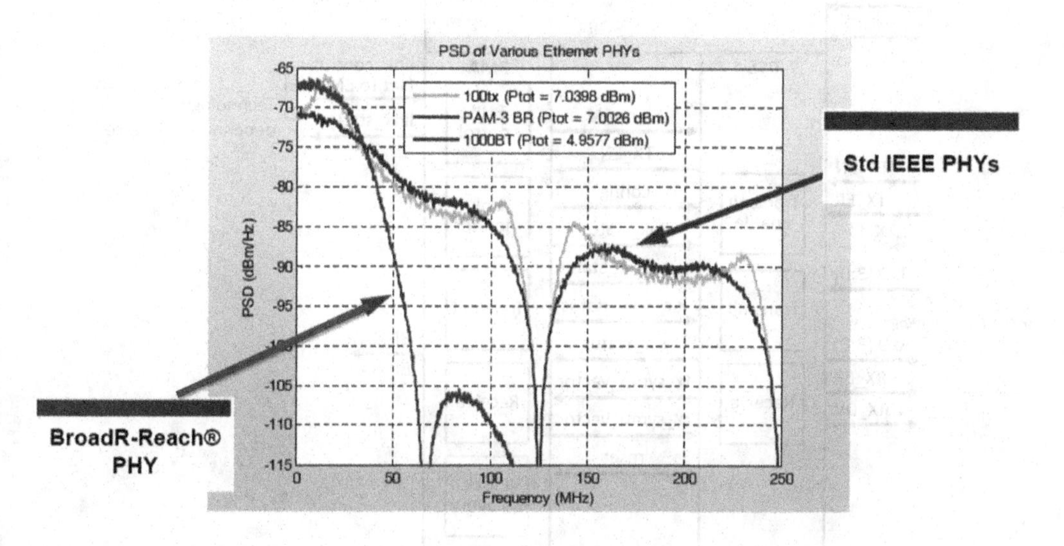

Figure 5.17 Power spectral density of 100BASE-TX, 1000BASE-T, and BroadR-Reach/100BASE-T1 transmit signals [45].
Figure reprinted with permission from Broadcom Inc., © 2012 Broadcom Corporation

BroadR-Reach/OABR/100BASE-T1 technology, which made this possible. Aggressive in-band filtering and a maximum cable length of 15 m are key elements.

The three main blocks of a 100BASE-T1 PHY are the PCS, the PMA, and management. The latter holds a serial interface (Management Data Input/Output (MDIO) and Management Data Clock (MDC)) that writes to and reads from the PHY register data. The standard [1] further segments the PCS into the three blocks "PCS transmit enable," "PCS transmit," and "PCS receive" (see also Figure 5.18). The PMA consists of "PMA transmit," "PMA receive," and "clock recovery," which are all described in more detail later in this section. Figure 5.19 provides an overview of the signaling in the PCS transmitter including transmit enabling. The description of the PCS in the text follows the numbering provided in Figure 5.19 (for the PCS transmitter) and Figure 5.23 (for the PCS receiver).

1. **PCS transmit enable:** The PCS transmit enable converts the TX_ER and TX_EN signals into tx_error_mii and tx_enable_mii. If the link_status is not OK or the PCS is reset, then tx_enable_mii and tx_error_mii are FALSE. The situation changes when the tx_mode is in data transmission mode (SEND_N, see Section 5.2.1.2, Table 5.1). Then tx_enable_mii equals TX_EN and tx_error_mii equals TX_ER, else they remain "FALSE."
2. **Aligner:** The aligner adapts the lengths of tx_error_mii and tx_enable_mii such that they are fitted for PHY internal use. Adaptations of the duration of the values might be needed in case SSD/ESD/IDLE symbols are added to the data stream, or

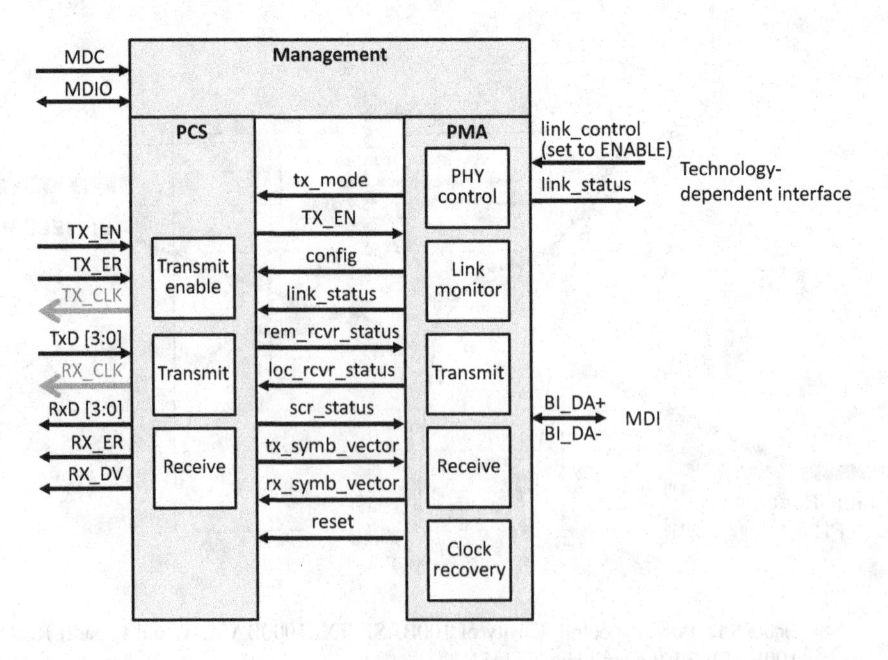

Figure 5.18 Building blocks of a 100BASE-T1 PHY [1], normally link_control is enabled with the completion of the auto-negotiation. As 100BASE-T1 does not support auto-negotiation, the value is set to enable at power-on

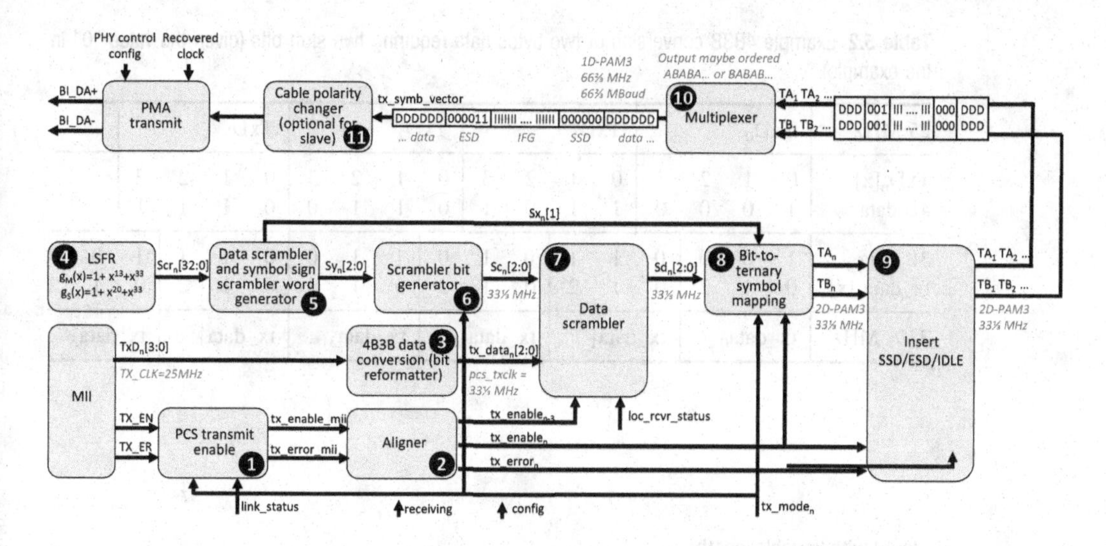

Figure 5.19 Overview of the PCS transmitter and transmit enable functions for 100BASE-T1

when stuff bits are used in the four bits to three bits (4B3B) conversion (see point 3 and Figure 5.20). The input data tx_error_mii and tx_enable_mii are changed into the output data tx_error_n and tx_enable_n.

3. **4B3B conversion (bit reformatter):** This block regroups four incoming bits TxD [3:0] into groups of three tx_data[2:0]. To keep the same data rate of 100 Mbps the clock rate needs to change for tx_data[2:0] from 100 Mbps = 4 × 25 Mbps to 3 × (25 Mbps*4/3) = 3 × 33⅓ Mbps = 100 Mbps. When the incoming data is not a multiple of three bits, one or two stuff bits have to be inserted (see Table 5.2 for an example). The value of the stuff bits is not specified in the standard. However, their value is not important as they are simply removed again on the receiver side. In case stuff bits are added at the end of the data portion of the Ethernet packet, the aligner has to adapt the lengths of the tx_error_mii and tx_enable_mii accordingly (see Figure 5.20 for examples).
4. **Linear Feedback Shift Register (LFSR):** The LFSR has the purpose of creating the pseudo random binary starting sequence used for randomizing/scrambling the transmit data. The 33 initial bits the LFSR holds are not specified, but are up to the implementer. These initial bits can have any of the 2^{33} possible values, except being all “0.” In that case the created output sequence would always stay “0” – 0 XOR 0 equals 0 – and defy its purpose of being pseudo random. For any other starting value the resulting output sequence is repeated after $2^{33} - 1$ register shifts. Note that this is not necessarily the case for every LFSR with the length 33, but due to the specific polynomials chosen.[11] With a frequency of 33⅓ MHz and thus 30 ns register shift, the sequence repeats itself every $(2^{33} - 1)*30$ ns = 257.69 s. In order to be able to reuse as much as possible from 1000BASE-T, 100BASE-T1 uses the same master and slave polynomials as 1000BASE-T (see Equation 5.10 and Figure 5.21). Both polynomials are always needed. In the master device, the

Table 5.2 Example 4B3B conversion of two bytes data requiring two stuff bits (given the value "0" in the example)

25 MHz	TxD_0				TxD_1				TxD_2				TxD_3					
$TxD_n[x]$	0	1	2	3	0	1	2	3	0	1	2	3	0	1	2	3		
4B data	1	0	0	0	1	1	0	1	0	1	1	0	0	1	1	1		
3B data	1	0	0	0	1	1	0	1	0	1	1	0	0	1	1	1	0	0
$tx\_data_n[x]$	0	1	2	0	1	2	0	1	2	0	1	2	0	1	2	0	1	2
33⅓ MHz	tx_data_0			tx_data_1			tx_data_2			tx_data_3			tx_data_4			tx_data_5		

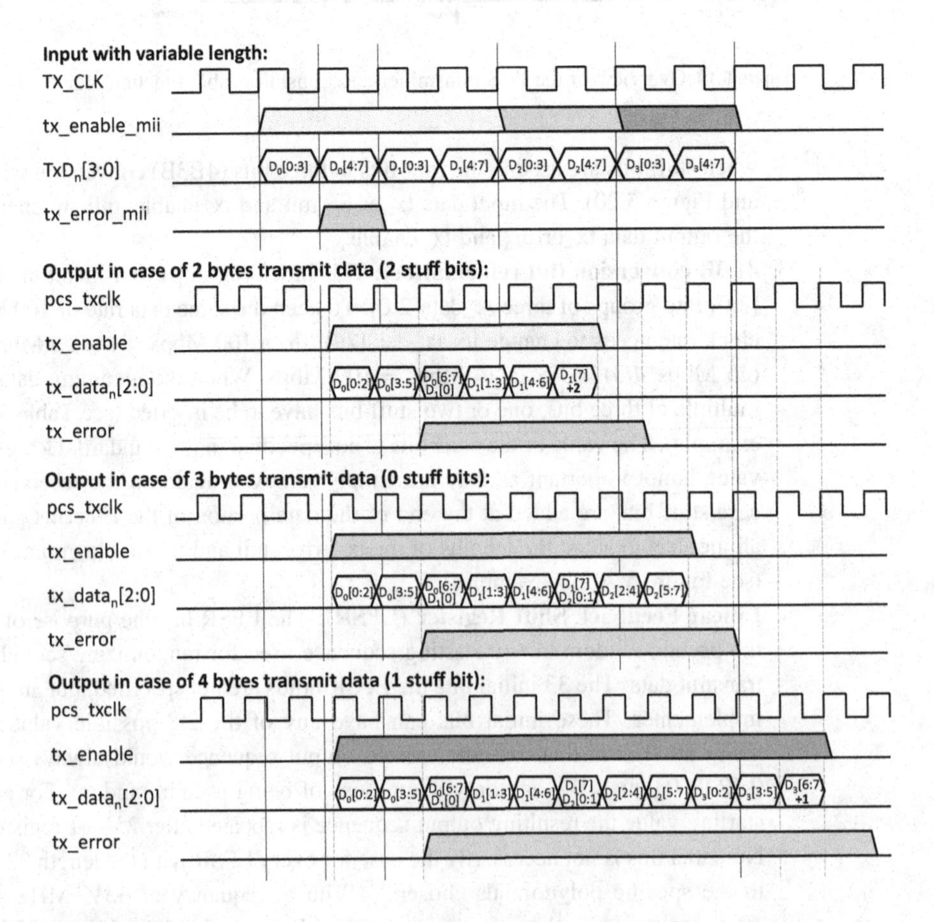

Figure 5.20 4B3B bit mapping and aligner extension of tx_enable and tx_error in case of different number of bytes $D_n[0:7]$ at the end of the packet [1]. In the example depicted, tx_error is set TRUE (= "1") to show how it is handled. Naturally, being able to use the data in the receiver requires tx_error to be FALSE (= "0")

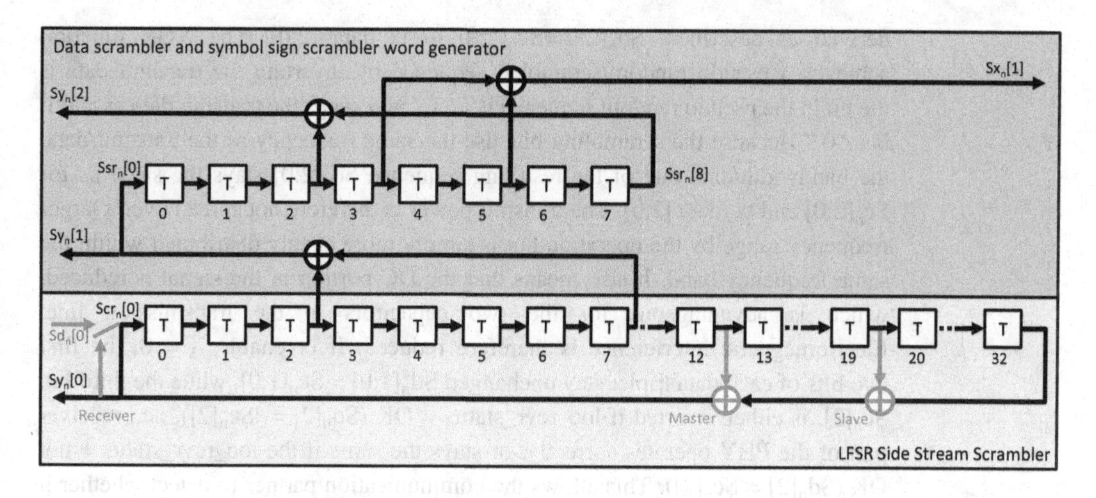

Figure 5.21 LFSR and data scrambler symbol sign word generator for 100BASE-T1/OABR (for an explanation of the receiver change see point 13)

master polynomial is needed for the transmit data and the slave polynomial for the receive data, and in the slave vice versa. Note that the scrambler for 1000BASE-T1 (see Section 5.3.1) has been further optimized and is shorter.

5. **Data scrambler and symbol sign scrambler word generator:** Figure 5.21 shows how the starting pseudo bit stream is generated in the LFSR and transformed into pseudo random, three bit words $Sy_n[2:0]$ by the symbol sign scrambler word generator. The value $Sx_n[1]$ is used for randomizing the IDLE symbols during tx_mode=SEND_ N (see also point 8 for bit-to-ternary mapping). This is a reduced version of how it is done for 1000BASE-T. Note that because 100BASE-T1 uses only one of the four $Sx_n[3:0]$ values provided in 1000BASE-T, the 100BASE-T1 specification refers to it as Sx_n only, without numbering the bit [1]. This book keeps the $Sx_n[1]$ designation in order to show the derivation from the 1000BASE-T standard. The second shift register shown in Figure 5.21 that depicts the function of the symbol sign word generator can be circumvented by applying Equation 5.11 directly to the LSFR output.

$$\begin{aligned} Sy_n[0] &= Scr_n[0] \\ Sy_n[1] &= Scr_n[3] \oplus Scr_n[8] \\ Sy_n[2] &= Scr_n[6] \oplus Scr_n[16] \\ Sx_n[1] &= Scr_n[7] \oplus Scr_n[9] \oplus Scr_n[12] \oplus Scr_n[14] \end{aligned}$$

Equation 5.11 Derivation of $Sy_n[2:0]$ and $Sx_n[1]$ from the LFSR polynomials ($\oplus$ denotes the xor function)[12]

6. **Scrambler bit generator:** The scrambler bit generator transforms or keeps the data depending on the tx_mode value. If the tx_mode is SEND_Z the output is all "0." In all other cases the output equals the input.
7. **Data scrambler:** If $tx\_enable_{n-3} = 1$, the data scrambler performs the bit-by-bit XOR combination between the transmit data and the pseudo random sequence

derived as described, $Sd_n[2:0]=Sc_n[2:0] \oplus tx\_data\ [2:0]$. The XOR function achieves a pseudo random/scrambled sequence by inverting the transmit data if the bit in the pseudo random sequence is a "1," and keeps the transmit data as is if it is a "0." Because the scrambling bits use the same frequency as the transmit data, the bandwidth/data rate of the resulting sequence $Sd_n[2:0]$ stays the same as for $Sc_n[2:0]$ and tx_data [2:0]. The transmit power is therefore not spread over a larger frequency range by the operation but is simply more evenly distributed within the same frequency band. It also means that the DC portion of the signal is reduced, which is advantageous for the two capacitors in the transmission line. Electromagnetic interference is therefore reduced. If $tx\_enable_{n-3} = 0$, the first two bits of each data triplet stay unchanged $Sd_n[1:0] = Sc_n[1:0]$, while the third bit, $Sc_n[2]$, is either inverted if loc_rcvr_status = OK ($Sd_n[2] = !Sc_n[2]$), i.e., receiver part of the PHY operates correctly, or stays the same if the loc_rcvr_status = not OK ($Sd_n[2] = Sc_n[2]$). This allows the communication partner to detect whether it can set its rem_rcvr_status to OK or not OK. The latter is relevant in the start-up phase, shown in Table 5.1 and Figure 5.25. Another important parameter during start-up is the scr_status, which indicates whether the scrambler in the receive path has synchronized or not.

To start with, all local_rcvr_status, rem_recv_status, and scr_status values are not OK. The master first sends IDLE symbols (SEND_I), while the slave is in SEND_Z, i.e., zero voltage is being put on the line. The slave uses the bit $Sd_n[0]$ of the received idle symbols to synchronize its master descrambler. When the descrambler is synchronized, the scr_status in the slave is set to OK. Once this happens, the slave will start sending IDLE symbols, too, and can recognize rem_rcvr_status of the master encoded in the IDLE symbols. As explained, the information is transmitted with $Sd_n[2]$. The master can then start synchronizing its own scrambler. When the receiver is the master and is ready (scr_status = OK represents only one element or loc_rcvr_status = OK), the transmitted $Sd_n[2]$ bit will be changed accordingly, in order to indicate to the slave the changed status (received as rem_rcvr_status OK). For more details see also point 13.

8. **Three Bits to Two Ternary conversion (3B2T):** This block translates the three bits $Sd_n[2:0]$ onto the ternary transmit symbols TA_n and TB_n used for PAM3. The ternary values of TA_n and TB_n (–1, 0, 1) are mapped directly onto the voltage levels (–1, 0, 1) V. The used 3B2T code does not guarantee that the signal is DC-free, nor that the clock is continuously available, nor does it use a Gray code (a Gray code normally ensures that successive code words only differ by one bit in order to limit the errors in the case of continuously changing signals [46] and thus reduces the Bit Error Rate (BER)[13]). Table 5.3 shows the 100BASE-T1 3B2T-mapping during data transmission SEND_N in contrast to an example Gray code like that used for 1000BASE-T1. The mapping $\{TA_n, TB_n\} = \{0, 0\}$, is not used for data, but reserved for the Start Stream Delimiter (SSD) and parts of the End Stream Delimiter (ESD). The mapping requires that three bits are always available. This is why, should the bit-length of the user data not de devisible by three, the data is extended by stuff bits, as explained in point 3.

Table 5.3 3B2T mapping used in 100BASE-T1 during normal transmission (tx_mode = SEND_N and tx_enable = 1) in contrast to a respective Gray code as used for 1000BASE-T1 [1, 11]

100BASE-T1 $T_A\downarrow$ $T_B\rightarrow$	−1	0	1	1000BASE-T1 $T_0\downarrow$ $T_1\rightarrow$	−1	0	1
1	101	110	111	1	011	111	110
0	011	SSD/ESD	100	0	010	SSD/ESD	100
−1	000	001	010	−1	000	001	101

Table 5.4 Bit to ternary symbol mapping for 100BASE-T1

	"inactivity" tx_mode = SEND_Z		"training," "idle" tx_mode = SEND_I or tx_mode = SEND_N, tx_enable = 0, $Sx_n[1]=0$		"idle" tx_mode = SEND_N, tx_enable = 0, $Sx_n[1]=1$		"data" tx_mode = SEND_N, tx_enable = 1	
$Sd_n[2:0]$	TA_n	TB_n	TA	TB_n	TA_n	TB_n	TA_n	TB_n
000	0	0	−1	0	−1	0	−1	−1
001	0	0	0	1	1	1	−1	0
011	0	0					0	−1
010	0	0	−1	1	−1	1	−1	1
SSD/ESD	0	0	Not used {00},{11},{−1−1}		Not used {00},{01},{0−1}		0	0
100	0	0	1	0	1	0	0	1
101	0	0	0	−1	−1	−1	1	−1
111	0	0					1	1
110	0	0	1	−1	1	−1	1	0

The exact mapping depends on a number of parameters. The first parameter is the tx_mode. If tx_mode=SEND_Z, TA_n and TB_n are mapped to 0, which, in the end, is the meaning of SEND_Z. If tx_mode=SEND_I (i.e., idle mode) the data is mapped according to Table 5.4. In this case neither tx_enable or $Sx_n[1]$ are taken into account and the scrambling might not be as effective. This potentially results in a slightly worse EMC performance. However, as explained in point 7, SEND_I represents the training mode during which the scrambler and other aspects of the receiver are adjusted and synchronized. Only after this has been successful can the receiver actually generate and use the value $Sx_n[1]$ out of its own descrambler. To counterbalance this and improve the robustness against EMC impacts, only six out of the nine possible $\{TA_n, TB_n\}$ combinations are used during training SEND_I. The specific selection allows better distribution of the transmit power across the used frequency band.

This is similar to when the system is idle in case of tx_mode=SEND_N, but tx_enable=0 (i.e., link acquisition has been completed, but the transmitter MII has no data to send and is therefore also in an idle state). Here the incoming bits are also mapped to only six $\{TA_n, TB_n\}$ combinations, while the scambling bit $Sx_n[1]$ now additionally decides on the exact values, in order to balance the power density better.

Finally, during normal transmission (i.e., when tx_mode = SEND_N and tx_enable = 1) eight of the nine possible ternary symbol combinations are used for $\{TA_n, TB_n\}$ (see also Table 5.4). The combination {00} is used for control purposes, i.e., before and after the actual transmit data, the SSD and ESD encoded in the {00} indicate to the receiver the beginning and end of transmit data (see point 9).

9. **Insert SSD/ESD/IDLE:** Depending on the values of tx_enable and tx_error, this block inserts control symbols into the data stream. With the end of the idle state and data ready to transmit, the value tx_enable changes from FALSE = 0 to TRUE = 1. This information comes from the MII interface where TX_EN changes when data is available for transmission. When tx_enable has changed, first of all three SSD PAM3 pairs [00 00 00] are inserted into the data stream. If tx_enable is still TRUE after their completion, the user data is inserted. As soon as tx_enable changes to FALSE the ESD is added. If an error was detected in the packet (the information comes with TX_ER from the MII) and tx_error is TRUE = 1 then the ESD information is [00 00 –1–1]. The PHY forward the packet regardless. The layer behind the MII, e.g., the switch, might decide to discard it. If tx_error is FALSE, i.e., no error has been detected before sending, the ESD is [00 00 11]. Figure 5.22 depicts the function.

   The insertion of additional symbols with SSD and ESD adds symbols to the data stream that the MII is not aware of. In order not to desynchronize the transmission

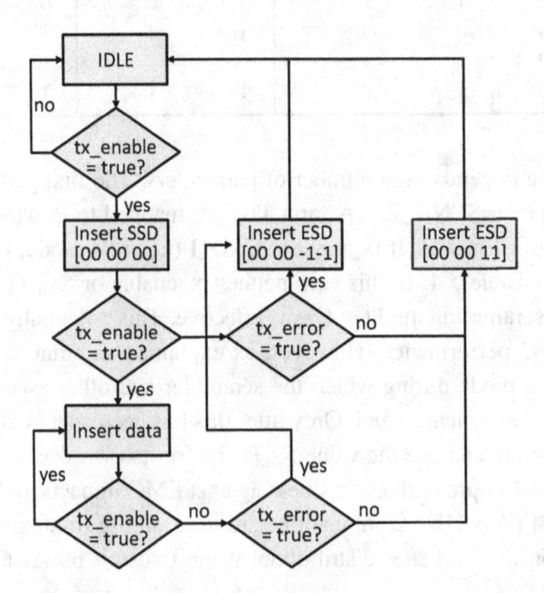

Figure 5.22 Insertion of control symbols in the "Insert SSD/ESD/IDLE" block. This happens only during SEND_N status. tx_enable defines whether the status is nevertheless idle or not.

or to change the data rate between MII and data on the channel, the following method is applied. Each Ethernet packet normally starts with a seven byte preamble (see Section 1.2.1 and Figure 1.5). In order to accommodate the SSD, the 100BASE-T1 (and also the 1000BASE-T1) PHY shortens this preamble. The SSD comprises six 1D-PAM3 symbols or three 2D-PAM3 symbols, respectively. Because three bits are mapped to two PAM3 symbols in the 3B2T conversion, the six PAM3 symbols represent 6*3/2 = 3*3 = 9 bits. Thus, the ternary symbols that represent the scrambled nine bits of the original preamble [101 01010 101] are replaced with the SSD symbols [00 00 00]. This is possible because in the switched Ethernet network, the original function of the preamble is no longer needed. The preamble was originally needed to allow synchronization at the beginning of every packet in a large network with bus topology (potentially including repeater hubs) that did not have a continuous connection. With the introduction of the switched architecture, the preamble, and also the length of the InterPacket Gap (IPG) have been kept in the standard for backwards compatibility reasons only, but without functional necessity. As a consequence, the preamble is shortened for the SSD and potential stuff bits and the ESD shortens the IPG. The receiver has to ensure that the original timing with preamble and IPG is restored (and SSD and ESD are removed), before passing the data on to the MII. Table 5.6 (see point 15) explains the principle realignment between the MII data and the transmit data.

10. **Multiplexing 2D-PAM3 to 1D-PAM3:** This block multiplexes the two parallel ternary symbols TA_n and TB_n into a one-dimensional data stream, before the PMA transmit block processes the data into BI_DA+ and BI_DA− and puts them onto the channel. With the multiplexer the symbol rate doubles from 33⅓ MBaud to 66⅔ MBaud and the symbol duration halves from 30 ns to 15 ns. It is not specified whether the output of the multiplexer starts with TA_n or TB_n. The receiver thus must include a function which can recognize the order of the symbols and acts accordingly (see point 14).
11. **Optional cable polarity changer:** The 100BASE-T1 specification includes an optional cable polarity changer for the slave. The slave can detect a polarity change during link acquisition in it's receive path, once it has completed its descrambler synchronization (see also point 14). After the master's idle data has been recognized, the cable polarity changer can detect a polarity change and correct it, if necessary. If this is indeed necessary, the slave also has to change the polarity of its transmit symbols, as the specification does not foresee a polarity change in the master. A polarity change means $1 \rightarrow -1$, $0 \rightarrow 0$, and $-1 \rightarrow 1$.

The standard specifies significantly fewer parts of the receiver than of the transmitter. To assure interoperability and compliance, the receiver has to be able to count on specific properties of the received signal, but how it handles them is a part of the Unique Selling Point (USP) of the implementer and thus not generally specified. The following therefore describes examples. In principle, the PCS receiver performs most

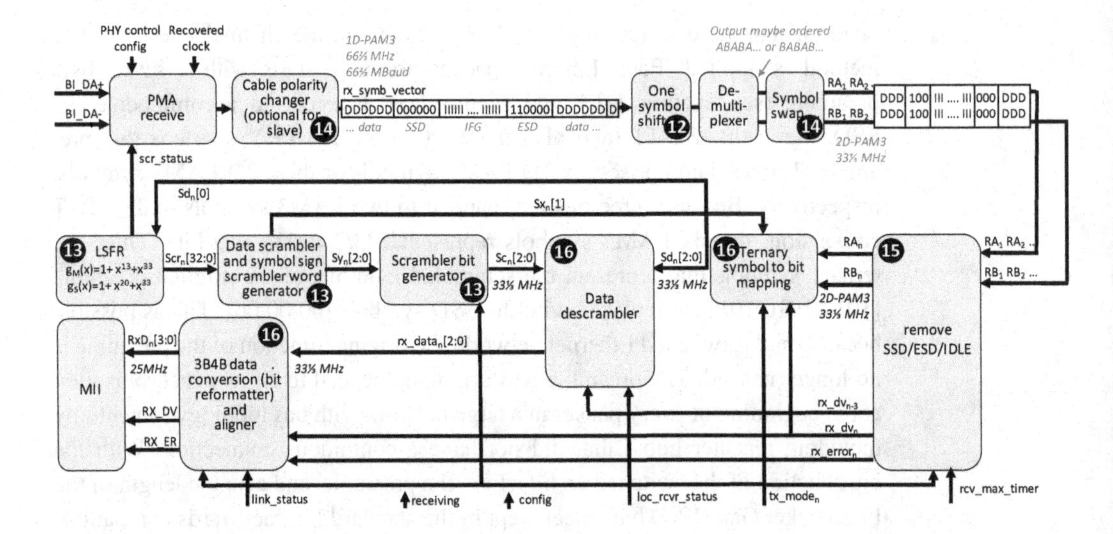

Figure 5.23 Example for elements of a 100BASE-T1/OABR PCS receiver

operations of the PCS transmitter in reverse order as is depicted in Figure 5.23. The different blocks are described in more detail in the following text.

12. **One symbol shift:** In order to receive data correctly, the PCS receiver must group the received symbols RA_n and RB_n into pairs with correct polarity, correct order (RA_n RB_n or RB_n RA_n), and the same timing RA_n and RB_n. Otherwise the ternary symbol to bit mapping will produce the wrong output. The receive PCS must also synchronize its scrambler, so that it can descramble the transmission data correctly. To start with the receiver has no information, but must detect all that is necessary from the data it receives during training mode.

    The very first thing, before it makes sense to start the process of scrambler synchronization, the symbol grouping needs to be correct and potentially corrected. For the symbol grouping, the receiver has to detect whether, independent of the symbol order, RA_n RB_n or RB_n RA_n, RA_n is grouped with RB_n and not accidently with RB_{n-1} or RB_{n+1} and vice versa (see Figure 5.24). The basis for the detection of the status of the symbol grouping is the 3B2T coding as shown in Tables 5.4 and 5.5. As can be seen in both tables, during SEND_I only six of the nine possible 2D-PAM3 symbols are used and only these symbols should be received. If the symbol grouping is wrong, the 2D-PAM3 symbols {00}, {11}, {–1–1} will be received. In this case, the symbol grouping must be shifted by one symbol.

13. **(De-)Scrambling:** As was explained before, the PCS receiver must first synchronize its scrambler before it can sensibly receive any data. The scrambler polynomials used are the same as defined for the transmitter LFSR, only that in the receive path the slave will use the master polynomial for reception and the master will use the slave polynomial. Additionally, in the receiver path the scrambler holds a selector that can open or close the feedback loop in the

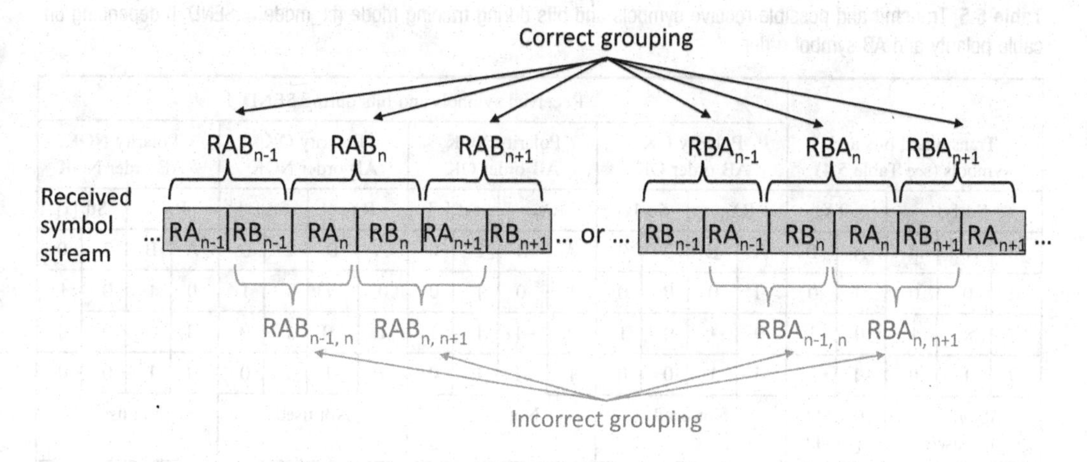

Figure 5.24 Possible grouping of 2D-PAM3 symbols during training at the receiver

LFSR (see Figure 5.21). Initially, the feedback loop is open, making the LFSR a shift register without feedback.

When the receiver is activated into training mode, the PCS receiver will start filling its register with the data $Sd_n[0]$ it receives from the master and that has been decoded in the 2T3B ternary symbol to bit mapping (see also Figure 5.21). Ideally, it takes 33 shifts (33 × 30 ns = 0.99 μs) to fill the register. However, to start with, the PMA must adjust properly (especially clock, equalizer, AGC) and the one-symbol shift must potentially be performed before $Sd_n[0]$ will represent usable values (for cable polarity and symbol order see point 14). Only then does the LFSR have the chance to successfully compare the scrambler output $Sy_n[0] = Scr_n[0]$ with the incoming data $Sd_n[0]$ and to synchronize. Once $Scr_n[0] = Sd_n[0]$ are continuously, i.e., long enough, identical, synchronization can be assumed. The standard does not describe when the synchronization has been completed. It is up to the implementer to decide when to set the scr_status to OK and to close the feedback loop. Once the LFSR is synchronized, the scrambler values $Sy_n[2:0]$ and $Sx_n[1]$ can be generated for 2T3B mapping in normal mode SEND_N and consecutive scrambler operations.

14. **Cable polarity changer and symbol swap:** The scrambler will not lock, when the selected symbol order RA_n RB_n or RB_n RA_n is not the same as the transmitted order. It is described in the following how this can be detected with help of the scrambling and that correct cable polarity is not necessary to do so. The cable polarity can be corrected once the symbol order is correct and the scrambler has locked. In order for the scrambler to lock, it has to receive $Sd_n[0]$ correctly, even if the polarity is changed and/or the symbol order was swapped. To explain how this can be done, Table 5.5 shows what happens during training mode (tx_mode = SEND_I) in the 2T3B conversion in the different (error) cases.

The leftmost columns in Table 5.5 show the 3B2T mapping of the transmission bit triplets into the 2D-PAM3 symbols during training mode

Table 5.5 Transmit and possible receive symbols and bits during training mode (tx_mode = SEND_I) depending on cable polarity and AB symbol order

Transmitted bits and symbols (see Table 5.4)					Received symbols and bits during SEND_I															
					Polarity OK AB order OK				Polarity NOK AB order OK				Polarity OK AB order NOK				Polarity NOK AB order NOK			
$Sd_n[x]$			TX_n		RX_n		$Sd_n[x]$		RX_n		$Sd_n[x]$		RX_n		$Sd_n[x]$		RX_n		$Sd_n[x]$	
2	1	0	A	B	A	B	2	0	A	B	2	0	A	B	2	0	A	B	2	0
0	0	0	−1	0	−1	0	0	0	1	0	1	0	0	−1	1	1	0	1	0	1
0	x	1	0	1	0	1	0	1	0	−1	1	1	1	0	1	0	−1	0	0	0
0	1	0	−1	1	−1	1	0	0	1	−1	1	0	1	−1	1	0	−1	1	0	0
SSD/ESD (not used)			{00}{11} {−1−1}		Not used				Not used				Not used				Not used			
1	0	0	1	0	1	0	1	0	−1	0	0	0	0	1	0	1	0	−1	1	1
1	x	1	0	−1	0	−1	1	1	0	1	0	1	−1	0	0	0	1	0	1	0
1	1	0	1	−1	1	−1	1	0	−1	1	0	0	−1	1	0	0	1	−1	1	0
							3)	1)			3)	1)				2)				2)

SEND_I as presented in Table 5.4. The following columns show what the receiver receives as RA_n, RB_n and consequently decodes as $Sd_n[0]$, $Sd_n[2]$ for the different scenarios. The first thing that can be noticed when comparing the columns marked 1) and 2) with each other is that $Sd_n[0]$ is actually not influenced by a polarity change, but only by a symbol order mismatch. In case of correct symbol order $Sd_n[0] = 1$ for $RA_n = \{1 \text{ or } -1\}$ and $Sd_n[0] = 0$ for $RA_n = \{0\}$, while it is the other way around for incorrect symbol order, i.e., $Sd_n[0] = 0$ for $RA_n = \{1 \text{ or } -1\}$ and $Sd_n[0] = 1$ for $RA_n = \{0\}$. This means that even with the wrong cable polarity, the scrambler can lock as long as the symbol order is correct. When starting, the slave can simply assume a certain symbol order and, if the scrambler does not synchronize within a certain time, it performs a symbol swap and restarts the synchronization effort of the scrambler.

With the scrambler synchronized the correct values $Sy_n[2:0]$ and $Sc_n[2:0]$ can be generated within the receiver. In training mode, the generated $Sc_n[2:0]$ should be the same as the received $Sd_n[2:0]$. This can be used to detect a polarity change. As shown in the columns marked 3), $Sd_n[2]$ is affected by a polarity change and will result in exactly the reverse value from what it should be if the assumed polarity is wrong. Once such a behavior has been detected, the polarity can be changed.

Note, that $Sd_n[2]$ is also used to indicate to the receiver the rem_rcvr_status (see also point 7) and that it might potentially be intentionally inverted by the transmitter. Using it to discover a polarity change is nevertheless possible because of the order of events. During start-up, the loc_rcvr_status in the master will never change to OK and the bit $Sd_n[2]$ will not be inverted, until the slave has started

Table 5.6 RA_n and RB_n mapping onto bits

	... Data					SSD			Idle					ESD			Data ...		
RA	D	D	D	...	D	0	0	0	I	I	...	I	I	1	0	0	D	D	D
RB	D	D	D	...	D	0	0	0	I	I	...	I	I	1	0	0	D	D	D
rx_data[2]	d	d	1	...	0	1	0	1	0	0	...	0	0	0	0	0	d	d	d
rx_data[1]	d	d	1	...	1	0	1	0	0	0	...	0	0	0	0	0	d	d	d
rx_data[0]	d	d	0	...	0	1	0	1	0	0	...	0	0	0	0	0	d	d	d
	... Data		SFD and preamble						IPG								Data ...		

transmitting SEND_I idle symbols (with the right cable polarity). The slave will only do so, once its scrambler has locked. However, as soon as the scrambler is locked, it can detect and correct a necessary cable polarity change. It will thus long have terminated its use of the $Sd_n[2]$ values for the cable polarity detection before the master might invert them to indicate his loc_rcvr_status as OK.

15. **Remove SSD/ESD/IDLE:** In normal transmission mode SEND_N and tx_enable=1, the transmitter replaces parts of the preamble and the IPG with the SSD and ESD (see point 9). The receiver must restore this. For the SSD, the block must reinsert the PAM3 symbols that represent the preamble and that, following 2T3B decoding and descrambling, result in the nine bits [101 010 101]. For the ESD it must be ensured that the state of the IPG is reinstalled (see Table 5.6 for the principle functioning) and that its values are replace by "0."

    With respect to the ESD, the standard foresees a measure to ensure that the receiver does not unintentionally stay in data mode if it misses the detection of the first two 2D-PAM3 symbols of the ESD [00 00]. The standard therefore introduces a state machine that uses the value rcvr_max_timer. If a timer with rcvr_max_timer expires before the change into idle state was initiated, the idle state in the PCS receiver is enforced.

16. **Data recovery and alignment:** The other activities in the PCS receiver are performed in reverse order. First, in the following 2T3B decoding, the rules as described in Table 5.4 are applied to obtain the values $Sd_n[2:0]$. Provided, tx_mode = SEND_N, these values are then scrambled with the correct scrambling triplets $Sc_n[2:0]$ that reverse the previously applied randomization by inversion if $Sc_n[x] = 1$ and keeping the data as is if $Sc_n[x] = 0$; just like it was done in the transmitter. The 3B4B conversion groups and reschedules the data from groups of three bits at 33⅓ MHz to groups of four bits at 25 MHz. The number of bits after 2T3B decoding must be a multiple of four. If there are additional bits left over, these bits are dropped as stuff bits and not used for the 3B4B bit mapping (reverse operation as shown in Table 5.2). The aligner adapts the lengths of rx_dv and rx_error, to forward them as RX_ER and RX_DV to the MII interface.

The PMA part of the Ethernet transceiver consists of PHY control, link monitor, transmit, and receive functions as well as clock recovery (see Figure 5.18). The 100BASE-T1 PMA transmit and receive functions are very similar to what has been explained with 1000BASE-T in Section 5.2.1.2 and Figure 5.15. The following description therefore focuses on the essentials.

In **PMA transmit**, the signals TA_n and TB_n coming from the PCS transmitter are first converted into analog signals. Other than for 1000BASE-T, no additional partial response operation is performed on the signals for 100BASE-T1 before the D/A. Following the D/A an internal filter can be used in order to ensure that the output signal meets the PSD mask defined in the specification. The hybrid then couples/decouples the transmit and receive signals onto one channel. An additional low pass filter can be used on the signal following the hybrid to improve its EME and EMI behavior. Such a filter would need to be effective from 33⅓ MHz on and can be PHY internal or PHY external.

PMA receive has the task of detecting the information that arrives and correctly converting it into PAM3 symbols that are then passed on with RA_n and RB_n to the PCS receiver. Because of the baseband transmission used with 100BASE-T1, the signals typically arrive at the receiver end of the transmission attenuated (smaller amplitude) and distorted (changed in shape) [47]. High and low pass filters, as well as amplitude correction (for AGC see description for 1000BASE-T in Section 5.2.1.2) can help to restore the received waveforms. However, for receiving data correctly it is extremely important that the clock runs at the right frequency and phase, in order to be able to sample the incoming waveform at the right time.

The PMA therefore also needs to perform **clock recovery**. The reference clock of 66⅔ MHz is set by the master link partner. The master obtains its clock value from a local clock source, making it available within its own system as well as imposing it onto the transmit signal. The slave then recovers frequency and phase from the incoming signal, and having done so, in return imposes the recovered clock onto its own transmit signal. The master knows the frequency the received signals have, because of its own clock source. But, it must still extract the correct phase from the incoming signals on its side, thus completing loop-timing. The loop-timing concept is the same as for 1000BASE-T (and in 1000BASE-T1). Note that because of the long scrambler sequence, baseline wander is not an issue for 1000BASE-T or 100BASE-T1 and compensational measures are not necessary (other than for 100BASE-TX).

The PMA receiver then performs the A/D conversion and echo cancellation that is needed because reflections of the own master transmit signal on the transmission channel can distort the received signal. As has been described for 1000BASE-T, the receiver performs some form of equalization (DFE, FFE) before the demodulator decides on the PAM3 signal values of RA_n and RB_n to be forwarded to the PCS. Because there is no convolutional encoding as with 1000BASE-T, 100BASE-T1 does not require a Trellis decoder for this. For 100BASE-T1, a simple slicer is sufficient for this task. Once the receiver perceives that it is able to do so correctly as well as continuously and the PCS has announced that its scrambler locked with scr_status = OK, the unit conveys this with the loc_rcvr_status = OK to the rest of the system and the connected unit.

In contrast to, e.g., a mobile communication channel, the changes the 15 m maximum wireline channel of Automotive Ethernet experiences are harmless. Nevertheless, it is advisable that the filter coefficients of a potential FFE and DFE can be adjusted dynamically. This ensures optimal reception in the case the channel changes because of, e.g., heat, humidity, aging, or mechanical strain. In the case of burst errors, 100BASE-T1 relies on a relatively high SNR. For systems like 1000BASE-T1 (see Section 5.3.1) that work with a significantly lower margin, an additional Forward Error Correction (FEC) is needed to improve the correct reception of data.

PHY control determines the procedure that enables the units to exchange data (and with it the tx_mode the transceiver is in, SEND_Z, SEND_I, or SEND_N) and initiates the changes from one tx_mode to the next. The main parameters used to initiate changes are scr_status, loc_rcvr_status, rem_rcvr_status, as well as two timers, maxwait_timer and minwait_timer. Most of the changes are initiated during the link-acquisition procedure, which has been explained in Section 5.2.1.2. While Table 5.1 in Section 5.2.1.2 shows the interrelation of the behavior of both units in sequence, Figure 5.25 shows the events for each unit individually, including the use of the timers.

Figure 5.25 shows that the main difference in behavior between master and slave is at the beginning. When a unit has been assigned master it simply goes directly into SEND_I state. On the other hand, the slave does so only when its scrambler has locked and the scr_status is OK. This means that the slave will remain in SEND_Z somewhat longer than the master, while the master can be expected to remain longer in SEND_I (which can be derived from Table 5.1 but not from Figure 5.25). Each unit remains in training/SEND_I for, at most, the time it takes to complete the training and set loc_rcvr_status to OK or for the time it takes for the maxwait_timer to expire; whichever is shorter. Should the maxwait_timer (200 ms $\pm$ 1% for 100BASE-T1)[14] expire before the loc_rcvr_status is OK, the link_status will go into not OK and the system will decide on a higher layer to restart the process and to go back to SEND_Z (hence the dotted line in Figure 5.25). When the unit, master or slave, has successfully reached loc_rcvr_status OK, it will check the status of the rem_rcv_status, i.e., whether the other unit they are communicating with is ready to receive, too. If the rem_rcvr_status is OK, data transmission mode SEND_N can begin. If the rem_rcvr_status is not OK, the system will remain in SEND_I to give the other unit more time to complete training (even though its own training was completed). If the rem_rcvr_status changes into OK, data transmission SEND_N will begin.

Data transmission SEND_N will continue for as long as there is data to transmit (TX_EN = TRUE) and the loc_rcvr_status and rem_rcvr_status are OK. If at any time the loc_rcvr_status changes from OK to not OK, e.g., because of a sudden burst of noise, the system will go back into SEND_Z once the unit has finished transmitting the last packet and TX_EN is set to FALSE. If the rem_rcvr_status changes to not OK,

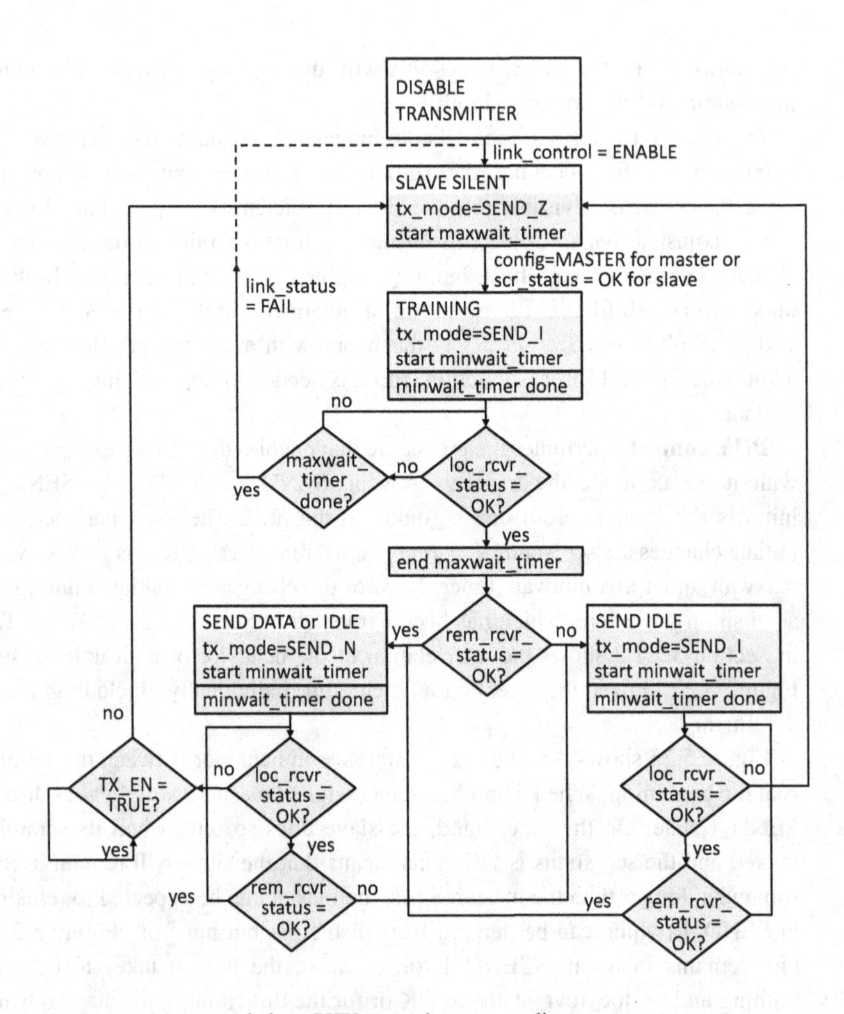

Figure 5.25 Master and slave PHY control sequence diagrams

the transmission goes from SEND_N to SEND_I in order to allow the rem_rcvr_status to recover. Whenever the unit changes into a new mode, the units run the minwait_timer, which determines the minimum time each unit has to stay in each new mode (1.8μs ± 10% for 100BASE-T1). This is added to ensure that the different states do not change too fast, as this could destabilize the whole system.

The **link monitor** function's purpose is to support the PHY control by determining the status of the underlying channel and by communicating the status with the link_status parameter. The state link_status = FAIL = link down can have different reasons. For one, link_status goes into fail if PMA is reset or if the link_control is not enabled.[15] Furthermore, as has been explained with the PHY control, the link_status goes into FAIL if the receiver has not been able to synchronize before the maxwait_timer expires. Should the link achieve synchronization and the loc_rcvr_status is OK during link_status = FAIL, the link monitor starts a stabilize

Table 5.7 Comparison of PHY parameters between 100BASE-TX, 1000BASE-T, 100BASE-T1, and 1000BASE-T1, based on best design practices, see e.g. [43]

	100BASE-TX	1000BASE-T	100BASE-T1	1000BASE-T1
Channel length	100 m	100 m	15 m	15 m
PHY transmission	Dual simplex	Full-duplex	Full-duplex	Full-duplex
X-level signaling	MLT-3	4D-PAM5	2D-PAM3	PAM3
	125 MBaud	125 MBaud	66.67 MBaud	750 MBaud
No. of twisted pairs	2 (Cat 5)	4 (Cat 5e)	1	1
Required Nyquist bandwidth	62.5 MHz	62.5 MHz	33.33 MHz	375 MHz
Error correction	n/a	Trellis Coded Modulation	n/a	Reed Solomon Coding
A/D conversion	5.5 bits ideal @ 125 MBaud	7 bits ideal @ 125 MBaud	7 bits ideal @ 66.67 MBaud	Up to 8 bits ideal @ 750 MBaud
DFE	16–24 taps	24 taps/channel	24 taps	Up to 128 taps
FFE	8 taps	12 taps/channel	8 taps	Up to 48 taps
NEXT cancellers	none	3×25 taps/channel	none	none
Echo canceller	none	160 taps	48 taps	150 taps
Critical path	• 3 input add • 3 input select • 1 slicer	• 3 input add • 5 input select • Branch metric compute • 4 input add-compare select	• 3 input add • 3 input select • 1 slicer • 3 input add-compare select	• 3 input add • 3 input select • 1 slicer • 3 input add-compare select
Normalized gate complexity	1	8	2	8
Additional features	Manchester coding provides spectral shaping	Partial response transmit filter	Transmit spectral shaping	Transmit spectral shaping

timer. If the loc_rcvr_status is still OK when the time has expired, the link_status parameter will be set to OK and remain in this state. It will only leave this state when the loc_rcvr_status changes into not OK and the link is not in training mode. In order for the link_status not to interfere with the training mode, a link_status OK can change into link_status not OK only when the maxwait_timer is not running (it does run during training).

Fast start-up is an important requirement for in-vehicle networking technologies. The 100BASE-T1 standardization project thus included the objective to achieve a valid transmission and receiving state from power on within less than 100 ms. The 100BASE-T1 specification thus explicitly specifies in the link monitor section that the time from power_on = TRUE to link_status = OK shall be smaller than 100 ms.

Table 5.7 compares different parameters for the different PHY technologies 100BASE-TX, 1000BASE-T, 100BASE-T1, and 1000BASE-T1. Some of the parameters presented depend on the specification; others on the implementation. They represent best practice values.

5.2.2 100 Mbps over 100BASE-TX

In 2012, Marvell and Micrel presented an alternative solution for using 100 Mbps Ethernet in the automotive industry [48, 49], which is sometimes called "QUIET-WIRE"; a registered trademark by Micrel [50] (now Microchip). Micrel had been an early development partner of BMW for the diagnostic Ethernet interface in 2008 (see also Section 3.1.3.4), had AEC-Q100 qualified some of their devices originally intended for industrial use, and thus had some experience with automotive requirements.

The proposed solution uses IEEE 802.3 compatible hardware, based on 100BASE-TX (see Figure 5.26 for the principle set up). The transmission is consequently dual simplex, i.e., using one wire pair for transmission and one wire pair for reception. Apparently, the original 100BASE-TX signal is passed through a different, better-adapted filter and transmitted via a better-balanced link [49, 51]. With this, the output signal ceases to be strictly 100BASE-TX compliant, but in return shows promising EMC results.[16] The original implementations shown used planar transformers which, together with two additional capacitors, also functions as a filter [49].

Pannell [49] shows example test results achieved with the 150 Ohm method; results from a DPI test have also been presented. As is explained in more detail in Section 4.1.3, the 150 Ohm method and the DPI test are a good starting point to assess the suitability of a semiconductor for automotive use. If it had been developed a few years earlier, the proposed solution might have had true automotive potential. As 100BASE-TX uses significantly shorter scrambling than is used for 1000BASE-T and 100BASE-T1, the solution is likely to show a different crosstalk behavior. Respective investigations have not come to the attention of the authors, though we expect that sufficiently good results can be achieved.

An interesting use case for the proposed solutions is the diagnostic interface as standardized in ISO 13400 (see also Section 3.1.3.3), which requires the use of 100BASE-TX. If this interface can have better EMC performance owing to the methods proposed with this solution, it might be possible to omit the disabling of the diagnostic interface during the runtime of the car that is necessary today.

5.2.3 100 Mbps Ethernet over Media Independent Interface (MII)

The xMII is an important element of Ethernet-based communication. It allows for the flexibility and scalability that makes Ethernet so attractive for the automotive industry.

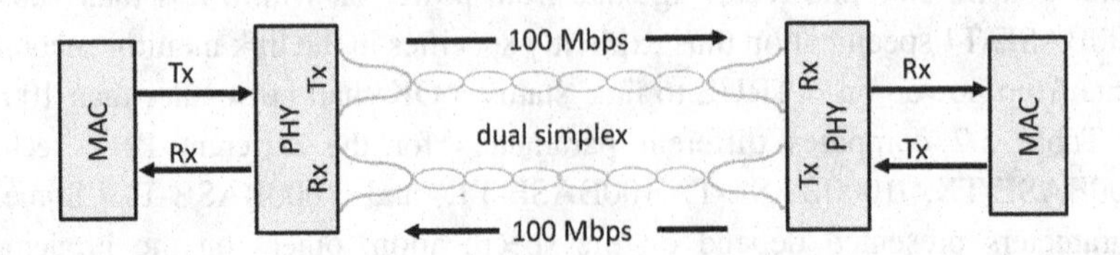

Figure 5.26 Key elements of the 100 Mbps over 100BASE-TX alternative

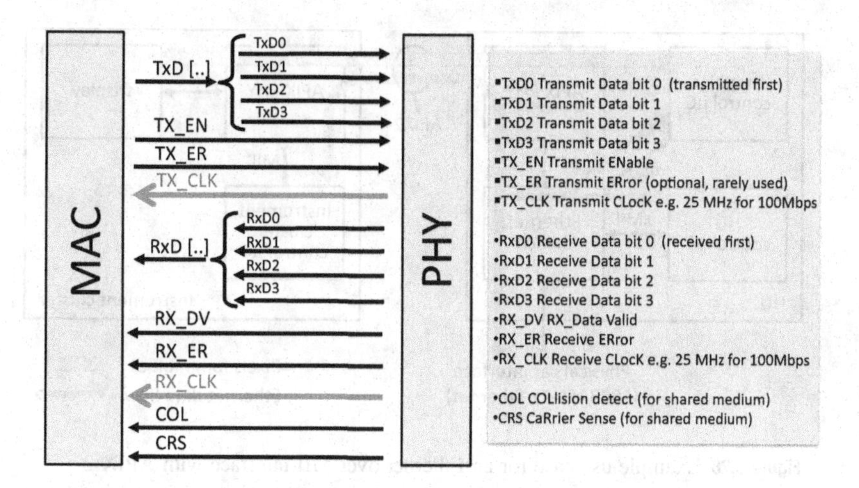

Figure 5.27 Definition of the MII interface

Because of the xMII interface, different PHY technologies, even ones that use different media or speed grades, can be attached to the same switch and can be part of the same Ethernet communication.

During the development of Ethernet for the diagnostic interface and the first investigations into the use of Ethernet in the automotive industry with unshielded cabling, an unexpected property was discovered at the MII interface: In 10 Mbps and 100 Mbps Ethernet systems, the PHY transceiver clock determines the clock for the communication with the MAC. It would have been more intuitive if the side of the communication that sends the data determines the clock, i.e., the PHY transceiver when passing on received data to the MAC and the MAC when passing on data for transmission to the PHY. In the case of a GMII, it is indeed organized this way (see also Section 5.3.1). When using an MII interface with 100 Mbps Ethernet, however, it is the PHY that determines the clock in both directions. The elements of the MII interface and the direction of the clocks are depicted in Figure 5.27. The MII transfers four bit words in parallel in each direction (see Figure 5.27), meaning that to achieve 100 Mbps the clock speed is 25 MHz. The MII was standardized with IEEE 802.3u and approved in 1995 [52].

This unexpected behavior of the MII clock can be used to enhance other transmission technologies with an Ethernet channel. As the PHY determines the clock, theoretically all clock rates are possible and it is not necessary to have synchronized clocks. This means that the 100 Mbps MII is suitable to connect communication systems with completely different clock rates, provided the amount of transmitted data can be handled by the communication system.

One example of an automotive communication technology exploiting this is the otherwise proprietary SerDes Technology Automotive PIXel link (APIX, see also Section 2.2.6). With the generation of APIX 2, the technology received an MII interface and enabled an additional bi-directional Ethernet channel [53]. There are

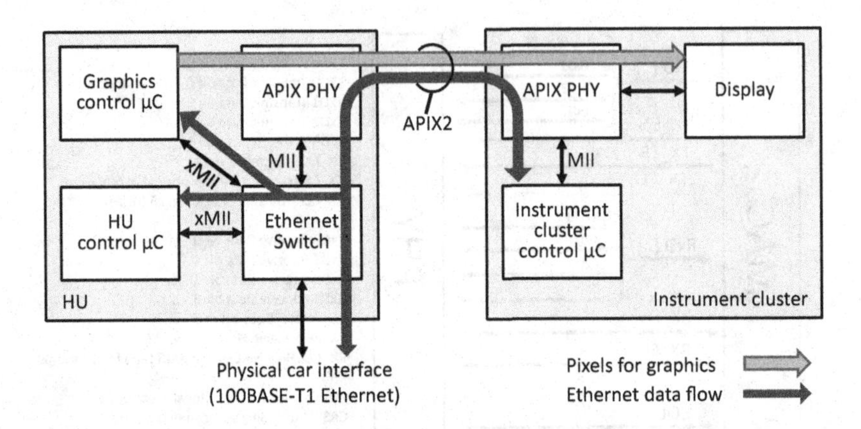

Figure 5.28 Example use case for an Ethernet over MII interface with APIX 2

various technical solutions possible to achieve this; the one used for APIX is sideband modulation. The clock rate at the MII interface can be adjusted by the low level settings of the APIX system and can be set to as high as 25 MHz. The advantages are obvious. It is possible to have a unidirectional single hop, Point-to-Point (P2P) connection for, e.g., video data at a data rate of several Gbps, while at the same time the control data of bidirectional Ethernet can be seamlessly integrated into the vehicle's communication network via the MII interface.

Figure 5.28 shows an example of a Head Unit (HU) that is connected to the instrument cluster. The task is to transmit graphics data at the same time and over the same physical connection as is used to connect the instrument cluster via the HU to the in-vehicle Ethernet network / vehicle backbone network. The latter is achieved via the included Ethernet channel that connects to the network via an Ethernet switch (in this case in the HU) just as any other Ethernet connection does. The micro controller in the instrument cluster runs a standard TCP/IP software stack (see also Section 7.3). Data to be transmitted over the additional Ethernet link might be the engine speed, the velocity of the vehicle, lists of data from the HU, etc. For the communication partners in the Ethernet network, the physical transmission technology is transparent. This is thus a good example for the "Ether" idea of Ethernet (see Section 1.2).

5.3 PHY Technologies for 1 Gbps

It was clear from the beginning that bandwidth requirements would keep increasing. At the time of initiation of the respective standardization project(s), autonomous driving was one of the driving use cases to require more data exchange in the car. More sensor data would need to be made available in more ECUs, redundancy would need to be provided, and computing "brains" would need to exchange data quickly. Furthermore, with each subsequent generation of mobile communication networks, the data pipes in and out of the car get bigger. Passengers employ this for their infotainment, but also the car manufacturer can profit from this with respect to remote software updates and the like. The in-vehicle network must support this with, ideally, a flexible layout that is not limited by bandwidth. Extending the Automotive Ethernet family with a 1 Gbps Ethernet PHY technology thus was inevitable. Section 5.3.1 discusses the PHY properties of the 1000BASE-T1 solution and Section 5.3.2 gives a brief overview of a technology for transmitting 1 Gbps Ethernet packets over Plastic Optical Fiber (POF) 1000BASE-RH.

5.3.1 Technical Description of 1000BASE-T1

In March 2012, the IEEE 802.3 accepted a Call For Interest (CFI) on Reduced Twisted Pair Gigabit Ethernet [54]. The automotive market, with its increasing demands on networking and bandwidth, was identified as the driving market for the technology. Owing to the standardized xMII interfaces, Gbps was the next speed grade to adopt. Because of cost and weight requirements, twisted pair was the targeted cable type.

During the SG phase, the automotive requirements were discussed in more detail. The whole range of topics – including temperature requirements, EMC, quiescence current, diagnostic capabilities, PHY latency, crystal accuracy, life time, wake-up time, and channel requirements – were presented in order to create a better understanding of the automotive requirements that were new in IEEE (see e.g. [55–57]). This also helped in creating suitable objectives. In November 2013, IEEE 802.3 approved the objectives and the SG to become a Task Force (TF) [58].

The final technology [10] differs significantly from 100BASE-T1 (and 1000BASE-T); not only in the way the signals are handled exactly (especially in the PCS), but also because the specification additionally includes (optional) auto-negotiation, (optional) Energy Efficient Ethernet (EEE), and an Operation, Administration, and Management (OAM) channel. Figure 5.29 gives an overview of the different building blocks. Instead of a separate block for PCS transmit enable as with 100BASE-T1, the 1000BASE-T1 PCS includes a separate block for OAM on the PCS side. The 1000BASE-T1 PMA holds the PHY control, link monitor, PMA transmit and receive, clock recovery, and an additional link synchronization block. For the signals exchanged between the PMA and PCS, 1000BASE-T1 includes extra information on the Low Power Idle (LPI) status needed when the optional EEE has been implemented. Also, 1000BASE-T1 not only uses the loc/rem_rcvr_status values but also loc/rem_phy_ready.

In this section first the elements of the PCS and then those of the PMA are explained, adding some information on auto-negotiation and EEE. It starts with the PCS transmitter elements, as shown in Figure 5.30.

1. **80B81B encoding:** The 80B81B function encodes the information coming from the GMII, i.e., the TxD, TX_EN, and TX_ER into groups of 81 bits (referenced "tx_co1" in Figure 5.30). This encoding depends on the tx_mode and whether the group contains data only or also control information. For normal data transmission in tx_mode = SEND_N and TX_EN = 1 and TX_ER = 0, 10 octets of TxD[7:0],

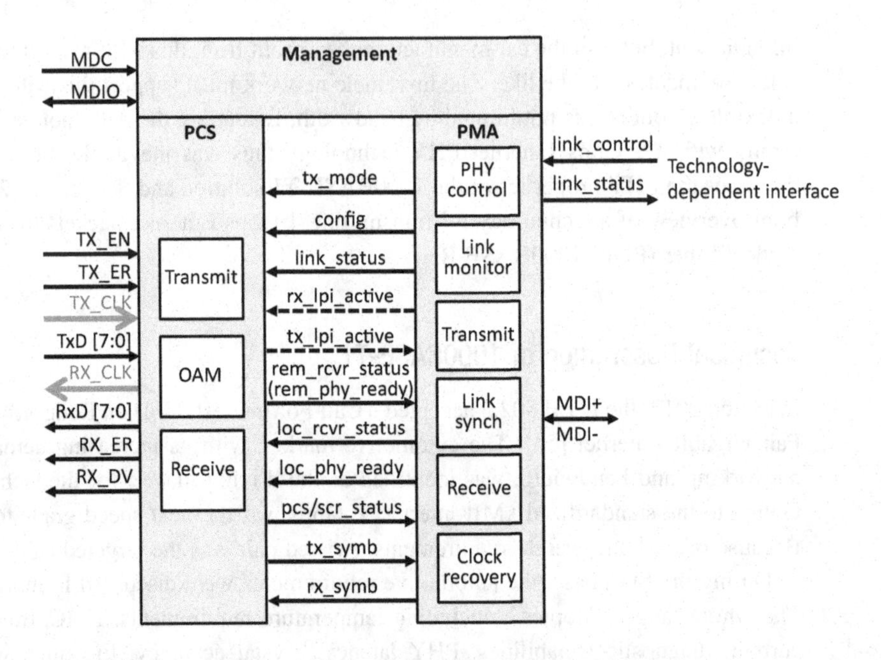

Figure 5.29 1000BASE-T1 building blocks (optional signaling for EEE in dashed lines)

i.e., 80 bits, are simply grouped together and prefixed by "0." If tx_mode = SEND_I, the 1000BASE-T1 specification is not specific. It is expected that the 80B81B function groups the same data it would use when the system is idle during SEND_N (i.e., TX_EN = 0 and TxD[7:0] = 0).

When the 81-bit group needs to contain control information (e.g., when TX_EN = 0 → normal idle, when TX_EN = 1 and TX_ER = 1 → error, when the loc_rcvr_status is not OK, or during LPI in the case of EEE) the prefix bit is "1." The first four bits following indicate where to locate the first 3-bit control code in the group "tx_co1," while the fifth bit indicates whether that code is the final control code in the block ("0") or whether more control information follows ("1"). Depending on the location indicated, the next bits are either the 3-bit control code, or data octets TxD[7:0] up to the control code, which may then be followed by further bytes of data.

2. **Aggregate and insert OAM:** Fort-five of the bit groups "tx_co1" are aggregated into one larger block of data (labeled "tx_co2" in Figure 5.30). Additionally, nine bits of the 1000BASE-T1 OAM are added, making each block consist of $45 \times 81 + 9 = 3654$ bits. The OAM bits may be used for exchanging messages intended for management, e.g., for monitoring the link health or supporting partial networking (see Section 6.3.3). Its deployment is optional; unless EEE is supported, in which case the OAM has to be used – at a slower pace – to monitor the link health. If OAM is not used in this PHY, nine "0" bits are added to each "tx_co2" block instead. If the link partner does not support OAM, the nine static bits are transmitted.

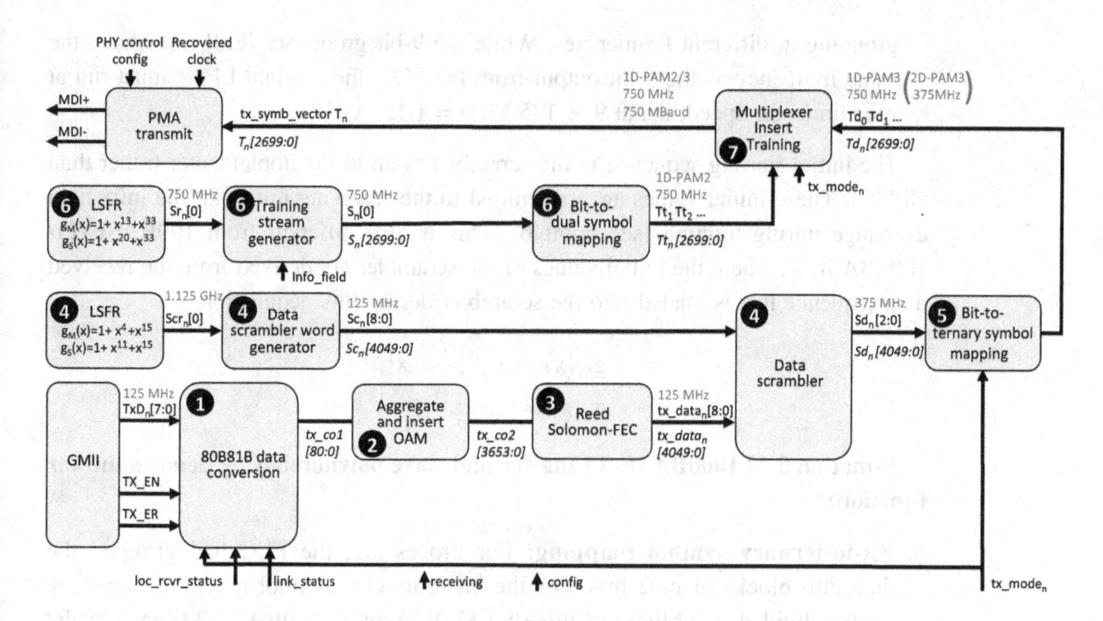

Figure 5.30 PCS transmitter elements in 1000BASE-T1 (without provision for link synchronization, auto-negotiation or EEE). The output of the block is depicted with different data groupings [x:0]. The text above the arrows indicates the frequency given, while the text below the arrows, in italics, describes the block of data that belongs together

An OAM frame consists overall of 12 bytes plus 12 parity bits, i.e., 12 × (8 + 1 = 9) bits. One 9-bit group (one information byte, one parity bit) is included in every block of data "tx_co2," meaning that 12 "tx_co2" blocks are needed to transmit one OAM frame. The first two bytes of every OAM frame have predefined uses (see [10] for details). The following eight bytes can be defined by the implementer, while the last two bytes contain a 16 bit CRC.

3. **Reed-Solomon FEC:** 1000BASE-T1 works with a lower SNR margin than 100BASE-T1. In order to maintain a target BER < 10^{-10}, 1000BASE-T1 requires an FEC (that 100BASE-T1 does not have). The FEC defined uses a shortened Reed-Solomon code that operates on 9-bit symbols. The code used is a (450, 406) code, meaning that it encodes 3654/9 = 406 information symbols and appends 450 – 406 = 44 parity symbols of nine bits each at the end of each block. It thus adds 44 × 9 = 396 bits to a block and the output block length of tx_data increases to 3654 + 396 = 4050 bits. The 44 parity symbols allow a correction of up to 22 symbol errors.
4. **Scrambling:** In order to improve the EMC performance, the block of tx_data [4049:0] is scrambled, i.e., bit-wise XORed with a pseudo random bit sequence. This sequence is generated with an LFSR using the master and slave polynomials as defined in Equation 5.12. This sequence does not undergo any further processing but directly feeds into the scrambler. The extra data scrambler word generator block in Figure 5.30 has only been introduced in order to point out the different bit

grouping at different frequencies. While the 9-bit group $Scr_n[8:0]$ must have the same frequency as the 9-bit output from the FEC, the original LFSR must run at nine times the speed, i.e. at 9 × 125 MHz = 1.125 GHz.

The initial starting sequence in the scrambler is up to the implementer (other than all "0"). These initial values are transmitted to the receiving unit with the info_field exchange during training (see point 6). This is also different from 100BASE-T1/1000BASE-T, where the initial values of the scrambler are derived from the received data sequence that is loaded into the scrambler during link acquisition.

$$g_M(x) = 1 \oplus x^4 \oplus x^{15}$$
$$g_S(x) = 1 \oplus x^{11} \oplus x^{15}$$

Equation 5.12 1000BASE-T1 master and slave polynomials ($\oplus$ denotes the xor function)

5. **Bit-to-ternary symbol mapping:** For processing, the FEC had grouped the data into blocks of nine bits. For the bit-to-ternary symbol mapping, these are further divided into blocks of three $Sd_n[2:0]$. As with 100BASE-T1, each triplet of bits is mapped onto two PAM3 symbols. Unlike 100BASE-T1, 1000BASE-T1 uses a Gray code for this (see Table 5.3 in Section 5.2.1.3). This Gray code mapping is used when the tx_mode is in SEND_N or SEND_I. The standard labels the two output PAM3 symbols T[0] and T[1], which defines their order and eliminates the necessity to detect their correct order in the receiver, as is needed for 100BASE-T1.

   Should the tx_mode be in SEND_Z, zero voltage will be put on the line, which corresponds to {T[0], T[1]} = {0,0}. Should the connection be in training, for which an extra tx_mode = SEND_T is introduced, an entirely different path and set of data is used, as explained in the next numbered item. The 4050 bits that are part of one block of data after FEC and scrambling make 2700 PAM3 symbols after the 3B2T conversion.
6. **Training mode:** The training mode of 1000BASE-T1 supports the receiving link partner in adjusting and in aligning to blocks of transmit data. Its processing during training is entirely different from the data and idle stream processing. The blocks for the symbols sent during training have the same length, i.e., they consist of 2700 symbols each. However, the symbols during training are PAM2 modulated, meaning that the bits are mapped directly onto the voltage levels. If the input bit $S_n[0] = 0$ then $Tt_n[0] = +1$. If the input bit $S_n[0] = 1$ then $Tt_n[0] = -1$. The overall bit sequence that comprises one block of training information thus consists of not only 2700 symbols but also 2700 bits.

   To generate this training data, a pseudo random bit sequence $Sr_n[0]$ is generated by an LFSR, which is then used to scramble the information that goes into the training stream $S_n[0]$. The polynomials for generating $Sr_n[0]$ are defined in Equation 5.13 (they are the same as the polynomials used for 100BASE-T1 and 1000BASE-T).

$$g_M(x) = 1 \oplus x^{13} \oplus x^{33}$$
$$g_S(x) = 1 \oplus x^{20} \oplus x^{33}$$

Equation 5.13 Master and slave polynomials for 1000BASE-T1 training (same as 100BASE-T1 and 1000BASE-T polynomials, $\oplus$ denotes the XOR function)

For the sequence generated during training $S_n[0]/S_n[2699:0]$, each 2700-bit block of training data is split into 15 partial frames of 2700/15 = 180 bits each. The first bit of each partial frame except the last is inverted. The first 96 bits of the 15th partial frame is XORed with the content of the info_field. Equation 5.14 describes the generation of the training stream $S_n[0]/S_n[2699:0]$.

$$S_n = \begin{cases} Scr_n \oplus info\_field_{(n \bmod 180)} & if\ 2520 \leq n \bmod 2700 \leq 2615 \\ Scr_n \oplus 1 & elseif\ (n \bmod 180) = 0 \\ Scr_n & else \end{cases}$$

Equation 5.14 Generation of training stream $S_n[0]$ before PAM2 modulation ($\oplus$ denotes the XOR function)

The info_field consists of 12 × 8 = 96 bits. The first three bytes always have the same predefined values that serve as a start-of-frame delimiter. The next three bytes contain the Partial PHY Frame Count which indicates the running number of the partial frame. The slave synchronizes its partial frame count to the master's. The one byte message that follows includes information on the PMA state (two bits that are 00 for training or 01 for countdown, see Figure 5.25, and used in the training symbol path only), the loc_rcvr_status, and whether the master is ready for the slave to transmit ("en_slave_tx") or whether the slave completed the timing lock. The next three bytes depend on the state during training. In the case of PMA = 00 (training state) the bytes include information on OAM and EEE capability and on the scrambler starting sequence for the data path. In the case of PMA = 01 (countdown state), the three bytes include the partial frame count that indicates when the system changes from PAM2 to PAM3 modulation. In both cases, the last two bytes of the info_field contain a CRC. The correct reception of the info_field is decisive for the start-up in 1000BASE-T1. It is therefore transmitted at least 256 times, to allow detection by the link partner.

7. **Multiplexing:** Depending on the tx_mode the multiplexer will put the training (tx_mode=SEND_T) or the idle/data streams onto the channel (tx_mode_SEND_I/SEND_N).

Unlike 100BASE-T1, the 1000BASE-T1 specification foresees a polarity change in the receive path only, independent from whether this is the master or slave unit. The PCS transmitter therefore does not need to include a polarity change unit; only the PCS receiver does.

There is one additional aspect that the PCS transmitter must take care of. It comes from the fact that the MII of 100BASE-T1 handles the clock differently than does the GMII of 1000BASE-T1. In the case of MII, both the receive and transmit clocks are determined in the PHY and passed up through the MII (see also Figure 5.27). Thus, there is no risk of misalignment in the PHY. In the case of GMII, this is different. Only the receive clock is determined by the PHY and passed up through the GMII. On the transmit side, the GMII passes a clock GTX_CLK down to the PHY. The PCS transmitter has to provide for a potential misalignment between GTX_CLK and the PHY internal (receive) clock. How this is done is left up to the implementer.

In principle, the PCS receiver just performs the tasks of the PCS transmitter in reverse order. Once the training has been completed and the data blocks are correctly aligned, the receiver demodulates the PAM3 symbols into bits, descrambles the data, decodes it with potential error correction, and extracts the OAM, potential control data (esp. RX_DV and RV_ER), and the data bytes RxD[7:0], which are then passed to the GMII. The 81B80B conversion removes the appended bit and marks a block as erroneous (RX_ER = 1), if a block is marked to hold control frames, either points to an invalid location or contains a control not defined, or if the FEC flagged a block as having errors that cannot be corrected.

However, in order to receive properly, the PCS receiver must first perform its share of the synchronization process. It must start with potentially correcting the polarity and locking its training descrambler. How this can be done is not defined and it is up to the implementer to determine how the receiver makes use of the incoming PAM2 training stream for this. Similar to what is being done in 100BASE-T1, the receive descrambler can load the received data into the LFSR registers until the output and input signals of the descrambler are the same. However, the 1000BASE-T1 implementation must also take the processing of the training data (as described in Equation 5.14) into consideration.

Once the descrambler has been locked, the PCS sets the scr_status to OK. With the descrambler locked, the PCS will also know when one 2700-bit data block starts, i.e., it will have achieved frame synchronization. It can then extract the information from the info_field essential to complete the process. In particular, it needs the starting input for the descrambler in the data path and the partial frame count for when the PHY control goes into countdown (see PHY control description later in this section).

On the transmit side, the **PMA transmitter** puts the PAM2/3 symbols onto the channel such that the electrical specifications such as the PSD mask limits can be met. This may be comprised of filtering, D/A, and subsequent analog filtering, as described for 100BASE-T1. Because 1000BASE-T1 uses a single pair UTP cable, a hybrid is needed to allow simultaneous reception and transmission on the same line and to be able to extract the PAM2/3 symbols from the incoming signals. On the **PMA receive** side for 1000BASE-T1, a loop-timing concept is foreseen for **clock recovery** (see Sections 5.2.1.2 or 5.2.1.3 for details). Like the 100BASE-T1 PMA, the 1000BASE-T1 PMA receive function includes equalization and cancellation of the echoes its own transmit signal imposes on the receive signal. Quite a few differences exist in the link acquisition process described by the PHY control (see also Figure 5.31). The **link monitor** of 1000BASE-T1 is not much different from the one in 100BASE-T1, except

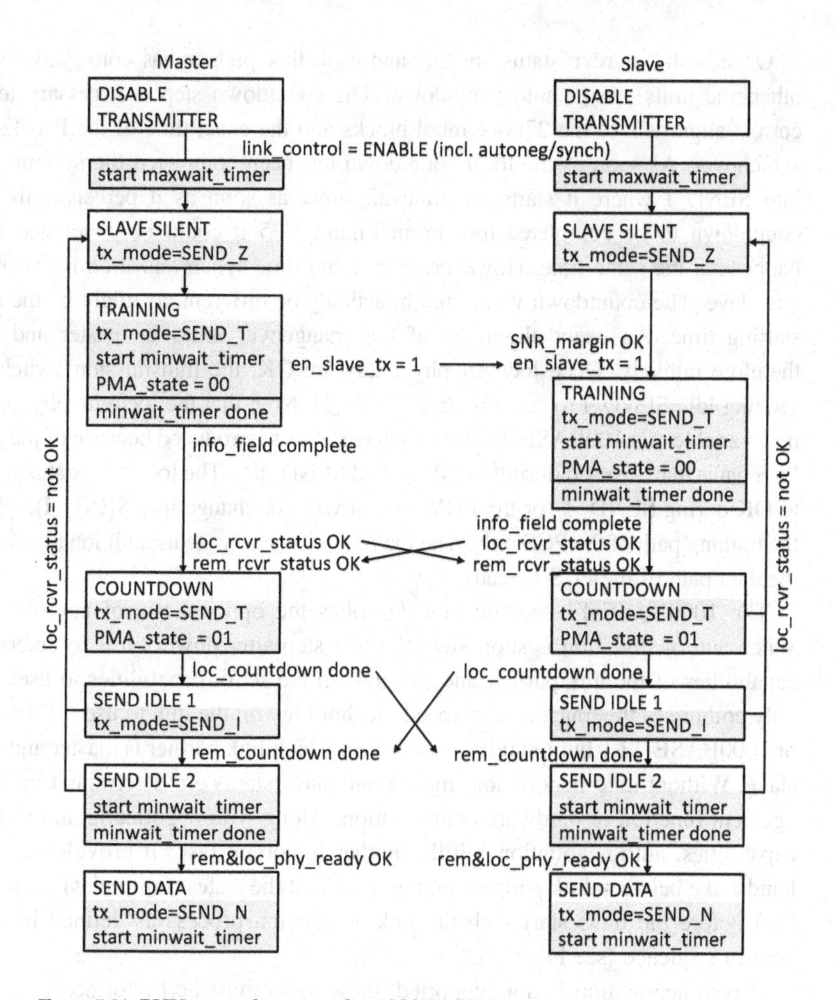

Figure 5.31 PHY control process for 1000BASE-T1

that the hysteresis used in 100BASE-T1 is replaced by the integration of the minwait and maxwait timers into the process.

Figure 5.31 depicts the flow charts for master and slave **PHY control**, including some correlation in timing they might have. The process is started by enabling link_control under the condition that the link synchronization (see more below) or, if available, the auto-negotiation has been completed. Both master and slave start in the slave silent SEND_Z mode. The master immediately changes into training mode SEND_T and starts sending the PAM2 training data. Now that it has started to transmit, the master adjusts to receive data, e.g., by adapting its echo canceller. Once the master is sufficiently prepared (which is up to the implementer to decide), it sends en_slave_tx = 1 to the slave within the info_field encapsulated in the training stream. The slave, which will have made use of the master's training stream to adapt its AGC, FFE, etc. to determine that it has sufficient SNR_margin, can then also go into SEND_T and starts sending its own PAM2 training data.

Once both loc_rcvr_status are OK and each link partner has conveyed this to the other, the units can go into countdown. The countdown step is necessary to ensure correct alignment to the 2700-symbol blocks and the exact time of the PAM2–PAM3 switchover. As soon as the local countdown has been completed the system switches into SEND_I where it starts the minwait_timer as soon as it perceives the remote countdown to be completed too. From Figure 5.25 it could be concluded that this happens at the same time. However, there is no time synchronization between master and slave. The countdown values might actually be different, as might be the absolute starting time. An exact alignment of the changeover times in master and slave is therefore unlikely. Once loc/rem_phy_ready are OK, the transmission switches from sending idle SEND_I to sending data SEND_N. Note that the loc/rem_phy_ready is a new variable that 100BASE-T1 does not have. It was introduced because of the different PCS paths that are used during SEND_T and SEND_I/N. The loc_rcvr_status will be set to OK during SEND_T (or the PHY control will not change into SEND_I), relying on the training path of the PCS only. The loc/rem_phy_ready status additionally ensure that the data path of the PCS is ready.

The 1000BASE-T1 specification describes the optional use of auto-negotiation. When auto-negotiation is supported, the first step after power on is to determine the capabilities of the link partner and to agree on the set of capabilities to use. This not only comprises the transmission speed / technology on the link to use – 100BASE-T1 or 1000BASE-T1 – but includes establishing which link partner is master and which is slave. Without auto-negotiation, master and slave status are determined by the management function or hardware configuration. Along with negotiating and aligning on capabilities, auto-negotiation fulfills another important task: It provides for the first handshake between link partners and ensures that the states (tx_modes) are synchronized before the units start with the link acquisition process as defined in the PHY control sequence (see Figure 5.25).

If auto-negotiation is not supported, these tasks must be performed in a different way. 1000BASE-T1 therefore specifies a separate **link synchronization** process. Link synchronization is used by the link partners to discover the other link partner and to synchronize the units into the same state SEND_S/SEND_Z; not only once at the beginning, but every time link acquisition needs to restart. This process provides for a proper handshake and prevents the units from ending up in a deadlock situation, where, e.g., because one unit lost the link and the other notices this with a time delay, the link partners try to reacquire the connection in different stages [59]. When auto-negotiation is available, the auto-negotiation process is restarted and thereby ensures a defined starting state. In SEND_S, link synchronization deploys a special pseudo random sequence derived by an LFSR using the master and slave polynomials as defined in Equation 5.15. These bits are PAM2 modulated before being forwarded to the PMA. This sequence is transmitted independently of any blocks of the transmit states SEND_T, SEND_I, and SEND_N, but repeats after every 255 bit symbols. The transmission in SEND_S is half-duplex. For the handshake, first the master transmits SEND_S, then the slave.

$$P_M(x) = x^8 \oplus x^4 \oplus x^3 \oplus x^2 \oplus 1$$
$$P_S(x) = x^8 \oplus x^6 \oplus x^5 \oplus x^4 \oplus 1$$

Equation 5.15 Master and slave polynomials of the link synchronization

1000BASE-T1 provides for the (optional) use of EEE (see also Section 6.3.1). The idea is to significantly reduce the power consumption of a PHY when it has a link established but no data to send. The LPI state can thus only be entered during normal data mode SEND_N (and if both units support it). Next to the normal data transmit state, EEE knows QUIET and REFRESH. Per specification, the LPI is initiated upon TX_EN = 0, TX_ER = 1, and the transmit data group TxD[7:0] = [00000001] = 0 × 01 at the GMII interface. The PHY then first sends a sleep frame, i.e., a whole block of 2700 symbols of low power idle symbols (TX_EN = 0, TX_ER = 1, TxD = 0 × 01). During QUIET, no voltage is put on the channel and the transmitter as well as the receiver can power down parts of their circuitry. In order not to lose synchronization, the PHY periodically sends refresh blocks, which are generated from the scrambler output in the data path that is PAM3 encoded. To wake the system up again, the PHY sends a full block of idles (i.e., zero data FEC encoded and scrambled) at the next possible wake window (see below) and thus end the LPI mode. The EEE process is similar to what has been defined for 10GBASE-T with the exception that 1000BASE-T1 uses a single-pair UTP cable and, hence, refresh uses the data path and that the data blocks in 1000BASE-T1 are longer than for 10GBASE-T [60].

1000BASE-T1 EEE also functions on the basis of 2700-symbol blocks. A period of quiet and refresh lasts 16 blocks = 16*2700*3/2 bits = 180 bits × 360 partial frames, with 354 partial frames QUIET and six partial frames REFRESH. This reduces the active transmit time to 6/360 = 1.66%. The refresh from the slave is sent at an offset of 360/2 + 15 = 195 partial frames from the master refresh. When sending wake blocks, this is possibly every second block boundary for the master and every alternating second block for the slave. The wake signal is an entire 2700-symbol block of idles. The wake up can be initiated by the GMII when data is available to send, or when the PHY perceives that it cannot maintain a good enough SNR during LPI. Once the system has ended the LPI, it can attempt to recover the link during normal power up mode.

5.3.2 Overview on 1000BASE-RH

The strict layering and availability of xMII interfaces is the basis for the scalability and attractiveness of (Automotive) Ethernet. Provided both ends of a link use the same technology and have the respective xMII interface, the technology as such is transparent to the network. Thus, many Ethernet standards have been developed for different media, like coaxial, twisted pair, twin-ax, backplane, and fiber (see also Section 1.1). One of the first 802.3 standards (IEEE 802.3d) was for a fiber-optic inter repeater link and standardized in 1987. 10BASE-F was standardized in 1993 [61].

Optical transmission, at least over Glass Optical Fiber (GOF), has the advantage of extremely low attenuation (IL) and delay and therefore supports higher data rates and/or longer distances [62]. However, GOF is not (yet) suited for automotive use. It is seen as too expensive and difficult to handle, while the long reach is not needed. The automotive industry has always preferred Plastic/Polymeric Optical Fiber (POF), which is mechanically less sensitive than GOF. For the IT/communications industry POF is not of interest, because the reach is significantly shorter than for GOF [63], while for automotive use the reach is sufficient. The key argument for using an optical Ethernet system in automotive is it is impervious to EMC in the harness [64, 65] (see also Section 2.2.4 for using POF with MOST). The price level falls between UTP and STP [66].

Costs and reliability are the main criteria when deciding between UTP and POF Ethernet. A few other considerations are listed here [66]:

1. The POF medium in automotive use traditionally supports temperatures from –40°C up to only +95°C. Some car manufacturers require up to +125°C. Higher temperature POF can be provided, however at a noticeably higher price level.
2. For optical systems it is not possible to transmit power over the data-line, as it is with electrical systems (see also Section 6.2.1). For optical systems, power always requires separate wiring.
3. Electrical PHYs have fewer restrictions when being integrated into switches or microcontrollers than optical PHYs. The reason for this is depicted in Figure 5.32.

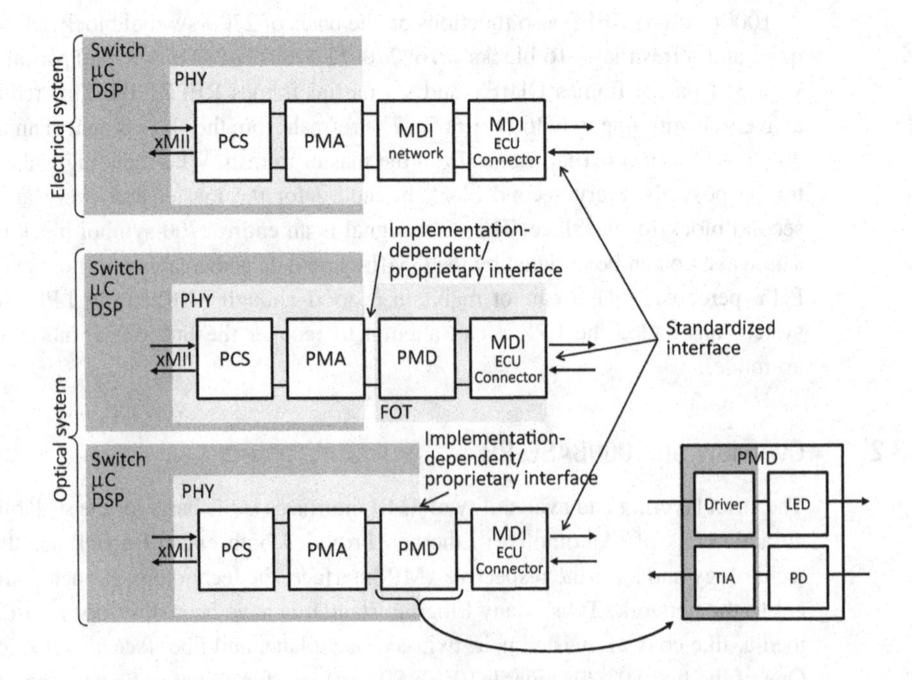

Figure 5.32 Transmitter elements and their possible integration in semiconductors in an optical system in comparison to an electrical system; TIA = TransImpedance Amplifier

The CMC and DC isolation capacitors that make up the MDI network between the output of an electrical PHY chip and standardized output of the MDI/connector are more or less transparent from an electrical point of view. This means that PHYs are developed toward the standardized output. Optical transmission, in contrast, requires a media conversion in the Physical Medium Dependent (PMD) element, whose Light Emitting Diode (LED) and PhotoDiode (PD) cannot be integrated in a CMOS semiconductor. Most commonly the PMD and the MDI are integrated in a Fiber Optical Transmitter (FOT). This means that the PHY is developed to a proprietary output and PHY and FOT vendors have to align their products during development, which limits the number of FOTs usable in each case. Thus, anyone integrating an optical PHY into a switch or microcontroller also restricts their parts to specific FOTs or related products.

4. Optical systems are more demanding in their mechanical handling than UTP systems. The bending radius is limited. The tensile force is strictly limited. The optical harness element needs to be pre-assembled, while it is essential it receives complete protection against contamination, cuts, or corrugations on the inner sheathing when being installed in the cars.
5. With regard to problems in the field, at least BMW engineers find the maintenance of optical solutions to be more difficult than for UTP but better than for STP cables.

In March 2014 a CFI was passed for 1 Gbps Ethernet transmission over POF [67], which mentions three application areas: Consumer networking, automotive networking, and professional networking. Under the assumption that Gigabit Ethernet over Plastic Optical Fiber (GEPOF) is more expensive than the electrical 1000BASE-T1 in the automotive industry, it is a potential alternative for special use cases where UT(S)P is not possible, either owing to link distance or to EMC criticality. In consumer/home networks POF is seen as an alternative when neither wireless technologies provide sufficient quality (limited reach owing to building material or size), nor Power Line Communication (PLC) or UTP cables are suitable (for EMC reasons). In professional networks POF is of interest in EMC sensitive areas or when galvanic isolation is needed.

Optical transmission systems started off using red LEDs – red LEDs were commonly available and lower priced than other colors[17] – and used a straightforward Non Return to Zero (NRZ) modulation. However, with this concept it is not possible to achieve a data rate of 1 Gbps over POF for two reasons: (1) The LED cannot emit light fast enough [68] and (2) The attenuation of POF is too large. For POF the attenuation varies between 10 and 1000 dB/km, depending on the material used [69]. Absorption and scattering are reasons for the loss, and they are caused by impurities in the fiber. With SI-POF, the maximum bandwidth with red LEDs and NRZ modulation is only about 75 MHz (End-to-End@–3dB) or 150 MHz with improved driver pre-emphasis [70].

1000BASE-RH therefore uses an intelligent multi-level coding; which includes PAM16 (for the header) and PAM8 (for the payload) modulation at 162.5 MBaud and Tomlinson Harashima precoding [71, 72] in order to mitigate error propagation in the

DFE [73]. As a consequence of the multilevel modulation, analog reception of the optical information is necessary. Like 100BASE-TX, this solution uses dual simplex transmission (in a "full-duplex" Ethernet network), i.e., one link for transmission and one for reception of data.

The resulting IEEE802.3bv/1000BASE-RH standard was published in March 2017 [73].[18]

5.4 PHY Technologies for 10 Mbps Ethernet

5.4.1 Background to 10BASE-T1S

It seems to be a natural instinct to always look for "more." However, in the discussions around the introduction of Automotive Ethernet with different car manufacturers, it became evident that for many use cases, a data rate of 100 Mbps actually provides too high a data rate. Well over 90% of the communication links in cars need far less than 10 Mbps (estimate for 2020 [74]). Using 100 Mbps Ethernet in such cases is not cost and energy efficient, especially as other, less complex in-vehicle networking technologies exist that support lower data rates such as FlexRay or CAN FD (see also Sections 2.2.2 and 2.2.5).[19] To really extend the use of Ethernet in the in-vehicle network, an Automotive Ethernet version is needed that is able to compete with those legacy technologies, while seamlessly integrating into the Ethernet network and, with that, reducing the number of technologies and gateways deployed [75].

A reduction of the data rate to 10 Mbps was an obvious starting point. It led directly to the first question: Does this reduction in data rate allow for developing a competitive 10 Mbps Ethernet technology (with competitive meaning less than 50% compared with 100 Mbps Ethernet)? First estimations confirmed that the effort in the PHY could be sufficiently reduced [76, 77].[20] This fostered the standardization of a respective PHY technology at IEEE 802.3, which in turn coincided with the industrial automation community also looking for a (one pair) 10 Mbps Ethernet solution. While their main concern was reach (1000 m), it was still fitting to have both efforts joined. A CFI to start a respective study group thus passed in July 2016 [78]. The respective IEEE 802.3cg task force took up its work in January 2017 [79].

In the adopted objectives [80], the otherwise common "full-duplex" requirement was intentionally omitted for the short reach link, thus not precluding the inclusion of mechanisms that support a bus topology, referred to as half-duplex within IEEE 802.3. As visualized in Figure 5.10, a bus topology allows for further cost reduction, because it reduces the number of PHYs, connectors, and (potentially) switches, and is thus of interest in a cost competitive environment. The resulting IEEE 802.3cg specification indeed includes an optional, so-called multidrop" mode for the short reach 10BASE-T1S ("S" for short) solution.[21] Multidrop is described in detail in Section 5.4.3, while the 15 m/25 m 10BASE-T1S PHY (simplifications) itself is described in Section 5.4.2. For the 1000 m reach a second PHY was developed, called 10BASE-T1L ("L" for long). It was just not possible to combine the requirements of cost efficiency and the need for better properties for the long reach PHY in one solution. As 10BASE-T1L is overkill for automotive use – especially in the light that a suitable solution exists with 10BASE-T1S – this is not described further in this section.

5.4.2 Technical Description of the 10BASE-T1S PHY

The main objective for the 10BASE-T1S PHY was to create a cost competitive Automotive Ethernet alternative to legacy busses. For 10BASE-T1S, there are two elements to achieve this cost competitiveness, a simplified PHY and the possibility of using a bus/multidrop topology. While the multidrop function itself is described in more detail in Section 5.4.3, the choice to include it in the specification nevertheless impacted the mandatory 10BASE-T1S P2P PHY mode. As multidrop only functions when one node transmits at the time, i.e., in half-duplex mode, the mandatory 10BASE-T1S P2P PHY is half-duplex. One consequence of this choice is that the available 10 Mbps data rate is shared among the units sharing the link segment – between two in the case of P2P, more in the case of multidrop – instead of being available for all nodes, as is the case for switched full-duplex.

This was a matter of much debate, as almost all recent IEEE 802.3 Ethernet PHY projects define full-duplex. As a consequence, an optional full-duplex mode was also defined, which allows for the simultaneous transmission of 10 Mbps data in both directions. Naturally, the two options – full-duplex and multidrop – are mutually exclusive.[22] The following description shows that a lot of simplifications in the PHY can be made because of half-duplex. This means that full-duplex leads to a more complex design and more costs for the 10BASE-T1S full-duplex mode. It is thus less of a focus for automotive use; at least to start with, and the following description focuses on the half-duplex mode.

Figure 5.33 shows the main building blocks for the 10BASE-T1S PHY. There is no PHY control, clock recovery is simplified, the transmit enable is replaced with a collision detect (see Figure 5.18 for a comparison with 100BASE-T1), and the link control is also not really needed in the envisioned use case.

When in half-duplex mode, the 10BASE-T1S link is electrically silent when there is no transmission. This eliminates the need for a link monitor that is required with the "active idle" used in full-duplex (10BASE-T1S or other speed grades). Additionally, the use of auto-negotiation – which would require the link monitor – is not enabled for multidrop. The link monitor is primarily needed for the full-duplex mode and is thus not here. Another effect of the potentially silent link in the half-duplex mode is that Energy Efficient Ethernet (EEE) is not needed (see Section 6.3.1 for details on EEE).

In 100BASE-T1, the PHY control is responsible for coordinating the training phase (depicted in Table 5.1 for 100BASE-T1). Because of the selection of Differential Manchester Encoding (DME) and the lower bandwidth of 10BASE-T1S, a training phase, and, with that, PHY control, is also not required. This also impacts the time needed for start-up, which is immediate. The impact on clock recovery and collision detect are discussed further later in this section.

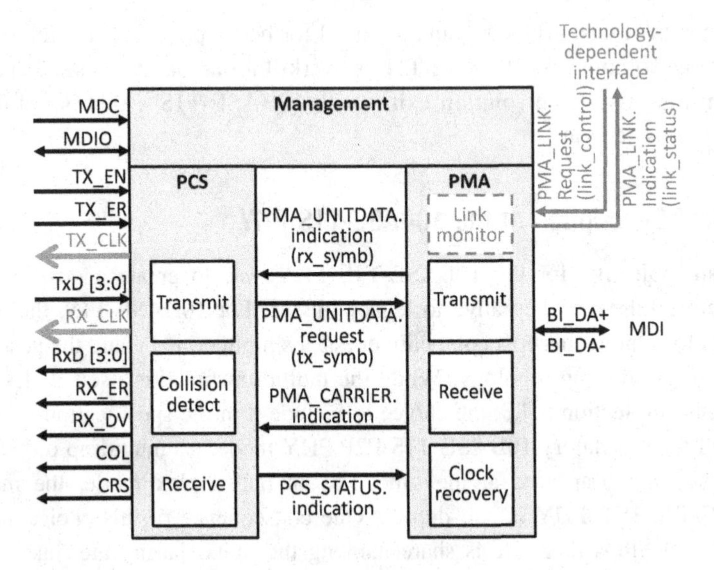

Figure 5.33 10BASE-T1S building blocks

Figure 5.34 compares the PMA functions needed in 10BASE-T1S with those used for 100BASE-T1 in an exemplary receiver design. When in half-duplex mode, there are no simultaneous transmit and receive signals. A hybrid, that separates transmit and receive signals immediately, is therefore not needed (in full-duplex mode it is). Because of the low speed and DME, echoes from the own transmission can be rejected without adaptive filters and echo cancelling, even in full-duplex mode. DFE or FFE are thus not needed either. Because of DME, there is no need for baseline wander correction, for complex A/C, nor for precise clock recovery. DME is intrinsically balanced and self-clocked, see also Figure 5.37 and the description of DME further later in this section.

Naturally, some sort of clock recovery is still needed, as the nodes have independent clocks that can drift with ±100 ppm precision. But because there is no echo to cancel, the clocks do not need to be phase locked, plus the DME is more robust and allows for more jitter.[23] The PLL can thus be replaced with something simpler and less costly, like a digital clock multiplier or similar. Also, the DME decoder itself performs a simplified clock recovery. The A/D conversion is still needed, but can be of lower complexity. Instead of, e.g., seven or eight bits needed for the A/D in 100 and 1000BASE-T1, two comparators of one bit each are sufficient (see also Table 5.9).

Figure 5.35 shows a block diagram of the PCS transmitter and receiver, as well as the MII and the DME encoding and decoding part of a 10BASE-T1S PHY. This figure presents the multidrop mechanism. It is shown as number 6 "Physical Layer Carrier

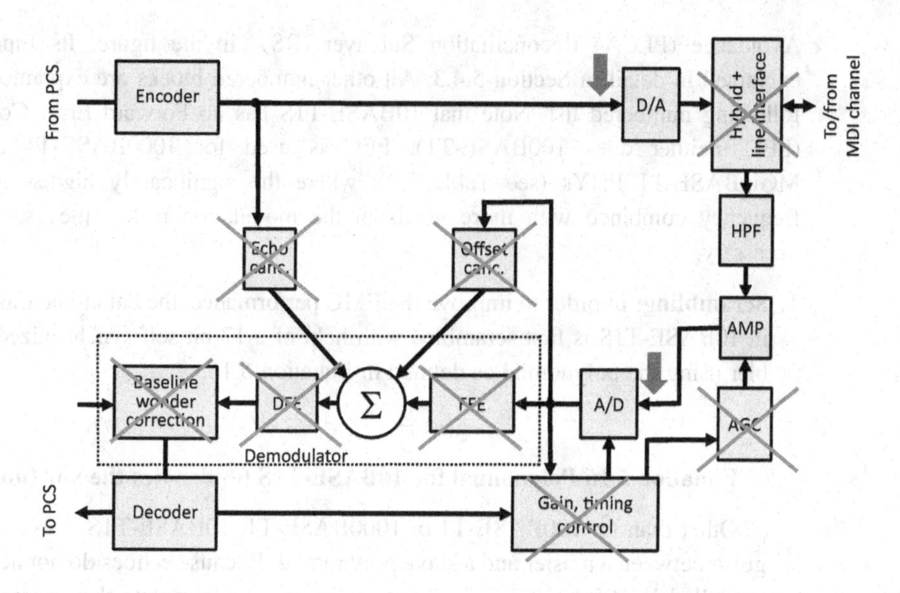

Figure 5.34 10BASE-T1S PMA blocks in comparison with 100BASE-T1; all blocks not crossed out are needed but generally less complex

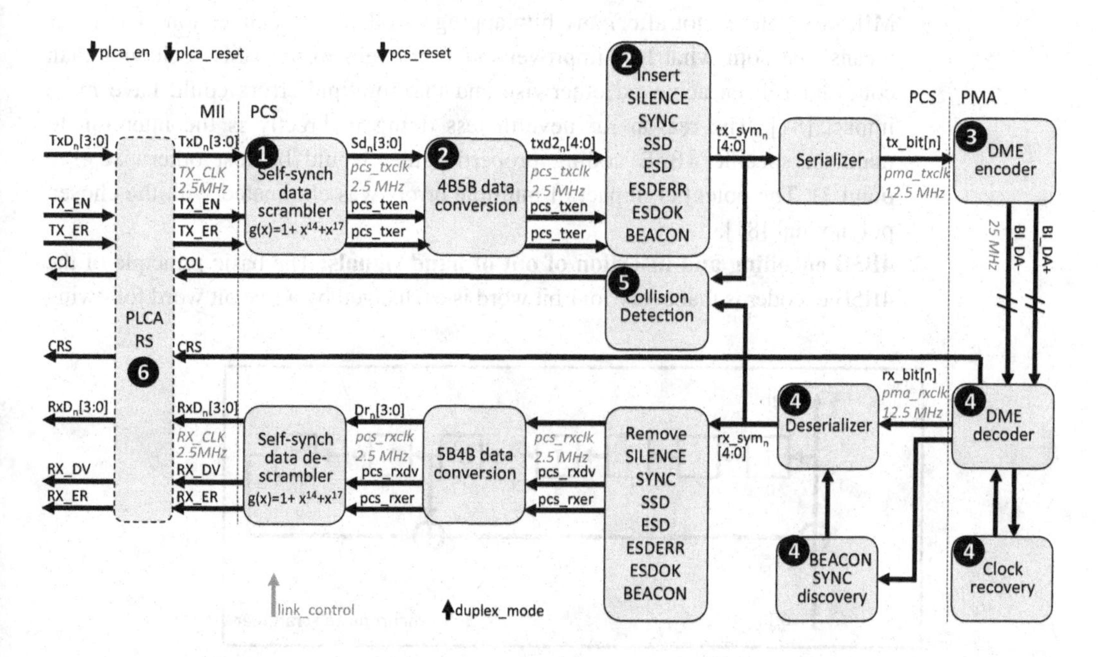

Figure 5.35 10BASE-T1S PCS transmitter and receiver elements in 10BASE-T1S plus MII and PMA part

Avoidance (PLCA) Reconciliation Sublayer (RS)" in the figure. Its function is explained in detail in Section 5.4.3. All other numbered blocks are explained in the following numbered list. Note that 10BASE-T1S has no Forward Error Correction (FEC, neither does 100BASE-T1). FEC is used for 1000BASE-T1 and the MGigBASE-T1 PHYs (see Table 5.9), where the significantly higher operating frequency combined with more levels in the modulation makes the use of FEC necessary.

1. **Scrambling:** In order to improve the EMC performance, the data to be transmitted in 10BASE-T1S is first scrambled with help of a 17-bit self-synchronized scrambler using the polynomial as defined in Equation 5.16.

$$g(x) = 1 \oplus x^{14} \oplus x^{17}$$

Equation 5.16 Polynomial for 10BASE-T1S (⊕ denotes the xor function)

Other than for 100BASE-T1 or 1000BASE-T1, 10BASE-T1S does not distinguish between a master and a slave polynomial. Because echoes do not need to be cancelled in 10BASE-T1S, it is not necessary to decorrelate the transmit directions, which means that the same polynomial can be used in the receive and transmit paths. Furthermore, the scrambling for 10BASE-T1S is done by a self-synchronized scrambler (see Figure 5.36); i.e., transmitter and receiver seeds do not need to be synchronized. 100BASE-T1 and 1000BASE-T1 generate separate sequences that are XORed into the data stream, which is not necessary here. Last but not least, scrambling is done immediately after receiving the data from the MII, i.e., before (not after) any bitmapping or other data conversion. The latter means that somewhat less improvement is brought to the EMC behavior than could have been achieved otherwise and that multiple errors could have more impact [81]. The reason for nevertheless doing it directly is the intention to maintain specific 4B5B coding properties that would be lost otherwise (see point 2). The potential impact of multiple errors was eliminated with the chosen polynomial [81].

2. **4B5B encoding and insertion of out of band signals:** The basic principle of the 4B5B encoder is that every four bit word is exchanged by a five bit word following

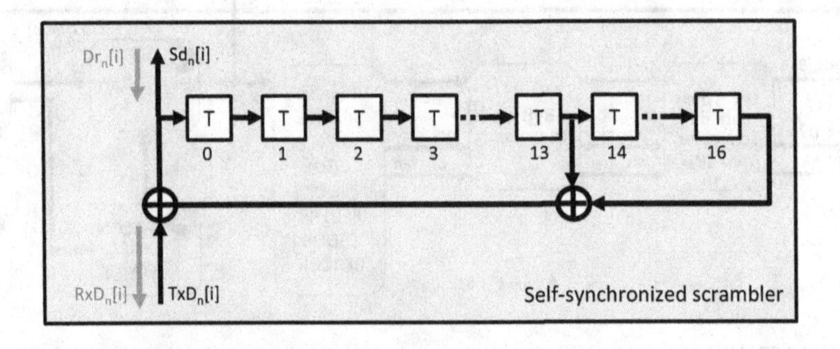

Figure 5.36 10BASE-T1S self-synchronized scrambler

Table 5.8 4B5B encoding and special symbols

Name	4B	5B
0	0000	11110
1	0001	01001
2	0010	10100
3	0011	10101
4	0100	01010
5	0101	01011
6	0110	01110
7	0111	01111
8	1000	10010
9	1001	10011
A	1010	10110
B	1011	10111
C	1100	11010
D	1101	11011
E	1110	11100
F	1111	11101

Name	4B	5B	Function
H	n/a	00100	SSD
I	n/a	11111	SILENCE
J	n/a	11000	SYNCH/COMMIT
K	n/a	10001	ESDERR
N	n/a	01000	BEACON
R	n/a	00111	ESDOK/ESDBRS
T	n/a	01101	ESD/HB

the mapping shown in Table 5.8. This 4B5B encoding was previously used in 100BASE-FX [82]. It was proposed and selected for 10BASE-T1S for various reasons. For one, the 4B5B encoding allows for additional code words (see the following paragraph). Next, it does not make the PLL necessary that was possible to avoid because of the DME. The 4B5B 4/5 clock ratio allows for a rate change with a standard 25 MHz crystal without PLL. Furthermore, the 4B5B encoding works well with the selected DME (see point 3), has good EMC properties and is easy to implement [P. Beruto, Reasons for selecting 4B5B, email correspondence, August 24, 2018].

Naturally, 5-bit code words allow for twice the number of combinations than four bit code words. This means that additional code words are available. Some of these are used in 10BASE-T1S (see right part of Table 5.8). As they are not added to the data section of the packet to be transmitted (but to the preamble and at the end of the packet, see description below), they do not add to the overhead. This is why they are called “out of band” signals. The special symbols are used in the following way:

When there is data to be transmitted, four words out of the preamble are replaced with two SYNCH and two SSD symbols, in order to facilitate synchronization. At the end of the packet an ESD is added – as is in 100BASE-T1 (see Figures 5.19 and 5.23) – plus an ESDOK (or an ESDERR, in case a transmit error occurred). BEACON, COMMIT, ESDBRS, and ESDJAB are special symbols needed when the system is used in a multidrop topology (see Section 5.4.3 for details). The special symbol SILENCE is an indication for the PMA to change its PMD state either to high impedance (multidrop mode) or 0 V differential operating voltage (P2P mode). The HeartBeat (HB) symbol is needed only when auto-

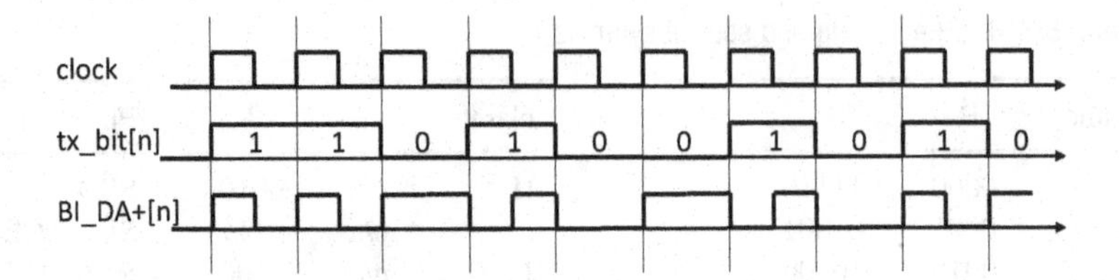

Figure 5.37 Basic functionality of Differential Manchester Encoding as used for 10BASE-T1S

negotiation is enabled (which is only supported as an option for the full-duplex 10BASE-T1S version; its removal is one of the simplifications of the half-duplex 10BASE-T1S).

3. **Differential Manchester Encoding (DME):** DME is a two-level, self-clocked modulation method. The encoding works such that, for every bit, one or two voltage level changes are performed. For 10BASE-T1S a "1" has two voltage changes and a "0" has one at the beginning of the bit (see also Figure 5.37).[24] Basing the modulation on voltage changes means that the system constantly changes voltage levels independent of the actual bit sequence sent, which makes it intrinsically balanced. It therefore shows good EMC performance, but doubles the frequency (to 25 MHz). Because any modulated symbol can start at high or low voltage state, the polarity of the cables is irrelevant. And, as has been mentioned before, DME provides low DC baseline wonder as well as robust clock and data recovery. One advantage of the increased frequency is that it lowers the complexity and cost for Power over DataLine (PoDL, see also Section 6.2.1) since the inductor size decreases with the frequency.

For 10BASE-T1S there is an additional parameter of differentiation compared with the other Ethernet PHY technologies: The functions that the actual PHY transceiver chip comprises. Figure 5.38 gives an overview. The 100 or 1000BASE-T1 PHYs are (normally) sold as either standalone PHYs that connect per xMII interface or are integrated within a switch or μC (see upper row of the figure). Both possibilities exist for the 10BASE-T1S PHY as well, and the stand-alone PHY can be used with all existing processors or switches that support an MII-Interface (provided the MAC that the MII connects to still supports the CSMA/CD function and has not eliminated it for efficiency reasons). Nevertheless, for 10BASE-T1S, there is more.

In principle, the costs of a (PHY) semiconductor consist of three parts: Silicon area, packaging, and testing. A less complex PHY, like 10BASE-T1S, can save silicon area and testing simply because it is less complex. However, when using the same 16 pin MII interface as 100BASE-T1 (see also Figure 5.27), the package size is determined by the MII interface (instead of the silicon area) and cannot be reduced.[25] To support the cost effectiveness of 10BASE-T1S, the following two additional product categories have been made available (see also lower row of Figure 5.27):

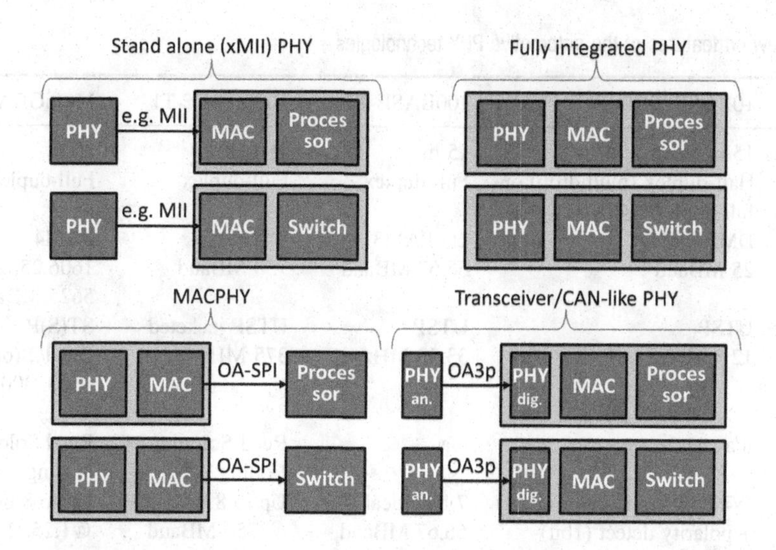

Figure 5.38 Function of possible 10BASE-T1S PHY types; the light grey boxes indicate the chip

1. The **MACPHY**: In this case the MAC is integrated with the PHY and a 4+1-pin Serial Peripheral Interface (SPI) is used to connect the PHY to the processor or switch (four pins for the SPI and one pin for the Interrupt ReQuest, IRQ). The MACPHY thus connects to many existing μCs, even smaller ones without an MII interface. Also, in switches, connecting the MACPHYs via SPI saves a significant number of pins. As the CSMA/CD function provided by the MACPHY, its availability in the connecting parts is not a concern. To ensure interoperability, the exact configuration of the here-called OA-SPI was defined in the OPEN Alliance [83].
2. The **CAN-like PHY** (transceiver): In this case the analog part of the PHY is separated from the digital part. The resulting analog transceiver is comparable in size and effort to a CAN-transceiver and connects with a 3-pin interface (called here OA3p) defined in the OPEN Alliance (see [84]) to a processor or switch. Both processor and switch need to integrate the digital PHY part with the MAC. Also, this is comparable to CAN, where the CAN-controller is part of the μC and where every new CAN-version requires the development of new processors, as is the case for this Ethernet transceiver.

Table 5.9 compares the different properties for the Automotive Ethernet PHYs (for details on MultiGBASE-T1 see Section 5.5). As expected, requirements and complexity increase with the transmission speed. With increasing frequency, the modulation, equalization effort, and overall complexity increase as well. The cable requirements become more stringent and the cable thus becomes more costly. Note that with the technical progress in semiconductor process technologies, the digital elements of the transceivers (DFE, FFE, critical path, ...) become less challenging to realize, while the burden shifts to the analog portion.

Table 5.9 Overview on features of the automotive PHY technologies

Technology	10BASE-T1S	100BASE-T1	1000BASE-T1	MultiGBASE-T1
Channel length	15 m (25 m)	15 m	15 m	15 m
PHY transmission	Half-duplex (multidrop) or full-duplex (P2P)	Full-duplex	Full-duplex	Full-duplex
X-level signaling	DME 4B5B 25 MBaud	2D-PAM3 66.67 MBaud	PAM3 750 MBaud	PAM4 1606.25, 2812.5, 5625 MBaud
Cable	UTSP	UTSP	UTSP jacketed	ST(S)P
Required Nyquist bandwidth	12.5 MHz	33.33 MHz	375 MHz	Baud/2 (or 750, 1500, 3000 HMz)
Error correction	n/a	n/a	Reed Solomon Coding	Reed Solomon Coding
A/D conversion	Window comparator (1 bit) + polarity detect (1bit)	7 bits ideal @ 66.67 MBaud	Up to 8 bits ideal @ 750 MBaud	Up to 8 bits ideal @1.25, 2.5, 5 MBaud
DFE	none	24 taps	Up to 128 taps	Up to 200/400/800
FFE	none	8 taps	Up to 48 taps	Up to 80/160/320
NEXT cancellers	none	none	none	none
Echo canceller	none	48 taps	150 taps	280/560/1120 taps
Critical path	• 2 input add-compare select	• 3 input add • 3 input select • 1 slicer • 3 input add-compare select	• 3 input add • 3 input select • 1 slicer • 3 input add-compare select	• 3 input add • 3 input select • 1 slicer • 3 input add-compare select
Normalized gate complexity	0.33	2	8	12/24/48
Additional features	DME is balanced and self clocked, PLCA	Transmit spectral shaping	Transmit spectral shaping	Transmit spectral shaping

5.4.3 10BASE-T1S Multidrop

5.4.3.1 Introduction to 10BASE-T1S Multidrop

In a switched Ethernet network, each communication link needs two PHYs to operate; one at each end. In a switched system with, e.g., $N = 5$ Ethernet nodes the system requires at least $(5 - 1) \times 2 = 8$ PHYs for all nodes to be connected (see also Figure 5.10). That some of the needed PHY functionality might be integrated with the switch still makes them a cost factor. When using a bus/multidrop topology, only five PHYs are needed; one for each node. This reduces the number of PHYs needed by 25% (in case of three units) to 42% (in case of, e.g., eight units). Such a structure is thus very attractive to reduce system costs further. This section explains how the multidrop functionality is realized with 10BASE-T1S.

In the 1980s, Ethernet at IEEE started with 10 Mbps Ethernet and half-duplex/multidrop using Carrier Sense Multiple Access with Collision Detection (CSMA/CD)

to organize the access of the different nodes to the channel (see also Sections 1.1 and 1.2.1). In CSMA/CD, every node simply listens whether the channel is available and only starts transmitting when it is not. Should a collision occur, because another node starts transmitting at the same time, each node goes into a random back-off, before retrying the transmission. In, e.g., a 10BASE-T Ethernet network, in which only three nodes want to transmit continuously, this reduces the overall network throughput to below 7 Mbps [85]. Additionally, even if the resulting average delay might be acceptable, in the worst case, a specific unit might have to back-off until the data is no longer of interest.

While this unpredictability is naturally undesirable and the Ethernet community (largely) moved on to switched full-duplex communication, CSMA/CD still needs to be supported in respective Ethernet systems to provide backwards compatibility. The two signals that indicate whether there is activity or a collision on the line – CRS (carrier sense) and COL (collision) – are part of every compliant MII and MAC implementation. That these signals are available, was exploited in the channel access method that specifies the multidrop mode of 10BASE-T1S, which is called Physical Layer Collision Avoidance (PLCA). It is explained in more detail in the following, how PLCA – in contrast to CSMA/CD – guarantees a maximum latency, a per-unit fairness, and optimum use of the available bandwidth.

5.4.3.2 Technical Description of 10BASE-T1S Multidrop

The basic principle of the 10BASE-T1S multidrop is that units get the chance to transmit in sequence of their node IDs in a Round Robin fashion [86]. For this, the unit with the node ID zero functions as a head node. In principle, any unit can be head node, however, it is likely to be the most important unit, e.g., the one linking the multidrop segment to the rest of the Ethernet network. The head node starts every cycle with a beacon that indicates to all attached (end) nodes to restart their PLCA timers and counters. The beacon is a physical layer signal that is handled by the reconciliation layer (and not a packet that is handled by the MAC).

Every unit then decides in the order of their IDs, whether they want to make use of their Transmit Opportunity (TO) or not. If a unit does not make use of its TO to transmit a packet, the unit with the next higher node ID can start its transmission once the TO of the previous unit has expired. As a TO with, e.g., 32 bit times is significantly shorter than even the minimum size Ethernet packet, the bandwidth is optimally used.[26] Once the cycle is completed a new beacon is sent and the round starts anew.

Figure 5.39 visualizes the principle difference between PLCA and a TDMA set up. A TDMA system classically uses time slots of predefined and generally equal length that is derived from the maximum packet duration. If packets are shorter than the predefined length or if slots are not used, these are wasted. Naturally, a fully loaded TDMA system with only cyclic traffic, in which every packet has the same maximum length that just fits into a transmission slot, and a comparable PLCA system have the same efficiency. However, as soon as packets have different lengths, the transmission slots are not always used, and/or there is irregular/non-cyclic traffic, PLCA will have a better performance with shorter delays and better bandwidth utilization (see e.g. [87]).

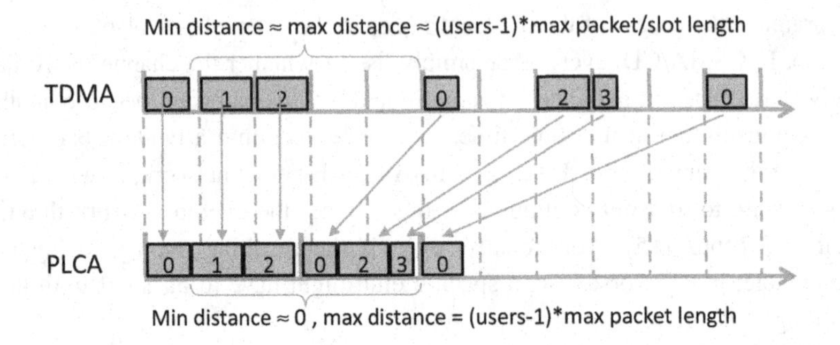

Figure 5.39 Principle difference between PLCA and TDMA

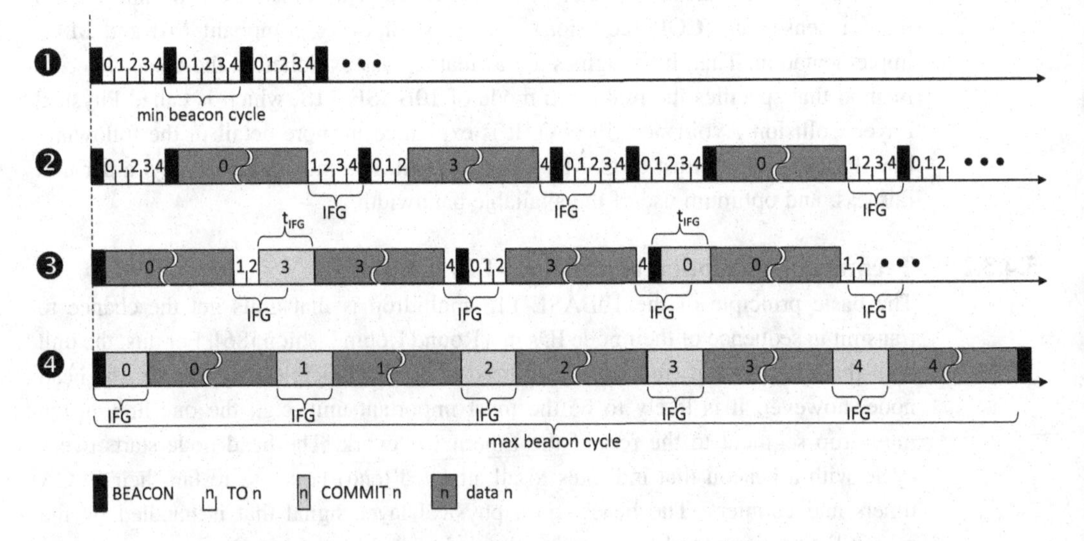

Figure 5.40 Example channel access scenarios for a multidrop link segment with five ECUs having each one node ID /TO; (1) No ECU has anything to send; (2) ECU 0 and 3 occasionally have a packet to send; (3) ECU 0 and 3 continuously have packets to send; (4) All ECUs have immediate messages to send (maximum size)

Figure 5.40 depicts the functioning of PLCA in four different exemplary scenarios in more detail.

1. In this scenario, no unit has any data to send. This means that the beacon is simply re-sent every time the TOs of all node IDs have elapsed. The duration of the minimum beacon cycle is thus (#TO+1) × 32 bit times, e.g., 19,2 μs = (5 + 1) × 3.2 μs in case of five units each with one TO.
2. In this scenario, only unit 0 and unit 3 send packets, while all other units remain silent. As soon as their transmit opportunity comes up, they simply transmit. As is indicated in the picture, immediate transmission is possible because the moment the units have packets ready to transmit, the channel is idle and the IPG from the

previous packet has elapsed. In this case, the throughput is comparable to that of a lightly loaded CSMA/CD system.

3. In this scenario, also only unit 0 and unit 3 send packets, while all other units remain silent. The difference to scenario 2 is that the MACs of units 0 and 3 have packets to transmit before the IPG elapses. Because it is not possible to start the transmission of a new packet during the IPG, the PHYs send a COMMIT in their TO, to indicate to others, that they intend to transmit as soon as this is possible. This happens only when the PHY knows that the MAC has packets to transmit. This again is only possible when the MAC had previously started forwarding a packet to the PHY before PHY and MAC noticed that the channel is busy after all because another unit was faster to start transmitting.

   As can be seen, the COMMIT symbol itself is repeated for the length of the IPG, which somewhat extends the original IPG. The reason is that unit 3 first has to wait to find out whether unit 1 or 2 want to use their TOs or not. Only once their TOs have passed unused can the counter of the IPG start (after which unit 3 will start its transmission no matter what, see [87] for a detailed example).

   Whether scenario 2 or scenario 3 happens simply depends on the point in time the PHY, or more precisely the PLCA in the PHY, learns from the MAC that there is data to transmit.

4. In this scenario, all units continuously have packets to send. As soon as a packet ends, the next unit commits to its TO and sends its packets as soon as the IPG elapsed. The COMMIT symbol is also extended (repeated) until the end of the IPG. Should every unit send maximum size packets, this scenario represents the maximum beacon cycle. The maximum beacon cycle is thus #TO times (MaxPacketDuration+IPGDuration), e.g., 6.15 ms = 5 × 1.23 ms in case of five units each with one TO and maximum size Ethernet packets (see Figure 1.5).

The key to PLCA is that the MII at the MAC interface and the MAC itself are not changed but function as they always have. In order to ensure this, the PLCA makes use of the Reconciliation Sublayer (RS) as part of the MII (tagged with No.6 in Figure 5.35). The principle task of the PLCA RS is to ensure that, when the ECU wants to transmit data, it does so at the assigned time only. The PLCA thus avoids collisions altogether (and with that improves bandwidth utilization, latency, fairness, and EMC performance[27]). Figure 5.41 details the function blocks and signals used in the RS.

In principle, all the PLCA RS does is delay data coming from the MAC until the TO of the unit. In order to achieve this, it slightly adapts the CRS and COL signals toward the MAC in the following way. When the MAC has data to transmit and the $CRS_{PCS} = CRS_{MII}$ indicates an unoccupied channel, it will forward the data to the RS. In the RS, the data is buffered until the unit's TO. When the TO is there and the CRS_{PCS} is still low and therefore the channel is still unoccupied, the RS will forward the data to the PCS and a regular transmission starts (see example point 2). However, during the wait for the TO, it might happen that another unit starts transmitting and CRS_{PCS} will indicate this. The PLCA RS will then stop the MAC from sending more data (to minimize buffering in the PHY) by pretending that there was a collision on the

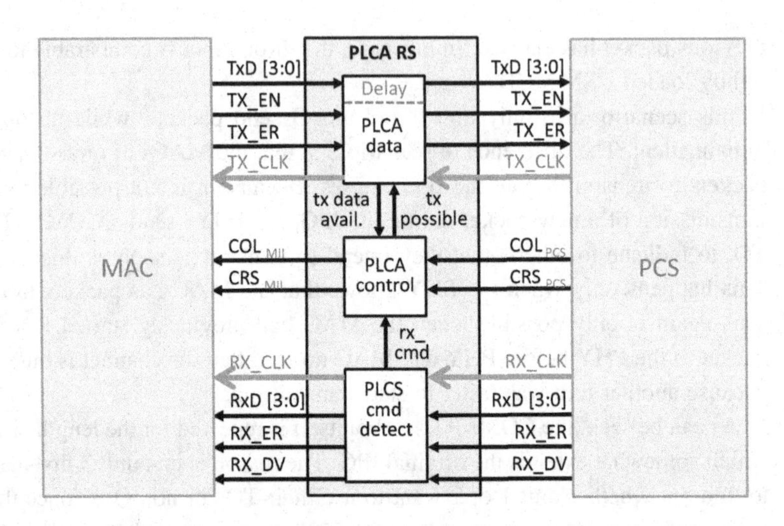

Figure 5.41 Function blocks of the RS enabling PLCA[28]

line with eth COL_{MII}, even though this was not the case. Because the data was in the RS buffer and not yet on the channel, the COL_{PCS} is different from the COL_{MII} in this moment. CRS_{MII} will simply, like CRS_{PCS}, indicate an occupied channel.

The COL_{MII} is set only long enough to make sure the MAC goes into random back-off with the packet it wants to send. The PLCA makes use of the fact that the maximum duration of the first random back-off is always shorter than the minimum Ethernet packet size and that the MAC is ready to resend the data before the transmission on the line has ended. The CRS_{MII} is not released when the channel is clear and CRS_{PCS} is low again, but artificially kept high until it is clear that no unit with a prior TO wants to commit (see also example point 3 or [87]). Once the unit's next TO is there, the CRS_{MII} is lowered. The MAC will then wait for the length of the IPG and start transmitting. During this wait time the PHY will send continuous COMMIT to ensure no other unit transmits.

To control the sending and reading of the BEACON and COMMIT signals, TX_EN, TX_ER, RV_DV, and RX_ER are used (see Table 5.10). The PLCA command detect block (see Figure 5.41) knows when to extract the correct rx_cmd, based on the RX_DV and RX_ER values and the RxD[0:3] received. The PLCA control resets its timers and counters at the beginning of a new beacon, makes the right control decisions, and initiates the transmission of BEACON and COMMIT accordingly with help from the TX_EN, TX_ER, and TxD[0:3]. The code words defined for the BEACON and COMMIT are of interest for the PLCA command detect block only. They are not forwarded to the MAC but are, instead, terminated in the RS.

The correct operation of PLCA assumes that every ID that identifies a TO is unique to the multidrop link [21]. However, the assignment of these IDs is not part of the IEEE 802.3cg specification and thus might not have been done correctly. The following list shows that the only error that would really disrupt the function of PLCA is when two nodes receive the ID 0:

Table 5.10 TX_EN, TX_ER, RX_DV, RX_ER control and encoding

TX_EN or RX_DV	TX_ER or RX_ER	TxD[0:3] or RxD[0:3]	Meaning
0	0	No relevance	InterPacket Gap (IPG)
0	1	Reserved	Reserved
		0001	Assert LPI (see Section 6.3.1)
		0010	PLCA BEACON
		0011	PLCA COMMIT
1	0	Any	Normal data transmission
1	1	Any	Transmit error

- PLCA- and non-PLCA-enabled units may be used on the same multidrop link. Naturally, the more non-PLCA-enabled units there are on the link the more the performance of the link converges to that of a CSMA/CD only system.
- If PLCA is enabled and there is no BEACON, e.g., because of an erroneous behavior of the unit with the ID 0, there will be no BEACON to indicate the start of a new PLCA cycle. However, if there is no BEACON within a certain time (13 ms according to the specification), the PLCA becomes "transparent" and allows any transmissions from the MAC, just as in plain CSMA/CD. Should node 0 recover and start sending BEACONs again, normal PLCA operation resumes.
- In case several identical units are attached to the same multidrop link without pre-configuration (a desirable scenario in production), it is to be expected that these units all have the same Node ID(s) (other than 0). In such a situation, these units will try to transmit at the same time, causing collisions and the MAC to go into (multiple) random back-off(s). The standard does not provide a solution for such a scenario. However, it is not too complex to envision methodologies that allows changing the ID(s) of a unit based, e.g., on the number of collisions.
- In case several units have the same ID 0, each of them will issue BEACONs at different times and all units will constantly reset their timers and a successful transmission becomes an unlikely event, especially for units with large IDs. Such a scenario should be avoided. At the same time, the head node is likely to be a node that is different from all other units and thus easier to pre-configure with a node ID of 0 while other units are (initially) configured with non-zero IDs.

5.4.3.3 Performance Parameter of 10BASE-T1 Multidrop

One of the advantages of the PLCA multidrop mechanism is that the maximum latency between two transmit opportunities of a unit is known with the number of TOs and the maximum packet size allowed. A network's maximum latency is a common design criteria. For example, the FlexRay cycle time or the maximum load of a CAN bus are defined with a maximum latency in mind. Naturally, other than in the static segment of FlexRay where fixed timeslots are assigned and preplanned, jitter

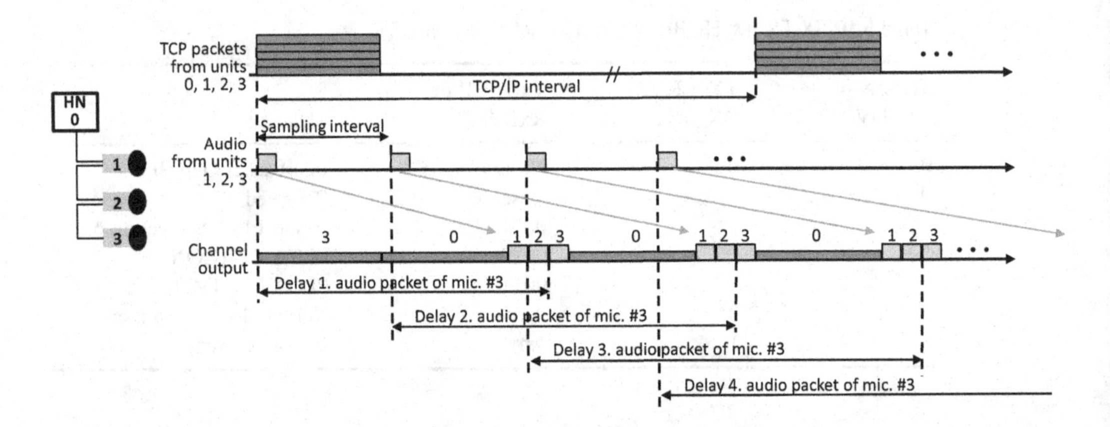

Figure 5.42 Example scenario with mixed sensor data and TCP/IP control traffic; the TCP/IP interval is significantly larger than the four audio sample intervals shown

cannot be avoided on a CAN bus or a PLCA segment, where there is no central control of the data being transmitted by all units.

In this context, there are two aspects of interest: (1) How can fairness on the channel with respect to the data rate be improved? and (2) How can the maximum latency be reduced? Both of these aspects concern quality of service, which is generally a layer two and above topic and is discussed in detail in Section 7.1. However, 10BASE-T1S provides some means to influence both aspects closer to the PHY, which is why they are (also) discussed in this section.

In PLCA, the fairness of channel access is established by TOs. If the bus is shared between units sending small, e.g., sensor data packets and units sending full TCP/IP control packets, the units sending the small sensor data packets will be able to transmit as often, but eventually receive less bandwidth. At the same time, the long TCP/IP packets might unnecessarily delay the sensor information. In a switched network the switch would simply put the sensor data in a higher priority queue than the TCP/IP control data. In a PLCA multidrop link each unit can have internal queues with priorities; however, on the PLCA bus a unit sending lower priority traffic than another unit sending higher priority traffic will get the same access (see also [88]). Figure 5.42 visualizes an example with three microphones attached to a head node, where the control traffic, especially from the head node, delays the microphone data (see also [89]).

Even if the average data rate on the channel is more than sufficient, during the time the head node sends its burst of TCP/IP packets (with an additional odd TCP/IP control packet from one of the other units) the delay increases beyond what is acceptable during that period. It was proposed during the standardization process to define the 10BASE-T1S PHY such that it would allow a multidrop link to behave exactly like a switched network with respect to IEEE 802.1Q priorities [90]. However, this was not accepted. Instead, the following possibilities exist to improve the delay at PHY level. Note, at layer two other additional possibilities exist to influence the timing behavior (such as the Time Aware Shaper, TAS, see IEEE 802.1Qbv in Section 7.1.4.4):

Table 5.11 Maximum cycle duration (ms) on a multidrop link depending on the number of TOs and packet sizes on the link

Payload bytes head node	Payload bytes end node	2*	3	4	5	6	7	8
1500	1500	2.47	3.70	4.93	6.17	7.40	8.64	9.87
1500	500	1.71	2.19	2.67	3.15	3.63	4.10	4.58
1500	128	1.41	1.60	1.78	1.96	2.14	2.32	2.50
1500	8	1.32	1.40	1.49	1.57	1.66	1.74	1.86
500	500	0,96	1,44	1,91	2,39	2,87	3,35	3,83
128	128	0.36	0.54	0.72	0.90	1.08	1.27	1.45
8	8	0.17	0.26	0.34	0.43	0.51	0.60	0.68

* with two TOs, two units full-duplex operation is also possible.

1. **Optimize the packet sizes** to the use case: The delays as shown in Figure 5.42 depend on the sampling interval of the audio and the size of the TCP/IP control packet. By increasing the sampling interval or by decreasing the maximum allowed TCP/IP packet size, the delays on the multidrop link can be significantly reduced [91]. When increasing the sampling time, the size of the audio packets increases, which naturally improves the fairness on the PLCA channel. But, of course, it must be allowable by the application. For limiting the maximum size of the (TCP/IP) control packets, a distinction must be made between the head node that connects the multidrop link to the rest of the network and the end nodes.

   In the end nodes, the length of the packets can be easily limited by ensuring at higher layers that the unit does not want to send more data per packet than some set limit. For the head node, which potentially relays packets from other parts in the network, this is not as simple. In theory, it is possible to break up large packets with IP fragmentation [92]. However, deploying IP fragmentation is not recommended. For one, it would require a router, i.e., a layer three bridge. For another, the method itself is not robust when packets are dropped or discarded. Another option is to filter the number of control packets that propagate onto the multidrop link. In the case of, e.g., service discovery (see also Section 7.4.4) only a minimal subset of control information might be needed. The disadvantage is that this would require some kind of gateway. For that reason, it might be a preferred solution to limit the packet size in the end nodes only. Table 5.11 shows the guaranteed maximum delay between the start of two packets in the same TO ID depending on the maximum allowed payload size and number of TOs on the multidrop link.

   As can be seen, by just limiting the payload size in the end nodes, cycle times below 5 ms can easily be achieved. Cycle times below 1 ms require limiting the payload size in the head node as well. Note that, should the control packet size be decreased in general, the Ethernet efficiency as such only experiences a small decrease. When decreasing the allowed payload size from 1500 bytes to 500 bytes it reduces the Ethernet efficiency from ~97% to 92%.

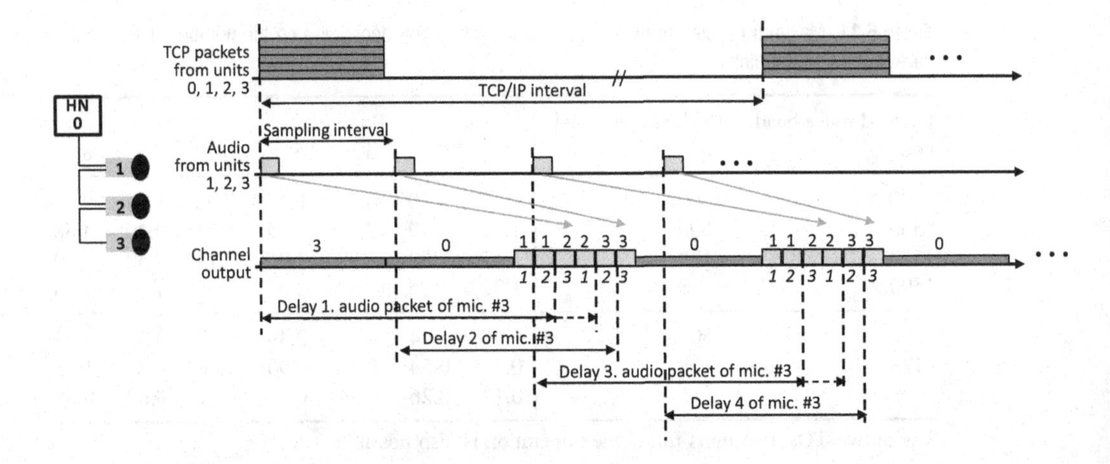

Figure 5.43 Example reduced latency with burst mode or multiple transmit opportunities (*italic*) allowing two transmissions each per beacon cycle

2. **Assign multiple TOs** to specific units within one beacon cycle: The 10BASE-T1S standard does not explicitly address this possibility, but it does not preclude it either. Units that have to send small packets at small intervals could receive multiple TOs within one cycle. The only attention that would need to be paid is that these TOs are not consecutive, but need to allow at least one other unit to transmit in between. The worst case delay can be significantly decreased with just one additional TO [91] (see also Figure 5.43).
3. **Enable PLCA burst mode**: The burst mode allows that individual units can transmit a configured number of additional packets within one TO. A burst mode enabled unit will terminate a packet not with ESD/ESDOK but with ESDBRS/ESDOK (see Table 5.8) and then COMMIT during the IPG before immediately sending the next packet. If it has fewer packets to send than the configuration allows it to send, it will simply send an ESD/ESDOK after the COMMIT [93]. Figure 5.43 shows the packet order when using the burst mode allowing for two packets per burst (upper row) to every sensor unit or when assigning two TOs to the sensor units within beacon circle (lower row in italic). When comparing the figure with Figure 5.42, it can be seen that the maximum delay reduces with either method more or less to the length of two max size packets. If these are reduced additionally, e.g., by limiting the packet length in the end nodes, the maximum delay decreases directly with it. Unfortunately, none of the three methods is just plug and play, but requires some thought and engineering.

Another aspect of interest is the achievable user data rates for 10BASE-T1S. As 10BASE-T1S addresses use cases that require lower data rates, which might also be addressed with CAN(-FD) or FlexRay, 10BASE-T1S needs to withstand the comparison in case packets with small payloads are transmitted. The overhead of CAN is ~6 bytes for an 11-bit identifier and ~8 bytes for a 29-bit identifier. For CAN FD it is

Table 5.12 Comparison of maximum effective throughput for CAN, CAN FD (both with 11-bit identifier and 50% load maximum), FlexRay, and 10BASE-T1S multidrop for different user payloads at 100% load[29]

User data rate [Mbps] = f (user payload)	CAN	CAN FD 2 Mbps	CAN FD 5 Mbps	FlexRay static only	10BASE-T1S
8 bytes	0.14	0.37	0.62	1.93	0.95
16 bytes	n/a	0.54	1.00	3.11	1.90
64 bytes	n/a	0.82	1.82	6.65	6.04
256 bytes	n/a	n/a	n/a	7.28	8.59
1500 bytes	n/a	n/a	n/a	n/a	9.73

~8.5 bytes for an 11-bit identifier and ~10.5 bytes for a 29-bit identifier. FlexRay has an overhead of 8 bytes plus a not-insignificant amount of "channel idle" (of which the exact length depends on the selected FlexRay low-level parameter). Ethernet, in contrast and independent of the speed grade, has 38 or 42 bytes overhead plus a minimum packet size of 46 or 42 bytes, respectively (see also Figure 1.5).

At first glance, sending an 8-byte payload with 42 + 42 – 8 = 76 bytes overhead seems like a lot. However, the overhead has a purpose: It allows gateway-free communication between all nodes in small or extensive Ethernet networks. Moreover, as can be seen in Table 5.12, the achievable user data rate of 10BASE-T1S is still significantly larger than what can be achieved with CAN or CAN FD, even for small payloads. Obviously, a fully loaded FlexRay used statically achieves the best, albeit also significantly reduced, net data rate. The overhead of FlexRay is smaller than that of Ethernet. Because there is no contention in the static segment, FlexRay can be used with a higher load than CAN(-FD). The loads of FlexRay and 10BASE-T1S are comparable when, in both cases, the units transmit as much as possible. As soon as packets of varying lengths are transmitted or the channel is not fully loaded, 10BASE-T1S is more efficient and the latencies are smaller. When the load is small, 10BASE-T1S packets will likely receive immediate channel access, just as in CAN(-FD).

Last but not least, at the time of writing, the IEEE 802.3cg specification had just been published and first silicon was being presented. Various topics for improvement and enhancement were being discussed to be specified either with the "multidrop enhancements" in IEEE 802.3da [94], in the OPEN Alliance, or simply to be realized in a vendor manner. While industrial automation had some different additional requirements, the following items are of interest for the automotive industry: Power over MultiDrop (PoMD, see also Section 6.2.1), wake-up and sleep (see also Section 6.3.2), topology, and termination detection. Termination detection is desirable to ensure that exactly two units are terminated, despite different topology options in production. Topology detection allows for physically locating a unit on the multidrop link with certain accuracy. This is of interest if, e.g., a number of identical sensors are used and not only the ID needs to be assigned, as was discussed in point 3 of Section 5.4.3.2, but also a specific IP address, i.e., which unit and ID has which physical location.

Table 5.13 Required data rate in Gbps for uncompressed data depending on the horizontal and vertical resolution (Hres, Vres), the bit depth, and the frame rate (fps) without overhead (10% expected)

			30fps			60fps		
	Hres	Vres	8 bit	16 bit	24 bit	8 bit	16 bit	24 bit
720p HD	1280	720	0.22	0.44	0.66	0.44	0.88	1.33
SXGA	1280	1024	0.32	0.63	0.94	0.66	1.33	1.99
1080p HD	1920	1080	0.50	1,00	1.49	1.00	1.99	2.99
	2880	1280	0.89	1.77	2.65	1.77	3.54	5.31
4k	3840	2160	1.99	3.98	5.97	3.98	7.96	11.94
8k	7200	2160	3.73	7.47	11.20	7.47	14.93	22.40
8k+	7680	4320	7.97	15.93	23.89	15.93	31.85	47.78

5.5 Technologies for 2.5, 5, and 10 Gbps

5.5.1 Background to MultiGBASE-T1

For a long time, the need for higher data rates in cars was driven by infotainment and driver assistance. At the beginning, it was audio and map data that required more powerful transmission technologies, then it was video and graphic data, i.e., camera and display usage. Table 5.13 shows data rates needed for various different resolutions and frame rates for uncompressed video and graphic data. It can be seen that, for better HD resolutions, a 1 Gbps data rate is easily exceeded, especially when considering that at least 10% overhead must be added to the numbers shown. When the standard development of MultiGBASE-T1 started, 4k and 8k resolutions were on the horizon and had to be addressed [24]. Table 5.13 shows that 4k and 8k resolutions generate multi Gbps data rates, but also that the data rates stay mainly below 10 Gbps.

Naturally, all video data rates can be significantly reduced with compression. In the first use case for Automotive Ethernet (see also Section 3.4.2), camera data was being compressed such that it was possible to transmit the video stream over a 100 Mbps Ethernet link, without any loss in functionality. However, with the shift from human vision to machine vision (especially in the context of autonomous driving), there was a demand for uncompressed data in order to have the lowest latency possible and to ensure that there is absolutely no loss of information due to compression artifacts.

Also, in the realm of infotainment, ever increasing data rates can be observed for the communication between the car and the outside world either in the form of mobile communication, different facets of WiFi, or digitized media broadcast. The development of mobile data communication over time is shown in Figure 5.44. It can be seen that the achievable data rates have continuously increased over the decades and services with data rates larger than 1 Gbps have started [95]. Car manufacturers want

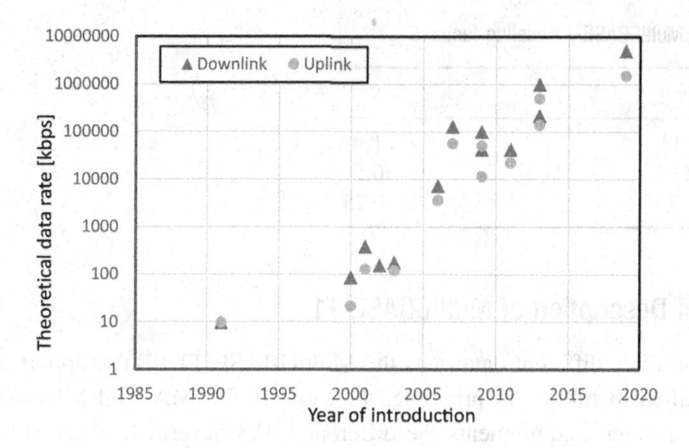

Figure 5.44 Development of (theoretical) data rates in mobile communication over time [96, 97]

to use those data rates for more services to their customers as well as for connecting the vehicle to the manufacturer's backend; and especially, again, in light of autonomous driving functions. Data to or from the antenna must be distributed inside the car with a matching data rate.

As for digital radio/TV/media broadcast, broadcast is being digitized worldwide in order to save energy and to use the bandwidth more efficiently [98, 99]. To achieve the latter, digital broadcast standards use compression methods. However, broadband reception methods and/or decompression potentially increase the data rate to be transmitted significantly.

Other reasons for requiring higher data rates are based on changes in the EE architecture. It might be that several 1 Gbps links need to be aggregated, e.g., for port mirroring or diagnosis, that data recorders require a significant amount of (raw) data to be recorded simultaneously, or that centralization makes the high data rate exchange of data between a few but powerful computing platforms necessary inside the car (see e.g. [100]).

The various existing Point-to-Point (P2P) technologies such as USB, HDMI, or proprietary SerDes links, are often too limited. Relaying data coming over these links requires a gateway, they cannot easily be integrated with the synchronized time of, e.g., IEEE 802.1AS (see Section 7.1.3.2), they might require expensive cabling (application specific), and/or make changes in the next generation difficult as they do not scale [24]. Also the duplication of sensor data routes for redundancy is not easy to realize with these technologies.

So, sooner rather than later, the in-vehicle network required an Automotive Ethernet PHY that supports a data rate of more than 1 Gbps and a respective CFI was passed in IEEE 802.3 in November 2016 [24]. With the objectives not specifying the media, the project invested a not insignificant amount of effort in selecting and defining it (see also Section 5.1.4). The results of respective investigations prior to the IEEE efforts have been published in, e.g., [101–104].

Table 5.14 MultiGBASE-T1 scaling factor *S*

Speed grade	*S*
10GBASE-T1	1
5GBASE-T1	0.5
2.5GBASE-T1	0.25

5.5.2 Technical Description of MultiGBASE-T1

Despite the three different data rates the MultiGBASE-T1 PHYs support, they have been specified to follow, in principle, the same PCS, PMA, and MDI descriptions [30]. The different requirements the different PHYs nevertheless have to fulfill are accommodated by a frequency scaling factor *S* (see Table 5.14), additional options for the different speed grades (e.g., in the error protection, see later in this section), or simply by different limit lines (see e.g. the RL limits in Section 5.1.4). The following text explains the differences when relevant.

Figure 5.45 shows the building blocks for the MultiGBASE-T1 PHYs. The depiction is very similar to the one of 1000BASE-T1 (see Figure 5.13) using the interface appropriate for speeds greater than 1 Gb/s – XGMII instead of GMII – and making two changes in the signaling between PMA and PCS; alert_detect and pcs_data_mode signals are used instead of a loc_phy_ready. The alert_detect may be used to signal a remote PHY to come out of the Energy Efficient Ethernet (EEE) Low Power Idle (LPI) mode (see also Section 6.3.1) without having to send data. The difference between pcs_data_mode and loc_phy_ready is simply the result of a somewhat different realization of the PHY control.

The optional XGMII transmits four bytes of TxD data in parallel, each byte associated with one control bit from TxC. If the respective control bit is 0, user data is being transmitted. If the control bit is 1, control data is being transmitted (such as IPG, interframe status, start/preamble, terminate, transmit error propagation). XGMII was originally designed to only support a data rate of 10 Gbps. However, this was changed with the development of 2.5 and 5GBASE-T (without "1"). In case only 2.5 or 5 Gbps are needed, the clock rate is simply lowered. In case the XGMII and PMA data rates are not synchronized, this is compensated for within the receive process with inserted or deleted idles or the removal of sequence ordered sets. Note that, while the standard mentions XGMII, real implementations will likely use a serial or reduced pin-count interface, such as, e.g., USGMII.

Figure 5.46 depicts the PCS elements for the MultiGBASE-T1 PHYs.

1. **64B/65B conversion:** The transmitter first of all groups two XGMII blocks of 32 bits of data TxD plus the four control bits TxC into blocks of 72 bits. These blocks of 64 data and eight control bits are converted into blocks of 65 bits, where the first bit indicates whether the block contains data only ("0") or also control information ("1"). The 64B/65B selected is a simplification of the 64B/66B that was first introduced with 10GBASE-R. 64B/65B just removes the second upfront added bit that was needed for optical transmission.

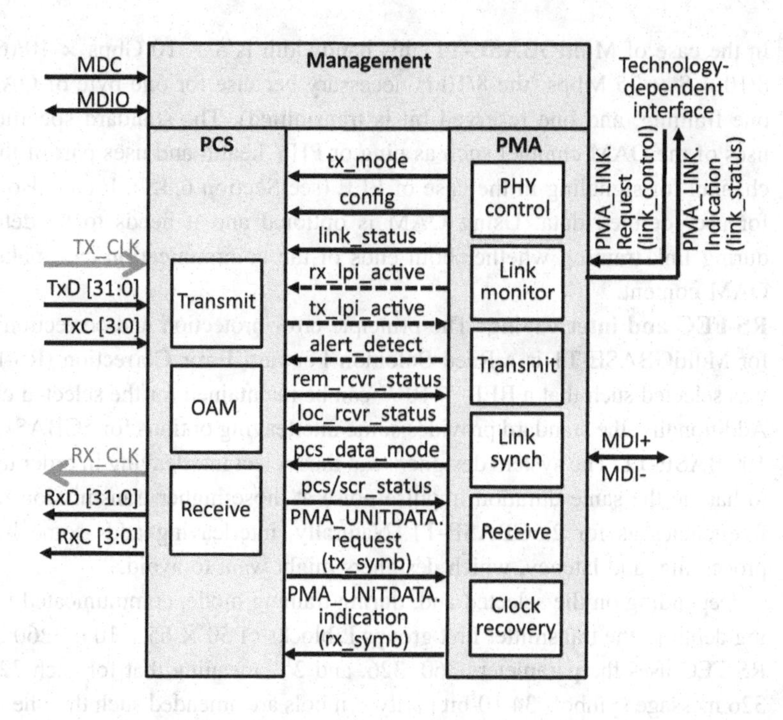

Figure 5.45 MultiGBASE-T1 building blocks (optional signaling for EEE in dashed lines)

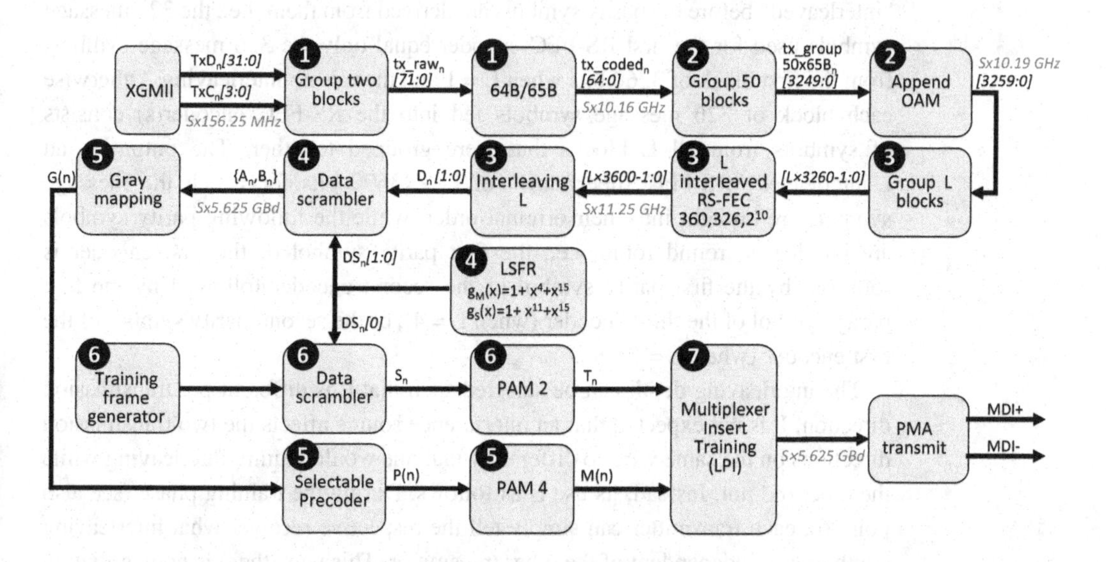

Figure 5.46 PCS transmitter elements in MultiGBASE-T1; the nomenclature is used as in the specification [30]

2. **Appending the OAM:** Following the 64B/65B conversion, 50 65-bit blocks are grouped and each block is appended by 10 bits of an Operation, Administration, and Management (OAM) channel. As in the case of 1000BASE-T1, OAM uses some excess bandwidth available with the selected coding and link parameters.

In the case of MultiGBASE-T1, this bandwidth is $S \times 10$ Gbps $\times$ 10/(6450) $\times$ 8/10 = S $\times$ 25 Mbps (the 8/10 is necessary because for one byte of OAM data, one framing, and one reserved bit is transmitted). The standard specifies some uses of the OAM channel such as ping or PHY health and uses part of the OAM channel for signaling in the case of EEE (see Section 6.3.1). It can also be used for user defined data. Using OAM is optional and it needs to be determined during link training whether both ends of the communication can make use of OAM content.

3. **RS-FEC and interleaving:** The principle error protection and correction method for MultiGBASE-T1 is a Reed Solomon Forward Error Correction (RS-FEC). It was selected such that a BER $\leq 10^{-12}$ can be maintained for the selected channels. Additionally, the standard provides some interleaving options for 5GBASE-T1 and 10GBASE-T1. The system designers can thus select interleaving in order to be able to handle the same duration impulse noise at those higher transmission rates and frequencies as for 2.5GBASE-T1. Naturally, interleaving adds some buffering, processing, and latency, which designers might want to avoid.

   Depending on the selected and, during training mode, communicated interleaving depth L, the transmitter first groups L blocks of 50 $\times$ 65 + 10 = 3260 bits. The RS-FEC uses the parameters 360, 326, and 2^{10}, meaning that for each 3260/10 = 326 message symbols 34 10-bit parity symbols are amended such that the resulting RS-FEC blocks have a length of 360 $\times$ 10-bits. Note that the message symbols are "interleaved" before the parity symbols are derived from them, i.e., the 326 message symbols used for the first RS-FEC encoder equal only the 326 message symbols from the first block of 3260 bits when L = 1 and there is no interleaving. Otherwise each block of 326 message symbols fed into the RS-FEC encoder(s) consists of symbols from all L blocks that were grouped together. The output is an L-interleaved "RS-FEC superframe" of L $\times$ 3600 bits for which the message symbols are brought into their original order, while the following parity symbols are interleaved round robin, i.e., the first parity symbol of the first encoder is followed by the first parity symbol of the second encoder followed by the first parity symbol of the third encoder (when L = 4), or the second parity symbol of the first encoder (when L = 2) etc.

   The interleaving depth can be selected such that it is different per transmission direction. It is not expected that an interference source affects the two transmission directions on the same wire so differently that one would require interleaving while the other did not. Instead, its use is as follows: During the training phase (see also point 6), each transmitter can simply tell the respective receiver what interleaving depth to use, independent of the other transmitter. This way, there is no negotiation necessary in which both sides need to agree on the interleaving depth. It might thus happen that the interleaving depth ends up being different in both directions, accidently, out of ignorance, or even on purpose. Table 5.15 shows the interleaving options for the different speeds.

4. **Data Scrambler:** Next, the data is scrambled with the help of a 33-bit LSFR in order to improve the EMC performance. The LSFR deployed is the same as for

Table 5.15 Interleaving options for MultiG-BASE-T1 PHYs

	Interleaving	2.5GBASE-T1	5GBASE-T1	10GBASE-T1
L=1	none	Default	Default	Default
L=2	2	Not defined	Optional	Optional
L=4	4	Not defined	Not defined	Optional

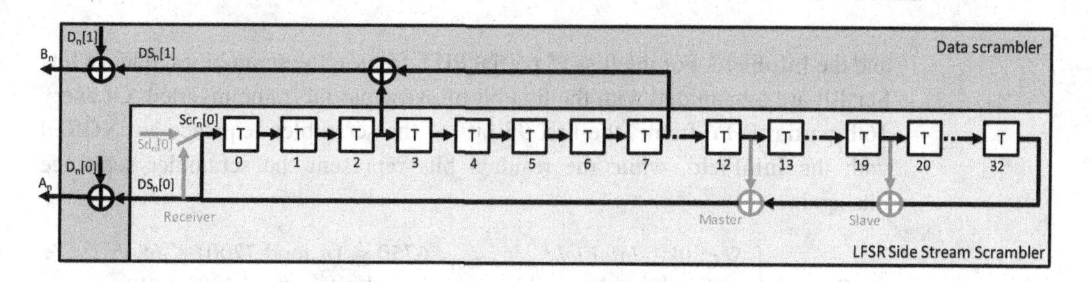

Figure 5.47 LFSR and data scrambler word generator for MultiGBASE-T1

100BASE-T1 and 1000BASE-T (see Equation 5.10). Figure 5.47 shows how the data stream is scrambled. With this scrambler two data bits D_n are processed in parallel. Note that the LSFR output $DS_n(0) = Scr_n(0)$ is also used for scrambling the training sequence (see also point 6).

5. **Gray coding, selectable precoder, PAM 4:** The next three steps change the bit-pairs $\{A_n, B_n\}$ created by the scrambler into PAM 4 symbols. First the bit pairs are Gray mapped to numbers G(n). Then there are four options to precode these numbers, before the output of the precoder P(n) is mapped to PAM 4 values M (n). Which one of the four precoder options used per communication direction is determined by the receiver (implementation) and communicated to the transmitter during the training phase. This means that different precoders might be used per transmission direction, which is perfectly acceptable since the precoder choice needs to match the receiver design, but does not affect the performance. Table 5.16 shows the respective transformation rules. Note that only one transformation rule is selected and used to create P(n); independent of the input G(n).
6. **Training data processing:** During training mode, 96-bit long "InfoFields" are transmitted at regular intervals to provide capability information and requests for remote transmitter settings (e.g., the precoder setting, interleaving depth, OAM, EEE) as well as receiver status information (e.g., on scrambler lock, polarity detect and correct, and RS-FEC frame boundaries, as well as the immanent modulation change over when switching data mode). For the training mode, partial PHY frames of 450 bits length are introduced. Because the training mode uses PAM 2 (with the levels −1 and +1), 7200 bits with 16 partial PHY frames correspond to the duration of one RS-FEC superframe with L = 4. Equation 5.17 shows how the training data S(n) is derived with the help of the LFSR, scrambler,

Table 5.16 Gray mapping, precoding, and PAM 4, one precoder option is selected and communicated during link training for all values of G(n)

A_n	B_n	G(n)	G(n) ➔ P(n), the same for all G(n)	P(n)	M(n)
0	0	➔ 0	P(n) = G(n) or	0	➔ –1
0	1	➔ 1	P(n) = (G(n) + P(n-1)) mod4 or	1	➔ –⅓
1	1	➔ 2	P(n) = (G(n) – P(n-1)) mod4 or	2	➔ +⅓
1	0	➔ 3	P(n) = (G(n) + P(n-2)) mod4	3	➔ +1

and the InfoField. For the first 15 partial PHY frames, the scrambler sequence bits $Scr_n[0]$ are transmitted with the first bit of every partial frame inverted. Of every 16th partial PHY frame, the first 96 bits of the scrambler sequence are XORed with the InfoField, while the residual bits represent the scrambler sequence bits again.

$$S(n) = \begin{cases} Scr_n[0] \oplus InfoField_{(n \bmod 450)} & 6750 \leq (n \bmod 7200) \leq 6845 \\ Scr_n[0] \oplus 1 & \text{elseif } (n \bmod 450) = 0 \\ Scr_n[0] & \text{otherwise} \end{cases}$$

Equation 5.17 Training frame scrambling ($\oplus$ denotes the xor function)

7. **Multiplexing:** The MultiGBASE-T1 PHY control knows the three transmission modes SEND_N, SEND_T, and SEND_Z, which determine the data that is forwarded by the multiplexer to the PMA. SEND_N represents the normal data mode and SEND_T the trainings mode. SEND_Z, transmit "silent," is used at the beginning of a link start-up as well as for the optional EEE, which can be activated asymmetrically for MultiGBASE-T, i.e., in one direction only (see also Section 6.3.1).

The processing in the receive path follows the same blocks in reverse. Naturally, it is crucial that the receiver aligns correctly to the RS-FEC superframes, for which it uses knowledge of the encoding rules and the PMA training phase.

5.6 Technologies for Other Data Rates

One of the main advantages of Automotive Ethernet is its scalability. An in-vehicle Automotive Ethernet network can comprise links with different data rates depending on the local requirements, while this is completely transparent to the higher layer implementation. After the completion of the 10BASE-T1S, 100BASE-T1, 1000BASE-T1, 1000BASE-RH, and 2.5, 5, 10GBASE-T1 standards the question therefore was "what other data rates are needed for Automotive Ethernet?."

By the time the matter of additional data rates was discussed in 2019, the industry had gained significantly more experience and expectations on what it takes to realize fully-autonomous vehicles. The main motivation for higher data rates thus came from exactly this use case. The CFI consensus presentation for >10 Gbps Electrical

Table 5.17 Expected multi Gbps data rates for different automotive use cases [105, 108, 110, 111]

	2.5 Gbps	5 Gbps	10 Gbps	25 Gbps	50 Gbps	100 Gbps	Asym.
Cameras	x	x	x	x			x
Other sensors	x	x	x	x			x
Displays	x	x	x	x			x
Redundant processing	x	x	x	x	x		
Back bone	x	x	x	x	x	x	
Data logging		x	x	x	x	x	
Smart antennas	x	x					

Automotive Ethernet PHYs [105], as a consequence, addresses the Tera bytes of data that autonomous cars need to process on average per day. Demanding communication needs arise from sensor aggregation of, e.g., camera, radar, and lidar data, from redundant processing, live software migration, black box data recorders, etc. This justified the objectives that request the support of 25, 50, and 100 Gbps data rates over 11m links with two inline connectors [106]; especially as autonomous test vehicles of the time already deployed 25 and 50 Gbps consumer grade Ethernet links [107].

An additional CFI for multi Gbps optical Automotive Ethernet PHYs passed just four months later [108]. It emphasizes that, for the same use cases, optical communication allows for galvanic isolation, superior EMC performance, lower weight, and longer link lengths. As the only optical IEEE Automotive Ethernet PHY developed until the time of this second CFI supported just 1 Gbps data rate, the respective objectives cover 2.5, 5, 10, 25, 50, and 100 Gbps for a 40 m link with four inline connectors [109].

Table 5.17 provides an overview of the anticipated data rates for different use cases. Naturally, with the number of use cases addressed, there is the expectation for a sufficiently large market to justify the development [24, 112]. However, what data rate and technology will truly be used is a different matter. In general, it can be expected that the higher the data rate and the more functionality supported, the higher the costs of the connection. The car manufacturers will thus decide on the actual use of these data rates with care.

The Automotive Ethernet solutions from 10BASE-T1 to 10GBASE-T1 all allow for fully symmetric data transmission with full networking capabilities. For the same data rates, such functionality is generally more costly and more power hungry than asymmetric transmission without networking capabilities, as provided by, e.g., SerDes solutions (see also Section 2.2.6). If networking capabilities are needed, Ethernet must be used. However, if the application does not need networking and is asymmetric as in the first Automotive Ethernet use case of connecting cameras (see also Section 3.4.2), the choice will be for the less costly solution that still meets the functional requirements of the application. To support a specific data rate symmetrically when only asymmetric behavior is needed will always be unfavorable.

Discussions to enable asymmetric Ethernet transmission thus started with the MultiGBASE-T1 standardization (see e.g. [113]). However, allowing for different speeds per direction in the PHY creates multiple PHYs, high speed receive and lower speed transmit in one PHY, and high speed transmit and lower speed receive in the other. For MultiGBASE-T1 this was not desired. Instead. MultiGBASE-T1 enables asymmetric EEE, so that at least the power consumption of the MultiGBASE-T1 PHYs can be improved for such use cases (see Section 6.3.1).

Nevertheless, this was the reason that, for the new automotive projects – 10G+ electrical (shortened IEEE 802.3cy) and the 1G+ optical (shortened IEEE 802.3cz) – enabling asymmetric data transmission was being discussed from the beginning. After all, more than 80% of all 10G+ use cases are expected to be asymmetric in nature [32]. At the time of writing it remained an open discussion as to whether the asymmetry would be sufficiently supported by an extended EEE scheme that would allow total shutoff in the transmission direction not in use (while the basic transmission capabilities remain fully symmetric), or whether asymmetric data transmission capabilities should be supported as well. One aspect of disagreement was whether the savings in silicon area – especially for the complex echo canceller that is needed for full-duplex asymmetric communication – would actually merit the effort of specifying an asymmetric system [114, 115].

Notes

1 Which was a matter of many serious debates during the review process of the IEEE 802.3cg specification. See the meeting minutes or comments on the drafts on the project homepage for details [122].

2 The measured emissions are somewhat closer to the limit line than was the case for the first ever automotive 100BASE-T1/OABR measurement shown in Figure 3.6. This is due to the fact that the transmit power was increased for these later measurements. As 100BASE-T1/OABR had sufficient margin to the limit line, transmit power was added in order to further improve the immunity.

3 Not all car manufacturers have published their limit lines. It is therefore not possible for the authors to claim more than "many."

4 At IEEE 802.3 these channels were mainly specified at the TIA, sometimes at the same time as the IEEE Ethernet technology. This table shows the timeline.

Ethernet name	IEEE number	Year of publication	TIA channel used	Year of publication
10BASE-T	802.3i	1990	CAT 3 (telephone wiring)	1991
100BASE-TX	802.3u	1995	CAT 5	1995
1000BASE-T	802.3ab	1999	CAT 5e	1999
10GBASE-T	802.3an	2006	CAT 6 (shorter distances), CAT 6A	2002, 2007
40GBASE-T	802.3bq	ongoing	CAT 7A	2009

5 In the IEEE specification, the 15 m automotive link segment is called "Type A," while the optional 40m link segment (for aircraft, trains, buses, and heavy trucks) is called "Type B." This book only addresses the 15 m link and thus does not make the distinction.

6 The PSAACRF limit line calculation was significantly simplified from the IEEE description in [10]. The equation given here leads to values very close to the originally proposed ones.

7 Car manufacturers generally like star topologies as they reduce the number of harness variations in case of options and often simplify the repair process. If in a star an option is not selected, that link segment is simply removed. If a middle node in a passive linear topology is removed the harness segment changes. If the communication on one link is broken in a star topology, there is only one unit attached to the link, which must be checked. In contrast, in a passive linear topology, more than one ECU might have to be checked. However, for a function such as topology discovery, a passive linear topology is actually advantageous. Topology discovery would allow, e.g., the assignment of a specific IP address to a unit on a specific position in the chain. This can in many cases simplify the production process.

8 First internal tests show that a significantly larger number of nodes is possible. At IEEE a multidrop enhancement study group was initiated to formalize this into a standard with support for at least 16 nodes [94].

9 1000BASE-T, in turn, apparently had its roots in 100BASE-T2, a technology which never became successful commercially [124].

10 In IEEE 802.3 nomenclature, 100BASE-TX is "full-duplex" (see also endnote 5 of Chapter 1 for an in-depth explanation). At IEEE 802.3, the definition of full or half-duplex at IEEE is based on what happens at the (x)MII interface and not on what happens on the medium. In order not to cause too much naming confusion, this book will use "dual simplex" for 100BASE-TX. The occasional addition "true" to full-duplex results from trying to be unambiguous in the naming and indicates when transmit and receive signals simultaneously use the same cable pair(s).

11 Not every LFSR results in a sequence of maximum length that is repeated after 2^n-1 register shifts. A suitable LFSR should have few XOR functions, should be long enough to be able to create an almost (pseudo) random sequence, and should have the right length for the transmission power to be spread relatively evenly over the power density spectrum. LFSRs that do repeat after 2^n-1 register shifts use so-called primitive polynomials [116, 119, 120].

12 The following equation explains the derivation of Syn[2] and Sxn[1] in Equation 5.11 using the shift register depicted in Figure 5.21.

$$\begin{aligned}
Sy_n[2] &= Ssr_n(3) \oplus Ssr_n(8) \quad with\ Ssr_n[0] = Scr_n[3] \oplus Scr_n[8] \\
\Rightarrow Sy_n[2] &= Scr_n[3+3=6] \oplus \underbrace{Scr_n[8+3=11] \oplus Scr_n[3+8=11]}_{=0} \oplus Scr_n[8+8=16] \\
&= Scr_n[6] \oplus Scr_n[16] \\
Sx_n[1] &= Ssr_n(4) \oplus Ssr_n(6) \quad with\ Ssr_n[0] = Scr_n[3] \oplus Scr_n[8] \\
\Rightarrow Sx_n[1] &= Scr_n[4+3=7] \oplus Scr_n[6+3=9] \oplus Scr_n[4+8=12] \oplus Scr_n[6+8=14] \\
&= Scr_n[7] \oplus Scr_n[9] \oplus Scr_n[12] \oplus Scr_n[14]
\end{aligned}$$

13 During transmission, the signals on the line generally get distorted, in amplitude and in shape. This means that, though the PAM3 symbol "0" is transmitted as 0 V, it might actually be received as 0.1 V = "0" if the disturbance is low, or as 0.6 V = "1" if the disturbance is high. While the Symbol Error Rate (SER) is independent of the 3B2T encoding used but solely depends on the noise added, it makes a difference for the BER.

With a Gray code, detecting a "1" instead of a "0" for one of the PAM3 symbols in the 2D-PAM3 conversion will result in one bit error. In the 3B2T coding of 100BASE_T1 used, it might result in two bit errors.

14 Note that this number changed from the BroadR-Reach to the 100BASE-T1 specification. For BroadR-Reach a much longer maxwait_time of 1406 ms ± 18 ms for the master and 656 ms ± 9 ms for the slave had been defined [121].

15 The link_control parameter is normally used by the auto-negotiation function to overrule other system functions in case of required changes initiated by it. If auto-negotiation is not activated/foreseen in the system (as is the case for 100BASE-T1), the link_control parameter can be expected to be enabled continuously as soon as the system has power.

16 As was described in Section 3.3, it had been BMW's initial intention to create a 100BASE-TX based solution that passes the automotive requirements. In 2007 this was not successful. The authors therefore believe that the general knowhow increase that has been generated in order to enable 100BASE-T1/BroadR-Reach for automotive EMC (plus some general technological progress) helped to find more suitable measures to also improve the EMC behavior of 100BASE-TX hardware.

17 Blue, green, and yellow light actually show better attenuation behavior for POF than red light [117]. However, blue light has only recently started to be commercially exploited (e.g., Blu-ray) and more uses might appear in the future.

18 The technology was first published at the VDE/DKE v 0885-763 in 2012 [118] before its standardization was launched at IEEE 802.3 with a CFI in March 2014 and the respective IEEE802.3bv alias 1000BASE-RH TF in January 2015.

19 As discussed in Section 2.2.2, CAN FD is designed for shared gross data rates of 2 Mbps and 5 Mbps. With a proposed maximum load of 50% and a header that is still sent at 500 kbps, the resulting data rate is significantly lower than 10 Mbps. Even if the next generation CAN development successfully realizes an effective line rate of 10 Mbps, the arbitration will still prevent loading the bus to much more than 50%. For FlexRay (see Section 2.2.5), the gross data rate of 10 Mbps cannot be fully realized. The experience at BMW showed that the system overhead as well as the actually chosen segmentation between static and dynamic sections and unused assigned transmission slots can easily reduce the effective data rate to 20% (i.e., to 2 Mbps). CAN and FlexRay are, however, at a similar order of magnitude and sufficient for many applications.

20 The authors recall that, in the beginning, the possibility to simply tone down a 100BASE-T1 PHY was also discussed. Experts from semiconductor vendors then assessed that this would yield too small a cost reduction to justify the development of a new PHY. It was therefore clear from the beginning that a 10 Mbps Automotive Ethernet PHY would require a new development.

21 For the topology, in which many ECUs are connected to and share the bandwidth of the same cable, the automotive industry normally uses the term "bus," as has been done in this book up to this point. As the IEEE specification [21] uses the term "multidrop" for the same setup with 10BASE-T1S, this term is used in the following sections.

22 The IEEE 802.3cg specification defines half-duplex P2P as mandatory. While it is possible to be standard-compliant and sell a 10BASE-T1S PHY supporting half-duplex P2P only, it is not very likely that this will be done. It is more likely that the 10BASE-T1S PHY will always either also support full-duplex or multidrop, as both variants significantly improve the value proposition (full-duplex increases the overall available data rate and multidrop allows for cost savings).

23 100BASE-T1 defines a root mean square jitter tolerance of 50 ps in master mode and 150 ps in slave mode [1]. 10BASE-T1S accepts 5 ns symbol-to-symbol jitter [21]. When converting the root mean square jitter into a symbol-to-symbol jitter (times ~6.7 at

BER = 10^{-10}), this leads to approx. 350ps in master mode and approx. 1 ns in slave mode. That is, 10BASE-T1S can tolerate about five times the jitter of 100BASE-T1.

24 The general principle of DME also allows this to be done the other way around. Also, the one voltage change can be defined to be at the beginning of a bit, or once during the bit transmission.

25 Alternatively, a 14 pin RMII can be used. However, this does not make much difference in the overall packet size.

26 Note that this is similar to the access method used in the (optional) dynamic segment of a FlexRay cycle (see Section 2.2.5.2).

27 Plain CSMA/CD could be very problematic in an automotive environment because of the EMC requirements. On the one hand, collisions and re-transmissions create undesirable peaks in the spectrum, which creates problems from an emission perspective. On the other hand, the noise levels can be so high (e.g., during a BCI test) that all PHYs would mistake the noise for a collision, preventing the MAC from transmitting and eventually discarding the packet. As PLCA avoids collisions by design, packets can be sent through a noisy link and rely on the ability of the receiver to properly decode it, just as in 100 and 1000 BASE-T1. This part allows for making robust implementations.

28 The figure is intentionally different from the one provided in the specification (see [21]). This picture visualizes the logic.

29 The exact throughput data for FlexRay depends on the low level parameters decided for in the application. The basis for the calculation shown here is a 3 ms cycle with 91 transmit slots when the payload contains 16 bytes. This defines the absolute channel idle times, which were used also in cases where the payload contains fewer or more bytes.

References

[1] IEEE Computer Society, *802.3bw-2015 – IEEE Standard for Ethernet Amendment 1 Physical Layer Specifications and Management Parameters for 100 Mb/s Operation over a Single Balanced Twisted Pair Cable (100BASE-T1)*, New York: IEEE-SA, 2015.

[2] B. Körber, S. Buntz, M. Kaindl, D. Hartmann, and J. Wülfing, *BroadR-Reach® Physical Layer Definitions for Communication Channel, v1.0*, Irvine, CA: OPEN Alliance, 2017.

[3] N. A. Wienckowski, "I Can Use CAN Wire for Ethernet, Right?," in *IEEE-SA Ethernet & IP @ Automotive Technology Day*, Detroit, 2014.

[4] K. Matheus, M. Kaindl, S. Korzine, D. Goncalvez, J. Leslie, Y. Okuno, N. Kitajima, M. Gardner, R. Orosz, M. Jaenecke, D. Kim, R. Mei, and D. v. Knorre, "1 Pair or 2 Pairs for RTPGE: Impact on System Other than the PHY Part 1: Weight & Space," January 23, 2013. [Online]. Available: http://grouper.ieee.org/groups/802/3/bp/public/jan13/matheus_3bp_01_0113.pdf. [Accessed May 14, 2020].

[5] S. Buntz, "IEEE RTPGE Information: Update on Required Cable Length," July 17, 2012. [Online]. Available: http://grouper.ieee.org/groups/802/3/RTPGE/public/july12/buntz_02_0712.pdf. [Accessed May 14, 2020].

[6] L. Winkel, "CFI 'Reduced Pair Gigabit Ethernet PHY' Industrial Environment," April 26, 2012. [Online]. Available: http://grouper.ieee.org/groups/802/3/RTPGE/public/may12/winkel_01_0512.pdf. [Accessed May 14, 2020].

[7] S. Carlson, "Reduced Twisted Pair Gigabit Ethernet Study Group: Objectives," November 15, 2012. [Online]. Available: www.ieee802.org/3/RTPGE/Approved_Objectives_1112.pdf. [Accessed May 14, 2020].

[8] C. DiMinico, "Reduced Twisted Pair Gigabit Ethernet Link Segment Definitions Decision Process," March 22, 2013. [Online]. Available: www.ieee802.org/3/bp/public/mar13/diminico_3bp_02_0313.pdf. [Accessed May 14, 2020].

[9] M. Tazebay, "A Modeling & Measurement Technique for RTPGE EMC Channel Analysis," January 24, 2013. [Online]. Available: www.ieee802.org/3/bp/public/jan13/tazebay_3bp_01a_0113.pdf. [Accessed May 14, 2020].

[10] IEEE Computer Society, *802.3bp-2016 – IEEE Standard for Ethernet Amendment 4: Physical Layer Specifications and Management Parameters for 1 Gb/s Operation over a Single Twisted Pair Copper Cable*, New York: IEEE-SA, 2016.

[11] M. Tazebay, "Immunity Analysis & Test Results via Bulk Current Injection Method for 1-pair UTP Channels," May 15, 2013. [Online]. Available: www.ieee802.org/3/bp/public/may13/tazebay_3bp_01_0513.pdf. [Accessed May 14, 2020].

[12] A. Chini and M. Tazebay, "RTPGE BCI Noise Analysis for Common Mode Termination & Grounding Effects," November 12, 2013. [Online]. Available: www.ieee802.org/3/bp/public/nov13/Chini_Tazebay_3bp_02a_1113.pdf. [Accessed May 14, 2020].

[13] G. Zimmerman, "Salz SNR Text and Procedure," August 25, 2015. [Online]. Available: www.ieee802.org/3/NGEBASET/public/archadhoc/zimmerman_3bzah_01a_0815.pdf. [Accessed May 14, 2020].

[14] A. Vareljian and H. Takatori, "Channel Qualification Based on Salz," September 2012. [Online]. Available: www.ieee802.org/3/bj/public/sep12/vareljian_3bj_01_0912.pdf. [Accessed May 14, 2020].

[15] M. Tazebay, "A Review of 1000BASE-T1 PHY Architecture for EMC Constrained Channels," in *IEEE-SA (4th) Ethernet & IP @ Automotive Technology Day*, Detroit, 2014.

[16] C. DiMinico and M. Tazebay, "IEEE P802.3bp (RTPGE) PHY Task Force Channel Definitions Ad Hoc Report," November 2013. [Online]. Available: www.ieee802.org/3/bp/public/nov13/ch_ad_hoc_3bp_01_1113.pdf. [Accessed May 14, 2020].

[17] M. Tazebay and S. Buntz, "Reduced Twisted Pair Gigabit Ethernet EMC & Noise Ad Hoc Report," November 12, 2013. [Online]. Available: www.ieee802.org/3/bp/public/nov13/EMCnoise_ad_hoc_3bp_01_1113.pdf. [Accessed May 14, 2020].

[18] W. Larsen, R. Mei, B. Moffitt, and T. Herman, "RTPGE Channel Requirements Proposal for 1-Pair Ethernet," July 10, 2013. [Online]. Available: www.ieee802.org/3/bp/public/jul13/herman_3bp_01_0713.pdf. [Accessed May 14, 2020].

[19] B. Körber, B. Bergner, D. Marinac, D. Dorner, F. Bauer, H. Patel, J. Razafiarivelo, M. Dörndl, M. Kaindl, M. Rucks, M. Nikfal, P. Gowravajhala, T. Müller, T. Wunderlich, V. Raman, W. Mir, and Y. Bouri, *Channel and Component Requirements for 1000BASE-T1 Link Segment Type A (STP), v1.0*, Irvine, CA: OPEN Alliance, 2019.

[20] T. Müller and B. Bergner, *Channel Component Requirements for 1000BASE-T1 Automotive Ethernet, v. 1.1*, Irvine, CA: OPEN Alliance, 2016.

[21] IEEE Computer Society, *802.3cg-2019 – IEEE Standard for Ethernet Amendment 5: Physical Layer and Management Parameters for 10 Mb/s Operation and Associated Power Delivery over a Single Balanced Pair of Conductors*, New York: IEEE-SA, 2019.

[22] C. Wechsler, C. Schanze, and O. Krieger, "10SPE Automotive PHY Channel Consideration – TIme Domain Simulation," May 21, 2017. [Online]. Available: www.ieee802.org/3/cg/public/May2017/wechsler_3cg_01a_0517.pdf. [Accessed January 11, 2020].

[23] M. Kaindl and K. Matheus, "Automotive Link Segment for 10SPE Including Multidrop," September 13, 2017. [Online]. Available: www.ieee802.org/3/cg/public/Sept2017/kaindl_matheus_3cg_01c_09_2017.pdf. [Accessed January 11, 2020].

[24] S. Carlson, H. Zinner, K. Matheus, N. Wienckowski, and T. Hogenmüller, "CFI Multi-Gig Automotive Ethernet PHY Consensus Presentation," November 9, 2016. [Online]. Available: www.ieee802.org/3/ad_hoc/ngrates/public/16_11/20161108_CFI.pdf. [Accessed May 6, 2020].

[25] IEEE 802.3, "Homepage of the IEEE 802.3 Multi-Gigabit Automotive Ethernet PHY Study Group," 2017 (closed). [Online]. Available: www.ieee802.org/3/NGAUTO/index.html. [Accessed January 11, 2020].

[26] IEEE 802.3, "Approved Objectives for IEEE 802.3ch," March 16, 2017. [Online]. Available: www.ieee802.org/3/ch/0317_approved_objectives_3NGAUTO.pdf. [Accessed January 11, 2020].

[27] E. DiBiaso and B. Bergner, "Media Coniderations – Insertion Loss and EMC," May 2017. [Online]. Available: www.ieee802.org/3/ch/public/may17/DiBiaso_3NGAUTO_01_0517.pdf. [Accessed January 15, 2020].

[28] O. Grau, "Some Remarks/Thoughts about Automotive Multigig Link Segments," May 2017. [Online]. Available: www.ieee802.org/3/ch/public/may17/grau_3NGAUTO_02_0517.pdf. [Accessed January 15, 2020].

[29] R. Vernickel, "Actual Cable Data Update," November 2018. [Online]. Available: www.ieee802.org/3/ch/public/nov18/vernickel_3ch_01c_1118.pdf. [Accessed January 15, 2020].

[30] IEEE Computer Society, *802.3ch-2020 – IEEE Standard for Ethernet Amendment: Physical Layer Specifications and Management Parameters for 2.5 Gb/s, 5 Gb/s, and 10 Gb/s Automotive Electrical Ethernet*, New York: IEEE-SA, 2020.

[31] B. Bergner and E. DiBiaso, "Robust Connectivity Solution for Next Generation Automotive Ethernet," *Ethernet & IP @ Automotive Technology Day*, Detroit, 2019.

[32] C. Mash, "Network Topology Analysis," July 2019. [Online]. Available: www.ieee802.org/3/B10GAUTO/public/jul19/mash_B10GAUTO_1_0719.pdf. [Accessed February 3, 2020].

[33] Broadcom, *YD/T 1947-2009: Technical Requirement of Physical Layer for Extended Reach Ethernet based on 2D-PAM3 and 4-D PAM5*, Beijing: Ministry of Industry and Information Technology, PRC, 2009.

[34] C. Mick, C. DiMinico, S. Raghavan, S. Rao, and M. Hatamian, "802ab, A Tutorial Presentation," March 17, 1998. [Online]. Available: http://grouper.ieee.org/groups/802/3/tutorial/march98/mick_170398.pdf. [Accessed November 27, 2013].

[35] IEEE Computer Society, *802.3-2008 IEEE Standard for Carrier Sense Multiple Access with Collision Detection (CSMA/CD) Access Method and Phyiscal Layer Specifications*, New York: IEEE-SA, 2008.

[36] F.-J. Kauffels, "1000 BASE-T: Gigabit Ethernet über Kupferverkabelung (1/2)," September 18, 2012. [Online]. Available: www.comconsult-research.de/1000-base-t-gigabit-ethernet-uber-kupferverkabelung-12./1000-base-t-gigabit-ethernet-uber-kupferverkabelung-12./. [Accessed February 27, 2014, no longer available].

[37] S. Patwardhan, "Gigabit Ethernet over Copper: Hardware Architecture and Operation," 2001. [Online]. Available: www.dell.com/content/topics/global.aspx/power/en/ps4q01_patward?c=u. [Accessed November 13, 2016, no longer available].

[38] W. Buchanan, "Gigabit Ethernet Over Category 5 U.T.P. Cabling," Master Thesis, unknown.

[39] B. Noseworthy, "Gigabit Ethernet 1000BASE-T, update," March 29, 2000. [Online]. Available: www.iol.unh.edu/services/testing/ge/knowledgebase/pcs.pdf. [Accessed 14, May 2020].

[40] IEEE Computer Society, *802.3az-2010 – IEEE Standard for Carrier Sense Multiple Access with Collision Detection (CSMA/CD) Amendment 5: Media Access Control Parameters, Physical Layers, and Management Parameters for Energy-Efficient Ethernet*, New York: IEEE-SA, 2010.

[41] O. E. Agazzi, J. L. Creigh, M. Hatamian, D. E. Kruse, A. Abnous, and H. Samueli, "Multi-Pair Gigabit Ethernet Transceiver." US Patent 2011/0064123 A1, March 17, 2011.

[42] Ethernet Alliance, "Ethernet Alliance Panel #2: Bandwidth Growth and The Next Speed of Ethernet," 2012. [Online]. Available: www.ethernetalliance.org/wp-content/uploads/2012/04/Ethernetnet-Alliance-ECOC-2012-Panel-2.pdf. [Accessed May 7, 2020].

[43] M. Tazebay and T. Suermann, "Deployed 100Mb/s One Pair OABR PHY," May 2012. [Online]. Available: www.ieee802.org/3/1TPCESG/public/tazebay_01_0512.pdf. [Accessed May 14, 2020].

[44] IEEE 802.3, "IEEE 802.3ba 40Gb/s and 100Gb/s Ethernet Task Force Homepage," June 19, 2010 (closed). [Online]. Available: http://grouper.ieee.org/groups/802/3/ba/index.html. [Accessed May 15, 2020].

[45] A. Abaye, "BroadR-Reach Technology: Enabling One Pair Ethernet," 2013. [Online]. Available: www.ethercat.org/2013/mobile_applications/files/04_EtherCAT_Mobile_App_Broadcom.pdf. [Accessed May 15, 2020].

[46] Wikipedia, "Gray Code," May 15, 2020. [Online]. Available: https://en.wikipedia.org/wiki/Gray_code. [Accessed May 16, 2020].

[47] Züricher Hochschule Winterhur, "Kapitel 5: Digitale Übertragung im Basisband," 2005. [Online]. Available: https://home.zhaw.ch/~rur/ntm/unterlagen/ntmkap51einbb.pdf. [Accessed November 5, 2016, no longer available].

[48] M. Jones, "Zwickau University of Applied Science EMC Lab Testing for Automotive Ethernet Physical Layer Testing," September 2012. [Online]. Available: http://grouper.ieee.org/groups/802/3/RTPGE/public/sept12/jones_01_0912.pdf. [Accessed May 14, 2020].

[49] D. Pannell, "Passing Automotive EMC with IEEE Standard PHYs," September 2012. [Online]. Available: http://grouper.ieee.org/groups/802/3/RTPGE/public/sept12/pannell_01_0912.pdf. [Accessed May 14, 2020].

[50] LegalForce Trademarkia, "Quiet-wire," July 24, 2013. [Online]. Available: www.trademarkia.com/quietwire-85360886.html. [Accessed May 14, 2020].

[51] L. Frenzel, "Networking Technologie Take Charge at Home and on the Road," November 8, 2012. [Online]. Available: http://electronicdesign.com/communications/networking-technologies-take-charge-home-and-road. [Accessed May 14, 2020].

[52] Computer Business Review, "Media-independent Interface for 100BASE-T Published," April 12, 1994. [Online]. Available: www.cbronline.com/news/media_independent_interface_specification_for_100base_t_published. [Accessed May 15, 2020].

[53] R. Kraus, "APIX2 – a Whole Flood of Data," *Elektronik*, pp. 2–5, July 2011.

[54] S. Carlson, T. Hogenmüller, K. Matheus, T. Streichert, D. Pannell, and A. Abaye, "Reduced Twisted Pair Gigabit Ethernet Call for Interest," March 2012. [Online]. Available: www.ieee802.org/3/RTPGE/public/mar12/CFI_01_0312.pdf. [Accessed May 6, 2020].

[55] H. Zinner, K. Matheus, S. Buntz, and T. Hogenmüller, "Requirements Update for IEEE 802.3 RTPGE," July 18, 2012. [Online]. Available: http://grouper.ieee.org/groups/802/3/RTPGE/public/july12/zinner_02_0712.pdf. [Accessed January 2, 2020].

[56] T. Hogenmüller and H. Zinner, "Tutorial for Wake Up Schemes and Requirements for Automotive Communication Networks," July 18, 2012. [Online]. Available: http://grouper

.ieee.org/groups/802/3/RTPGE/public/july12/hoganmuller_02a_0712.pdf. [Accessed May 14, 2020].

[57] T. Hogenmüller and H. Zinner, "Tutorial for Lifetime Requirements and Physical Testing of Automotive Electronic Control Units (ECUs)," July 18, 2012. [Online]. Available: http://grouper.ieee.org/groups/802/3/RTPGE/public/july12/hoganmuller_01a_0712.pdf. [Accessed May 14, 2020].

[58] IEEE 802.3, "Approved Minutes, IEEE 802.3 Ethernet Working Group Plenary Grand Hyatt, San Antonio, TX USA," November 12–15, 2012. [Online]. Available: www.ieee802.org/3/minutes/nov12/minutes_1112.pdf. [Accessed May 6, 2020].

[59] P. Wang and M. Tazebay, "1000BASE-T1 PHY Synchronization Method & Start-up Process and Link Failure Case Analysis," July 15, 2014. [Online]. Available: www.ieee802.org/3/bp/public/jul14/wang_3bp_01_0714.pdf. [Accessed May 16, 2020].

[60] J. Graba, "EEE Proposal for 1000BASE-T1," 16 July 2014. [Online]. Available: www.ieee802.org/3/bp/public/jul14/Graba_3bp_01a_0714.pdf. [Accessed May 16, 2020].

[61] Wikipedia, "IEEE 802.3," Wikipedia, April 28, 2020. [Online]. Available: http://en.wikipedia.org/wiki/IEEE_802.3. [Accessed May 6, 2020].

[62] F. Idachaba, D. U. Ike, and O. Hope, "Future Trends in Fiber Optics Communication," July 2, 2014. [Online]. Available: www.iaeng.org/publication/WCE2014/WCE2014_pp438-442.pdf. [Accessed May 14, 2020].

[63] Y. Tsukamoto, "Plastic Optical Fiber (POF) Technology for Automotive, Home Network Systems," May 2014. [Online]. Available: www.ieee802.org/3/GEPOFSG/public/May_2014/Tsukamoto_GEPOF_01_0514.pdf. [Accessed May 14, 2020].

[64] B. Bergkvist, "Automotive Grade Gigabit Ethernet Links Robustness against UWB Pulses," *Automotive Ethernet Congress*, Munich, 2019.

[65] R. Kawabuchi and D. Yumoto, "Comparative Analysis on the EMC Characteristics of Optical and Electrical Ethernet," *Automotive Ethernet Congress*, Munich, 2019.

[66] T. Lichtenegger, K. Matheus, V. Raman, A. Engel, S. Kobayashi, R. Orosz, M. Jaenecke, and N. Serisawa, "Comparison of Electrical and Optical Transmission for Gbps Ethernet v.1.0," May 13, 2014. [Online]. Available: www.ieee802.org/3/GEPOFSG/public/May_2014/Pardo_GEPOF_1_0514.pdf. [Accessed May 14, 2020].

[67] C. Pardo, T. Lichtenegger, A. Paris, and H. Hirayama, "Gigabit Ethernet over Plastic Optical Fibre, Call For Interest," March 27, 2014. [Online]. Available: www.ieee802.org/3/GEPOFSG/public/CFI/GigPOF%20CFI%20v_1_0.pdf. [Accessed May 6, 2020].

[68] KDPOF, "Simple Introduction to Gigabit Communications over POF, Revision 1.1," 2014. [Online]. Available: www.kdpof.com/wp-content/uploads/2012/07/Easy-Introduction-v1.1.pdf. [Accessed May 14, 2020].

[69] Thorlabs, "Graded-Index Polymer Optical Fiber (GI-POF)," not known. [Online]. Available: pofto.com/downloads/ieee/pof.branches.v6.doc. [Accessed May 15, 2020].

[70] B. Grow, C. Pardo, N. Serisawa, and T. Lichtenegger, "Tutorial: Gigabit Ethernet Over Plastic Optical Fiber (GEPOF)," November 3, 2014. [Online]. Available: www.ieee802.org/3/GEPOFSG/public/Tutorial/tutorial_GEPOF_1d_1114.pdf. [Accessed May 14, 2020].

[71] M. Tomlinson, "New Automatic Equalizer Employing Modulo Arithmetic," *Electronics Letter*, Vols. 5–6, no. 7, pp. 138–139, March 1971.

[72] H. Harashima and H. Miyakawa, "Matched-Transmission Technique for Channels with Intersymbol Interference," *IEEE Transactions on Communications*, vol. 20, no. 4, pp. 774–780, August 1972.

[73] IEEE Computer Society, *802.3bv-2017 – IEEE Standard for Ethernet Amendment 9: Physical Layer Specifications and Management Parameters for 1000 Mb/s Operation Over Plastic Optical Fiber*, New York: IEEE-SA, 2017.

[74] K. Matheus, "10Mbps Ethernet, the Missing Link for a Homogeneous In-Vehicle Network," in *Automotive Ethernet Congress*, Munich, 2018.

[75] K. Matheus, "Die Zukunft von Vernetzungstechnologien im Fahrzeug – Ergebnisse einer Umfrage," February 2016. [Online]. Available: www.hanser-automotive.de/zeitschrift/archiv/artikel/die-zukunft-von-vernetzungstechnologien-im-fahrzeug-ergebnisse-einer-umfrage-1299263.html. [Accessed May 15, 2020].

[76] C. Gauthier, "PHY Feasibility: The Impact of Signalling on Power and Area," September 13, 2016. [Online]. Available: http://grouper.ieee.org/groups/802/3/10SPE/public/Sept2016_Interim/gauthier_10SPE_01a_09132016.pdf. [Accessed May 15, 2020].

[77] M. Tazebay, "Feasibility Framework for 10SPE Automotive," in *IEEE-SA Ethernet & IP @ Automotive Technology Day*, Munich, 2016.

[78] L. Winkel, M. McCarthy, D. Brandt, G. Zimmerman, D. Hoglund, K. Matheus, and C. DiMinico, "10Mb/s Single Twisted Pair Ethernet Call for Interest," July 26, 2016. [Online]. Available: www.ieee802.org/3/cfi/0716_1/CFI_01_0716.pdf. [Accessed May 24, 2020].

[79] G. Zimmerman, "IEEE 802.310 Mb/s Single Twisted Pair Ethernet (10SPE) Study Group Closing Report," November 10, 2016. [Online]. Available: www.ieee802.org/3/minutes/nov16/1116_10M_stp_close_report.pdf. [Accessed May 6, 2020].

[80] G. Zimmerman, "Adopted Objectives," November 10, 2016. [Online]. Available: www.ieee802.org/3/10SPE/objectives_10SPE_111016.pdf. [Accessed May 15, 2020].

[81] P. Beruto and A. Orzelli, "TIS Scrambler Proposal," March 28, 2018. [Online]. Available: www.ieee802.org/3/cg/public/adhoc/beruto_3cg_scrambler.pdf. [Accessed August 23, 2018].

[82] IEEE Computer Society, *802.3u-1995 – IEEE Standards for Media Access Control (MAC) Parameters, Physical Layer, Medium Attachment Units, and Repeater for 100Mb/s Operation, Type 100BASE-T (Clauses 21–30)*, New York: IEEE-SA, 1995.

[83] H. Zweck, M. Miller, T. Baggett, P. Beruto, and T. Beilitz, *OPEN Alliance Serial 10BASE-T1x MACPHY Interface, v1.0*, Irvine, CA: OPEN Alliance, 2020.

[84] P. Beruto and M. Muth, *OPEN Alliance 10BASE-T1S Transceiver Interface Interoperability Specification, v1.0*, Irvine, CA: OPEN Alliance, 2019.

[85] P. Beruto and A. Orzelli, "PHY-Level Collision Avoidance – Adendum #1," August 17, 2017. [Online]. Available: www.ieee802.org/3/cg/public/adhoc/8023cg_PLCA_addendum_01.pdf. [Accessed January 19, 2020].

[86] Wikipedia, "Round-Robin Scheduling," December 10, 2019. [Online]. Available: https://en.wikipedia.org/wiki/Round-robin_scheduling. [Accessed January 22, 2020].

[87] P. Beruto and A. Orzelli, "802.3cg Draft 2.0 PLCA (Clause 148) Overview," July 9, 2018. [Online]. Available: www.ieee802.org/3/cg/public/July2018/PLCA%20overview.pdf. [Accessed January 22, 2020].

[88] D. Pannell, "802.3cg Supporting QoS on PLCA: A Problem Statement," October 4, 2018. [Online]. Available: www.ieee802.org/3/cg/public/adhoc/cg-pannell-QoS-for-PLCA-ProblemStatement-1018-v01.pdf. [Accessed January 26, 2020].

[89] A. Meier, "Analysis of Worst Case Latencies in a 10Mbps Ethernet Network with PLCA," January 2018. [Online]. Available: www.ieee802.org/3/cg/public/adhoc/Analysis_of%20Worst_Case_Latencies_for_10M_Eth_with_PLCA_Alexander_Meier_3....pdf. [Accessed January 26, 2020].

[90] D. Pannell, "Priority Support for PCA," September 2018. [Online]. Available: www.ieee802.org/3/cg/public/adhoc/cg-pannell-Priority-for-PLCA-0918-v05.pdf. [Accessed January 26, 2020].

[91] K. Matheus, "Fairness Considerations for PLCA, Example Microphone Use Case," April 11, 2018. [Online]. Available: www.ieee802.org/3/cg/public/adhoc/matheus_cg_1_0418.pdf. [Accessed January 26, 2020].

[92] Wikipedia, "IP Fragmentation," December 12, 2019. [Online]. Available: https://en.wikipedia.org/wiki/IP_fragmentation. [Accessed January 26, 2020].

[93] P. Beruto and A. Orzelli, "PLCA Burst Mode Fixes," January 2019. [Online]. Available: www.ieee802.org/3/cg/public/Jan2019/beruto_3cg_burst_mode_fixes_revC.PDF. [Accessed Janaury 26, 2020].

[94] IEEE 802.3, "Homepage of the IEEE 802.3 10SPE Multidrop Enhancements Study Group," 2019 (continuously updated). [Online]. Available: www.ieee802.org/3/SPMD/index.html. [Accessed January 29, 2020].

[95] Wikipedia, "5G," 3 February 2020. [Online]. Available: https://en.wikipedia.org/wiki/5G/5G. [Accessed February 3, 2020].

[96] Wikipedia, [Online]. Available: www.wikipedia.org.

[97] MakeUseOf, "EDGE, 3G, H+, Etc: What Are All These Mobile Networks?," February 2016. [Online]. Available: www.makeuseof.com/tag/edge-3g-h-etc-mobile-networks/. [Accessed May 16, 2020].

[98] Wikipedia, "List of Digital Television Deployments by Country," April 12, 2016. [Online]. Available: https://en.wikipedia.org/wiki/List_of_digital_television_deployments_by_country. [Accessed May 25, 2016].

[99] Wikipedia, "Digital Radio," May 9, 2020. [Online]. Available: https://en.wikipedia.org/wiki/Digital_radio. [Accessed May 15, 2020].

[100] K. Barbehön, "Mitten im Umbruch zur Digitalisierung: Wie die Zukunft der E/E Architektur von IT-Standards beeinflusst wird," in *21st International Congress on Advances in Automotive Electronics*, Ludwigsburg, 2016.

[101] G. Armbrecht, "Plugs and Cables for Data Transmission up to 12 Gbps," in High-Speed Datenübertragung in Automobilnetzen der Zukunft, Erlangen, 2015.

[102] H. Adel, "Methodology for Cable Performance Optimization," in High-Speed Datenübertragung in Automobilnetzen der Zukunft, Erlangen, 2015.

[103] M. Kuijk, "Qualification of Coax as Physical Layer for In-car Networks," in MOST Forum, Esslingen, 2014.

[104] T. Kimura and M. Dittmann, "Glass Fiber Solution for Multi-gigabit per Second," in MOST Interconnectivity Conference Asia, Tokyo, 2015.

[105] S. Carlson, C. Mash, C. Wechsler, H. Zinner, O. Grau, and N. Wienckowski, "10G+ Automotive Ethernet Electrical PHYs, Call for Interest Consensus Presentation," March 12, 2019. [Online]. Available: www.ieee802.org/3/cfi/0319_1/CFI_01_0319.pdf. [Accessed December 7, 2019].

[106] S. Carlson, "Greater than 10 Gb/s Automotive Ethernet Electrical PHYs Objectives," November 2019. [Online]. Available: www.ieee802.org/3/B10GAUTO/B10GAUTO_OBJ_DRAFT_03_1119.pdf. [Accessed February 3, 2020].

[107] H. M. Kadry, "B10G 100Gb/s Use Cases," January 14, 2020. [Online]. Available: www.ieee802.org/3/B10GAUTO/public/jan20/kadry_3B10GAUTO_01_0120.pdf. [Accessed February 3, 2020].

[108] C. Pardo, H. Goto, T. Nomura, and B. Grow, "Call for Interest Automotive Optical Multi Gig PHY Consensus Presentation," July 2019. [Online]. Available: www.ieee802.org/3/cfi/0719_1/CFI_01_0719.pdf. [Accessed February 3, 2020].

[109] C. Pardo, "Optical Multi Gig Ethernet for Automotive Objectives," January 2020. [Online]. Available: www.ieee802.org/3/OMEGA/public/jan_2020/objectives_OMEGA_01a_0120.pdf. [Accessed February 3, 2020].

[110] M. Eek, "Input – Automotive Use Cases and Requirements Multi Gigabit," January 2020. [Online]. Available: www.ieee802.org/3/OMEGA/public/jan_2020/eek_OMEGA_01_0120.pdf. [Accessed February 3, 2020].

[111] K. Matheus, *Survey Results SerDes*, Munich: Distributed among participants of the survey, 2018.

[112] C. Pardo, "Optical Multi Gigabit Ethernet for Automotive, Market Size Estimation," September 2019. [Online]. Available: www.ieee802.org/3/OMEGA/public/sep_2019/cpardo_OMEGA_01_0919.pdf. [Accessed February 3, 2020].

[113] T. Souvignier, M. Tu, and J. Graba, "Asymmetrical Data Transmission," July 11, 2018. [Online]. Available: www.ieee802.org/3/ch/public/jul18/souvignier_3ch_01a_0718.pdf. [Accessed February 23, 2020].

[114] G. Zimmerman, "Technical Feasibility – EEE for Asymmetry," November 2019. [Online]. Available: www.ieee802.org/3/B10GAUTO/public/nov19/zimmerman_3B10G_01_1119.pdf. [Accessed February 3, 2020].

[115] H. Sedarat, "Optimized Asymmetric Operation, Technical Feasibility," January 2020. [Online]. Available: www.ieee802.org/3/B10GAUTO/public/jan20/sedarat_3B10G_01_0120.pdf. [Accessed February 3, 2020].

[116] C. Stroud, "Linear Feedback Shift Registers (LFSRs)," October 2004. [Online]. Available: www.eng.auburn.edu/~stouce/class/elec6250/LFSR.pdf. [Accessed October 24, 2016].

[117] O. Ziemann, J. Krauser, and P. Zamzow, *POF Handbook*, Berlin: Springer, 2008.

[118] VDE Verband der Elektrotechnik Elektronik Informationstechnik e.V., *VDE V 0885-763*, Frankfurt: VDE, 2012.

[119] F. Ruskey, "Information on Primitive and Irreducible Polynomials," May 23, 2011. [Online]. Available: http://theory.cs.uvic.ca/inf/neck/PolyInfo.html. [Accessed October 14, 2016, no longer available].

[120] T. Ward and T. Molteno, "Table of Linear Feedback Shift Registers," October 26, 2007. [Online]. Available: http://courses.cse.tamu.edu/csce680/walker/lfsr_table.pdf. [Accessed November 5, 2016, no longer available].

[121] Broadcom, "OPEN Alliance BroadR-Reach (OABR) Physical Layer Transceiver Specification for Automotive Applications v.3.2," June 24, 2014. [Online]. Available: www.ieee802.org/3/1TPCESG/public/BroadR_Reach_Automotive_Spec_V3.2.pdf. [Accessed May 14, 2020].

[122] IEEE 802.3, "IEEE P802.3cg 10 Mb/s Single Pair Ethernet Task Force Homepage," 2019 (closed). [Online]. Available: www.ieee802.org/3/cg/index.html. [Accessed January 3, 2020].

6 Automotive Ethernet and Power Supply

In-vehicle power supply and energy consumption are engineered carefully. A reliable power supply is a prerequisite for the functioning of modern cars. At the same time, consumers are not only increasingly aware of the fuel consumption (costs), but also legislation worldwide is setting more stringent targets for CO_2 reduction.[1] In 2013, transportation was responsible for 23% of the worldwide CO_2 emissions and thus the second largest CO_2 producing sector after electricity and heat generation (which had a 42% share [1]). Within transportation, road vehicles account for 75% of the emissions, i.e., about 17% of the overall emissions. Over the last decades, CO_2 emissions in all sectors have grown significantly and, for road vehicles in particular, by 66% between 1990 and 2013. This shows that vehicle manufacturers have actually reduced the emissions per vehicle, as the number of registered vehicles about doubled in the same timeframe [2, 3]. However, more reductions are required and thus all elements inside the car are scrutinized with respect to their power efficiency.

There are many different factors that determine the energy consumption and CO_2 emission of a car [4]. Not all have to do with the car itself but are a result of other aspects such as driving style, weather, environment (city, country, highway, mountains, ...), etc. Those that are directly related to the car have to do with the energy needed by the engine and the energy needed for the electric and electronic components. Huge improvements have been made in respect to the efficiency of combustion engines; in fact, these are the main contributions to CO_2 emissions reduction. Nevertheless, the heavier the car, the more fuel the same engine needs for the same performance. With the wiring harness being the third heaviest component in a car [5], reducing weight in the harness directly impacts the power consumption. After a general introduction of the power supply system in a car (see Section 6.1), Section 6.2 describes possibilities with regards to Automotive Ethernet for reducing weight in the wiring harness.

While, overall, the car manufacturers have been able to make cars more fuel efficient – in Germany the fuel reduction per car between 1991 and 2012 is about 20% [4] – the amount of vehicle electrics and electronics has increased significantly, and with it the need for electric power. While very large power consumers include air-conditioning and exhaust purification, many new functions in the area of comfort and driver assistance also add to the power use. The in-vehicle networking technology can impact this in various ways; for one, with the complexity of the communication technology itself. With the specifications written (see Chapter 5), it is up to the semiconductor vendors to make competitive Automotive Ethernet products that consume as little power as possible when a link is up. Section 6.3 instead focusses on methods applicable to Automotive Ethernet that save power by reducing the power consumption as much as possible during the time a link and/or ECU attached to the link is actually not active: Section 6.3.1 discusses Energy Efficient Ethernet (EEE) and Section 6.3.2 wake-up and sleep mechanisms.

6.1 The Power Supply Network

The key purpose of the power supply network is to reliably provide power whenever needed. One aspect of reliability is that power is supplied at a constant level without voltage drops, as this might make ECUs go into reset. Drivers of cars will have noticed that most vehicle electronics are paused during engine start. This is because engine start is the most critical time for voltage drops. However, there are more situations in which voltage drops are possible. Figure 6.1 gives an overview on the elements relevant for the power supply network. The different elements are explained in detail in the following.

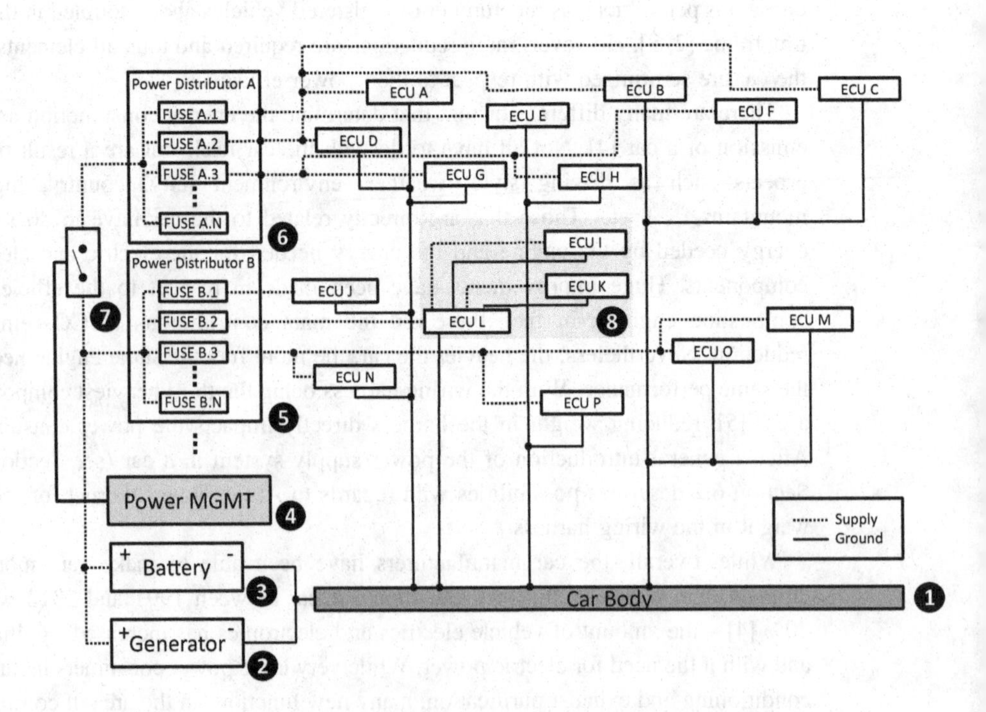

Figure 6.1 Example of a typical power distribution in a car. In this picture, "ECU" is used synonymously for all electrically powered units in the car, whether they are ECUs, sensors, or actuators

- Traditionally, cars are built with a conductive **metal body (1)**. This is very important, because the continuous metal body of the car is a reference point to which the negative pole of the car's battery is connected. The ground (negative pole of the battery) is thus distributed all over the car's body, and the ECU's ground can be connected to it via the most direct path available. By contrast, the positive pole requires distribution in a more complex fashion and must be well isolated from the car's body.

  With more digitization of vehicle functions and thus more electronics in vehicles, power consumption increases and with it the currents that need to be distributed. This causes new challenges as poor design can lead to EMC disturbances in the form of very strong magnetic fields due to the high currents needed. Twisting power supply cables where both the ground and positive cables are used, helps to reduce impairments with respect to EMC. Additionally, the body structure is changing gradually. This is mainly due to new materials being used with mixed-in plastic or bonded carbon fibers. These changes affect the current flow and can have a negative impact on the overall EMC behavior of the vehicle.
- **Electric generator (2)** is used to recharge the car's battery. The charging system in modern cars is extremely optimized. This is especially so in hybrid or electric vehicles where the power supply for the standard device uses a high-voltage system of approximately 400 V via DC/DC converters. A charge controller ensures that the electric generator charges the **battery (3)** correctly and makes the system stable. This maintains the power supply during vehicle operation.
- The **power management system (4)** ensures that energy is distributed and supplied to the ECUs. It must take care that the ECUs that do not require power in certain use scenarios are switched off from the power supply. This is also referred to as "clamp management."

  Most car designs distinguish between three different types of clamp systems. Units on "Clamp 15" receive power only when the engine has been switched on, which includes most units when the customer is driving. Units on "Clamp 30" are directly connected to the battery and always have power, even when the car is parked and not running (output of distribution box 5 in Figure 6.1). To consume as little quiescent current as possible, most car manufacturers will put only systems such as the alarm or door locks on Clamp 30. "Clamp 30g" is also directly connected to the battery but is switched (output of distribution box 6 in Figure 6.1). This might be used for functions that are usable for a certain time even when the car is switched off. Drivers might want to finish a telephone call they are having via the in-built hands-free system or the kids want to continue watching a movie in the Rear Seat Entertainment (RSE) while the parent buys fuel, etc.[2]

Depending on the complexity of the vehicle, a number of **on/off switches (7)** are used for this (Figure 6.1 shows only one). These switches are traditionally classic relays. Today, specialized semiconductors can be used for this purpose. In either case, if an ECU is switched off from the power supply, it cannot be switched on again based on anything that happens on the wire of the IVN-technology. Since the ECU is no longer supplied with power, no electronics are active that can recognize wake-up patterns or other activity on the communication link.

Thus there is a strong interdependency between the sleep/wake-up of the power supply and IVN-technologies. Using the power supply and the IVN to simultaneously control sleep/wake-up has destabilized vehicle functions in the past. This has two consequences: (a) The deployment of wake-up solutions provided today with traditional IVN-technologies is limited and (b) The power supply/wake-up systems are designed to be very hierarchical and need to be organized centrally. Note that battery drainage can be a result if shut down sequences of ECUs are not planned carefully. Additionally, with strict targets on CO_2 emissions, using wake-up and shut down of ECUs in cars is more important than ever (see also Section 6.3.2).

- The **power distributors (5, 6)** ensure the distribution of the positive supply to the ECUs. As a general rule, fuses are used on the distribution lines to prevent electrical fires and to protect the wires from overload. Power distributors are often spatially organized. There might, e.g., be a distributor for the front of the car, another for the rear, and one for the engine compartment as well. At the time of writing, semiconductors that can be used to replace fuses were, unfortunately, still relatively expensive. They have many advantages such as actively driving the power supply lines and diagnosing current faults through the use of active measurements. Overall, the power supply will be more individualized and adjusted to the actual need in the future. The increase of the number of fuses is also an indication of the increased amount of electronics in the vehicle. For example, between 1995 and 2015 the number of fuses in BMW cars increased from 18 to 70 [6].
- Figure 6.1 additionally shows the option of **one ECU** (L) **supplying the power to other ECUs** (I and K) **(8)**. One option for realizing this is PoDL (see also Section 6.2.1). If the power is provided over the almost perfectly symmetrical wire pair of the IVN-technology, this has a positive impact on the vehicle EMC.

Figure 6.2 illustrates the same scenario as in Figure 6.1 but with the focus on the communication. In this example the ECUs H, L, and M contain Ethernet switch ICs. The rest of the communication nodes are connected via PHYs only. Without the knowledge of the power supply network in the background it is very difficult to optimize the communication system in terms of start-up, shut-down, or wake-up. Simply speaking, it is not possible to wake up a communication node when this node has no power. In other words, the behavior of a communication systems in terms of start-up, shut-down, or wake-up depends on the basic principles of the car's power distribution.

The power supply network also impacts the EMC. It is a common misunderstanding that optical communications systems such as MOST or optical Ethernet technologies such as 1000BASE-RH preclude EMC issues. Naturally, electromagnetic radiation does not interfere with or emit from an optical fiber. However, ECUs connected to the IVN by an optical communication system need a power supply. If

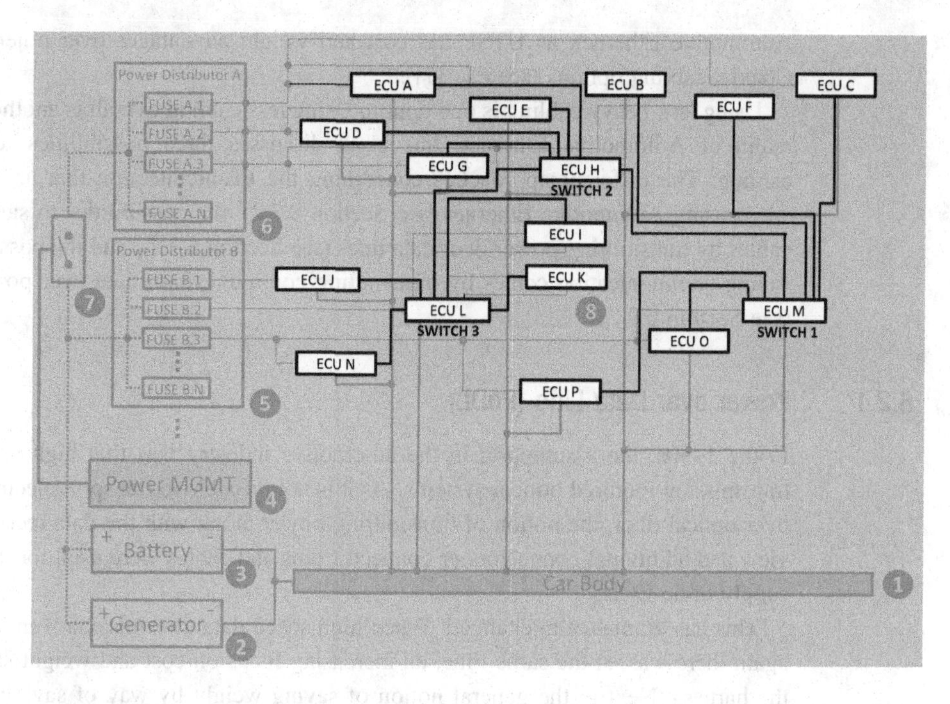

Figure 6.2 Example of the interrelation between the power distribution concept and an Ethernet communication network

the design of the ECU does not preclude this, high frequency clocked ICs might well couple onto the power lines from which ElectroMagnetic Emissions (EME) then disturbs the system.

EME from the power supply system is then particularly harmful, as the power supply systems are generally not designed to be symmetric in ground and positive. Communication systems using coaxial cables face similar issues, as coaxial cables are asymmetric by definition. Transient currents can occur on the shield as a result of this asymmetry and the difficulties of correctly bonding the shield to, e.g., the ground of the car's body. These transient currents are very critical for common mode noise. In order to suppress any impact the transient currents can have on a system, additional ferrite beads may be added to address this issue.

6.2 Saving Power by Saving Weight

As was said, the wiring harness is the third heaviest component in a car [5]. Its size has also increased significantly over time; for BMW, e.g., from 1.6 km in 1995 to 3.8 km in 2015 [6]. Internal investigations performed at BMW in 2016 showed that the power supply network has the largest share, followed by discrete wiring[3] and then by the cabling for the communication technologies. Any reduction counts. The possibility of using 100BASE-T1 with UTSP cabling was thus an important motivation to introduce

Automotive Ethernet, as UTSP has cost and weight advantages over other STP or Coaxial cabling options (see e.g. [7]).

Using less-heavy cabling is one option. Using less cabling overall is another. In the realm of Automotive Ethernet, this book discusses three possibilities to reduce cabling: The different approaches concerning the EE-architecture that are possible when using Automotive Ethernet (see Section 8.3.2), the possibilities to save power cables by transmitting power over data lines (see Section 6.2.1), and the possibility of saving communication cables by transmitting communication data over power lines (see Section 6.2.2).

6.2.1 Power over Data Line (PoDL)

In the 1990s, the assumption in the automotive industry was that high speed data transmission required optical systems. As it is not possible to transport electric power over optical fiber, the notion of transmitting power along with the data receded from view and additional connectors or connector pins and cables were used for the power supply of an ECU.

This has dramatically changed. When high speed data transmission went electrical again there was, at the same time, an increasing focus on cost and weight savings in the harness. Next to the general notion of saving weight by way of saving cabling, transmitting power along with the data was particularly investigated for sensors that are located at the far ends of a car (the required power is generally low and the savings in cabling and connectors are noteworthy). To save the power supply cables is also particularly attractive in the case of spatial restrictions in small openings or at connectors. The possibility of transmitting power over 100BASE-T1 Ethernet was thus investigated right from the beginning.[4,5]

The fact that the IEEE 802.3(af) Power over Ethernet (PoE) specification had existed since 2003 facilitated this, even if PoE was defined for two pair solutions such as 100BASE-TX. It ensured awareness, expertise, and motivation when in 2013 a task force was approved to define the transmission of power over one pair Ethernet technologies such as 1000BASE-T1 (and later also 100BASE-T1 and others) [8]. The resulting "Power over DataLine (PoDL)" IEEE 802.3bu specification was thus finalized in 2016 [9].

The principle behind PoDL is comparably simple, because the data is transmitted differentially. 100BASE-T1, e.g., achieves this with the help of a capacitor that blocks the DC part of the current. To transmit power as well, the common current is coupled onto the cables at the transmit side with help of a suitable circuit and coupled out again at the receive side (Figure 6.3 shows the principle set-up of PoDL with Automotive Ethernet). The standard works for the different PHY types with the same PoDL hardware, but with different coupling network bandwidths. Type A is cost optimized for 100BASE-T1, Type B is cost optimized for 1000BASE-T1, and Type A+B works with both.

The PoDL specification differentiates between Power Sourcing Equipment (PSE) that puts power on the link and the Powered Device (PD) that draws power from the

Table 6.1 Power classes defined for PoDL IEEE 802.3bu

	12 V un-regulated PSE		12 V regulated PSE		24 V un-regulated PSE		24 V regulated PSE		48 V regulated PSE	
Class	0	1	2	3	4	5	6	7	8	9
Voltage [V]	5.5–18		14–18		12–36		26–36		48–60	
Current [A]	0.1	0.22	0.25	0.47	0.10	0.34	0.21	0.46	0.73	1.3
PD power [W]	0.5	1	3	5	1	3	5	10	30	50

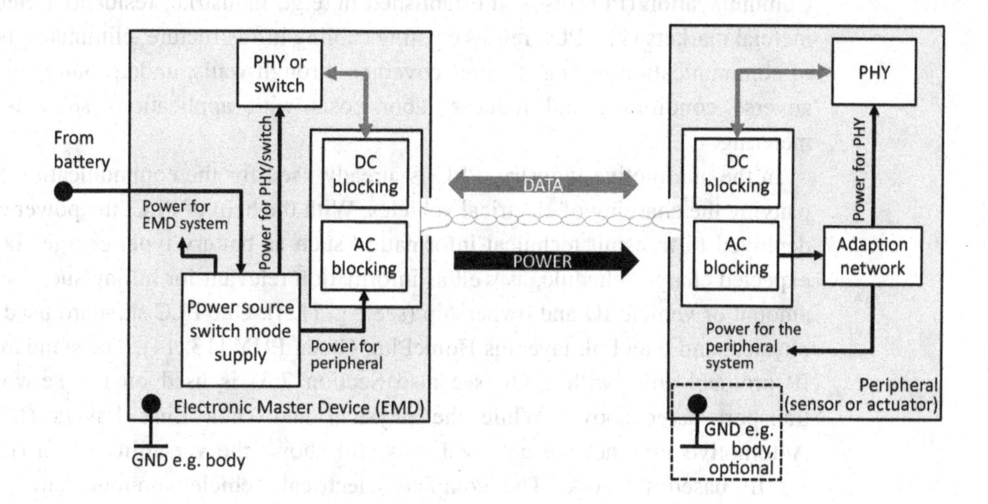

Figure 6.3 Example implementation for PoDL in cars

link [10]. When switched on, the PSE first of all tests the link. If the result is as expected, the power is turned on. The specification additionally offers a simple serial protocol with which it is possible to adjust the power in a plug & play scenario to different PDs. However, with the predefined networks and fast start-up requirements in the automotive industry, it is likely not deployed for automotive use. Table 6.1 shows the power classes, voltages, and power limits defined for each class.

PoDL has some additional advantages. With PoDL it is, e.g., possible to control the voltage drop on a link such that there always is the optimal voltage available at the peripheral sensor. Nevertheless, simple applications might live with power directly from the general power supply system. The voltage drop will be limited depending on the wire gauge and power consumption of the peripheral. The necessary electronic components must be selected according to the current required and, in the simplest case, consist of an inductivity. Unfortunately though, the inductors need additional space, which, in some use cases, is not available.

Another big advantage of PoDL is that it can further optimize the EME and EMI. The power supply wires are sources of electromagnetic interference due to conducted coupling. With the right termination and a good balance of the data wires, the current flow can be a closed loop right back to the source. In this case, only one ground level is used, and ground shifts between different locations in the car are avoided.

6.2.2 Data over the Power Supply Network

Every ECU needs power and thus some form of power supply. For the vast majority of units this is done via separate power supply cables. So, if every ECU is connected via cables anyway, why not use those cables for communication as well? Naturally, saving cabling is attractive not only in the automotive industry. Power Line Communication (PLC) is well established in, e.g., industrial, residential, and commercial markets [11]. PLC reuses existing cabling infrastructure, eliminates the costs of communication wiring, ensures coverage through walls, underground, and other adverse conditions, and reduces labor costs with applications such as smart metering.

In the automotive industry, PLC is already used for the communication accompanying the charging of electrical vehicles. With the help of PLC, the power cable is deployed to transmit technical information such as battery type, charge state, and expected charge schedule, as well as information relevant for billing such as charge amount or vehicle ID and ownership (see e.g. [12]). One PLC standard used on the physical and data link layers is HomePlug Green PHY [13, 14]. The standard TCP/IP protocol suite with IPv6 (see also Section 7.3) is used on the network and transport layer above. While the physical and Data Link Layers (DLL) of Automotive Ethernet are not used, this still shows the versatility of an (Ethernet &) IP based network. The complete electrical vehicle communication during charging standard is specified in the ISO 15118 series "Road vehicles – Vehicle to grid communication interface" [15].

ISO 15118 describes the PLC between the car and the outside world, which identifies the car as yet another IP node in the vast communication network. Because of the appeal of the idea of reusing the power supply cabling, many investigations have been performed to enable PLC also for communication between ECUs inside the car (see e.g. [16]). Latest efforts even include the requirement that such communication integrates with the Automotive Ethernet network. However, because of the previously-described asymmetrical design of the power supply system and additional problems with special power supply filters, PLC is not (yet) economical inside the car and has, to the authors' knowledge, not been brought to the road in a series production car.[6]

6.3 Saving Power by Reducing the Electrical Power Consumption

Electrical power is saved by building power efficient ECUs as such, but also by switching off complete ECUs or parts of ECUs that are momentarily not needed, e.g., communication, during runtime. The ability to switch CAN-based ECUs on and off is of such great importance to the automotive industry that new standardization efforts were undertaken and new CAN transceiver products were developed [17]. FlexRay and LIN also provide comparable mechanisms. This subsection thus describes methods available for Automotive Ethernet. Section 6.3.1 discusses the use of Energy Efficient Ethernet (EEE) for Automotive Ethernet. Section 6.3.2 covers wake-up and sleep mechanisms.

6.3.1 Using Energy Efficient Ethernet (EEE) in Cars

The need for specifying Energy Efficient Ethernet (EEE) arose from the introduction of full-duplex Ethernet. On a half-duplex link, units only transmit when they have data to send. In full-duplex mode, units having no data to send transmit idle symbols in order to constantly track the transmission conditions and to keep the PHYs synchronized. Naturally, this significantly increased the power usage, as suddenly all Ethernet nodes connected to a network constantly transmitted. In 2010, the IEEE thus published the IEEE 802.3az standard for Energy Efficient Ethernet (EEE) [18].[7]

The principle behind EEE is to have as little activity on the link as possible when there is no traffic and to have the link return to full operating speed again as quickly as possible when there is traffic [19]. To achieve this, the IEEE introduced the so-called Low Power Idle (LPI) mode. When two units attached to a link observe that the traffic consists only of idle packets, the units can agree to go into LPI instead. During the LPI, a defined period of zero transmission, e.g., 20 ms, is followed by a defined period of active refresh, e.g., 200 μs, that ensures that the PHYs stay synchronized. Because of the regular refresh, a link can be reactivated quickly when there is again data to send. Maximum 16.5 μs start-up times were specified for 100 Mbps Ethernet and 30 μs for 1 Gbps Ethernet [18].

When the project started, the target for EEE was to reduce power consumption of the PHY in the range of 50% [20]. The specified EEE solution can, depending on the ratio of zero and refresh transmission, actually achieve even up to 90% power savings on the link [19]. The developed solution takes the perspective from the link utilization that might be independent from the utilization of the compute unit(s) being connected by the link. Simple examples from the IT world are the following. Someone downloads a document from the Internet, then reads it without any network interaction before the next page is requested. Someone else writes a document locally, before it is printed on a network printer or forwarded to a colleague via email. The compute unit might be used continuously, while the communication is needed only sporadically (see [21] for examples that show the burstiness of related traffic patterns). To be able to save power on the communication link separately from saving power on the computer is thus a relevant advantage, even if the possibility to be able to save power beyond the PHY was always also intended and is desired [21, 22].

The automotive perspective is different. When an ECU is needed, it generally also needs to communicate. Or, put differently, when there is no need for communication, there is no need for the ECU. A method that seemingly focusses on power savings on the communication link thus appears to be insufficient. In automotive, the primary goal

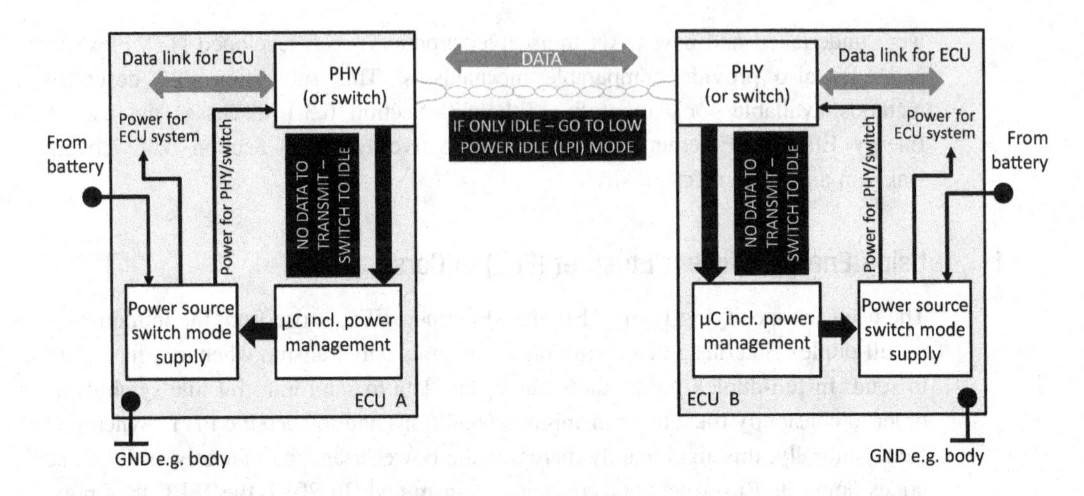

Figure 6.4 Example implementation for using power efficient Ethernet in cars

is to be able to switch the "complete" ECU on an off. The aforementioned mechanism specified for CAN, LIN, and FlexRay allow exactly that (see also Section 6.3.2).

Figure 6.4 shows a principle concept of how EEE may be used to control the power supply of the complete ECU. It would require that the PHY notifies the power management of the ECU when it goes into LPI, so that the complete ECU can be switched off (provided it really is no longer needed) and on again when the communication resumes. Such proceedings require that the LPI signaling is made available to higher layers (as IEEE 802.1az allows) and not isolated and terminated in the PHY (which seemed to be the case in automotive implementations at the time of writing [G. Zimmerman, email correspondence, April 11, 2020]). Naturally, the PHY itself would still need to be powered even if the rest of the ECU is asleep in order to send the regular refresh symbols. This would then be similar to some of the wake-up and sleep methods described in Section 6.3.2.

For the first Automotive Ethernet projects finalized at IEEE 802.3, EEE was thus not much of a focus. The first project, 1000BASE-T1 (see also Section 5.3.1), includes EEE as an optional feature [23]. 100BASE-T1 (see Section 5.2.1) took over the OPEN Alliance BroadR-Reach (OABR) specification. As the OABR specification does not consider EEE, EEE was not part of the IEEE 802.3 objectives for that project either [24]. In the 10BASE-T1S half-duplex mode (see Section 5.4), EEE is not required. Besides, transmission only happens when there is data to send. For the optional 10BASE-T1S full-duplex mode EEE is also optional [25]. For MultiGBASE-T1, EEE was again included in the objectives as an optional feature [26].

One feature of EEE that moved more into focus with MultiGBASE-T1 is the possibility to deploy EEE asymmetrically. The larger the data rate, the more power is needed for just the communication and the more likely the data transmission is highly asymmetric [27]. It is expected that for many uses of MultiGBASE-T1, data transmission will be predominantly in one direction. It is thus advantageous from a power consumption perspective that the LPI mode can be entered separately per direction [19].

The asymmetric LPI mechanism specified for MultiGBASE-T1 is based on the one specified for 2.5-10GBASE-T PHYs [19]. It allows entering LPI per transmission direction, independent from the communication in the other direction. However, MultiGBASE-T1 went beyond what had been specified for the MultiGBASE-T PHYs and additionally enabled a “slow wake” [28]. On top of the power saved in transmission, the slow wake allows for power savings in reception and in needing to be ready for a fast restart [29]. Normally in LPI mode, while the transmitter can stay turned off between sending refresh signals, the receiver needs to stay awake and ready to restart fast all the time, in order not to miss when the communication resumes. When in “slow wake,” wake is only possible directly following a refresh. This means that the receiver only needs to stay awake and ready at defined intervals and can save power in between.

If supporting more power savings with the asymmetric EEE was seen as essential for the MultiGBASE-T1 speeds, this is even more the case for speed grades larger than 10Gbps (see Section 5.6). For these new efforts it was thus discussed whether EEE can be taken even further to allow for almost 100% power saving by completely shutting off one transmission direction with a “Link Suspension” mode. This would require shutting off signaling completely while still enabling fast reactivation [19]. At the time of writing, it was yet to be decided whether and how this scheme will be adopted.

6.3.2 Wake-up and Sleep

Even during runtime, not all functions available inside a car are needed all the time. In order to save maximum energy it is thus desirable to put ECUs currently not needed to sleep and to wake them up again when they are needed. Just reducing the power consumption of the PHY transceiver only – as is possible with EEE – would waste chances for more power reduction. The power consumption of the transceiver IC is generally significantly smaller than that of the processor/μC and the rest of the ECU. Reducing power consumption with the help of wake up and sleep mechanisms thus reduces CO_2 emissions, improves the lifetime of ECUs, and increases the operating reach of, e.g., electrical vehicles.

For the automotive wake-up and sleep mechanisms there are, in principle, three choices for sleep states, each resulting in a different level of power saving and each requiring a different means for wake-up. Additionally, sleep and wake-up can each be selective or global; in different combinations with the wake-up and sleep choice (see Table 6.2 for an overview on the choices and different combinations possible). Because the power savings during “sleep” are the target of introducing wake-up and sleep mechanisms to start with, the three sleep states will be explained first. Then the wake-up means and the resulting different wake-up and sleep methodologies are explained in general before the specific solution for Automotive Ethernet is discussed.

Table 6.2 Possible combinations of wake-up and sleep choices discussed

Sleep state	Means for wake-up	Hardware requirement	Wake-up type	Sleep type	Applicable for
Deep Sleep, no current draw	Power supply	PoDL/intelligent power switches	Global Selective	Global Selective	All technologies (power switches), PoDL for Ethernet
ECU Sleep, quiescence current of ECU	Wake-up line	Wake-up circuit or respective system base chip, extra wiring	Global	Selective	Independent of IVN technology
Light Sleep, transceiver draws current	Activity on communication link "Wake-on-LAN"	A capable technology and transceiver chip	Global Global Selective	Global Selective Selective	LIN, CAN, FlexRay Ethernet CAN PN

In principle, there are the following three different "sleep" states:

1. "**Deep Sleep**": The ECU is cut off the power supply. In this case the ECU cannot consume any power.
2. "**ECU Sleep**": The complete ECU is put to sleep but the power supply is still "on." When in ECU sleep, an ECU will thus still draw some quiescent current, which is targeted to be in the range of some low 10 s of μA (10–20 μA for BMW, 30–50 μA according to [30]). Any wake up circuitry that might be needed in this scenario should ideally not increase this.
3. "**Light Sleep**": Most of the ECU, especially the μC, is asleep, but the transceiver is still powered [31]. This generally means that the transceiver IC needs a separate power supply circuit in the ECU. Following [30], in the light sleep case the current draw should not exceed 750 μA in total.

Sleep can be selective or global. Global sleep means that the complete section of the network considered is asleep (which always means global wake-up, see the following points 1–3). Selective sleep means that only some units of a specific network are asleep, while others are active. Selective sleep can be combined with a global or selective wake-up (see also the following points 1–3).

The corresponding means for wake-up are:

1. Wake-up **via power supply** by intelligently controlling power switches. This can be done globally for a number of ECUs connected at once (see e.g. the Clamp 15 example in Section 6.1) or selectively, when the power supply of specific units is controlled individually. In the latter case the power control is generally performed by a separate ECU. An example is the surround-view ECU, which switches the cameras on by controlling their power supply as soon as the car goes into reverse gear. Using PoDL (see Section 6.2.1) for this purpose is also possible and a special form of wake-up via power supply.

While wake-up and sleep via the power supply achieves the largest power savings, it is not always doable or practical. PoDL has stringent current limits, not every unit communicates with just one other ECU, nor is controlled switching very efficient if done on a large scale. Furthermore, wake-up might take somewhat longer in the case of full power up, which is also not desirable.

2. Wake-up **via wake-up line** is a means to switch complete ECUs on and off by controlling the ECU internal power supply with a special circuitry or system base chip. The wake-up line is a separate wire requiring a separate connector pin at the ECU. Generally, a number of ECUs are connected to the same wake-up line. The pulsed or continuous wake-up signal wakes all units attached to the wake-up line and is thus always performing a global wake-up. Each ECU wakes its own μC internally, which then decides, at a higher level, whether the ECU is really needed for the communication or whether it can go back to sleep. Going back to sleep is thus always selective. In theory, any connected unit can be empowered to trigger a wake-up over the wake-up line. In practice it is generally a selected number of units which are enabled to do so.

   Using a wake-up line is robust, simple, and independent of the communication technology and topology used. At the time of writing, BMW was still using the wake-up line for CAN-FD and Automotive Ethernet (albeit preparing to use the Automotive Ethernet alternatives). Naturally, the disadvantage of the wake-up line is the weight and cost of the extra wiring and circuitry needed to enable it.

3. **Wake-on-LAN** wake-up means that an ECU is woken up by (a certain type of) activity on the communication link. The key is that the communication technology has to be enabled to support this type of wake-up in the PHY transceiver IC and that the selected mechanism in the PHY is robust against unwanted wake-up in the presence of interference [31]. In this scenario, the transceiver needs to be continuously powered, even if the rest of the ECU is asleep. This generally requires a separate power circuit for the PHY. Furthermore, the PHY transceiver needs to be empowered to wake-up the rest of the ECU. This is often solved by an extra pin on the transceiver IC that triggers the power circuitry of the rest of the ECU. The automotive industry is using a number of different Wake-on-LAN methods, each employing its own combination of selective or global wake-up or sleep.

   The most straightforward method is global wake-up and global sleep, meaning any traffic in the considered network causes a wake up. As long as there is communication, all units connected stay awake independently of their own participation in the communication. LIN, FlexRay, and CAN WUP,[8] e.g., follow this **"Wake-on-Traffic"** principle [30]. As always, complete networks are either on or off and, hence, Wake-on-Traffic misses some power saving potential.

   The next step is therefore to initiate a global wake-up but to let units that are not needed to go to sleep selectively even if there is traffic in the network. This **"Pseudo Partial Networking"** [30] follows the same idea as the wake-up line. As with all cases with selective sleep, the in-vehicle networking technology with which this is used needs to be robust enough not to be woken by normal traffic in the network if the unit went back to sleep. The decision to stay awake or to go back to sleep is generally taken in the µC. Depending on the system, the wake-up message might already contain information that can be evaluated to that end, or the content of the traffic on the channel must be observed. The Automotive Ethernet wake-up and sleep, as described later in this section in more detail, follows this principle. A comparable CAN variant also exists (see e.g. [30]).

Last, but not least, wake up and sleep can be selective. In this case, the wake-up message sent by the unit initiating the wake up and received by the other units in the network includes data that the receiving PHY evaluates. Depending on the content, the PHY will initiate a wake up of the ECU or not. For CAN this has been specified as "**Partial Networking**" [17]. Of the different Wake-on-LAN methods, Partial Networking is optimal in terms of power savings. However, such methodology takes application decisions to the PHY level, which in itself might add other challenges and complexity.

Table 6.2 provides an overview on the different wake-up and sleep combinations. As IEEE802.3bw 100BASE-T1 comes without wake-up and sleep mechanisms, wake-up and sleep for Automotive Ethernet was specified in the OPEN Alliance [32]. Next to providing a mechanism that can be extended to other speed grades –especially 1000BASE-T1 – the OPEN Alliance specification had the objective to cover the complete wake-up process in ISO/OSI layer one without requiring additional components or any modification to the existing network architecture [33]. The solution was to be as simple as possible. Furthermore, the wake-up time of the complete network was to be less than 250 ms, including link start-ups.

In principle, the method selected follows a Wake-on-LAN concept with global wake-up and selective sleep (see also Figure 6.5). From a power consumption perspective, it might seem disadvantageous that all units need to wake up upon a single wake-up impulse in the network, even if many of them are not needed and can/will go back to sleep. The important advantage of this type of approach is that

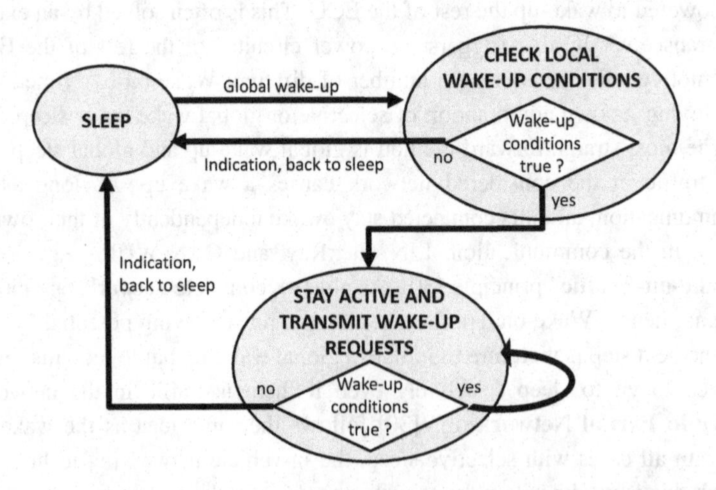

Figure 6.5 The basic state-machine for the OPEN Alliance wake-up/sleep

knowledge of the different use cases, i.e., what device is needed to fulfill a particular function, does not have to be exposed to the lower layers of the communication stack. It is the application that knows whether an ECU is needed for a specific function or not. To include such information at the PHY layer would mean to break with ISO/OSI layering. This, in return, could lead to an avalanche of unforeseen and unwanted interdependent reactions that might jeopardize the complete system design.

There is one core difference between applying a Wake-on-LAN concept to Ethernet and, e.g., to a CAN network. All units in a CAN network are connected in a bus structure to the same wire. In an Ethernet network, most units do not share the same wire, but are connected via switches. The only exception would be end nodes on a 10BASE-T1S segment. The switch simplifies the sleep of single attached units, as this attached unit will unlikely be disrupted by any traffic not intended for it. At the same time, a wake-up signal has to be able to propagate through switches over passive and active links.

Wake up mechanisms thus additionally depend on the network topology: Is it a single node with isolated communication that needs to be put to sleep, the communication of one network (represented by one physical bus), or a (large) network connected by switches (or even gateways)?

The specified OPEN Alliance wake-up mechanism discusses four main functions: Wake-up reception and signaling, wake-up transmission, wake-up forwarding, and sleep [33]. The switches play a special role. Naturally, they always need to be powered in order to be able to forward wake-up messages, even if the ECU the switch is part of is not needed. Next, their placement in the topology also has to consider the desired or necessary order of node wake-up. Furthermore, the layer one wake-up detection has to be passed through a layer two device. For this the MDIO and/or an extra pin needs to be used and a switch should be configurable to forward the wake-up on a per port basis [34].

To allow for a better understanding of the specified system behavior, Figure 6.6 depicts a start-up in the example system used in this section. The left upper diagram (1) shows the initial state in which only ECUs I, L, and N are active and communicating. The upper right diagram (2) shows a wake-up triggered inside the function of ECU C, which is then forwarded into the network (3). As switch 1 is the first unit to be woken up it sends wake-up pulses via all of its ports (4). One of the units that receives the wake-up from switch 1 is switch 2 (integrated in ECU H). This switch again forwards the wake-up through all of its ports (5). The next switch to receive the wake-up request is switch 3. Switch 3 has active and passive links connected. It sends wake-up pulses over the links not yet active (6) and wake-up requests messages over the active links.

At this point the whole system is awake (6). Now, each node decides, via higher layers in the communication stack, whether its functionality is needed in the use case that triggered the wake-up (7). If this is not the case, the units decides to shut down following their specific shut down sequence. This means that nodes that are not needed go back to sleep and the sequence ends in the next stable state (8). In the example of Figure 6.6 the units that go back to sleep are A, B, E, F, G, K, O, and P.

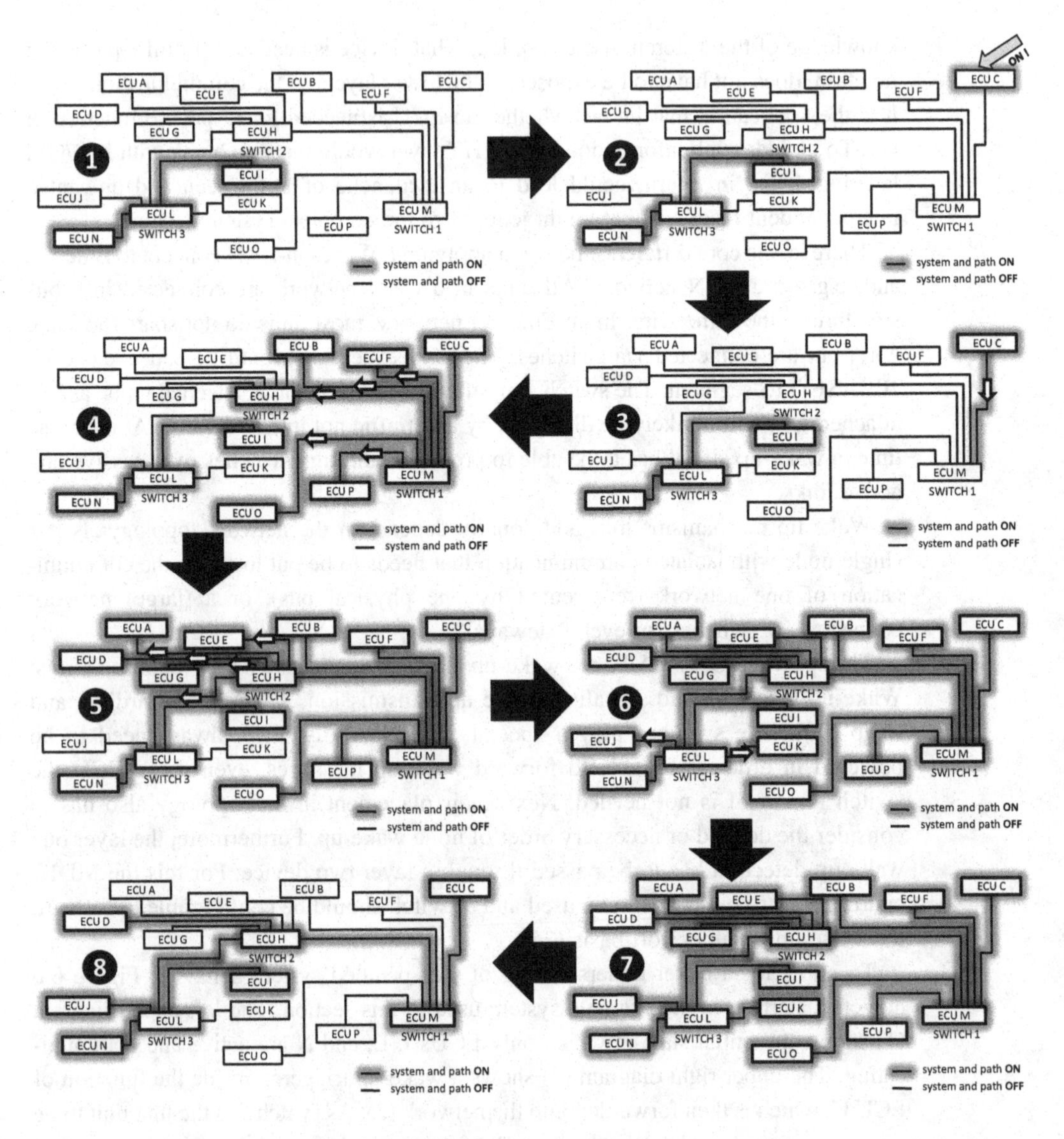

Figure 6.6 Start-up sequence following a wake-up request

Notes

1 An important milestone was set with the so-called Paris agreement, concluded by the United Nations Framework Convention on Climate Change (UNFCCC) in December 2015. Its goal is to limit global warming to below 2°C (ideally to 1.5°C) above preindustrial level [38]. Prior to the meeting, every country was asked to submit commitments on how to reduce greenhouse gas and carbon output in their countries. Based on 146 submittals, it was concluded that the proposed methods are not sufficient to achieve the target, but that even more stringent actions are needed [37].

2 Which function is on which power supply clamp can vary from car manufacturer to car manufacturer, especially with respect to those functions that are provided when the engine is switched off. Opening electric windows, adjusting mirrors, and use of the GPS navigation system are good examples for functions that are handled differently by different OEMs and which are sometimes available when the car has been unlocked, sometimes when the key has woken up the car, and sometimes when the engine has been started.
3 The whole purpose of the communication technologies was, and still is, to reduce the amount of discrete wiring. With the increasing complexity in the EE-architecture cars could no longer be built if all communication was done via discrete wires (see also Section 2.1). There are various reasons why there is still so much discrete wiring: Discrete wiring might have lower hardware costs, there might not be an efficient technology available for the specific use case, or simply no one has looked into it yet.
4 Transmitting power over 100BASE-T1 turned out to be feasible without major challenges. However, the cameras in the first BMW 100BASE-T1 use case (see Section 3.3) received separate power lines. The reasons were: Risk minimization, non-availability of smaller connectors, and severe spatial restrictions within the first camera that would have made it extremely difficult to place the necessary inductor.
5 The trend is not unique to Automotive Ethernet. Other communication technologies such as SerDes links (see Section 2.2.6) or the MOST cPHY (see Section 2.2.4.2) support the transmission of power with the data. For both technologies it is an economic must. Systems using STP cabling generally allow the inclusion of the power supply cables with the communication cables within the same shield and thus within the same connector. However, with the more cost-efficient coaxial cabling, this is not possible. Not only is an additional wire pair needed, but so is a completely different connector, when the power supply is separate. This can make such solutions economically unattractive. Because of the asymmetry of the coaxial cables, though, special attention must be paid to the EMC effects when power is transmitted with the data. For UTP, neither issue exists. A separate power supply can simply use separate pins on the same connector, without significantly increasing the cost and, if power is transmitted over the UTP wires, it finds a very symmetric cable.
6 EMC is also an obstacle to overcome when deploying PLC in other industries, as the use of electrical appliances or even the coupling of signals from radio broadcasters can degrade the communication performance [11]. In fact, the CFI for the 1000BASE-RH project explicitly mentioned POF and 1000BASE-RH (see Section 5.3.2) as an alternative to PLC as well as placing copper communication cables next to the power supply cables in homes, because of the robustness of POF to electromagnetic disturbances on the line [39].
7 Before EEE was finalized, apparently various companies already sold Ethernet equipment that allowed reducing power consumption. These methods were marketed under the name "Green Ethernet" [35, 36].
8 CAN WUP is explicitly named because, for CAN networks, different wake-up and sleep mechanisms have been specified. Details for the CAN Wake-Up Pattern (WUP) can be found in [40].

References

[1] International Energy Agency, "CO2 Emissions from Fuel Combustion; Highlights," 2015. [Online]. Available: www.iea.org/publications/freepublications/publication/CO2EmissionsFromFuelCombustionHighlights2015.pdf. [Accessed April 2, 2016, no longer available].

[2] S. C. Davis, S. W. Diegel, and R. G. Boundy, "Transportation Energy Data Book: Edition 30 – 2011," June 2011. [Online]. Available: http://info.ornl.gov/sites/publications/files/Pub31202.pdf. [Accessed May 16, 2020].

[3] statista, "Number of Passenger Cars and Commercial Vehicles in Use Worldwide from 2006 to 2014 in (1,000 Units)," 2016. [Online]. Available: www.statista.com/statistics/281134/number-of-vehicles-in-use-worldwide/. [Accessed May 16, 2020].

[4] VDIK; TÜV Nord; VDA, *Facts and Arguments About Fuel Consumption*, Berlin: VDA, 2014.

[5] S. Carlson, T. Hogenmüller, K. Matheus, T. Streichert, D. Pannell, and A. Abaye, "Reduced Twisted Pair Gigabit Ethernet Call for Interest," March 15, 2012. [Online]. Available: www.ieee802.org/3/RTPGE/public/mar12/CFI_01_0312.pdf. [Accessed May 6, 2020].

[6] K. Matheus, "New Speed Grades: Automotive Ethernet Manifests Itself as a Future Proof In-Vehicle Networking System," in Automotive Ethernet Congress, Munich, 2017.

[7] K. Matheus, M. Kaindl, S. Korzine, D. Goncalvez, J. Leslie, Y. Okuno, N. Kitajima, M. Gardner, R. Orosz, M. Jaenecke, D. Kim, R. Mei, and D. v. Knorre, "1 Pair or 2 Pairs for RTPGE: Impact on System Other than the PHY Part 1: Weight & Space," January 23, 2013. [Online]. Available: http://grouper.ieee.org/groups/802/3/bp/public/jan13/matheus_3bp_01_0113.pdf. [Accessed May 14, 2020].

[8] IEEE 802.3, "Approved Minutes, IEEE 802.3 Ethernet Working Group Plenary, Dallas, TX," November 11–14, 2013. [Online]. Available: www.ieee802.org/3/minutes/nov13/1113_minutes.pdf. [Accessed May 6, 2020].

[9] IEEE Computer Society, *802.3bu-2016 – IEEE Standards for Ethernet – Amendment 8: Physical Layer and Management Parameters for Power over Data Lines (PoDL) of Single Balanced Twisted-Pair Ethernet*, New York: IEEE-SA, 2016.

[10] D. Dwelley, "A Quick Walk Around the Block with PoDL," November 2015. [Online]. Available: www.ieee802.org/802_tutorials/2015-11/PoDL_tutorial_1115.pdf. [Accessed March 22, 2020].

[11] Markets and Markets, "Power Line Communication Market by Offering (Hardware, Software, and Services), Frequency (Narrowband, and Broadband), Application (Energy Management and Smart Grid, and Indoor Networking), Vertical, and Geography – Global Forecast to 2023," November 28, 2019. [Online]. Available: www.marketsandmarkets.com/Market-Reports/power-line-communication-plc-market-912.html. [Accessed March 29, 2020].

[12] D. Grossmann and H. Hild, "Smart Charging – A Key to Successful E-Mobility," July 2014. [Online]. Available: https://vector.com/portal/medien/cmc/press/Vector/EV_Smart_Charging_ElektronikAutomotive_201407_PressArticle_EN.pdf. [Accessed May 16, 2020].

[13] Wikipedia, "HomePlug," March 16, 2020. [Online]. Available: https://en.wikipedia.org/wiki/HomePlug/HomePlug. [Accessed March 29, 2020].

[14] M. Shin, H. Kim, H. Kim, and H. Jang, "Building an Interoperability Test System for Electric Vehicle Chargers Based on ISO/IEC 15118 and IEC 61850 Standards," *Applied Science*, vol. 6, p. 165, 2016.

[15] Wikipedia, "ISO 15118," May 31, 2019. [Online]. Available: https://en.wikipedia.org/wiki/ISO_15118. [Accessed March 29, 2020].

[16] F. Nouvel, P. Tanguy, S. Pillement, and H. Pham, "Experiments of In-Vehicle Power Line Communications," April 11, 2011. [Online]. Available: http://cdn.intechopen.com/pdfs/14996/InTech-Experiments_of_in_vehicle_power_line_communications.pdf. [Accessed May 16, 2020].

[17] ISO, *ISO/DIS 11898-6:2013 – Road vehicles – Controller Area Network (CAN) – Part 6: High-speed Medium Access Unit with Selective Wake-up Functionality*, Geneva: ISO, 2013.

[18] IEEE Computer Society, *802.3az-2010 – IEEE Standard for Carrier Sense Multiple Access with Collision Detection (CSMA/CD) Amendment 5: Media Access Control Parameters, Physical Layers, and Management Parameters for Energy-Efficient Ethernet*, New York: IEEE-SA, 2010.

[19] G. Zimmerman, "Technical Feasibility – EEE for Asymmetry," November 2019. [Online]. Available: www.ieee802.org/3/B10GAUTO/public/nov19/zimmerman_3B10G_01_1119.pdf. [Accessed February 3, 2020].

[20] S. Kerner, "Energy Efficient Ethernet Hits Standards Milestone," June 17, 2009. [Online]. Available: www.internetnews.com/skerner/2009/07/energy-efficient-ethernet-hits.html. [Accessed May 16, 2020].

[21] H. Barrass, M. Bennett, W. W. Diab, D. Law, B. Nordman, and G. Zimmerman, "IEEE 802 Tutorial: Energy Efficient Ethernet," July 16, 2007. [Online]. Available: www.ieee802.org/802_tutorials/07-July/IEEE-tutorial-energy-efficient-ethernet.pdf. [Accessed April 22, 2020].

[22] D. Chalupsky, E. Qi, and I. Ganga, "A Brief Tutorial on Power Management in Computer Systems," March 13, 2007. [Online]. Available: www.ieee802.org/3/eee_study/public/mar07/chalupsky_01_0307.pdf. [Accessed April 22, 2020].

[23] IEEE Computer Society, *802.3bp-2016 – IEEE Standard for Ethernet Amendment 4: Physical Layer Specifications and Management Parameters for 1 Gb/s Operation over a Single Twisted Pair Copper Cable*, New York: IEEE-SA, 2016.

[24] IEEE 802.3, "1TPCE Objectives," July 17, 2014. [Online]. Available: www.ieee802.org/3/bw/public/20140717_V3_Objectives.pdf. [Accessed March 30, 2020].

[25] IEEE Computer Society, *802.3cg-2019 – IEEE Standard for Ethernet Amendment 5: Physical Layer and Management Parameters for 10 Mb/s Operation and Associated Power Delivery over a Single Balanced Pair of Conductors*, New York: IEEE-SA, 2019.

[26] IEEE 802.3, "Approved Objectives for IEEE 802.3ch," March 16, 2017. [Online]. Available: www.ieee802.org/3/ch/0317_approved_objectives_3NGAUTO.pdf. [Accessed January 11, 2020].

[27] C. Mash, "Network Topology Analysis," July 2019. [Online]. Available: www.ieee802.org/3/B10GAUTO/public/jul19/mash_B10GAUTO_1_0719.pdf. [Accessed February 3, 2020].

[28] IEEE Computer Society, *802.3ch-2020 – IEEE Standard for Ethernet Amendment: Physical Layer Specifications and Management Parameters for 2.5 Gb/s, 5 Gb/s, and 10 Gb/s Automotive Electrical Ethernet*, New York: IEEE-SA, 2020.

[29] S. Benyamin and G. Zimmerman, "EEE Improvements for Highly Sparce Traffic," April 2019. [Online]. Available: www.ieee802.org/3/ch/public/apr19/Benyamin_3ch_01a_0419.pdf. [Accessed April 10, 2020].

[30] N. A. Wienckowski, "Automotive Wakeup Methods," September 2017. [Online]. Available: www.ieee802.org/3/ch/public/sep17/Wienckowski_3ch_01a_0917.pdf. [Accessed April 3, 2020].

[31] T. Hogenmüller and H. Zinner, "Tutorial for Wake Up Schemes and Requirements for Automotive Communication Networks," July 18, 2012. [Online]. Available: http://grouper.ieee.org/groups/802/3/RTPGE/public/july12/hoganmuller_02a_0712.pdf. [Accessed May 14, 2020].

[32] P. Axer, C. Hong, and A. Liu, *OPEN Sleep/Wake-up Specification 2.0*, Irvine, CA: OPEN Alliance, 2017.

[33] R. Sappia, "Enabling Power Efficient and Interoperable Behavior in a Multi-vendor Environment of 100BASE-T1 Components," in Automotive Ethernet Congress, Munich, 2016.

[34] P. Axer, "Automotive Wakeup and Sleep," in Automotive Ethernet Congress, München, 2019.

[35] Cisco, "Cisco Green Routers," January 27, 2009. [Online]. Available: https://web.archive.org/web/20090618035001/http://greenethernet.com. [Accessed May 16, 2020].

[36] Wikipedia, "Energy-Efficient Ethernet," February 2, 2020. [Online]. Available: https://en.wikipedia.org/wiki/Energy-Efficient_Ethernet. [Accessed May 16, 2020].

[37] UNFCCC, "New UN Report Synthesizes National Climate Plans from 146 Countries," November 17, 2015. [Online]. Available: www.buildup.eu/en/news/new-un-report-synthesises-national-climate-plans-146-countries-ahead-cop-21. [Accessed May 16, 2020].

[38] Wikipedia, "Paris Agreement," April 2, 2016. [Online]. Available: https://en.wikipedia.org/wiki/Paris_Agreement. [Accessed May 12, 2020].

[39] C. Pardo, T. Lichtenegger, A. Paris, and H. Hirayama, "Gigabit Ethernet over Plastic Optical Fibre, Call for Interest," March 27, 2014. [Online]. Available: www.ieee802.org/3/GEPOFSG/public/CFI/GigPOF%20CFI%20v_1_0.pdf. [Accessed May 6, 2020].

[40] ISO, *ISO/DIS 11898-2:2016 – Road Vehicles – Controller Area Network (CAN) – Part 2: High-speed Medium Access Unit*, Geneva: ISO, 2016.

7 Protocols for Automotive Ethernet

One of the reasons for the automotive industry to adopt Ethernet-based communication as an in-vehicle networking system is the chance for synergies, i.e., the possibility of reusing protocols that have been developed and tested in other industries. Across the various protocol layers for the various applications it therefore needs to be carefully investigated whether to adopt, adapt, or to add protocols. Figure 7.1 gives an example overview of a typical Automotive Ethernet protocol stack. This chapter discusses four areas that require special care: Audio Video Bridging (AVB) and its successor Time Sensitive Networking (TSN, see Section 7.1), Virtual LANs (VLANs) and switch configuration (see Section 7.2), the Internet Protocol (IP, see Section 7.3), and what is needed in terms of command and control (see Section 7.4). Additionally, Section 7.5 addresses security in respect to an Automotive Ethernet network.

Note that the described solutions make no claim to be complete; it might well be possible to use other protocols with the Automotive Ethernet PHY transceivers. However, the solutions described in this chapter represent a solution that works and that can be adopted by those wanting to deploy Automotive Ethernet.

7.1 Quality of Service (QoS), Audio Video Bridging (AVB), and Time Sensitive Networking (TSN)

Ethernet as such, i.e., the PHY and MAC layers as defined by IEEE 803 at the time, provide best-effort communication only. The introduction of switches reduced the collision domain and improved the determinism of each individual link. The various units connected to a switch no longer needed to contend for the same medium at potentially the same time, which in the case of collisions meant they had to go into random, i.e., non-deterministic, back-off periods. However, in a switched network, data of different sources with different destinations might still have to be sent over the same link at the same time. It is therefore at layer two in the switch – often also referred to as a (multiport) bridge[1] – that Quality of Service (QoS) requirements can be supported effectively. Today, the respective protocols and procedures are mainly standardized within IEEE 802.1.

This book uses the term QoS for requirements and solutions that influence the flow of data such that it can be received at a defined quality [2]. These requirements and

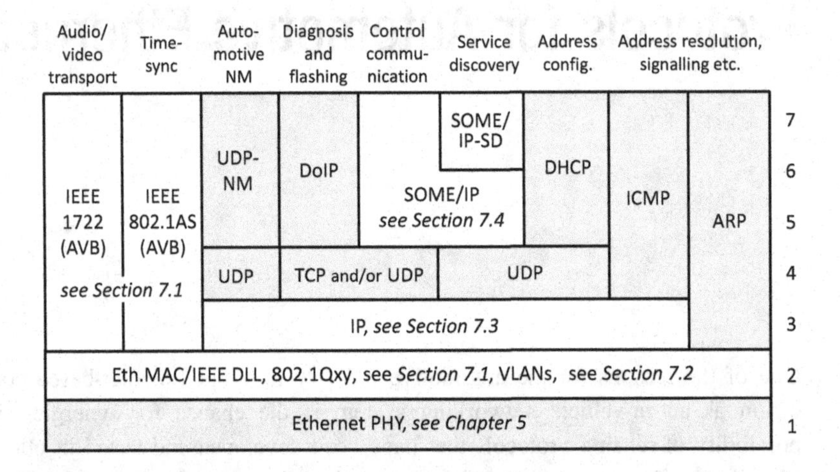

Figure 7.1 Protocol overview for Automotive Ethernet [1]

solutions can vary significantly depending on the use case and area of focus (see e.g. [3, 4]). It is therefore important to start with some background information on the origin of the Audio Video Bridging (AVB)/Time Sensitive Networking (TSN) standardization activity (Section 7.1.1). To better understand the deployment of various QoS protocols for the automotive audio video use case (Section 7.1.3), Section 7.1.2 first highlights the differences between the originally envisioned use cases and the automotive use cases. While addressing audio and video transmission for entertainment provides the origin for the series of IEEE standards, standardization has moved on to also support safety critical applications in an Ethernet network. Section 7.1.4 describes this second set of TSN standards and their use in the automotive industry.

7.1.1 How Audio Video Bridging (AVB) Came to Ethernet

In July 2004, the IEEE 802.3 group accepted a Call For Interest (CFI) on "Residential Ethernet" in order to investigate the use of Ethernet for time-sensitive Audio and Video (AV) applications [5]. Apparently, more than a year previously, discussions on the need for more Consumer Electronics (CE) centric Ethernet networking had started simultaneously in different groups of industry players, who then aligned the standardization in the IEEE [5, Y. Kim, email correspondence, August 12, 2013].

At the time, the Internet in combination with audio – and later video – compression formats was drastically and irreversibly changing the consumer behavior with respect to music consumption. Even if the "share it with all for free" Napster platform had only lived from May 1999 to February 2001 [6],[2] it initiated a change: The PC/laptop/mobile device replaced the home hi-fi and CD collection as the center for consumer entertainment. These new devices are able to serve simultaneously as storage, rendering device, synthesizer, sound mixer, and media server, and the PCs and laptops considered in 2004 always had an Ethernet interface [7]. The CFI on Residential Ethernet addressed the specific quality requirements of AV transmission in an

Ethernet LAN, and thus broadened the market potential of Ethernet into the consumer space.

Next to being widely deployed in PCs and laptops, Ethernet offered plug & play, high data rates, and network management in terms of neighbor discovery, virtual network support, and traffic prioritization [5]. Nevertheless, even with priority, there was/is neither timing guarantee nor a reference time to which the receiver can relate. Buffering data can help to overcome jitter up to a certain point. AV applications such as VoIP or IP-TV rely on buffering in order to improve the quality. However, determining the correct buffer size is no easy task: Buffers that are too small to bear the risk of buffer overflow, dropped packets, and quality degradation; while buffers that are too large are costly and introduce additional latency. So, Ethernet did not support end-point synchronization, timing support, bounded latency support, or bandwidth allocation [5]. The goal of the IEEE 802.3 effort was thus to develop mechanisms for better supporting AV applications in an Ethernet network by providing the appropriate mechanisms.

Very quickly after the respective study group had been set up by IEEE 802.3, it became apparent that the proposed solutions were better suited for standardization in IEEE 802.1 [8], and by the end of November 2005 the effort was officially moved [9]. In September 2011 (plus a latecomer in August 2013), the following set of standards associated with first generation Audio Video Bridging (AVB, AVBgen1) were completed (see also Section 7.1.3):

- IEEE 802.1Qav, "Forwarding and Queuing Enhancements for Time-Sensitive Streams" (traffic shaping), January 5, 2010.[3]
- IEEE 802.1Qat, "Stream Reservation Protocol," September 30, 2010.
- IEEE 802.1AS, "Timing and Synchronization for Time-Sensitive Applications," March 30, 2011.
- IEEE 1733, "Protocol for Time-Sensitive Applications in Local Area Networks" (AVB adaptation of RTP), April 25, 2011.
- IEEE 1722, "Transport Protocol for Time Sensitive Applications in a Bridged Local Area Network" (layer two transport protocol), May 6, 2011.[4]
- IEEE 802.1BA, "Audio Video Bridging (AVB) System" (overall system configuration, profiles) September 30, 2011.
- IEEE 1722.1, "Device Discovery, Connection Management, and Control Protocol for 1722™ Based Devices" (control mechanisms and service discovery), August 23, 2013.

These standards leave the implementer with a variety of options. In 2008, a number of companies participating in AVB standardization at IEEE thus started working on the formation of an industry alliance that would market AVB and aid interoperability [R. Kreifeld, email correspondence, 2013]. The Avnu Alliance that was launched in August 2009 [10] now offers the respective certification programs. Concerning the proliferation of AVB in the market, AVB-enabled silicon started to emerge in 2012 (see e.g. [11, 12]), which, from an automotive perspective, was just the right timing. Having finished the AVB standards listed above, the IEEE started directly with standardizing AVGgen2 in order to be able to support safety critical applications (see also Section 7.1.4). As these no longer comprise audio and video applications, the AVB effort was officially renamed Time Sensitive Networking (TSN) in November 2012 [13].

7.1.2 The Audio Video Bridging (AVB) Use Cases

Before going into the details of specific applications, here are some general remarks on the fundamental differences between the quality requirements of AV consumption (including speech) and those that traditional Ethernet was designed for:

- While most data applications require every single packet to arrive intact, the occasional **packet loss** can go unnoticed by the user in the case of AV transmissions. That not all information is needed is emphasized as such by the fact that many AV compression formats are "lossy," meaning that not all information can be recovered during decompression. Moving Pictures Expert Group (MPEG) with MPEG-1 or 2 Audio Layer III (MP3) are well known examples. Even though it is in general more critical to lose a packet of compressed data than a packet of uncompressed data, an additional and occasional loss of a packet containing compressed data is not necessarily perceived as a quality degradation [14].
- The situation reverses in the case of **delays**. In general, delays in the milliseconds range or even of a few seconds are not discussed when it concerns a file transfer or building up a website. AV applications, in contrast, have very stringent timing requirements [15]:
  - The **absolute delay** must be small in the case of live streaming data. A musician, e.g., tolerates only a 10 ms delay between initiating a sound and expecting to hear it [16, 17].
  - The data must be **synchronized**. In a home (or concert hall) sound and image might travel different paths with different delays. For quality replay, the delay between sound and picture replay needs to be smaller than ± 80 ms ("lip sync" requirement) [18]. Standard home stereo sound needs synchronization between the different streams of less than ± 1 ms [16]. For high-end surround sound the synchronization requirement reduces to less than ± 10 μs [18], or ± 1 μs in the professional environment [8].
  - No matter where the AV content is stored or replayed, there should be **no noticeable jitter**. Sudden interruptions or delay variations in AV streams can occur when competing traffic is consuming too much of the available data rate. Buffers can only partially compensate for this. For some applications the absolute delay is strictly limited and, in general, the larger the buffer, the higher the costs.

7.1.2.1 In Homes/Consumer Devices

Since the introduction of audio and video (AV) compression formats, AV streams are turned into strings of data packets that can be stored on and replayed on various consumer devices such as PCs, laptops, tablets, phones, etc. While traditional home entertainment consisted of units with a clear media-to-function relation (record player, tape recorder, CD-player, amplifier, etc.) and analog communication between them, the transformation of entertainment data into packets allows for/requires bidirectional networking between units, which Ethernet inherently supports. The general observations made above apply to all AV applications.

Additional requirements, e.g., on timing in consumer devices, come from gaming applications, which require a response time of less than 50 ms for human activity and less than a ±80 ms difference between video animation and audio. Other home related AV applications with different requirements yet again are home surveillance and healthcare [5].

In contrast to the professional audio and automotive use cases, requirements that prevail in consumer applications are (a) the support of ad-hoc/plug & play capability (no IT administrator) and (b) the lowest cost [5]. As consequence, aspects such as discovery (which is addressed at higher layers by UPnP[5] and DLNA[6]), self-configuration, and a high level of compatibility and interoperability are very important [10]. Furthermore, in a home, media might be shared over a variety of networks that include IEEE 802.11 WLAN/WiFi and Coordinated Shared Networks[7] (CSN) [19], as well as Ethernet.

7.1.2.2 In Professional Audio

Typical application areas for networked professional audio equipment are concerts/live shows, recording studios, conference centers, theme parks, houses of worship, art installations, or any other place where live sound is used professionally [15, 20]. This emphasizes one of the fundamental differences relative to the consumer use case: In professional audio, good quality perception is a core purpose. A network deployed for professional audio has to be absolutely reliable, with no audio defects, video dropouts, or other artifacts [15]. Furthermore, the timing requirements are very stringent: As has been said, for musicians the delay, e.g., between the microphone and the earphone of an artist, needs to be smaller than 10 ms [16, 17]. Allowing 8 ms for processing means that the network delay cannot exceed 2 ms [M. Johas Teener, email correspondence, 2013]. Professional audio also has very stringent requirements on speaker synchronization, which needs to be within a few microseconds [8].

As with all industrial products the use of a new technology/concept needs to provide for direct or indirect cost savings (or for new functionalities that are expected to result in monetary advantages, see also Figure 3.10). The starting point of professional audio networks is a set-up that comprises a huge amount of high-quality, single-purpose, unidirectional, analog (or even digital) audio cabling using different technologies. Furthermore, a separate set of extensive wiring is used for the respective video infrastructure, and yet another set of cabling for control (of amplifiers and loudspeakers), which might use an Ethernet infrastructure [8, 15, 20–22]. This is not only expensive with respect to the wiring, but also difficult to maintain, and invites the development of proprietary solutions at higher layers, which seems to have prevailed for a long time [5]. Being able to handle such a setup requires very specialized know-how [15].

Thus, the attraction to using a single network, i.e., the Ethernet infrastructure, for all data that needs to be networked in the professional AV applications, is obviously large. In pre-AVB times, this was too cumbersome [8, 15]. So when AVB activities started, it was only natural that these were supported by professional audio companies from the start [16] in the IEEE [5] as well as when establishing the Avnu Alliance [R. Kreifeld, email correspondence, 2013].

An important difference between the residential/consumer and automotive uses of AVB is that the extent of the professional audio network can be significantly larger, as measured in meters as well as the number of nodes. However, in contrast to the consumer use case, the professional AVB network can be expected to be professionally set up and controlled.

7.1.2.3 In Cars

Ever since Ethernet started being discussed for automotive use, the Quality of Service (QoS) capabilities of Ethernet and the potential of the AVB solutions have been investigated (see also Section 3.2). With Ethernet coming from the IT and CE industries, Ethernet was first considered for "similar" in-vehicle infotainment applications. So, while today, Ethernet is also being discussed for in-vehicle control applications, the focus at the beginning was on enabling AV applications. By the time the Avnu Alliance was set up in 2009, automotive applications were identified as one of the target areas for AVB [23].[8]

In-vehicle AV consumption was not one of the original use cases addressed with the standardization of AVB at IEEE. Nevertheless, infotainment is an important quality element for car users; after all, the stringent automotive Electromagnetic Compatibility (EMC) requirements (see also Section 4.1) have been installed to also ensure unblemished audio consumption while driving. Nevertheless, in the hierarchy of applications inside cars, infotainment is always secondary in relation to driving and safety. This is the most important difference to the consumer and professional audio use cases discussed before and its consequences impact the in-vehicle use of AVB (see the following sections). Furthermore, the automotive industry has an additional timing constraint: The AV system needs to be fully operational for some applications within two seconds of power on [24]. Neither in the consumer domain, nor in professional audio does such a (stringent) start-up requirement exist. As is true for the professional audio domain, the in-vehicle AV network is professionally set up beforehand, even if various car models exist and the exact layout depends on the options the customer selects.

Naturally, the automotive audio use case itself differs from the ones that can be expected in homes or even professional environments. The high-quality expectations from car customers and the complexity of handling the various different audio sources in cars had once even led to the development of a new in-vehicle networking technology (MOST, see Section 2.2.4.1). An example of audio use cases and their hierarchy inside the car is presented in Table 7.1. As can be seen, a significant amount of the complexity is not handled at the network interface but is organized at higher layers. From an

Table 7.1 Example audio hierarchy in an automotive audio network

Layer	Functional block	Features and functions
High	Human Machine Interface (HMI)	Customer interface for: Volume control, source changes, additional control interfaces (e.g., changes of volumes for audio interrupts sources like jingles and alarms)
Mid	Audio management system	Fixed system behavior: Controls the mixing stages in the audio sink by special control commands. An example is the audio output in case a navigation audio guidance message or park control beep occurs at the same time as the driver is listening to the radio or making a phone call via the in-built hands-free system. The solutions here are generally car manufacturer specific
Low	Audio network interface	Network resource management: Is responsible for the availability of the requested bandwidth. The source needs to know when to allocate bandwidth and the sink knows when and how to connect to the source data

automotive perspective, it is important to maintain the separation between application-specific requirements and the QoS functions that the network can provide based on AVB. The separation of the ISO/OSI layering model should be maintained.

Furthermore, costs and resources distinguish the use cases. The AVB functionality requires hardware capabilities in the Ethernet semiconductors and processing power from a separated micro-controller (μC) or from an Ethernet switch. In professional audio it can be assumed that all processing resources needed will be provided. After all, audio function and quality are their primary concerns. In the CE industry costs, in principle, need to be low, though the resources available and customer expectations are likely to vary significantly, depending on the monetary value of the CE device. However, in 2020 a laptop or even a tablet typically had much greater resources available than a typical ECU inside a car that is optimized for hardware costs, space, and processing power.

While costs are important, the question of where and how to implement the AVB functions in an automotive network has more facets. If the AVB functions are embedded on a micro-controller that is integrated into the switch, which supplier will provide the software for it, the Tier 1 or the semiconductor supplier? If the AVB software comes from the semiconductor supplier, but the Tier 1 is responsible for the function of the ECU, who, if there is a malfunction, can diagnose it? Who is responsible? If resources of the ECU's main μC are used, how can it be ensured that network functionality is never impaired by other ECU functions, especially during start-up or reboots? As a first approach, Matheus et al. [25] proposed to use only an absolute minimum of AVB features for the automotive networks and, with this, initiated an important discussion on automotive AVB. But the result was non-standard compliant, and was, therefore, reworked [26]. This again was relevant input for the actual solutions described in Section 7.1.3.

7.1.2.4 Direct Comparison of Use Areas

The description in Sections 7.1.2.1–7.1.2.3 showed that the three use cases have very different requirements. What they have in common is that all would like to realize high-quality AV streaming in a (mainly) Ethernet-based network. Additionally, the use case and the network are restricted to a certain purpose, size, and physical location, even if a concert hall network may have significantly larger dimensions than a LAN inside a family home or a LAN inside a car. All three use cases can live with the maximum delay guaranteed for 100 Mbps Ethernet AVB traffic of 2 ms over 7 hops on ISO/OSI layer two.[9]

Table 7.2 directly compares the main properties and requirements of the AVB use cases. The main difference is the low-cost, multi-vendor plug & play requirement of the consumer use case, in contrast to the very high-quality requirements of the professional audio use case, or the secondary nature of AV in automotive and its stringent start-up requirement.

7.1.3 First Generation AVB Protocols and Their Use in Automotive

The AVB specifications facilitate QoS guarantees for streaming data within an "AVB cloud," i.e., a group of networked devices all supporting the core AVB functions either in the role of forwarding switches (which in the IEEE context are referred to as bridges, see introduction to Section 7.1) and/or as end nodes. The basic QoS requirements are that the streams can be rendered in sync with each other, that network delays are not noticeable in the application, and that the network resources are available for as long as the application needs them [8]. AVB distinguishes between "Talkers," which are the source of the streaming data, "Listeners," which are the consumers of those streams, and the AVB capable switches in between. The implementation of AVB requires that the underlying Ethernet network runs at least at 100 Mbps full-duplex, that the Ethernet payload does not exceed the maximum size of 1500 bytes, and that flow control/pause packets (see Section 1.2.1) are disabled. The following subsections describe the AVB mechanisms in more detail.

7.1.3.1 Basic Transport

The IEEE 1722 [28] specification describes the transport of AV data. It leverages concepts from the IEC 61883 standards on digital interfaces of consumer AV equipment and thus FireWire/IEEE 1394 [21]. The key property of IEEE 1722 is that it identifies Ethernet packets carrying AV content on layer two and not on higher layers. This allows bypassing higher layer protocols (see Figure 7.2), thus reducing the processing time needed and making the latency more predictable.[10]

In principle, IEEE 1722 transports two types of content: Streaming data and data for controlling IEEE 1722. Figure 7.3 depicts the respective packet structures: An Ethernet frame/packet carrying an IEEE 1722 packet and its content, which has its own header for the AV content units. The Ethernet packet has to include the otherwise

Table 7.2 Comparison of the requirements and properties of the different AVB application areas

Criteria	Home/consumer devices	Professional audio	Car AV
AV application scenarios	Multiple source/sink AV replay in the home, home surveillance [5]	Recording studios, concerts/ live shows, conference centers, theme parks, houses of worship, art installations [15, 20]	Simultaneous audio streams of different priority, synchronous replay of AV, camera data for driver assistance
Importance of AV quality	Expectations are likely to correlate with the price paid for the equipment	**The core purpose**	Entertainment and comfort are important under normal circumstances but **driving (safety) is requirement No. 1**
Variability of network setup	Ad hoc, **plug and play**, no IT admin [5], requires self-configuration, service discovery, etc. [19]	Setup can change but is carefully planned from event to event	Known number of predefined variations per car model (limited plug and play from passengers)
Network technologies used	Ethernet, WiFi/ WLAN, Coordinated-Shared-Networks (CSNs) [19], Bluetooth	Ethernet	Ethernet
Service rejection (insufficient enough data rate)	Acceptable	Not acceptable	Not acceptable
Synchronization accuracy required	Stereo sync ~ 1 ms/10 μs [16, 18]	**~ 1 μs** [8]	Similar to consumer devices
Maximum network delays	50 ms [17]	2 ms application delay [M. Johas Teener, email correspondence, 2013]	80 ms for lip sync [17], 2 ms for echo cancellation
Start-up requirements	None	None	**AV system fully operational within 2 s** [24]
Link length	< 200 m [5]	Can be long, but < 1 km expected	≤ 15 m [26], 3.5 m average [27]
Available processing resources	Depends, larger on computers, smaller on mobile devices	As large as needed	Generally shared with other functions, rather small
Costs	Very low costs [5]	Function before cost, savings in cabling	Needs to be competitive (see Section 3.4.2)

optional IEEE 802.1Q header. The priority information (as defined in IEEE 802.1Q and used in IEEE 802.1Qat/SRP) is essential for the functioning of the AVB QoS concept. Using VLANs (see also Section 7.2.2) is, in principle, orthogonal to AVB and an independent consideration. However, to be able to receive streams, Listeners must be members of the Talker's VLAN [28]. The IEEE 1722 Ethertype is 0x22F0. If the IEEE 1722 packet is shorter than the required 42 bytes minimum Ethernet payload length, the Ethernet MAC will pad the packet automatically, as with any other protocol transmitted via Ethernet.

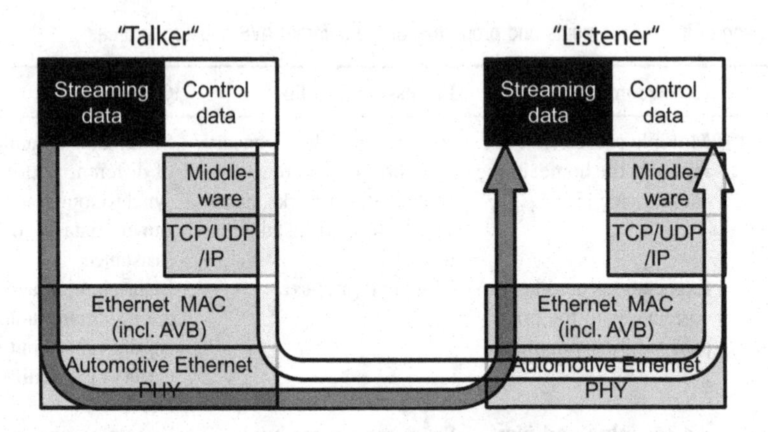

Figure 7.2 IEEE 1722 streaming data and application control data in the automotive industry

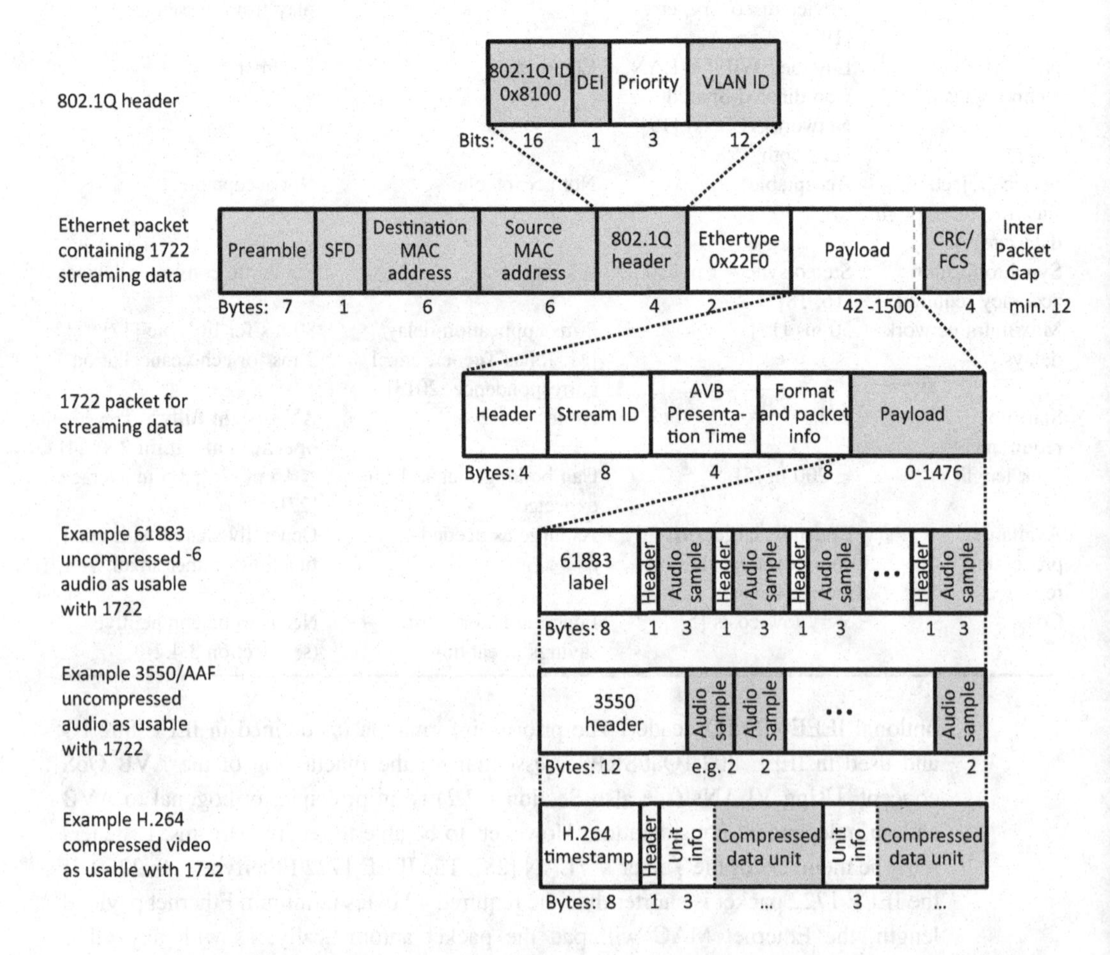

Figure 7.3 The IEEE 1722 packet format with example 1722 payloads; DEI is the (not commonly used) Drop Eligible Indicator, which allows to mark packets that can be dropped in the case of congestion

An IEEE 1722 streaming packet consists of a header, the stream ID, the Presentation Time, payload information, and the payload itself (see Figure 7.3). The header defines what type/format of AV data to expect. It also includes the sequence number in order to allow Listeners to identify missing packets. The stream ID unambiguously defines a specific data stream and is derived from the Talker's MAC address (see Section 7.1.3.3). The field for payload information is directly related to the format of the data inside the payload.

A very important part of AVBTP 1722 is the Presentation Time. It defines the time a received packet must be presented to the Listener application, i.e., when it should leave layer two at the receiver. The Talker sets the Presentation Time depending on the time the packet left the application buffer inside the Talker and the expected worst-case forwarding latency across the network ("Max Transit Time"). The default value for the Max Transit Time in the case of the highest priority streaming "Class A" traffic is 2 ms; in the case of the next lower priority, "Class B" traffic, it is 50 ms. The standard allows this value to be different/negotiated, but it does not define how this should be done. Standard plug & play equipment can thus be expected to use the standardized default value(s). The Presentation Time is represented in nanoseconds (ns) as the remainder when dividing the absolute time by 2^{32} ns. The concept of the Presentation Time is a good example of the close interrelationship between the AVB standards, as the Presentation Time can only work in a time-synchronized network (see Section 7.1.3.2 for IEEE 802.1AS). Once time synchronization has been established the Presentation Time can be used as feedback and correction for the synchronization. The Max Transit Time is also one of the values that determine the required buffer size of the Listeners (see also Section 7.1.3.3), because the Max Transit Time determines the possible jitter that is compensated by buffering the AV data.

The IEEE 1722 standard also provides an important mechanism for QoS in Automotive Ethernet. In terms of its use, the following considerations are important:

1. **Supported data formats:** The IEEE 1722-2011 specification covered mainly FireWire/ISO 61883 headers,[11] but not, e.g., the formats discussed in the automotive industry for camera use cases: MJPEG and H.264 (see also ISO 17215). This was changed with the IEEE 1722-2016 release [28], which makes using these formats a lot more efficient (plus supporting CAN, LIN, FlexRay). Should yet more formats be needed, the payloads defined for the Real-time Transport Protocol (RTP) in a number of RFCs related to IETF RFC 3550 [29] can be used with IEEE 1722 without modification.
2. **Use of the Presentation Time:** Figure 7.4 shows an example in-vehicle network consisting of an Audio Video Source (AVS), a Head Unit (HU), a Rear Seat Entertainment (RSE) system, and an Amplifier. The goal is to have a lip-synchronous replay of the content on two different displays and two speakers. While the concept of Presentation Time as defined in IEEE 1722 is important, it has its limits in this use case. IEEE 1722 specifies the time the data shall be presented to the system beyond layer two. In the example this means the time the data is passed on to the AV codecs, and not, as would be desirable, the time of playing the data on the displays and speakers. The example scenario is even more critical, as after

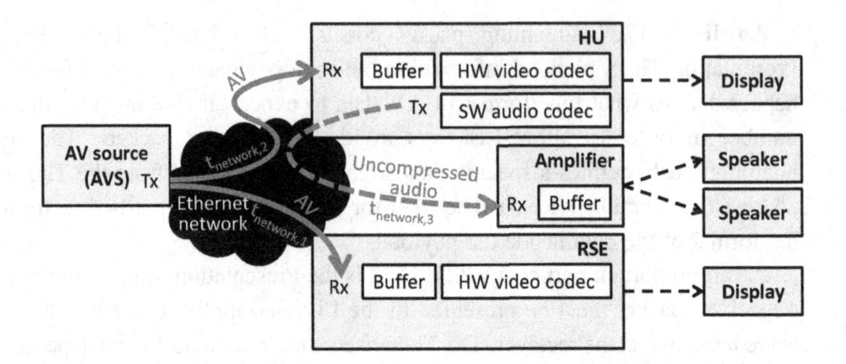

Figure 7.4 Different audio and video paths in an automotive in-vehicle network that are a challenge for achieving lip sync

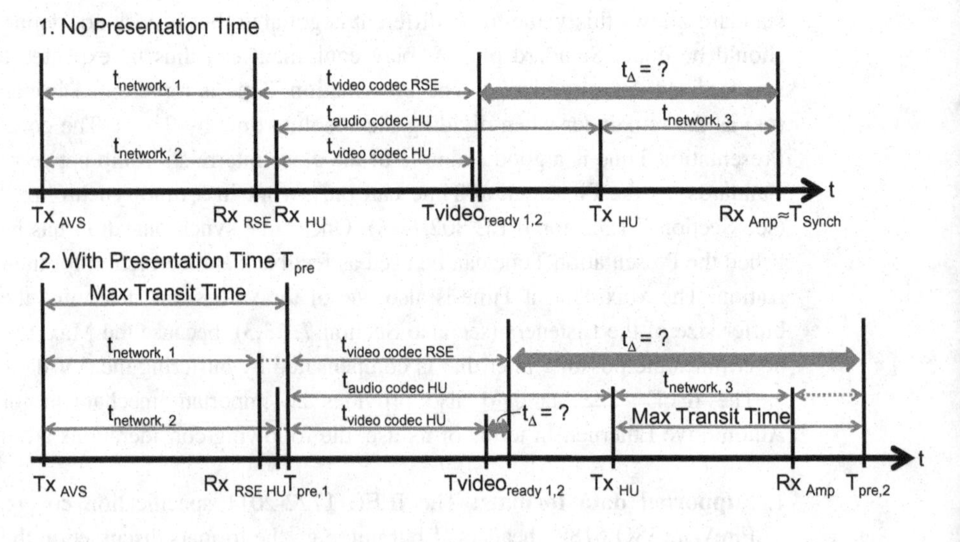

Figure 7.5 Timing behavior with and without Presentation Time T_{pre}, assuming T_{pre} is derived from the maximum delay in the network Max Transit Time

decoding the uncompressed data is reinserted into the network. This transmission is completely independent from the first and not considered in the original Presentation Time provided by the AVS. The Presentation Time as originally defined is thus not sufficient for ensuring synchronous replay in this scenario. Even if only the HU and RSE video replay was considered, the Presentation Time would only help if the processing delay caused by the video decoding in the two units differed only marginally.

Figure 7.5 depicts the consequences for the timing behavior. In the upper half of Figure 7.5, the Presentation Time is not used and the codecs start processing the data the moment they receive it (Rx_{RSE}, Rx_{HU}). In the example, the audio codec is implemented in software and is, therefore, slower (without this being decisive). In the end, neither of the two image streams are ready at the same time ($Tvideo_{ready,1}$, $Tvideo_{ready,2}$), nor is the audio, which arrives significantly later at the speakers (Rx_{Amp}). The lower half of the figure shows the same scenario under the assumption that the maximum delay possible, the Max Transit Time, is used to derive the Presentation Time ($T_{pre,1}$, $T_{pre,2}$). As can be seen, this does not help to improve the synchronization between any of the output files.

Instead of a Presentation Time that defines when to present the data to the application beyond layer two, it would thus be desirable to have a time available that defines when to present the AV information to the customer [24]. One approach would be to set the Max Transit Time to a value other than the default values of the standard. The standard, in principle, allows for this. However, it needs to be assured that all used units support the use of a different value.[12]

3. **Dynamic versus static use:** IEEE 1722 expects either locally administered unicast addresses or the use of multicast addresses, which can be statically or dynamically allocated. In order to support the dynamic allocation of multicast addresses the IEEE 1722 specification defines a MAC Address Acquisition Protocol (MAAP) [28]. Within a specifically reserved range of multicast addresses the MAAP can dynamically establish which addresses can be used for a new stream, while defending the address from other uses once it has been selected. As has been described with the use cases in Section 7.1.2, the automotive scenario is not particularly dynamic. Once a car model has been designed, the number of different Automotive Ethernet network topologies for this car is limited. At the same time, start-up is critical and therefore all dynamic negotiations disadvantageous. A static pre-configuration is thus preferred for the IEEE 1722 address allocation.

7.1.3.2 Basic Synchronization/Timing

The main purpose of IEEE 802.1AS-2011 [30] is to synchronize all nodes in an AVB cloud to a common reference time. The standard mandates a precision of ±500 ns for two end nodes that have fewer than 7 AVB nodes in between, which means that direct neighbors have to synchronize with nanosecond precision [16, 21]. IEEE 802.1AS – also referred to as the "generalized Precision Time Protocol (gPTP)" – is a simplified extension of the IEEE 1588 specification [31], which was started at the end of the 1990s, first completed in 2002 [32], and updated in 2008 (PTPv2). The main difference between IEEE 1588 and IEEE 802.1AS is that gPTP assumes that all communication between time-aware systems is done using IEEE 802 MAC PDUs and addressing only, while IEEE 1588 supports various layer two and layer three to four communication methods.

Like in most time synchronization approaches, one node in an IEEE 802.1AS network functions as a "Grandmaster," to whose clock all other clocks synchronize. Which unit needs to be the Grandmaster is not standardized. Ideally, the Grandmaster is the node with the best suited clock in the AVB cloud. The standard consequently addresses the following two topics: (a) How to select the Grandmaster and (b) how to correctly synchronize to its time throughout the AVB network.

The Grandmaster can be pre- or auto-selected with help of the Best Master Clock selection Algorithm (BMCA). The BMCA is a distributed algorithm: Every Grandmaster-capable node receiving a respective "announce" message compares the information of the current best Grandmaster with its own clock-related quality values. If the eight differently rated values of its own clock yield a better result than that of the current Grandmaster announced in the message, the unit having done the comparison announces itself as the new best Grandmaster. The process is repeated until the truly best Grandmaster in the network has been found. The "announce" messages are sent cyclically and the Grandmaster can change anytime the Grandmaster selection data changes, for example the actual Grandmaster leaving the network, a unit having access to a better clock,[13] or a new Grandmaster suitable unit joining.

As a side effect, the BMCA also determines the "clock spanning tree," i.e., the paths on which synchronization related messages pass through the network. This is done by labeling all ports in the AVB network as follows:

- "Slave port" (the port on which the last message from the Grandmaster was received).
- "Master port" (ports on which the message was passed on).
- "Disabled port" (ports that connect to non-1AS capable nodes).
- "Passive port" (ports that lead to redundant paths in the AVB cloud).

Normally, an ECU supporting AVB without a switch has only one port. Unless this ECU provides the Grandmaster – in which case the one port is Master – such nodes have one Slave port only.

In order to achieve synchronization, every unit needs to know the delays caused by the propagation of messages in the network. IEEE 802.1AS thus defines so-called pDelay measurements, in which every node in the AVB cloud learns the propagation delays between itself and its direct AVB neighbors to which packets might be sent. Additionally, the pDelay measurement is also suited to determine whether a direct neighbor is actually 1AS capable. An important tool for the pDelay determination is time-stamping: The IEEE 802.1AS Ethertype (0x88F7) triggers sampling of the local clock at the ingress and egress, respectively, of the PTP message from the MAC [21].[14] To achieve the aspired nanosecond precision in the time-stamping, it is necessary to implement the time-stamping in hardware instead of software [33]. The pDelay measurements are cyclically repeated.

Last, but not least, the time to synchronize needs to be made known in the network. This is done in two steps, with "sync" and "follow_up" messages along the clock spanning tree in the network. The two-step synch mechanism was explicitly introduced to allow realization in software. At the time the specification was developed, the hardware needed to realize an accurate one-step mechanism was perceived as too costly [M. Turner, email correspondence, April 20, 2020]. Figure 7.6 gives an overview of the mechanisms.

The class of MAC addresses starting with 01-80-C2 is reserved for MAC control protocol messages. Packets with such a MAC address are never forwarded but only exchanged between direct neighbors. IEEE 802.1AS messages use one of these

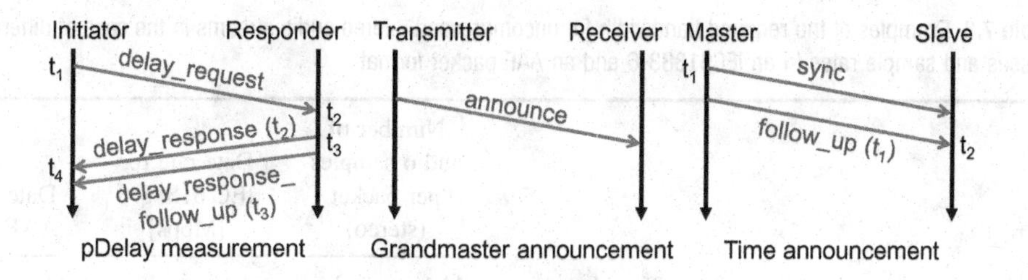

Figure 7.6 Flow charts of major IEEE 802.1AS functions

specific multicast MAC addresses (01–80-C2–00–00–0E) [30]. In consequence, it is not foreseen that such IEEE 802.1AS Ethernet packets include the optional IEEE 802.1Q header, as a VLAN information would potentially collide with the purpose of the multicast address used.

For automotive use, it is sensible to preselect the IEEE 802.1AS Grandmaster (and the clock spanning tree) and to choose an ECU every car is equipped with to be the Grandmaster. It may sometimes be the case that an optional ECU might actually provide a better quality clock (better accuracy, less jitter). Nevertheless, dynamic selection of the Grandmaster would not only unnecessarily strain the start-up time, it would also lead to more effort in the qualification and testing of the network. More variants (of, e.g., who the Grandmaster is) create more possibilities for malfunctions, and robustness is essential in the automotive industry.

The layout of in-vehicle networks is predesigned, known, and in principle does not change during use. Some links and ECUs might be disabled when not in use (see wakeup and sleep/partial networking in Section 6.3.2), like a surround view system that is initialized with the selection of the reverse gear and switched off when the velocity of the car has exceeded a certain limit. Nevertheless, the link lengths and the locations in the network do not change from one time the car is used to another. Thus, the pDelay values will not change (much) every time the car is started. Start-up can be speeded up further, if the last pDelay values are learned and stored.

7.1.3.3 Basic Stream Reservation

The IEEE 802.1Qat Stream Reservation Protocol (SRP) [34] supports allocating bandwidth for individual application and traffic streams within the AVB cloud. The main idea is that Talkers announce the availability of streaming data to all units in the AVB cloud. If Listeners would like to receive the stream, they also announce this. As a consequence, all switches the data must pass through in order to get from the Talker to the Listeners evaluate the availability of the needed bandwidth. If available, the specific streaming bandwidth is guaranteed. If not available, the reservation request is declined. By default a maximum of 75% of the available bandwidth can be reserved, though system designers can, with care, increase or decrease this number depending on the actual requirements. If a set-up allows, e.g., up to 50% of bandwidth to be reserved for Class A traffic, but actually only 30% have been reserved, Class B traffic can reserve the remaining 20% should the available bandwidth in Class B not be sufficient.

Table 7.3 Examples of the required bandwidth for uncompressed, stereo audio streams in the case of different traffic classes and sample rates in an IEC61883-6 and an AAF packet format

Class	Frequency of packets [kHz]	Time between packets [ms]	Number of audio samples per packet (stereo) 44.1 kHz	48 kHz	Date rate for IEC 61883-6 [Mbps] 44.1 kHz	48 kHz	Date rate for AAF [Mbps] 44.1 kHz	48 kHz
A	8	0.125	12	12	7.04	7.04	5.76	5.76
B	4	0.25	23	24	4.93	5.06	3.58	3.65
C (64 samples)	0.75/0.689	1.333/1.451	128	128	3.16	3.44	1.78	1.93

Note: Traffic Class C is not part of IEEE 802.1Qat [34]. The class with 64 samples at 44.1 kHz or 48 kHz has been published in [35]. The calculation includes 16-bit sampling, 30 bytes overhead for the Ethernet packet, 24 bytes overhead for the 1722 packet, 8 bytes + 1 byte overhead per sample for IEC 61883 or 12 bytes and no per sample overhead for AAF

Also IEEE 802.1Qat uses special Ethertypes. 0x22EA is used for the actual reservation with the "Multiple Stream Reservation Protocol (MSRP)." 0x88F5 (MVRP) and 0x88F6 (MMRP) identify control packets for necessary information and registration associated with it.[15]

At a minimum the information needed for the stream reservation is a unique stream ID, which is generated from the Talker's MAC address and a 16 bit number the Talker assigns to the stream, and quality data about the stream itself. This includes the traffic class, the packet rate, and the length of every packet sent. Table 7.3 gives examples for the bandwidth needed for uncompressed stereo audio data for the defined traffic Classes A, B, and a new traffic Class C.[16] This new traffic class has been generated in order to meet today's DSPs and DMAs typical process block rate of 32 or 64 audio samples at either 44.1 kHz or 48 kHz [35]. As can be seen, transmitting audio samples with the originally defined IEC 61883-6 packet format consumes significantly more bandwidth than using the newer AVTP Audio Format (AAF). The Avnu automotive profile thus recommends the use of the AAF and does not support IEC 61883-6 [35]. Table 7.3 also shows that the higher the frequency of packets the larger the bandwidth consumed.

The use of SRP in the automotive industry poses three main challenges:

1. Table 7.3 shows that the average number of samples streamed in a Class A packet is very small (for the assumed simple stereo audio use case) and packets are sent at short intervals. Having significantly more overhead than payload is not only a waste of bandwidth, it also results in an unnecessarily high processing load in all involved devices (while admittedly Class A realizes the shortest latency). Seen from this perspective, it is more advantageous to use longer packets; ideally they use the complete maximum MAC packet size [24]. To somewhat improve the ratio, it is thus advantageous to introduce new traffic classes, like the one identified with "C" in Table 7.3.

2. A denial of an IEEE 802.1Qat reservation request inside a car is not acceptable. It would be disastrous if, e.g., a driver assist function using camera data failed because a rear seat passenger was watching a High-Definition (HD) video. Even rejecting an audio stream from a passenger's mobile device is critical, as it would likely be perceived as a malfunction of the car [24]. It is therefore essential that the AVB network and the expected transmission rates, including the ones from consumer devices, are carefully planned upfront and that these considerations are reflected in the network design (see also Chapter 6).
3. Last, but not least, like the dynamic selection of the Grandmaster, a dynamic reservation of streams at system start-up potentially takes too long. Inside cars, the applications using the Ethernet network as well as their data rate requirements are known; this includes the multimedia applications passengers might bring in. After all, every car only seats a certain number of passengers who each can use only a limited number of devices. It is thus in principle possible to envision a static reservation of streams. However, the information on the reserved streams needs to be provided to all nodes involved. One possible solution is to pre-store data that can be accessed with every start-up. There are two ways to generate the pre-stored data. One is to run the SRP protocol once as part of the network set-up during the manufacturing process and to store the outcome. The other is to separately design the data in the development process and store different tables for every car/option combination in every AVB ECU.

7.1.3.4 Basic Traffic Shaping and Quality of Service (QoS)

The idea of IEEE 802.1Qav [34] is to improve the actual quality of all AV transmissions in the AVB network by ensuring that the packets of each individual stream are evenly distributed over time instead of one stream blocking all others for a while. Even if, e.g., a stream cannot use more than the assigned 10% of the available bandwidth, it makes a significant difference for (the delays of) the rest of the streams (and the network as such), whether this stream uses all bandwidth for 1 minute and then sends nothing for 9 minutes or whether it sends something for 0.1 ms and then nothing for 0.9 ms.

To achieve this is relatively straightforward at the source of streams, the Talkers. The traffic classes assigned with the streams determine the frequency of the packets sent; every 125 μs with Class A, every 250 μs with Class B, and every 1 ms with Class C. It just needs to be assured that the Talkers do indeed send the packets at this rate. For live streams like camera or microphone data or for CD audio the data is generated continuously at the application data rate anyway, so all that needs to be done is to package the data at the assigned packet frequency. In other cases of pre-stored streaming data, this can be quite different. The transmitting node will likely send as much data as possible at once, with the idea that the receiver will buffer the data until it is played. This unnecessarily strains the bandwidth available in the network at the switches and causes the risk of dropped packets owing to buffer overflow. Larger buffers can mitigate this risk, but increase the costs of the receiver [8] and increase the latency. So, in the case of pre-stored streaming data, pacing the output of the Talker can make an essential difference to the AV quality in the network.

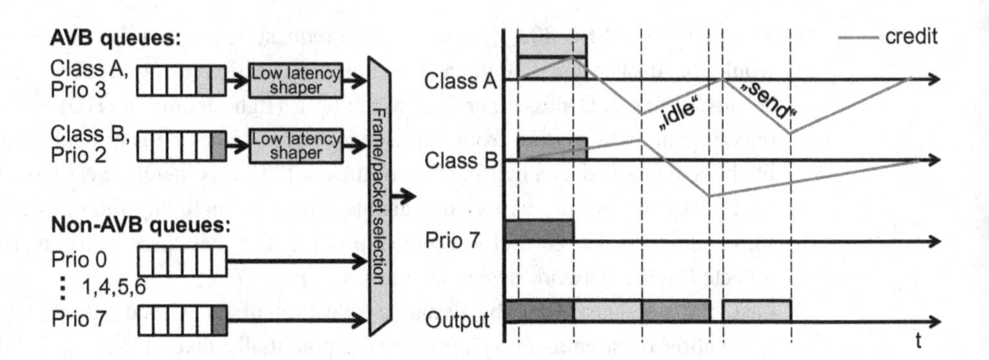

Figure 7.7 Principle of credit-based shaping and queuing with IEEE 802.1Qav

In the switches, the situation is more complex. A switch potentially needs to handle AV prioritized streams of multiple Talkers and traffic that passes through from connected non-AVB units. The issue with the latter is the following: Within an AVB cloud the Stream Reservation (SR) traffic classes are matched to some of the eight quality values provided with IEEE 802.1p; by default Class A traffic has priority 3 and Class B traffic has priority 2. Traffic that passes through the AVB cloud from connected non-AVB units might use the same priority values. These priority values then need to be changed at the switch where the traffic enters the AVN cloud and restored in the switch where it exits.

A simple switch only evaluates the Ethernet header to decide through which output port(s) the data is sent and into which priority queue of the output port(s) the data needs to go. All data with the same priority goes into the same queue, which generally works on a First In First Out (FIFO) basis, independent from the source of the data. SR traffic has priority over non-SR traffic.

The functioning of the AVB credit-based shaper is visualized in Figure 7.7. If SR traffic enters the priority queue of a port that is currently busy it collects credit at an "idle" rate, which is equal to the overall bandwidth currently reserved for the respective SR traffic class. As soon as the port is no longer occupied and the credit is equal to or larger than 0, the priority packet is transmitted. During transmission the credit reduces at a "send" rate which equals the transmission rate of the link minus the idle rate. If at the end of the transmission the credit is still equal to or larger than 0, more packets from the same queue can be transmitted. If at the end of the transmission the credit is still equal to or larger than 0 and the priority queue is empty, the credit is set to 0. When the credit for a queue is negative and no packet is ready to send, the credit increases according to the idle rate up to 0 but not higher. Only once the credit is 0 or higher can a packet be transmitted.

The main concern for the automotive industry with IEEE 802.1Qav is that the immediate reception of safety critical control data is more important than the AV quality. To simply use the highest priority SR Class A for control data is not a solution, as the traffic shaping might actually delay the transmission of a safety critical message [36, 37]. The use of the highest priority non-AVB traffic class for critical

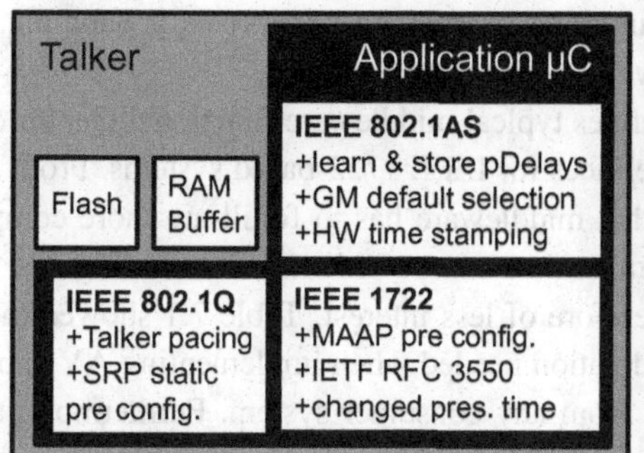

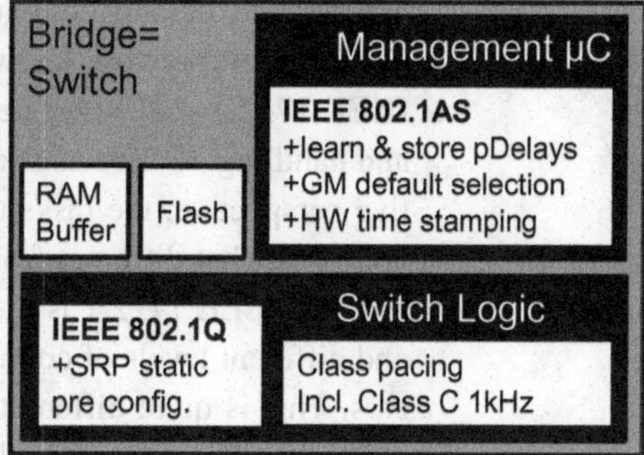

Figure 7.8 Proposed elements of an automotive AVB implementation for automotive AV ECU(s) either in the function of a Talker (Listener) or within the switch semiconductor

control data would guarantee an average throughput of 20% or more (depending on how much has been reserved for the AVB queues). However, AVB-queue packets within their reserved bandwidth always pass first, and the critical control data might be delayed. The simplest solution would thus be to ensure that safety critical data does not share the same link and direction as AV traffic. For topologies in which link sharing cannot be avoided, a number of new TSN standards address the requirements for safety critical control traffic in an Ethernet network (see Section 7.1.4).

Apart from the principal concern addressed in the previous paragraph, the advantages of shaping as such have been discussed at length for automotive use cases. Two aspects need to be observed: For one, not all MAC-implementations in the µCs used in the Talker support shaping in hardware, while it requires significant effort to realize it in software. Furthermore, not every shaping algorithm is suitable for the use cases that require support. If multiple streams pass through a switch, a different algorithm might be optimum for each. The solutions thus require careful design.

Figure 7.8 summarizes the AVB elements proposed in Sections 7.1.3.1–7.1.3.4 for use in Automotive Ethernet. Note that there is one design aspect, which has not been explicitly addressed in this section, but which is very important and needs careful attention when designing the network and selecting the means for QoS: Ensuring that the switch buffer has the right size to not drop packets while keeping the costs at bay!

7.1.3.5 Other First Generation AVB Protocols

The following specifications are part of AVB (gen1), but at the time of writing were of minor relevance for automotive implementations:

- The Real-time Transport Protocol (RTP) describes how to transport audio and video (AV) streams over layer three/IP-based networks. The standard **IEEE 1733–2011** [38] describes how to map RTP time to IEEE 802.1AS Presentation Time [8] and thus how to let layer three AV data benefit from layer two AVB mechanisms [39]. This increases the flexibility of the technologies used within an AVB network.
- As Sections 7.1.2 and 7.1.3.1–7.1.3.4 indicated, AVB supports various use cases and ways of setting up the AVB network. **IEEE 802.1BA-2011** [40] defines

profiles and default configurations in order to support easy handling, especially where expert network knowledge cannot be expected.

- **IEEE 1722.1–2013** [41] defines typical middleware functionalities for discovering and handling devices and services for IEEE 1722-based systems. From an automotive perspective, the tasks that middleware has to fulfill are more complex than – and somewhat different from – what is expected when connecting AV multimedia devices. IEEE 1722.1 is therefore of less interest. Table 7.1 showed the hierarchy and different levels of coordination needed when implementing AV applications in cars. This is quite different from any consumer system. Furthermore, the middleware should cover not only AV services but also be useable throughout all in-vehicle Ethernet ECUs, independent from their use case, size, operating system, etc. A functioning solution for an automotive middleware that can be used in an Ethernet-based network is thus discussed in detail in Section 7.4.

7.1.4 Time Sensitive Networking (TSN) for Safety Critical Control Data

Quality of Service (QoS) for Ethernet-based AV applications in the car is important. After all, a very high percentage of drivers use some form of audio entertainment while driving [42]. Additionally, new driver assist functions like automated stop and go in traffic jams increase the interest in in-vehicle video entertainment. Nevertheless, the most important requirement in a car is safety. When Ethernet is used for safety critical applications – and, with the prospect of autonomous driving, Ethernet will be used for such applications [43] – the respective communication needs to fulfill stringent requirements with respect to reliability and timing.

The AVBgen1 standards discussed in Section 7.1.3 satisfy important QoS requirements for audio and video transmission in an Ethernet network. Their standardization had been initiated to ensure QoS in a plug & play environment. To support time critical control traffic was not one of the use cases. With the completion of the AVBgen1 standards the scope of the QoS effort within IEEE 802.1 was thus broadened from AV data only, to more types of data with QoS requirements and more stringent requirements. The project was renamed "Time Sensitive Networking (TSN)" in November 2012 in order to reflect the new scope [13].[17]

The new TSN standards cater to quite a variety of requirements and the degrees to which these requirements can be fulfilled. On the whole, the inclusion of control traffic added not only a vast number of use cases that can be supported but also a vast number of choices with varying complexity. In cases of standards with many options, the industries generally adopt profiles that help with the selection of options and make interoperability easier. While the automotive industry was fast in starting the creation of an automotive audio and video profile out of the AVBgen1 specifications in Avnu [35], it has been more hesitant in adopting the AVBgen2/new TSN standards. At the time of writing the IEEE was still at the beginning of developing a TSN automotive profile specification in IEEE P802.1DG [44], following many workshops and tutorials hosted by Avnu on the topic years earlier [45]. Table 7.4 gives an overview on the specifications with potential interest for automotive use cases that are briefly

Table 7.4 Overview on AVB and TSN specifications as provided by IEEE with respect to automotive use (see also [47, 48])

	Transport	Time sync	Stream reservation	QoS/latency	Safety (seamless redundancy)	Ingress Policing
AVB (AVBgen1)	1722-2011	802.1AS-2011	802.1Qat-2010	802.1Qav-2009		
TSN (AVBgen2)	1722-2016	802.1AS-2020	802.1Qcc-2018	802.1Qbv-2015	802.1Qca-2015	802.1Qci-2017
				802.1Qbu&	802.1CB-2017	
				802.3br-2016	802.1AS-2020	
				802.1Qch-2017		
				802.1Qcr-2020 (est.)		

Note: Additional useful standards that were available prior to the AVB effort are IEEE 1588 and 802.1Q and p

introduced in this chapter. Note that TSN itself consists of a significantly larger number of standards and standard projects (see e.g. [46]).

7.1.4.1 Enhanced Transport

It became apparent, even with the first implementations of AVB in the automotive industry, that the data types supported in the IEEE 1722-2011 specification are not sufficient. IEEE 1722-2016 thus incorporates more AV formats common in the automotive (and other) industries [28]. Moreover, the new IEEE 1722 specification supports encrypted packet formats, UDP/IP encapsulation, and the tunneling of typical automotive in-vehicle networking messages from technologies like LIN, CAN (FD), FlexRay, and MOST. Furthermore, the 2016 version of 1722 allows the distribution of additional, application dependent clock and event information, which can be useful for some use cases. At the time of writing, IEEE was in the process of enhancing the 1722 specification with further formats and interface protocols as well as timing related features [49].

7.1.4.2 Enhanced Synchronization/Timing

A robust and precise time base is crucial for time-sensitive applications in a network. In addition to an initial synchronization to the clocks, time synchronization requires that the clocks stay synchronized not only during normal operation, but also when there are disruptions (such as the addition of units, removal of units, or malfunctioning units). The 2011 specification offers limited choices when, e.g., the Grandmaster suddenly becomes unavailable (e.g., because of an ECU reset). If BMCA is implemented, a new Grandmaster is selected which, however, takes time (>200 ms [50]). Without BMCA, as it is specified in the Avnu Automotive Profile [35], switches simply continue to transmit the most recent valid values they have received (called "(bridge) holdover").[18]

The 2020 specification now offers redundant clock spanning trees, redundant Grandmasters, and the simultaneous use of multiple gPTP time domains [51]. A redundant clock spanning tree can recover immediately, if a lost connection is the reason for the loss of the Grandmaster (provided, of course, the network has redundant paths, which is not yet common in cars, see also Section 7.1.4.5). The redundant Grandmaster means that a second Grandmaster, synchronized to the first Grandmaster, is in hot standby. In case the first Grandmaster fails, the second Grandmaster can take over seamlessly.

The use of multiple gPTP time domains simply means that a number of different time domains can co-exist within one system. The clocks belonging to one domain, are synchronized within the domain, but not necessarily to the clocks of other domains. This refines the options if timescales vary for different applications.[19]

7.1.4.3 Enhanced Stream Reservation

In an AVBgen1 setup, Talkers and Listeners announce and reserve streams independently of each other, while all switches in the distribution path check individually if they can accommodate the reserved traffic or not. This distributed organization based on local knowledge works, but is time consuming. The enhanced SRP specified in IEEE 802.1Qcc improves this by enabling centralized network management. This allows for optimizations in the form of reduced management traffic for reservation and configuration as well as by being able to use the knowledge of the best, most reliable, and fastest paths through the network.

Specifically, 802.1Qcc allows, e.g., for configurable SR classes and streams, better description of stream characteristics, support for layer three streaming, deterministic stream reservation convergence, and UNI (User Network Interface) for routing and reservations [52]. The required centralized knowledge of the network is supported by IEEE 802.1Qca (see also Section 7.1.4.5) and is likewise needed for Time Aware Shaping (TAS)/IEEE 802.1Qbv (see Section 7.1.4.4).

7.1.4.4 Enhanced Traffic Shaping and Quality of Service (QoS)

In AVBgen1, improved network quality is achieved by applying traffic shaping locally to the Talker, so that it cannot overload the network or consume bandwidth unfairly. This improves the transmission conditions and their time sensitive data (audio or video) is not unpredictably delayed. For audio and video transmission, the relative latency is crucial and can thus be better controlled. For time-critical control data, the situation is different, as the key performance parameter is absolute latency. Still, even for absolute latency, the requirements can vary and it is worthwhile in any such discussion to clarify the exact requirements that must be fulfilled. Does the information have to arrive within a specific maximum latency (while it might arrive earlier, too)? Does the information have to arrive at exactly the same time interval (and not earlier nor later)? Does the information have to arrive as fast as possible? Or is it simply necessary that the receiver knows the exact time the data was generated, without having particularly stringent requirements on how long it takes to arrive?

When defining the goals for the AVB and TSN standardization projects, a maximum latency was targeted: 100 μs over five hops for TSN/AVBgen2 and 2 ms over seven hops for AVBgen1. However, as Table 7.5 shows, the latency goals of the different methodologies described in this section (and Section 7.1.3.2 for 1AS and Section 7.1.3.4 for Qav) can be differentiated further.

Table 7.5 Different methodologies to control QoS and latency

Latency < max. value	Latency always the same*	Minimize the latency as much as possible	Knowing when the data was generated
802.1Qav	802.1Qbv	Cut-through switching	802.1AS
802.1Qcr	802.1Qch	802.1Qbu & 802.3br	

* Naturally, if the latency is always the same, it is also always smaller than a maximum value (that is larger than the minimum possible latency). Still, if the latency is always the same, there is no jitter, while if it is smaller than a maximum value, there may be jitter. For time-critical control traffic this makes a difference and the different types of latency are different concepts.

A very basic method for reducing delays in switches is to implement cut-through switching. "**Cut-through switching**" allows an incoming packet to be sent out before the packet has been completely received. This can, in principle, be done as soon as the destination address has been recognized. However, even with cut-through switching, any incoming packet has to wait for packets currently being transmitted on the same egress port to be completed, even if the waiting packet has a higher priority. For a 100 Mbps Ethernet link these blocking packets might cause a delay of up to about 122 μs, for a 1 Gbps Ethernet link this might take up to about 12.2 μs, at every switch on the path from source to destination. Note that while cut-through switching is commonly discussed (and is available in products) it is not part of the TSN standards.

"**Time Aware Shaping (TAS)**"/IEEE 802.1Qbv enable bridges and end stations to schedule the transmission of packets based on timing derived from IEEE 802.1AS, which thus always needs to be implemented to allow TAS to work. TAS is especially useful in engineered networks, when time critical information is sent at regular intervals [46]. It basically introduces a circuit-switched/TDM channel into the otherwise packet-based communication; a methodology also used in the Industrial Ethernet technologies Profinet or TTEthernet [53].

In order to block lower priority traffic during pre-programmed, regular time windows, TAS adds time gates to each queue at a port and makes switches aware of which cycle times to apply. Each port is then associated with a gate control list that determines gate operation. A guard band of the size of the largest possible interfering packet must be enforced before each reserved time slot. Otherwise, any packet starting its transmission just before the reserved time slot might spill into the reserved timeslot and delay the transmission of packets that are intended for the timeslot [54] (see also the cut-through switching example). In combination with the preemption mechanism discussed further later in this section, the length of the guard band can be reduced and the efficiency of the transmission thus increased. With TAS and cut-through switching a minimal switch latency of 1 μs can be guaranteed regardless of packet size [55], when using, e.g., 100 Mbps Ethernet.

While IEEE802.1Qbv has been promoted to ensure low and deterministic latencies, several factors present in automotive networks may make TAS cumbersome from the OEM perspective. First of all, using TAS requires detailed planning. The size of the packets in each queue needs to be known, otherwise TAS is not efficient. Like in any system with reserved timeslots, timeslots, and thus bandwidth are wasted, when not needed. Furthermore, the limited precision of the underlying synchronization protocol, jitters, and delays may accumulate along a path, due to variable switching delays and other TAS flows joining, in different orders over time, on the same egress port. This leads to larger TAS transmission windows, resulting in long and varying end-to-end latencies in practical implementations. A way to combat this is to have several TAS classes, but then the overhead of the multiple guard bands and the overall complexity of the system increase as well.

At the application layer, software that generates the signals must be statically scheduled and symbiotic with the TAS network schedule otherwise even longer latencies and jitter may be experienced. Such a symbiotic relationship would necessitate an operating system that is synchronous with network scheduling. While this is feasible and is done, it represents a lot of effort and overhead for a technology – Ethernet – that was introduced for exactly not needing this type of planning and overhead.

The IEEE 802.3 specification on Interspersing Express Traffic (IET)/IEEE 802.3br [56] defines how the transmission of long, low priority packets can be interrupted in the PHY and how high priority, time-sensitive traffic is inserted. The required "**pre-emption**" methodologies at layer two are filed under IEEE standards numbering 802.1Qbu [57]. IET and pre-emption can be deployed without any higher level organization in the network, as long as both ends of a link agree to do so (e.g., with help of the Link Layer Discovery Protocol (LLDP) or static configuration).

In order to enable preemption, a preemption status table is needed per port, which sets each priority class as either "preemptable" or "express." While implementing preemption does not require changes to the PHY, packet integrity is preserved, and MAC operation remains the same, it does, essentially require that two MAC functions be associated with a single PHY; with one being paused during preemption while the other transmits a packet.

The minimum fragment size is 64 bytes, and fragments must be reassembled to its original packet before the packet can be passed onto other links in the network. The MAC merge sublayer adds a 4-byte CRC. The worst case delay in the case of IET and pre-emption is thus 123 bytes, when a packet of such length has just started transmission (123 + 4 = 127 < the 128 bytes needed to make two fragments). Note that IET and pre-emption as such cannot guarantee any specific latency or delay values. They simply provide a methodology to reduce latencies for certain traffic in mixed traffic environments with long low priority packets.[20] See also [47] for a comparison of different performance parameters between IEEE 802.1p, 802.1Qav, 802.1Qbv, and 802.1Qbu/802.3br.

IEEE 802.1Qch introduces **cyclic queuing and forwarding** to ensure deterministic latencies with closely bounded upper and lower limits. This is done by introducing a buffering mechanism especially into the switches that stores any data received in one cycle until the following cycle, when the data is then transmitted via the appropriate port and queue [58]. The overall latency is the cycle time multiplied by the number of hops. While this results in a fixed latency, this is not the smallest latency possible, but might even add latency in order to achieve the cyclic alignment. A positive consequence is that buffer overflow and thus packet losses are significantly reduced if not eliminated altogether.

To work, IEEE 802.1Qch needs the gates and cycle times provided by the TAS/802.1Qbv. It needs the possibility to redirect incoming packets to certain gates provided by IEEE 802.1Qci (see Section 7.1.4.6). Furthermore, it needs the respective buffer space and a common time base in the network. Packet preemption can help in reducing the cycle times. Regular and periodic traffic can profit from this, if the packet size is known.

Last but not least, IEEE 802.1Qcr enables **"Asynchronous Traffic Shaping (ATS)"** [59]. The goal of IEEE 802.1Qcr is to guarantee low latencies – smaller than those achievable with IEEE 802.1Qav or a not optimally used 802.1Qch – without the need for time synchronization and planning, as is necessary for, e.g., the TAS/IEEE 802.1Qbv. At the same time, it is more bandwidth efficient than synchronous mechanisms, especially when traffic types with different requirements share the same link/network (e.g., on a backbone). Furthermore, it functions even in the case of time synchronization errors.

All this is achieved by the asynchronous operation of the ATS. It is the end stations' applications that define the packet transmission time, while every switch optimizes the traffic flow locally, based on the knowledge it has accumulated and relying on its own clock. For optimization, every switch deploys an Urgency-Based Scheduler (UBS) which prioritizes urgent over relaxed traffic by per-class, hierarchical queuing and per-stream reshaping. Traffic bursts are thus smoothed with help of a token bucket that allows for transmission only when a sufficient number of tokens have been accumulated.

At the time of writing, IEEE 802.1Qcr was the latest of the QoS related TSN specifications to be completed. It defines very promising concepts that address a number of concerns of the other options. Its acceptance and adoption for Automotive Ethernet networks, however, remains to be seen.

7.1.4.5 TSN Supported Redundancy

For safety critical control and fail-safe operation systems, it is not only important that data arrives with small latencies under normal circumstances. It is also important that it arrives without loss in case of unforeseen disruptions (see also Section 8.5). Redundancy is an important concept discussed in this context. In the 2020 Automotive Ethernet architectures on the road, redundancy was not yet common. However, with upcoming EE architectures for autonomous driving some redundancy is probable and it is therefore worthwhile to explore the options.

One redundancy option already discussed in Section 7.1.4.2 is the redundant spanning tree and redundant Grandmasters for ensuring the robustness of the time base as specified in IEEE 802.1AS-2020. IEEE 802.1Qca and IEEE 802.1CB then enable redundant paths through an Ethernet network. Naturally, a key prerequisite for making use of this is that the network topology actually allows for alternative paths.

IEEE 801.1Qca uses central knowledge of the physical topology and its availability status to provide an alternative path in the case the currently-used path fails. IEEE 801.1Qca defines how to set up, modify, and tear down the respective TSN streams

[60]. IEEE 802.1CB enables "seamless redundancy" [61]. Critical packets are duplicated and sent on alternative paths in the network such that there are no (significant) delays or retransmissions necessary in case of, e.g., congestion on one of the paths. In the unit where the different copies merge back onto the same path, the duplicate packet arriving later is removed from the network. A sequence number in a dedicated tag ensures that the receiver can deliver the packets in the right order, even if they have arrived out of order. The concept of this is similar to what has been specified for AFDX.

7.1.4.6 TSN Supported Ingress Policing

In an Ethernet network, faulty or malicious data flows can propagate through the network and thus block other traffic from being transmitted or forwarded, disrupt bandwidth and latency guarantees, cause buffer overflows and packet losses, and can simply overload switches and microcontrollers, which have to handle more traffic than anticipated [62]. To counteract this, IEEE 802.1Qci defines a means for ingress policing and filtering, i.e., to meter incoming data streams on a per packet basis at the switch entrance in order to prevent inundating a switch with excess or erroneous traffic [43, 63]. In this IEEE 802.1Qci is seen as a fundamental mechanism to make an Ethernet network more dependable [48].

When performing ingress policing, every arriving packet is associated with a stream by examining certain attributes of a packet (e.g., receive port, VLAN ID, addressing, priority, etc.). A packet's stream ID is then used to retrieve a set of parameters that define, for example, the maximum packet length and the maximum data rate that is allowed. If any of the limitations are violated, a switch may: Increment a counter (to keep track of such events), mark a packet as being eligible for being dropped (e.g., in case of congestion), drop the packet immediately, or drop all packets associated with the stream [62]. These capabilities require some additions to the MAC function and some additional state machines (that are also used for IEEE 802.1Qbv). IEEE 802.1Qci is one of the most interesting TSN standards for the automotive industry [64].

7.1.4.7 Automotive Use and Acceptance

As can be seen, TSN offers a variety of specifications and, within each specification, a variety of choices for supporting the transmission of safety critical control data within an Ethernet-based in-vehicle network. The designers of the networks can choose from the TSN specifications as if from a toolbox, depending on the exact requirements that need to be fulfilled [48]. The good part is that the specifications are more independent from each other than their sheer number and volume implies. Strong dependencies exist between IET and pre-emption (IEEE 801.1Qbu and 802.3br), and TAS requires a synchronized time (IEEE 802.1Qbv and AS). Packet preemption improves TAS (IEEE 802.1Qbu, Qbv, and 802.3.br), and redundant paths require configuration (IEEE 802.1Qca, CB). The state machines used for IEEE 802.1Qbv are also needed in IEEE 802.1Qch and Qci. Ultimately, designers can make individual choices to suit their needs (see e.g. [48], for combination options).

Table 7.6 gives an overview of which network participant needs to be involved to enable the functionality of the respective TSN specification and where the network

Table 7.6 Overview on (network) impact of the different TSN specifications [65]; the "Backbone" represents the network among switches

	AS	Qci	Qav	Qbv	CB	Qcr	Qbu/br	Qch	Qcc	Qca	1722
Sender	x		x	x		x	x		x		x
Forwarding unit/switch	x	x	x	x	x	x	x	x	x	x	
Receiver	x	x					x		x		x
Support System				x				x			
SW Stack	x					(x)			x	x	x
Config.	x	x	x	x	x	x	x	x	x	x	x
MAC	x	x	x	x		x	x	x			
PHY	(x)										
Involvement	Almost all	Individual	Talker, ideally data path	Backbone data path	Edge switches of redundant paths	Backbone, Talker	Backbone data path	Backbone data path	Almost all	Backbone	Individual

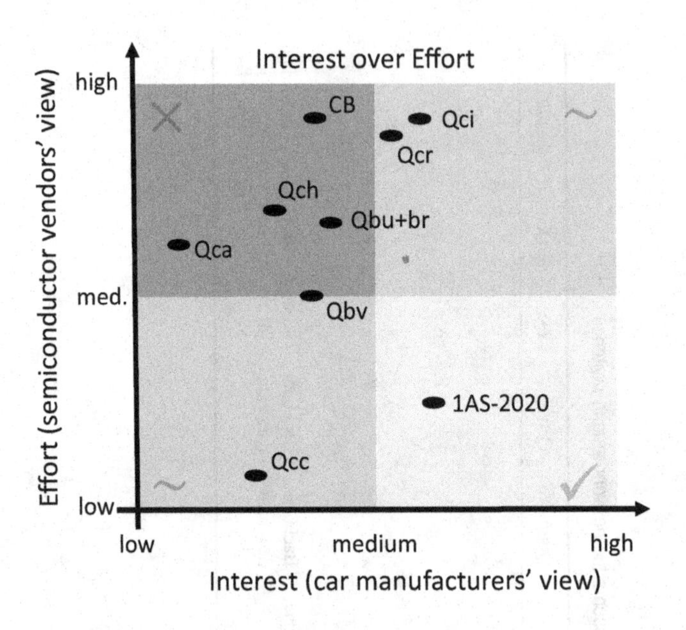

Figure 7.9 Effort versus interest of different TSN specifications [64]

components need to be modified. As is not surprising, all specifications (except 1722) require the involvement and configuration of the switches. Most specifications have no value as standalone features, but require other units in the network – or at least those in the data path – to also support the feature. Some functions additionally require that the function is enabled in the sender and/or receiver in order to be of value.

At the time of writing, many car manufacturers had just started bringing Ethernet into their vehicles. Those who had just introduced Ethernet as individual links or only as small networks, did not yet require any QoS for their networks. Those who had introduced larger networks had gained experience with the AVBgen1 protocols, especially in the infotainment domain. At the time of writing no clear picture had yet emerged regarding which of the new TSN functionalities would be deployed or be essential for the TSN automotive profile IEEE P802.1DG. Figure 7.9 show the results of a survey conducted in the context of a number of Avnu automotive TSN workshops in 2017. The goal was to rate effort/costs of a specification versus the potential interest of use. As can be seen, IEEE 802.1AS-2020, Qci, and Qcr had the highest interest (all with large uncertainty factors), while for Qci and Qcr the effort was also rated as high.

7.2 Switches and Virtual LANs (VLANs)

In an Ethernet network, the (physical or logical) topology may have many shapes and sizes. This is a decisive difference to a CAN(-FD), LIN, or FlexRay network, where a limited number of units may share the same bus. The reason for it being different is the Ethernet (layer two) switches that connect different units and segments of the network.

Table 7.7 Overview of means to improve robustness in an Ethernet switch

	QoS/AVB/TSN (Section 7.1)	VLANs (Section 7.2.2)	Switch configuration (Section 7.2.3)
Too much traffic coming in	(++) Ingress policing (e.g., IEEE 802.1Qci)	(++) VLAN configuration (enforce membership on ingress)	
Too much traffic going out	(++) Shaping and Queuing (e.g., IEEE 802.1Qat, Qcr)	(++) Limit VLAN configuration (+) Only forward packets with VLAN tag	(0) (Semi)static unicast forwarding configuration (++) Multicast configuration (++) Discard unknown (multicast) addresses

Because of the switches, the available data rate is not simply shared among all participants, but might actually be additive, depending on the traffic flow in the system. In other words, ten participants on a network are not limited to using just 1/10th of a network's per-segment bandwidth. This advantage and flexibility, however, also requires different methodologies compared with legacy networks to ensure that the network is not overloaded. While the QoS mechanisms of Section 7.1 can help to solve the majority of the problems, this section presents basic features available to increase the robustness of an Ethernet network. Section 7.2.1 explains the fundamentals, Section 7.2.2 what can be achieved with VLANs, and Section 7.2.3 what can be achieved when using the switch configuration appropriately.

7.2.1 Switch Robustness

In an Ethernet network there are two important properties that can have an impact on the network's robustness: (a) Communication can be flooded and (b) there is no default control in an Ethernet network on how much traffic a network participant can transmit [66, 67]. Flooding occurs not only with all unrestricted broadcast and unconfigured multicast messages, but also when the addressee of a unicast message is not (yet) known. So, flooding can happen any time. An ECU, in theory, should be designed such that it does not send too much traffic. However, aside from design errors or simple malfunctions that might cause a unit to send too much data, this may also be the outcome of a malicious attack.

Both too much flooding and too much overall traffic being sent by even just one participant can overburden an Ethernet network and result in a denial of service or other degradations of the network. Additionally, broadcast messages as such can be received anywhere in the network. The following thus investigates the means available at the switch level to support the following two tasks: (1) Stop too much traffic at the ingress of switch and (2) Stop too much traffic at the egress of a switch. Table 7.7 gives an overview on the countermeasures available.

Keep in mind that some of the discussed mechanisms for limiting flooded packets mainly work for non-security situations. These mechanisms may help to achieve robustness against random malfunctions. Regarding security (see also Section 7.5), the attacker can carefully choose to use the defence mechanisms against the network. For example, when limiting broadcast traffic by policing or shaping, an attacker can use this feature to force the dropping or regular packets as well. To overcome this, ingress policing (see Section 7.1.4.6) is most effective (on traffic that can be associated with a stream with a high degree of granularity).

7.2.2 Virtual LANs (VLANs)

A powerful way to structure an Ethernet network is virtualization. On top of the physical Ethernet network, different virtual networks may be created, each with possibly different subset topologies and QoS configurations. For Ethernet this feature is called Virtual LANs (VLANs) as defined in IEEE 802.1Q. VLANs were developed in the context of very large IT networks in order to be able to handle and segment the network, as well as to limit flooding. They are also useful in smaller Automotive Ethernet networks.

A VLAN describes a virtual Ethernet network in the physical Ethernet network, which is identified by a VLAN ID (see also Figure 7.3 or Figure 1.5 in Section 1.2.1). This means that a VLAN enabled switch will pass on packets with a particular VLAN ID only between switch ports of the same VLAN and not to, from, or between others, even if units belonging to other VLANs are physically connected to the same switch or if the message is a broadcast message. This consequently can be used to limit a broadcast domain, thus limiting the flooding of broadcast and other packets.

VLANs are often used to not only limit broadcast domains but also to isolate traffic. Depending on the design, the isolation can be between critical/uncritical traffic, internal/external traffic, or it can isolate the traffic flows of different application areas or security zones. An example of separating different VLAN traffic flows is shown in Figure 7.10 and is described further in the following paragraphs. VLANs can also be used to select appropriate ingress policing and to discard packets that do not belong to one of the VLANs assigned to a packet's receive port.

For packets arriving without a VLAN tag, the switch can be configured to discard the packet or to add a VLAN tag based on available packet information such as port, protocol, fixed header fields, etc. [67]. Due to these isolating properties, VLANs can also be used as a powerful building block for security. Note though, VLANs do not provide sufficient security in themselves (see Section 7.5 for options for Automotive Security).

Figure 7.10 shows an example of what traffic isolation in an in-vehicle Ethernet network could look like. In the example an ECU called "Car infrastructure switch" has been selected as the major unit to perform traffic isolation. This unit can be physically located anywhere inside the car. The VLAN filtering rules, defined during the development process, are applied per port. The depiction also shows where adding

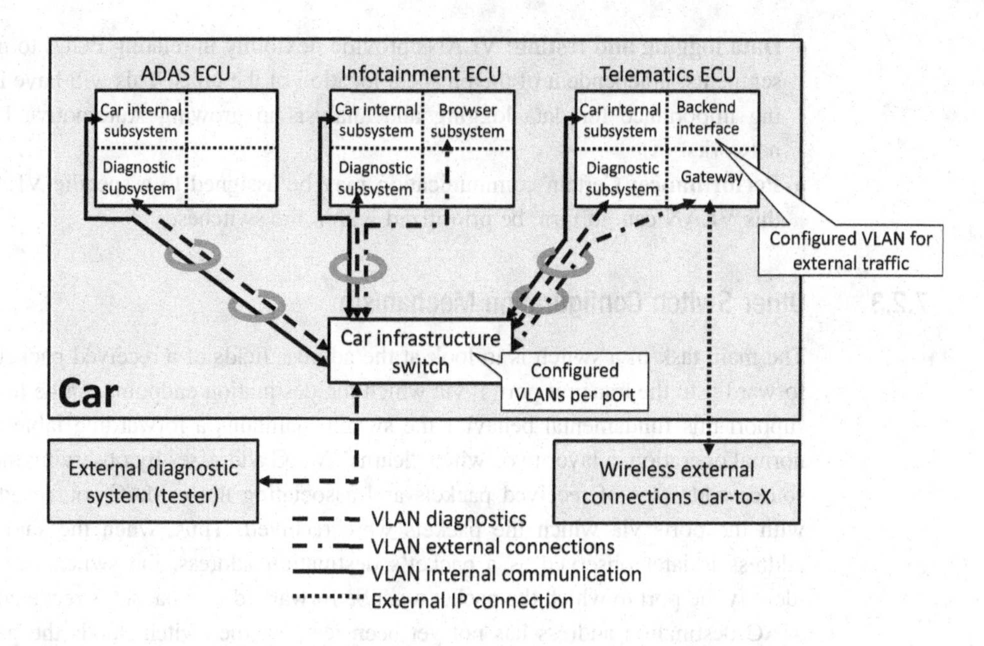

Figure 7.10 Example use of VLANs including software compartmentalization

or removing a VLAN tag makes sense: For both external interfaces. When e.g. diagnosing the car via the standardized diagnostic interface, the diagnostic packet simply receives the respective diagnostics (VLAN) tag, when they are received by the car infrastructure switch and are then distributed via the diagnostics VLAN inside the car. Within the car, this traffic is always seen as "diagnostics traffic" and never as "car internal traffic," which makes effective isolation quite easy and results in a quite efficient firewall. The external tester is unaware of the tagging of its packet's by the car's network and is unaffected by it (i.e., packets sent to the external tester by the car's network are stripped of their VLAN tags prior to transmission onto the car/tester link).

The situation is the same for the other external interfaces that most modern cars will have, e.g., for connecting to the Internet. In this case, it is possible to work identically. Traffic that enters or leaves the car via one of the many radio interfaces can be tagged as coming from outside the car, while the tag is removed when data from the car leaves via the same interface. Inside the ECU such tagged traffic is handled in an isolated area only. This ensures that a browser application has no access to car internal data, even if it passes along the same wires. The strict separation of traffic is crucial and car manufacturers should ensure this with the help of a respective development process.

VLANs are unprecedented in cars. Their implementation offers a powerful tool to the designers of the Automotive Ethernet network and will also provide competitive advantages for those who do it well. It is therefore unlikely that the automotive industry will standardize how to implement VLANs on a general basis. Two additional aspects to consider with respect to VLANs are the following:

- **Data logging and testing:** VLANs provide flexibility in relating ECUs to network segments, independent of the physical location of the units. This will have increasing importance for data logging and analysis in growing Automotive Ethernet networks.
- **Performance:** Certain communication may be assigned to a specific VLAN and this VLAN can, in turn, be prioritized within the switches.

7.2.3 Other Switch Configuration Mechanism

The main task of a switch is to look at the address fields of a received packet and to forward it to the transmit port(s) via which the destination endpoint can be found. To support this fundamental behavior the switch maintains a forwarding table. During normal operation, a layer two switch "learns" MAC addresses by observing the MAC source addresses of received packets and associating those MAC source addresses with the ports via which the packets were received. Thus, when the same MAC address is later observed as a packet's destination address, the switch can readily identify the port to which the packet must be forwarded. If a packet is received whose MAC destination address has not yet been learned, the switch floods the packet to every transmit port except the port via which the packet was received. This ensures that the packet makes it to its intended destination regardless of whether or not its destination address is known to the switch. Sending broadcast or unknown multicast packets has a similar flooding effect.

To prevent the flooding of packets with unknown unicast addresses, static or semi-static configuration of unicast addresses is often recommended. In this solution, the Ethernet Switch is configured with all unicast addresses in the network or they may be learned in very limited circumstances, e.g., just once with the first start in the factory. In order to limit the flooding of unknown addresses, the Ethernet switches would then need to be configured to discard all packets with unknown source or destination addresses. Unfortunately, the usefulness of this in a car is limited, despite the network being preconfigured and more or less static in its configuration. It would make testing and replacing units during service more complex, since, e.g., the tester addresses cannot be known a-priori and partial networking becomes difficult.

For multicast traffic, configuration of multicast addresses is very helpful. Instead of flooding the multicast packets within the packet's VLAN, each Ethernet Switch is configured to only forward these packets to the specified outgoing switch ports. This is especially helpful if VLANs are used to define domains and not traffic types. In addition, configuring Ethernet switches to discard packets with unknown multicast destination addresses is highly recommended. This feature is already common for multicast addresses configured by SRP.

Another important option is to configure ingress policing for Multicast and Broadcast traffic. Additionally, in some Ethernet switches a specific variant for this is present as Denial of Service (DoS) prevention. With carefully configured rates, this is useful as well.

7.3 The Internet Protocol (IP)

The Internet Protocol (IP) is the fundamental protocol associated with enabling and using the Internet. However, it is only one of several protocols that represent the "Internet protocol suite" or "TCP/IP protocol suite"; a combination of protocols that enables vendor- and operating system-independent communication between networking-enabled electronic devices [68].[21] The first version of the protocol(s) was published in 1974 as RFC 675, "Specification of Internet Transmission Control Protocol (TCP)" [69]. In 1981, with the fourth version, TCP and IP were, for the first time, separately described in RFC 793 [70] and RFC 791 [71]. The User Datagram Protocol (UDP) was standardized in 1980 (RFC 768 [72]) and is also part of the TCP/IP protocol suite. Today, most networking uses the TCP/IP protocol suite. Thus, cars must support it if they are to be handled as nodes in the (worldwide) network. Cars must support it as part of the Ethernet-based communication in in-vehicle networking.

In general, there are no diverging functional requirements when implementing the respective protocols for automotive applications. After all, the possibility of reusing standard-compliant implementations of protocols such as TCP, UDP, and IP is one of the reasons for adopting Ethernet-based communication in the automotive industry in the first place. One aspect that must be taken into account when implementing the TCP/IP protocol suite is the footprint of the software. Small ECUs, such as cameras integrated into the side mirror, have very little processing power, RAM, and/or flash memory available in optimized DSPs or µCs. It is therefore important to pay attention to the implementation of the software [73]. The skillful implementation of the TCP/IP stack gives a competitive advantage to those capable. Naturally, the right configuration, suitable APIs, and the right programming guidelines (see e.g. MISRA [74]) must also be taken into account. However, this is not the topic of this book.

The portion of the TCP/IP protocol suite that does leave some structural choices in the automotive industry is the use of IP. The core functions of IP are addressing, i.e., identifying and locating units (called "hosts" in IP), and routing packets from a source address to a destination across, well, anything, from within a small network to across multiple networks and around the world. In the public Internet, IP addresses thus must be globally unique. To ensure this, Regional Internet Registries distribute IP addresses, while the Internet Assigned Numbers Authority (IANA) publishes the availability of addresses. Additionally, there are IP address ranges available for closed/private networks [75]. Anybody can use these, as long as the units using them have no direct access to the global network.

An Internet Protocol version 4 (IPv4) address consists of 32 bits, which originally only identified the network and the host in the address classes. It became evident quickly that the original concept was not sustainable and various methods were developed to either make routing more efficient and/or to make the address range stretch further. Examples are subnetting, Variable Length Subnet Mask (VLSM), and Classless Inter-Domain Routing (CIDR) [76].[22] Despite these efforts, the predominantly-used IPv4 has run out of addresses [77] and everyone designing networks using IP today – which includes the automotive industry – must consider integrating IPv6 support (see Section 7.3.2). IPv6, first published in 1995 [78], uses 128 bits instead of 32 bits for its addresses, so there is reasonable hope that the address space will last longer.

Last, but not least, the question of security is often discussed in the context of IP. Because IP is the connecting element in the worldwide network, it is seen as an entry point for undesirable elements. Section 7.5 provides some basic considerations for security in the automotive industry, including a brief description of IPsec, a security protocol for use at layer three.

7.3.1 Dynamic versus Static Addressing

Modern IT systems need to be very flexible. They have a changing number of hosts and routes in the network. One of the key elements to support this is the dynamic setting of IP addresses. Deploying the Dynamic Host Configuration Protocol (DHCP) and Domain Name System (DNS) servers is the current state of the art. Automotive Ethernet networks are entirely different. An in-vehicle network is an almost closed system. Even if the number of active nodes in an in-vehicle network may vary (see also Section 6.3), the maximum number of nodes is known and limited up front. Furthermore, in contrast to IT networks, cars might be started and parked several times a day (see also Section 8.3.1). Fast start-up is therefore very important, which makes a static IP configuration the natural choice. However, there are some use cases in which dynamic IP addressing makes sense in cars, too.

The following possibilities exist to assign IP addresses inside cars. The list shows how Automotive Ethernet adds, with IP addressing (and VLANs, see Section 7.2.2), another design dimension to the electronics development inside cars.

- **Static:** The IP addresses are assigned during the development and every ECU with the same function, e.g. Head Units (HU), always receives the same address, independent from the car they are built into. As the same address is obviously used repeatedly between cars, it is selected from a specific address range. The private address pool would be a designated source for it, but does not have to be so. In the case of static addressing, it is absolutely important that two ECUs with the same function, i.e., same IP address are never built into one car.
- **Pseudo-dynamic (branding):** In this case the ECU is delivered without an IP address, but receives a then-static IP address during the assembly. This process is needed in case the same part is assembled multiple times inside a car. The cameras of the surround view system provide an example as exactly the same camera is placed at different locations inside the car. So, for assembly and also for repair, the cameras are delivered without an IP address. This reduces costs for logistics and storage. This branding procedure is standardized with the automotive camera interface of ISO 17215 (Part 4). In ISO 17215 the address is reassigned at every start-up, depending on the switch port the camera is connected to.
- **Dynamic:** This is required when the car or parts of the car communicate with external components/the world outside the car, e.g., the diagnostic tester (see ISO

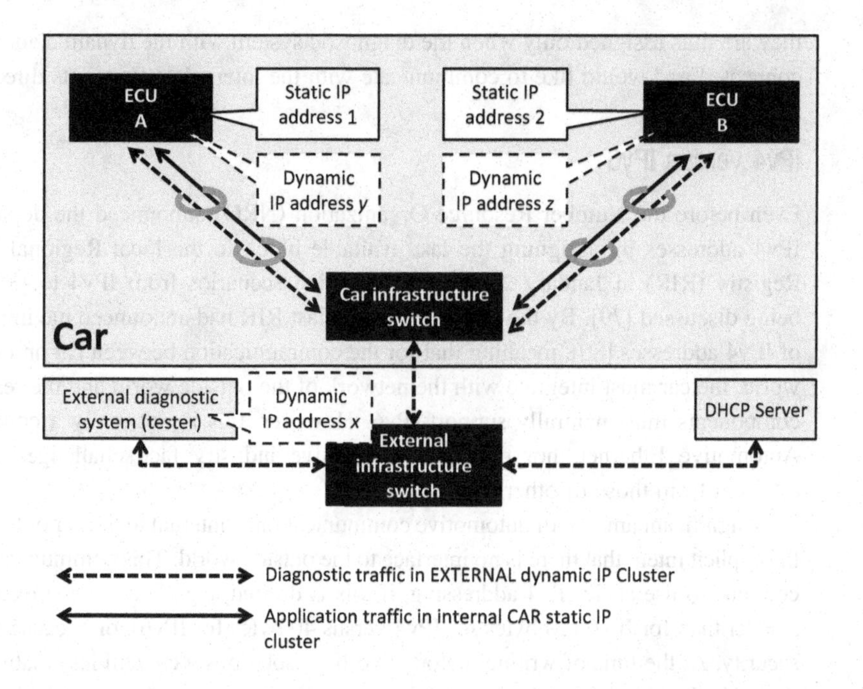

Figure 7.11 Example use-case for multiple IP addresses (solid line for regular car internal traffic, dashed line for diagnostic traffic)

14300). In this case, it is no longer possible to use addresses from the car's internal address space. Instead, the ECUs directly connecting the car to the outside world, "port ECUs," use the IP address(es) that an externally located DHCP server assigns to them. If it is intended that more ECUs behind the port ECUs communicate directly with the external world, it is possible to implement a dynamic/static address translation with a Network Address Translation (NAT) in the port ECUs. Another is to request more temporary IP addresses from the external DHCP server.

- **Multiple:** In this case, an ECU accommodates and uses several IP addresses. This is the case if ECUs want to use various address spaces. One example is the diagnostics use case (see Figure 7.11). The external tester must connect to a unique IP address inside the car. The static, internal addresses are not unique. For the duration of the testing, additional IP addresses are thus assigned to the internal units by the DHCP functionality of the external network that the tester is part of. If the internal units had only one IP address, they would need to replace it every time they are connected to the outside world. This would make communication (and with that the car's function) impossible during the switch over. Even if the test duration is short, this is undesirable.

Figure 7.11 shows an example of the use of multiple IP addresses. As can be seen, where appropriate, separation of traffic can be achieved with multiple IP addresses assigned to the communication partners on layer three, and not only with VLANs on layer two. In the example shown, only the dynamic addresses *y* and *z* are needed and they are thus assigned only when the diagnostic system with the dynamic address *x* is connected and would like to communicate with the internal components directly.

7.3.2 IPv4 versus IPv6

Even before the Number Resource Organization (NRO) announced the depletion of IPv4 addresses by assigning the last available block to the local Regional Internet Registry (RIR) in January 2011 [77], migration scenarios from IPv4 to IPv6 were being discussed [79]. By the end of 2019, the last RIR had announced having run out of IPv4 addresses [80], meaning that for the communication between car and external world, the car must integrate with the network of the outside world and the respective components must naturally support IPv6. However, this is not really a concern for Automotive Ethernet, nor does the automotive industry face challenges that are different from those of other use cases.

A significant amount of automotive communication is internal to the car only and it is the explicit intent that there is no interface to the outside world. This communication can continue to use static IPv4 addressing, if this is desired, e.g., because the overhead is smaller than for IPv6 (20 bytes for IPv4 versus 40 bytes for IPv6), or because of better security. At the time of writing, automotive IPv6 solutions were still less mature. With the multiple addressing scheme an Automotive Ethernet network can support both.

7.3.3 Routing versus Switching

Ethernet-based communication generally provides for two ways to pass packets through a network: Using the MAC addresses in the Ethernet packet header at ISO/OSI layer two, or using the IP addresses provided in the header of the IP packet (that can be found in the payload of the Ethernet packet) at layer three. The first method is called switching (throughout this book), the second routing. When switching is possible, it seems to be the simpler choice. With switching, packets are forwarded using the addresses of a packet's outermost header (Ethernet), whereas routing must examine the IP header that follows the Ethernet header. Switching can thus offer somewhat lower latencies and is performed in hardware within dedicated switch semiconductors that are currently readily available to the automotive industry.

Routing is used to interconnect separate layer two networks (i.e., LANs). This is done to minimize the scope of a LAN's "broadcast domain," to allow a mix of different layer two technologies (i.e., something other than Ethernet), and to build networks of massive scale (e.g., the global Internet). Routing is also well established and commonly used in the IT world and in data centers. These networks benefit from the efficiency and stability of layer three, for which well-proven methods and protocols exist.[23,24] As the in-vehicle network is always limited in size, as routers seem more complex and expensive (software), and as dedicated automotive router chips are not (yet) available, the car industry has, until now, shied away from using routing instead of switching. However, the following list gives examples of three well-established features and methods enabled by IP that might be worthwhile to investigate for automotive deployment:

- The **Time-To-Live (TTL) field** of IP ensures that any single packet cannot make it further than all of the way across the network. If a packet's TTL expires, the packet is discarded; guarding the network against infinite loops. This feature safeguards a network that may have many redundant routes. While the automotive network sees little redundancy today, it might play a larger role when the concept of networks is truly realized. A feature such as TTL will be of use in the automotive industry then, too.
- The **Explicit Congestion Notification (ECN)** is useful when best-effort traffic is mixed with time-critical and/or synchronous data. It functions as follows: When a router determines that a packet was present in its queues longer than some specific time limit, the router sets the ECN field to reflect this. This allows the recipient to notify the source about the issue, which, in turn, can reduce its transmission rate and, consequently, reduce congestion. Reducing congestion has the beneficial effect of reducing network latency (no queued packets means no delays). And, if congestion is allowed to grow to excess, packets may be dropped. These dropped packets must be dealt with by a higher-layer protocol such as TCP (or even the application itself). This, unfortunately, leads to greatly reduced network throughput. Remember that TSN reserves QoS for prioritized traffic (see Section 7.1). As cars adopt more distributed computing over time, providing some minimum guarantees and greater efficiencies for best-effort traffic may also become relevant.
- **Remote Direct Memory Access (RDMA) over Converged Ethernet (RoCE) v2** is an important tool for distributed computing that runs on top of Ethernet, IP, and UDP [81]. It implements a large part of the protocol stack in hardware and thereby saves power and time as well as frees up CPU load. It may be beneficial, when software is virtualized in cars and arbitrarily distributed among the ECUs.

Some more aspects to consider are, e.g., the possibilities to tunnel legacy busses such as CAN. CAN may be tunneled directly on Ethernet (e.g., with IEEE 1722) or on top of UDP. When SOME/IP is used to tunnel signal-based communication (see also Section 8.6, Note 8), it does exactly that: It uses UDP and IP. IEEE 1733 (see Section 7.1.3.5) can be used to perform the same functionality as IEEE 1722, but in a routed instead of a switched system. Alternatively, IEEE 1722 packets could be encapsulated with IP and UDP, if need be (remember, though, that it was the intention to use IEEE 1722 in order to avoid having to use IP and higher layer protocols, see Section 7.1.3.1). PTP is generally best run on top of Ethernet directly. In a routed system this could be solved by configuring a VLAN that spans the entire network for PTP.

7.4 Middleware and SOME/IP

7.4.1 Definition of "Middleware"

This section starts with the disambiguation of the term "middleware." The term originates in the development of complex software systems and addresses all functions that are needed for a "service" to allow data exchange between otherwise decoupled software components. This data exchange often passes through a network and it is the task of the middleware to ensure that the network itself is transparent to the software components exchanging the data. As is shown in Figure 7.1, middleware ("SOME/IP") operates at the higher layers of the ISO/OSI layer model. It organizes the transport of complex data (messaging) and moderates function calls (Remote Procedure Calls, RPCs) between the software components.

Using a middleware has the important advantage that these functions are implemented once, instead of with every application in the system. This significantly simplifies qualification and trouble-shooting. In times where in-vehicle software is becoming ever more complex, middleware allows for a separation of concerns and thus to manage the software more confidently. This makes distributed development feasible, owing to the much better testability of the modules.

The amount of software in cars today can be huge [82] and is still increasing along with the distribution of functions and systems inside cars (see also Chapter 9 for an outlook). These distributed functions can use various processes within one ECU, but they might also be spread over various processes in different ECUs. With the thus increasing complexity, simply placing messages onto the network under the assumption that the correct function will receive them is no longer sufficient. RPCs provide the right methods and are required to control the distributed functions. Additionally, different ECUs might use different software architectures (and Operating Systems, OSs). This means that middleware also has the important role of bridging between Portable Operating System Interface (POSIX) capable Unix-like operating systems such as Linux or QNX and Classic AUTOSAR systems, which are all used in the automotive industry.

7.4.2 The History of SOME/IP

When starting the development of Automotive Ethernet, the intention was to reuse one of the many middleware solutions available at the time; preferably one following an open source licensing model free of licensing fees. Various approaches were analyzed and some solutions, such as Etch [83–85] or Google Protocol Buffers [86] (serialization only) for middleware or Bonjour [87] for Service Discovery (SD), were investigated more closely. In principle, both solutions could have been modified to fit the small processing capacities available. However, two issues remained unsolved:

- Classic **AUTOSAR** provides many software modules, which incorporate some of the middleware functions and are configured with the help of a separate tool chain. In order to avoid **incompatibilities**, the reuse of existing (IT) middleware solutions would have required bypassing the AUTOSAR modules.
- **The licensing** of the existing (IT) middleware solutions was **not** quite **as needed**. Although the licensing of the respective implementations of the investigated solutions was open, the essential patents needed for adapting the solutions were not. Instead, those patents were protected and owned by large IT companies, with unknown consequences.

While, in theory, it would have been possible to make one of the technically suitable solutions useable by changing the architecture of Classic AUTOSAR, this was not possible in combination with the licensing issue. It was thus decided to develop a new solution. To reduce the risk of running into licensing issues with the new solution, the IPR situation was taken into consideration, while at the same time the new solution was being published as state-of-the-art technology. Naturally, the solution was developed to be directly usable with Classic AUTOSAR systems. The Scalable service-Oriented MiddlewarE over IP (SOME/IP) specification has thus been an integral part of AUTOSAR since Classic AUTOSAR version 4.1. Additionally, SOME/IP is provided as a GENIVI library. More public information is available on [88].

With the successful introduction of SOME/IP in series production cars at BMW in 2014, the discussion on the right middleware to be used with Automotive Ethernet did not stop. Two additional concepts that were often discussed are Type Length Value (TLV) support for SOME/IP and Data Distribution Service (DDS).

The goal of TLV support for SOME/IP was to create more flexibility by allowing more compatible changes to the interface, such as, e.g., the reordering of parameters. The TLV extension is very similar to using the already existing concepts of arrays and unions for transporting an array of unions. The parameters are matched by the union type and not by the position in the message [89]. While this allows for some more flexibility than regular SOME/IP, the following disadvantages come with it: Lower bandwidth efficiency due to the additional type and length field per parameter, lower performance due to the higher flexibility in the format which makes optimization more difficult, and higher complexity that also has to be managed during development and testing. Furthermore, compatible APIs are missing. To use the additional compatibility of the TLV extension the APIs up to the applications need to support this or an additional adaption layer needs to be built.

So, overall, the additional flexibility possible with the TLV extension seems to be rather expensive (especially for small ECUs). Depending on the update strategy and the vehicle development process, the advantage compared to regular SOME/IP should be considered carefully.

Data Distribution Service (DDS) is middleware that is used in, e.g., military or air traffic control systems [90]. The DDS specification has a very strong focus on behavior and APIs. With DDS, the Real-Time Publish-Subscribe protocol (RTPS) defines the transportation of data and subscription messages between DDS instances. RTPS defines most of the headers but does not define the format of exchanged messages and leaves this aspect (i.e., the serialization) to the application [91]. In 2017 DDS introduced Remote Procedure Call over DDS (DDS-RPC) support by adding rules for how to emulate a request–response method call over DDS [92].

The main differences between DDS and SOME/IP are based on the different target applications and markets. While SOME/IP was designed for automotive applications, DDS was not. DDS can be seen as more "plug and play" in the way that it was envisioned that "OEMs" would buy DDS as an application and install it on every node. A strong focus of the specification is on defining APIs for these applications, which can then be easily written by the implementer. When using AUTOSAR this is of limited advantage since AUTOSAR already defines comparable APIs. It might be even disadvantageous, as this makes the compatibility between different implementations harder to achieve.

Compared with SOME/IP, DDS has various disadvantages. To start with, it has a lower efficiency because of the high overhead of the RTPS. Then, the DDS default timings (30 second cycle) were not designed for systems in which a fast start-up is needed. The timings would need to be adapted if fast start-up is to be achieved. Most importantly, DDS is missing full AUTOSAR compatibility. It is currently only possible to run DDS on Adaptive AUTOSAR. Classic AUTOSAR cannot be supported, which limits the use of DDS in cars tremendously. Last but not least, at the time of writing, the automotive DDS ecosystem (e.g. standardized test cases, tools, etc.) seemed to be very weak compared to the one of SOME/IP. Overall, DDS seems to be well suited for prototypes, while SOME/IP is much stronger for use in production cars.

7.4.3 SOME/IP Features

SOME/IP was designed to support the following features needed for automotive use cases:

- **Service-based communication** approach: SOME/IP provides a powerful mechanism to structure the complex configuration present in modern automotive systems.
- **Small footprint**: SOME/IP is highly optimized compared to other solutions. While other solutions use highly inefficient text-based encodings (e.g., JSON, XML) or TLV-based encodings (e.g., Google Protocol Buffers or Etch), both are avoided in SOME/IP for maximum efficiency. In contrast to CAN messages, SOME/IP is strictly byte-aligned to allow for highly efficient zero-copy implementations, while giving up a small amount of efficiency of the overall message size – a rather necessary tradeoff, when considering the massively-higher bandwidth of Ethernet-based systems. Even SOME/IP's Service Discovery was designed to be much more efficient for its use cases than comparable solutions (e.g., Bonjour) by using optimized binary formats and taking into account automotive message and lifecycle patterns.
- **Compatibility with Classic AUTOSAR**: Due to the header structure of SOME/IP, it is compatible to Classic and Adaptive AUTOSAR. At the time of writing, no other solution was able to achieve this.
- **Scalability** for the use on very small to very large platforms: SOME/IP was designed with the smallest and largest automotive ECUs in mind.
- **Flexibility** in respect to different operating systems used in the automotive industry, like AUTOSAR, OSEK, QNX, and Linux.
- **Adaptability to OEM-specific requirements:** SOME/IP is highly customizable to the OEM specific need. This includes, for example, the timings used.

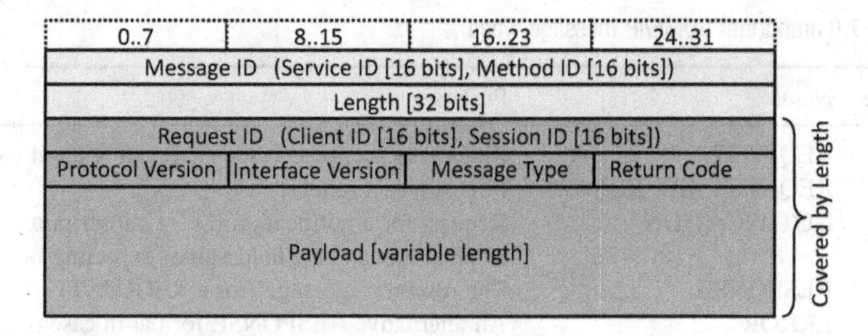

Figure 7.12 The header format for SOME/IP

7.4.3.1 The Header Format

Figure 7.12 shows the SOME/IP header. The individual elements are explained in the text in the following bullet points.

- **Message ID:** The first 16 bits of the Message ID identify the service used (Service ID). The service provides the overall structure for the middleware communication. An example of a service could be "*CD_Player*" (the complete example is given in the notes[25]). Each service needs to have a unique Service ID, which the system integrator assigns. A service can consist of a set of methods, events, and fields, which are identified in the Method ID. The 16 bits for the Method ID represent the lower half of the Message ID. An example of a method could be "track_number. set." In comparison, CAN provides only a small subset of what is possible with service-based communication. However, the idea behind the SOME/IP message IDs is similar to that of CAN message IDs, in the sense that a message is identified and analyzed into logical syntactic components (see Section 2.2.2.2). It is therefore possible to treat the SOME/IP message IDs with the same process structure, which just needs to be enhanced/adapted for SOME/IP.
- **Length:** The length field uses 32 bits to specify the number of bytes succeeding it including the payload and the remaining part of the header.
- **Request ID:** The Request ID allows a client to differentiate between multiple calls of the same method. The first 16 bits of the Request ID are called Client ID and identify a specific client. For example, if a user would like to set the track in the CD-player (server) from the Head Unit (Client A), this would have a different Client ID than if a user of the Rear Seat Entertainment (RSE) (Client B) would like to set the track in the same CD-player. The second 16 bits of the Request ID represent the Session ID. If, e.g., Client A sends the message to set the track in the CD-player multiple times, each of these messages receives a different Session ID. When generating a response message, the server always has to copy the Request ID from the request to the response message. This allows the client to map a response to the correct request. The Request ID is an inheritance from AUTOSAR's Client/ Server communication.
- **Protocol Version:** An 8-bit field which identifies the SOME/IP protocol version. At the time of writing, SOME/IP is on version 1.

Table 7.8 Important SOME/IP message types

Value	Name	Purpose
0x00	REQUEST	Request expecting a response (even without payload)
0x01	REQUEST_NO_RETURN	Fire & forget request
0x02	NOTIFICATION	Request for a notification (i.e., a subscription for an event call back or a field value) expecting no response
0x80	RESPONSE	The response message (for a REQUEST)
0x81	ERROR	An alternative RESPONSE format in case of errors

- **Interface Version:** These eight bits identify the major version of the service interface. The interface definition and version numbering are up to the designer. In case additions are made and new versions are defined, this field in the header allows the automatic detection of version incompatibilities in the design.
- **Message Type:** This field differentiates between the different possible types of messages. With SOME/IP version 1 the values shown in Table 7.8 were defined.
- **Return Code:** The eight bits of the Return Code signal whether a request was successfully processed. This includes errors generated by the middleware and errors the application wants to signal.
- **Payload:** The payload field contains the parameters of the SOME/IP message. In the case of the example, this might be "10," if that represents the value the track should be set to. The size of the SOME/IP payload field depends on the transport protocol used. For UDP, the SOME/IP payload can contain 0–1400 bytes. The decision to limit the payload length to 1400 bytes was taken in order to allow for future changes to the protocol stack, such as using IPv6 or adding security protocols. Since TCP supports the segmentation of payloads, larger payload sizes are automatically supported. With SOME/IP Transport Protocol (TP) segmentation, larger payload sizes are also supported for UDP. The serialization of parameters, i.e., the order of the values in the payload and in which order to place the least to most significant bits, is also specified in SOME/IP.

7.4.3.2 The Service Concept and the Supported RPC Mechanisms

SOME/IP defines a service by its Service Interface, i.e., the activities of client and server, based on the defined communication principles. In this, a Service Interface is comparable to a MOST FBlock (see also Section 2.2.4.2). Figure 7.13 gives an overview of the different communication principles supported by SOME/IP. A Service Interface may include (a) methods with response (request/response) or without response (fire & forget); (b) events, i.e., a message from the server to the client when something happens; (c) fields, which get, set, or notify of a property or status; or (d) event groups, which are logical groups of events and fields used for publish/subscribe handling.

- **Request/Response:** Describes a method with Request and Response messages. The Request is a message from the client to the server calling a method.

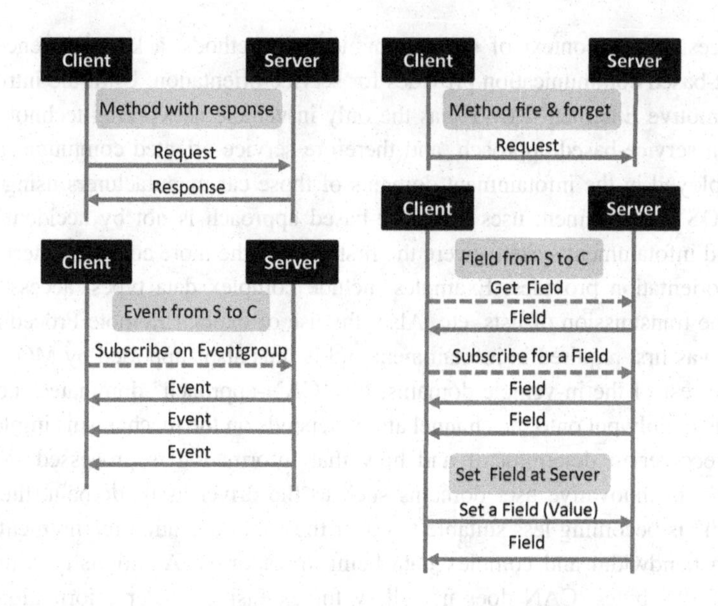

Figure 7.13 Communication principles supported by SOME/IP; C stands for "client," S for "server"

The Response is a message from the server to the client transporting the result of the method invocation.

- **Fire & Forget:** Describes a method with just Request messages. As in the Request/Response case, the client invokes a method at the server. In contrast to the Request/Response case, the client does not expect a Response.
- **Events:** In this case the server sends messages with specific information to the client, either cyclically or when there is a change (event). Prior to this, the client will have told the server that it wants to receive the information, i.e., will have "subscribed." As the server expects no response from the client, this could also be seen as a "Fire & Forget" communication principle from the server side. The Event message pattern is similar to the most common pattern on regular CAN.
- **Fields:** Fields represent "properties" that can be accessed remotely. The current value of a Field can be retrieved by calling the optional "getter," the value of the Field can be set by calling the optional "setter," and the optional "notifier" sends an event on change or cyclically. Compared with regular events, Fields represent an underlying property which is available at all times that the parent service is alive, while the Event is only valid at the time the event is happening. Thus, a Field can be seen as a kind of software variable that can be addressed from an external interface. Fields are similar to "Properties" in MOST.

7.4.4 Service Discovery (SD)

Automotive Ethernet is often just seen as a new in-vehicle networking technology that simply allows for higher data rates. However, there are a few more important differences. In the context of the communication methods, a key difference is that Ethernet-based communication provides for service-orientation. Until the introduction of Automotive Ethernet, MOST was the only in-vehicle networking technology supporting a service-based approach, and therefore service-oriented communication was only deployed in the infotainment domains of those car manufacturers using MOST. That MOST/infotainment uses a service-based approach is not by accident though. High-end infotainment systems were the first to need the more complex interfaces that service-orientation provides. Examples include complex data types, access to databases, the transmission of lists, etc. Also, the use of explicit Remote Procedure Calls (RPCs) was first required in infotainment and is thus also supported by MOST.

In the rest of the in-vehicle domains, the "CAN-approach" dominated, i.e., information is mainly put onto the channel and it depends on the mechanisms implemented in the receiver to determine if and how that information is processed. However, especially in innovative user domains such as the driver assist domain, the "CAN-approach" is becoming less suitable to cover the communication requirements due to the high bandwidth and complex data being transported. Additionally, with a data payload of 8 bytes, CAN does not allow for extensive header information, which limits the adoption of new concepts such as RPC or SD. With Automotive Ethernet being used in the driver assist domain and more service-orientation diffusing into other in-vehicle areas, the authors are convinced that a general shift to service-oriented communication is necessary to meet the in-vehicle communication requirements of the future.

Supporting the use of the standardized TCP/IP protocol stack and allowing for service-based communication are key properties and key advantages of SOME/IP. Compared with CAN, the TCP/IP stack needs additional configuration parameters (e.g., IP addresses, port numbers), which would typically be statically configured to ECUs. However, when looking at the overall lifecycle of car models, it may be advantageous to have more flexibility here (e.g., allowing the merging or splitting ECUs) without needing to touch the configuration of all other ECUs. SOME/IP-SD allows this configuration at start-up.

Furthermore, when using not only broadcast or hard-coded multicast communication but the highly scalable unicast communication, additional aspects must be taken into account: With unicast, it makes sense to address communication partners only if they are really available, otherwise it would be possible that, in a given car configuration, unneeded traffic is consuming bandwidth on links that might be needed for other use cases. Depending on the Ethernet Switch configuration, even flooding could occur for unicast packets.

With SOME/IP-SD, SOME/IP also supports a mechanism that determines if a service is available or not. The following list reflects a number of situations in which the seemingly static network is faced with increasing dynamics, but for which SD provides a solution:

- **Dynamics during start-up:** One of the more complex tasks in the system design of cars is the start-up. Each ECU in a car can show a different start-up behavior. Some

ECUs will start-up quickly; others will be slower. Some ECUs start even if the voltage is as low as 3.5 V; for others, a start-up voltage of 8 V is not sufficient. This means that, during start-up, functions are available at different times. Without SD, a hard limit needs to be set that defines the time at which all functions are expected to be available. This must be defined according to the function or ECU needing the longest start-up time. With the above-mentioned different option/combinations, the time limit would either be different for every car or always the longest. In contrast, with SD available, every function/ECU can announce its availability when ready and, in general, user functions can be available earlier. This significantly simplifies the start-up process. During start-up, SD has another advantageous side-effect in a switched Ethernet network: The switches can learn the addressing tables directly from the SD messages. For CAN-based systems some OEMs have chosen to encode the availability of the data into the data itself; therefore, emulating the Service Discovery functionality by increasing the complexity of the serialization.

- **Dynamics because of customer variants:** Most car manufacturers offer options for their customers to choose from when buying a car. As a rule of thumb, the bigger and more premium a car is, the greater the number of options that can be selected. A large number of options means an even larger number of combinations of options. So, as a consequence, many car manufacturers build individual cars according to the specific customer's requirements. Without SD, each ECU needs to be statically configured with respect to the availability of functions of other ECUs in the car. With SD available, ECUs can establish on their own picture of which other functions/ECUs are available in a car, without requiring any option-combination specific pre-configuration. This is significantly less error prone. Therefore, the more complex the car, the larger the advantages of SD are.
- **Dynamics in the event of failures:** In a network that functions with Fire & Forget messages only, it is not always directly evident when a communication partner has failed. An ECU not detecting any related messages on the link might simply assume that a certain event has not happened or that a value has not changed. In contrast, with SD active in the background, an ECU will know immediately when a service/another ECU is no longer providing a certain functionality. Failures are thus detected better and the respective failure modes can be activated within a certain timeframe.

  Additionally, SD can be used with individually adjustable "Time-To-Live (TTL)" values, which indicate how long an entry is valid. A user of that value expects an update once the TTL has expired. If it fails to arrive, the user can detect the faulty behavior of the communication partner and can start specific error processing. This improves the stability of a network, but, of course, cannot replace messages in case cyclic data is missing for a safety critical application. Cyclic messages with "Application Cyclic Redundancy Checks (CRCs)" are normally used for an end-to-end safety application.

- **Dynamics in the case of partial networking for energy efficiency:** Because of the increasing size of the in-vehicle network and the increasing number of ECUs, energy efficiency is ever more important. As is pointed out in Section 6.3.3, it is of interest to fully power only those ECUs that are needed at a particular moment in time. This needs to take different scenarios into account. A customer might want to finish a call via the built-in hands-free system, despite having arrived at their destination and parked the car. The car should then be smart enough to deactivate all ECUs on the network that are not needed, like the engine control or drive train. This example shows that, with partial networking, the in-vehicle network can be expected to change dynamically. In the changing environment, active ECUs have to know which functions are still available and which are not. Without SD this can be realized with timeouts. Like in the start-up scenario, however, this makes the system slower than with SD. With SD, the knowledge of availability is more immediate.

The more complex an in-vehicle network becomes, the better it is to have service-based communication and SD available. Without service-based communication, the complexity of an Automotive Ethernet network is much higher, as has just been explained. When implementing SD in a network, there are two principal approaches, a centralized and a decentralized one:

- In the **centralized approach**, one ECU monitors and maintains the service information of the network. Each participant sends its respective information just to this one ECU, and each participant requests the respective information just from this one ECU. This approach was taken by MOST.
- In the **decentralized approach** all communication partners apply the following rules: Each communication partner offers its available services to all other units via multicast or broadcast, and each communication partner requests available services from all other units via multicast or broadcast. If two communication partners find each other, they can establish a respective one-to-one connection. Important advantages of the decentralized approach are: Minimal start-up delays, which then mainly depend on the start-up time of the physical network; no specific third-party ECU is needed; support for partial networking; and multiple sources of data, meaning that no single component has to handle all the data but the load and risk in the event of an error are distributed. In other words, there exists no single point of failure. SOME/IP has therefore taken this decentralized approach.

7.5 Network Security

In the Digital Age cyber security has become a major concern. Ever since the first PC virus in the 1980s, the number of threats and attacks has continued to grow with significant (financial) impact [93]. With the increasing digitalization and increasing connectivity of cars, it is an understandable concern that cars might also become the target of attacks. Next to mere inconvenience (car cannot be used), intervention on privacy (use is monitored), and monetary losses (car was stolen, repair is needed), the potential of attacks into cars has another dimension: Personal safety. An attack that succeeds in tampering with the basic driving functions such as acceleration, breaking, and steering might well put lives at risk. It has been shown that the threat is real (see e.g. [94, 95]). Attacks therefore have to be prevented as effectively as possible.

Considering the potential consequences and the deviousness of its source – after all security is needed in order to protect against the malicious criminal energy of other humans – the topic is vast (see [96] for reading recommendations). In order to come to an effective security solution, this first of all requires explicit consideration and an analysis of the security threats and attack surfaces at the automotive system level. Then a comprehensive protection strategy must be developed in order to minimize the risk of a security attack. Such a strategy generally pursues a layered approach. The basic idea is that no implementation is perfect. Vulnerability is caused by software bugs, configuration errors, weak network design, and the like. But, if there are various layers of security and attackers have overcome one, they still have to tackle several others. This approach is standard in the IT-industry and is also appropriate for security in the car industry [97].

This section therefore focuses on the aspects relevant for and special to Automotive Ethernet and is structured as follows: Section 7.5.1 starts by discussing requirements unique to the automotive industry, Section 7.5.2 examines different attack vectors, and Section 7.5.3 gives an overview of available solutions. This section shows that Ethernet-based communication in the automotive industry benefits from the 15+ years head start the IT industry has in security compared with the automotive industry [98].

7.5.1 Security Requirements in the Automotive Industry

Before discussing a layered security structure in more detail, it is necessary to emphasize that (again) the circumstances in the IT industry vary from those found in the car. In IT, the network is generally plug & play and often unique per location. IT networks can be huge with manifold resources and frequent updates. And IT networks often ignore the security of the LAN itself due to physical security. Traffic traversing untrusted links (e.g., the Internet) is protected, while some local traffic is not protected.

Cars, by contrast, have a fixed topology of limited size and resources (memory, compute power, . . .). Each model is designed once and built often, with a long product life cycle and limited opportunities for (security software) updates. If one car is successfully attacked, all cars of the same model are potential targets. Furthermore, attack patterns can change over time and become more effective. A data encryption method that is state of the art and that protects a car well today might be successfully attacked in 10 years, when the same car model is still on the road. The processing power 10 years in the future is likely to attack longer keys more easily, while it would have been too time consuming to attack such keys at the time of development. In addition, physical security is somewhat more limited than in IT data centers, especially considering attackers with vehicle access (e.g., in rental cars).

Last but not least, an IT network is generally not restarted various times per day, and, when it is restarted, it does not need to be fully operational within two seconds. A car does (see also Table 7.2). So, security methods adopted inside the car can neither use the same resources nor need as much time to complete, as IT security methods. This has to be taken into consideration when adopting security standards from the IT industry.

7.5.2 Overview of Attack Vectors

Network security generally aims to achieve Confidentiality, Integrity, and Availability (short CIA). This means the goal of network security is to prevent an attacker from gaining access to confidential data, to ensure that data cannot be changed and is from an authentic source (integrity), and that the services are available as intended [99].

A car, first of all, generally allows easy physical access. An attacker can, e.g., cut the cable harness to break the overall system. The first protection of a car is therefore to limit the physical access, such that an attacker can only manipulate the (as few as possible) ECUs and sensors on the surface of the car, as well as the communication leading to them. Because some physical access is unavoidable, availability in these areas can never be achieved completely. The availability of all other ECUs and their communication can be protected, provided the right mechanisms also decouple the attack surface. To attack confidentiality and integrity is more demanding (e.g., requires more complex tools). However, the target can be any ECU or communication line, and those that need protection need to be protected individually.

Independent of whether an attacker is external (with physical access to an Ethernet link) or internal (by having gained control of an ECU), a car needs to be protected from the following generic attack categories:

1. **Reading communication** (confidentiality). For example, reading the navigation history or the driver monitoring camera.
2. **Replaying, changing, or injecting communication** (integrity). For example, capping the car speed message to circumvent certain limitations.
3. **Selectively removing communication** (integrity, availability). For example, removing safety critical messages, such as a crash message.
4. **Denial of Service** (DoS) attacks (availability). For example, flooding the network with broadcast messages (layer one/two), flooding with Internet Control Message Protocol (ICMP) messages (layer three), or trying to exhaust the TCP state of another ECU (layer four).
5. **Attacking vulnerabilities of ECUs** in order to control software or hardware of an ECU (confidentiality, integrity, availability). For example, exploiting a buffer overflow in a protocol stack to take over control of the software stack and the ECU. This is typically only possible when implementation bugs are present. For a buffer overflow this could be (partially) missing checks of a length field. Keep in mind that complexity of the protocol stack raises the probability of exploitable bugs in the implementations.

The first three of them depend on the access the attacker has and they can be performed on all protocol layers. The last two often focus on communication protocols. This is especially important, when comparing network security solutions and the layer at which they are implemented (see Section 7.5.3.1). Specific manipulation of protocol headers and data are possible at different network layers to implement or enable the generic attacks described in the previous list. Examples, based on the layer of the protocol, are:

- **Layer 1** (physical layer) attacks are historically very uncommon. The only somewhat related attack is the manipulation of the wiring harness.
- Typical **Layer 2** (data link layer) attacks include manipulation of MAC addresses, changing the VLAN ID (i.e., VLAN hopping), or changing other parameters such as a packet's priority. The manipulation of the MAC addresses allows circumventing filtering mechanisms implemented to stop attacks (e.g., preconfigured routing tables). Helper protocols such as ARP to translate IPv4 to MAC addresses may also be attacked. This would allow spoofing in order to inject communication or to reroute communication (in order to access the data). MAC address tables in switches may be attacked by generating vast numbers of MAC source addresses either as DoS attack or to force traffic to be flooded to switch ports other than those originally intended. This is a common attack in order to read specific data via the port the attacker gained access to.
- **Layer 3** (network layer) attacks include changing the address (IP spoofing) and taking additional control by carefully crafted IP fragments, so as to circumvent filtering.
- **Layer 4** (transport layer) attacks include attacks against TCP, such as blind injection of data into TCP streams or terminating other TCP streams. Another favorite TCP attack is to send large numbers of SYN packets – SYN packets are part of the three way handshake to start a TCP connection to a server and overwhelming the TCP state machines of the target [100].
- **Layers 5–7** (session, presentation, and application layers): These attacks are manifold and include circumventing access control.
- **Almost all layers**: Attacks include trying to use bugs in software implementations of protocols to control the software stack, thus gaining privileged access.

7.5.3 Network Security Solutions and Mechanisms

With the vast number of different attacks possible, it is impossible to protect communication with a single security mechanism. It is thus common to implement a layered security system, which uses a combination of different mechanisms.

At the time of writing, the industry had not only not yet converged on algorithms to use for security with Automotive Ethernet, but there also was no respective industry-wide standardization activity. Various efforts existed that discussed other, specific aspects of security in the automotive industry. One example is AUTOSAR secOC [101]. Another outcome was produced by the US-based Society of Automotive Engineers (SAE), which was working on recommended practices for in-vehicle cybersecurity [102]. The focus of the resulting SAE J3061 is on the integration of cyber security in automotive processes and not on specific protection mechanisms.

Furthermore, members of the German Association of the Automotive Industry (VDA) had initiated the ISO 21434 standardization project in order to cover procedural aspects of providing automotive security [103] and, in the Japanese car industry, the Japan Automotive Software Platform and Architecture (JASPAR) had set up a security group in their organization [104]. The following description will thus focus on elements the authors see as relevant in the discussion.

7.5.3.1 Security Solutions along the ISO-OSI Layering Model

Since the communication protocols are structured in the ISO-OSI layer model, the same can be done for the network security solutions. Every security solution is based in a certain layer and protects that layer plus the ones above (but not the ones below). A network security solution at layer four cannot protect against attacks at layer two. Figure 7.14 shows examples of security solutions in relation to the layer of the Ethernet-based communication stack. The depicted solutions are briefly described in the following list. Table 7.9 gives an overview on the protection level that can be achieved with the different solutions discussed.

- The AUTOSAR SECure Onboard Communication (**AUTOSAR SecOC**) has been developed in order to provide a resource-efficient and practical security mechanism that seamlessly integrates into the AUTOSAR communication [101] and that, being at AUTOSAR level, can be used with all networking technologies supported by AUTOSAR (CAN (FD), FlexRay, Ethernet, LIN). It provides end-to-end authentication and integrity based on message authentication codes and freshness values (counters or timestamps). For efficiency in computation and bandwidth consumption it assumes symmetric keys [105], though neither asymmetric keys nor encryption are precluded.

  Unfortunately, only a mere core feature framework is currently defined by AUTOSAR, which lacks for example the key exchange, session establishment, and a concrete mechanism for freshness. These highly critical parts need to be filled in by the OEM – small mistakes here may lead to a completely unsecure solution and significant issues. Another problem is that the size of CAN, LIN, and FlexRay

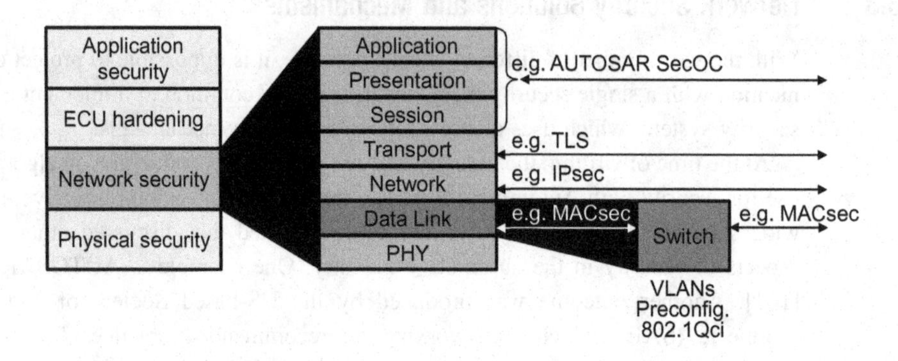

Figure 7.14 Layered automotive security approach (see e.g. [97, 105]) and related mechanisms

Table 7.9 Protocols protected by the different security standards

	SecOC	TLS/ DTLS	IPsec	MACsec
Allows protection of payload?	Yes	Yes	Yes	Yes
Allows protection of application protocol (above TCP/UDP)?	Partially	Yes	Yes	Yes
Allows protection of TCP and UDP headers?	No	No	Yes	Yes
Allows protection of IP headers?	No	No	Partially	Yes
Allows protection of VLAN-Tags?	No	No	No	Yes
Allows protection of Ethernet headers?	No	No	No	Yes
Allows protection of SOME/IP-SD?	No	Partially	Partially	Yes
Allows protection of ARP/NDP?	No	No	No	Yes
Allows protection of ICMP?	No	No	Yes	Yes
Allows protection of IGMP?	No	No	No	Yes
Allows protection of (g)PTP?	No	No	No	Yes
Allows protection of IEEE 1722	No	No	No	Yes
Allows protection of Unicast messages?	Yes	Yes	Yes	Yes
Allows protection of Multicast messages?	Yes	No	(Extension)	Yes
Allows protection of Broadcast messages?	(Yes)	No	No	Yes

messages is very small. While this is not an issue for Automotive Ethernet, the difficulties in integrating SecOC with the other networking technologies might also inhibit its use with Ethernet.

With SecOC being located in the application layer, the protection of other protocols is very limited. In the best case, SecOC can protect the SOME/IP message content. TCP, UDP, IP, VLAN-Tags, (g)PTP, and Ethernet can still be manipulated. The same applies to many helper protocols, like SOME/IP-SD, ARP, and the Neighbor Discovery Protocol (NDP) used with IPv6. SecOC has the unique capability to protect messages being passed, e.g., from CAN to Ethernet. However, the overall protection of Automotive Ethernet with SecOC is small.

- One of the most common security protocols in the IT world is called **Transport Layer Security (TLS)**, which was previously known as Secure Sockets Layer (SSL). This protocol ensures privacy and data integrity between two communicating applications such as HTTP, IMAP, SMTP, etc., by providing encryption and authentication mechanisms [106]. The protocol supports a number of different methods for encryption, cryptographic key exchange, and authentication, which are negotiated between client and server. TLS 1.2 was specified in the RFC 5246 in 2008 [107] and was replaced by TLS 1.3 in RFC 8446 [106] in 2018. TLS 1.3 closes several security problems encountered in recent years, simplifies the protocol, and, design-wise, gets close to IPsec in the way authentication and encryption are combined (no more authentication before encryption). Using TLS 1.2 requires careful consideration and configuration and is not recommended for future projects. TLS runs on top of TCP and does not work with UDP. For use with UDP a variant of TLS called Datagram TLS (DTLS) exists [108]. Note that TLS can only recognize, but not recover, when a blind injection attack destroys the state of the TCP connection.

TLS and DTLS can protect only TCP- or UDP-based communication and, with that, also protect the application protocol used on top of TCP. This could, e.g., be SOME/IP or HTTP. The protection of, e.g., TCP, UDP, IP, VLAN-Tags, (g)PTP, and Ethernet is still not possible. The same applies to most helper protocols such as ARP and NDP. In addition, TLS and DTLS only allow the protection of unicast communication.

- To supplement IP at layer three, Internet Protocol SECurity (**IPsec**) was developed. IPsec can be used for two different use cases. The first use case is to ensure end-to-end privacy, authenticity, and integrity. This use case is not really that common today but is for example used inside enterprise networks to secure the communication between servers. The second and more common use case is VPNs, since IPsec allows site-to-site as well as end-to-site VPNs. With a site-to-site VPN, for instance, branch offices can be connected, while end-to-site VPNs are usually common in remote or home worker scenarios. Both VPN scenarios are commonly used across the (in principle) unsecure Internet. IPsec uses various mechanisms to achieve this such as encryption and the addition of a header element containing a message authentication code; all directly integrated at layer three of the ISO/OSI layering model and thus transparent for the higher-layer applications [109]. IPsec was developed in conjunction with IPv6, but can be and is used with Ipv4 as well. It was standardized by the IETF in a number of RFCs (see [110] for an overview). IPsec covers more corner cases than MACsec (described in the next bullet point).

  The subvariant, IPsec AH (Authentication Header), enables transparent authentication only, and no encryption. Additionally, systems that do not understand IPsec can skip the AH header and inspect the rest of the packet.

  IPsec can only be used for IP-based communication, but it is the first solution in this list that can protect a significant number of protocols since it is implemented at layer three. However, IPsec cannot protect, for example, VLAN-Tags, (g)PTP, and Ethernet as well as helper protocols such as ARP and NDP. While multicast protection was not originally defined with IPsec, (experimental) extensions for this exist today.
- Finally, at layer two, the IEEE 802.1 standardized IEEE 802.1AE,[26] whose latest amendment, 802.1AEcg, was released in 2017 [111] and which is generally referred to as MACsec. MACsec offers P2P encryption and authentication between directly connected nodes at layer two. It is performed for every hop (also protecting the VLAN tag) and not end-to-end like IPsec or SecOC. This means that basically every Ethernet switch has to change the authentication and encryption for every switch transition (provided encryption is used).[27] Its authentication algorithm is especially of interest for the automotive industry, as it provides a good level of integrity in a single infrastructure with mixed security domains that the in-vehicle network represents. However, to be usable in automotive, MACsec requires, like many of the TSN features discussed in Section 5.1.4, hardware support in controllers and switches and thus adds costs to the semiconductors.

The main advantage of MACsec is that all communication is protected. This includes unicast, multicast, and broadcast messages of all protocols; also for the layers above, thus limiting the attack surface against external attackers significantly. Using MACsec in combination with ingress filtering and policing (IEEE 802.1Qci) limits DoS attacks.

Comparing the different Network Security solutions can be done based on different criteria. Some people believe that an implementation at a higher layer is more "end-to-end," which of course is not true when referring to the implementation in different systems (e.g. AUTOSAR classic). Solution at lower layers, however, may protect more protocols and messages. While SecOC can only protect very limited messages, TLS can protect everything in a TCP connection or UDP association. IPsec can, in addition, protect many of the helper protocols and aggregates multiple connections and associations. MACsec can protect the most traffic since it allows for protecting multicast traffic as well as AVB and TSN based protocols. For additional comparisons of protocols, see [112]; for detailed discussion of use cases and automotive requirements, see [113], and, for guidance, refer to [114].

7.5.3.2 Efficient Key Management

As a basic principle of cryptography, the same key should not be used for different functions and devices, and a key should not be used for a long period of time [115]. Additionally, key exchange, and data authentication/encryption require different keys. If the key for data authentication/encryption is compromised, for example, the key for key exchange is not affected and a new key can be exchanged. To limit the amount of data for which a key is used, every key is assigned a specific lifetime for which it is valid and/or an amount of data to be protected, depending on its use. Moreover, communication can be divided into separate groups with separate connection keys according to the respective car domain or other functional aspects.

The distribution of the keys to the ECUs in the car is a very complex task. While in IT the level of automation for such processes may be limited, these processes need to be fully automated in the automotive industry. The installation of keys needs to work in distributed plants world-wide as well in different service places. It needs to work even in places of service with very limited trust by the OEM (e.g., independent repair shops or when customers repair cars on their own).

Examples of properties that need to be considered:

- **Lifetime of the key**: Cars are sold and are then used for a very long time compared with IT systems. In addition, the security and keys in the car need to work for the lifetime of the car without maintenance in the meantime. Updating certificates as in the IT industry is not always possible.
- **Connectivity of cars**: While there is a clear trend for cars to always be connected, an OEM cannot limit its solutions to only operate when the car is connected. Reasons for not having a connection might be regional coverage gaps or customers not desiring this connectivity for legitimate reasons. Such cars must still work and be secure.

- **Individual keys:** To limit counterfeit units and fleet-wide attacks, keys should always be car specific. Breaking the key of a single unit may be worth it, if it enables attacking the whole fleet of cars. Breaking the key of every single car is less likely to be worth the attack.

When using certificates (e.g., X.509 based on IETF 5280) special care must be taken regarding the validity of the certificates (not-before and not-after) as well as the use of online revocation methods, such as is supported by TLS or IPsec. On the one hand, the certificate should not be limited in its usage too much, especially when the car might not always have a correct time and date set. No limitation might be best. On the other hand, online revocation does not work without connectivity and cannot fulfil the extreme fast start-ups as required for many use cases of a car.

For the use inside the car, symmetric algorithms might be of benefit in case smaller units can only fulfill lower computational requirements having to use current hardware without good acceleration for asymmetric cryptography. While being the most efficient solution, pure symmetric cryptography cannot achieve all cryptographic properties (e.g., Perfect Forward Secrecy, PFS). So, a symmetric solution might somewhat limit the security. In addition, the more mature security solutions may be based on asymmetric cryptography, since this is what most IT solutions use. So, the usage of asymmetric keys or certificates inside the car for, e.g., key exchange might allow for the highest security.

Asymmetric solutions are especially useful, when the exact communication peer is not known before the communication and revocation mechanisms are needed, such as in the case when an IT backend is compromised. Therefore, the use of asymmetric keys for connecting to the OEM's backend is often the preferred solution since it allows the scaling of servers without the need to share keys between IT systems and revocation. Typically, TLS-based solutions with certificates or different VPN protocols are used here.

7.5.3.3 Other Security Aspects

Access control and filtering are similar mechanisms to limit the communication between systems. Communication might only be allowed, if the communication partner is considered trusted and the communication protocol used has been allowed before. This is called access control and it works best if the identity is protected cryptographically through the use of a network security protocol. If, for example, IPsec protects your IP header, the IP addresses and port number can be trusted as much as the other ECU can be trusted. Access control is often implemented very close to the applications.

Additionally, certain identifiers in the header of a packet – e.g., VLAN IDs, IP addresses and port numbers – should be checked to make sure that only identified traffic is passed through the network. This is typically achieved by packet filters or firewalls. An Ethernet switch can be configured to filter the packets on an incoming Ethernet port, so that the attached ECU can only send packets using its own IP address but not the IP address of someone else. Also, ECUs can filter what traffic they accept. For example, a strict separation between ECU internal and external traffic such that traffic from the ECU ingress cannot be forwarded directly onto the link between internal applications, and traffic from the link between internal applications cannot directly be forwarded to the ECU egress.

Cryptography is obviously very important for network security. There are a variety of algorithms available for different purposes such as key exchange, peer authentication, message authentication, message encryption, etc. Their details are not decisive in the scope of this book and will therefore not be discussed (see [116] for suggestions on respective publications). Their implementation is crucial in the automotive industry. It needs to be effective and fast. In comparison with other in-vehicle networking technologies, the implementations need to cope with significantly higher data rates when used with Ethernet. Pure software-based implementations are not efficient enough, neither in terms of processing time nor in the use of resources. Hardware support is therefore needed, which is typically provided with the help of a dedicated Hardware Security Module (HSM). An HSM efficiently executes cryptographic functions and securely stores cryptographic keys [117]. As a consequence, the implementation of crypto algorithms offers opportunities for suppliers to differentiate their products and is not discussed further here.

Once the communication in the network has been secured, the ECU itself, i.e., its software and electronics, must be protected. This concerns the ECU implementation, the processors chosen, the partitioning of software on the processor (or onto different processors) and the like. Also, ports that are currently not active can be deactivated in order to better protect the ECU. At the highest level, the application can comprise, e.g., plausibility checks and data-use policies in addition to yet more authentication and encryption. An application can be made to only accept expected data or to only accept certain data (such as control messages) in specific application states. Anomalies can be detected if, e.g., cyclic messages are received more often than expected or if sensor data contains undefined information.

Notes

1 The terms bridge, switch, and router are not always unambiguously used. This book shall use the terms similarly to what is being described in [128]. All terms – including hubs and repeaters – refer to units in a communication network that pass on data. What differs is how and on what basis they do this.

The use of the term **router** is comparably consistent. A router forwards data packets on the basis of ISO/OSI layer three addresses, which in today's networks are IP addresses. "Routing" describes a functionality. The term in itself does not say whether a router is a standalone box or a function integrated into a micro-controller. Generally, routers are used for large-scale communication: To pass on data between different Wide Area Networks (WANs), between different Local Area Networks (LANs), and also between LANs and WANs. Routers can also be used to pass on data inside LANs, though there need to be good reasons for doing so, as this can also be done on a layer below via what in this book shall be referred to as "switches."

The use of the term **switch** is not so consistent. In this book, a switch forwards packets in an Ethernet network based on the ISO/OSI layer two addresses, i.e., the hardware/MAC address conveyed by an Ethernet packet (see Section 1.2.1). A switch is directly related to the Ethernet technology and, at least in in-vehicle networking, a new concept. The impact of this concept on the in-vehicle EE architecture and topology choices is severe and the subject of a separate chapter (Chapter 8). The switching function is generally realized in hardware in a special switch semiconductor. The semiconductor can be "switch only" (which is rare), a "switch with integrated PHYs" (which is most common), or a "switch integrated into a System on Chip (SoC)." If this book talks about switches, the default meaning is the (part of the) semiconductor that provides the respective function. In the IT industry, the term "switch" often refers to a standalone networking product, which actually has quite a market in itself (see Table 1.3 in Section 1.3). This switch is a box with a number of RJ-45 ports that allows the connection of various devices via Ethernet and that will direct the traffic between them (based on the layer two address). In Chapter 8, this book will also consider a separate ECU containing the switching function, which is then a "switch box" or "standalone switch." Sometimes the function of a switch is extended to layer three as a "layer three switch," which somewhat blurs the distinction and can be confusing. In this book, layer three forwarding is "routing," and layer two forwarding is "switching." Note that the IEEE 802.1 specifications never utilize the term "switch" as used here, but only "bridge."

Following [128], a **bridge** also passes on traffic based at layer two. This means it can handle different MAC control algorithms and can thus bridge traffic between different IEEE technologies without needing the IP address. In the IEEE 802.1 terminology this distinction makes sense; for Automotive Ethernet it is of minor relevance. Occasionally the term "bridging" is also used on layer three; after all, various different technologies can be bridged (e.g., with the help of IP addressing). Sometimes the term bridge is also used when translating between different protocols (e.g., in the context of SerDes). When using the term "bridge" in this book, it will always be used in combination with the layer or technology it bases its functionality on – like "layer two bridge" – in order to avoid confusion.

Last, but not least, there are repeaters and hubs. Both function at layer one, i.e., they have no intelligence that allows them to forward a packet based on addressing. A **repeater** is a simple one-to-one device that amplifies and retimes the signal to increase the range. Repeaters can be of interest in the automotive industry, too, when, e.g., Ethernet PHYs designed for 10 or 15 m links are being used in trucks or buses. **Hubs** are multiport repeaters that take the input data from one connected device and broadcast it to all attached other devices, similar to an active star coupler in FlexRay. They make sense in case of a shared medium only. In Automotive Ethernet, only 10BASE-T1S uses a shared medium. However, hubs have not yet been discussed in this context.

2 Since then, Napster has been relaunched more than once. Most recently, in June 2016, the music streaming service Rhapsody rebranded itself as Napster [132]. This shows that the memory of the revolutionary change Napster initiated in music consumption around the turn of the century is still attributed with sufficient marketing potential 15 years later; even if users must accept the change from "for free" to "free-of-adds" [124].

3 IEEE 802.1 Qat "Stream Reservation" as well as the IEEE 802.1 Qav "Traffic Shaping" were both incorporated in the IEEE 802.1Q revision of 2011 [34].

4 The date listed here reflects the first specification release that was relevant for the car industry. In the meantime, the next version of IEEE 1722 had been released on March 4, 2016. At the time of writing, further amendments to the specification were being prepared [49].

5 Universal Plug and Play (UPnP) describes a set of protocols that allow for vendor-independent, distributed media management, discovery, and control in an IP-based network that consists of consumer devices such as computers, printers, Internet gateways, audio rendering units, mobile devices, etc. The first version of UPnP was published as ISO/IEC 29341 in December 2008 [134] and updated/extended in 2011. The UPnP forum, which drives and markets the developments, was founded in October 1999. Since January 2016, the Open Connectivity Foundation (OCF) has taken over the assets from the UPnP Forum [123].

6 The Digital Living Network Alliance (DLNA), founded in June 2003, has the goal of ensuring the interoperability between applications of networked consumer devices that involve images and AV data [122]. For this, the DLNA provided design guidelines based on higher layer standards such as UPnP and certification programs. The guidelines were based on standards such as UPnP for media management, discovery, and control. In 2016 the DLNA homepage listed about 1500 DLNA certified products. In 2017, the DLNA dissolved, moving the certification activities to a newly formed service company [120].

7 Coordinated Shared Network (CSN) is a generic term for a network in which the media is shared on a contention-free, time-multiplexed basis. The network access in a CSN is coordinated by one unit designated or elected as the network coordinator, which also might be the interface to e.g., an Ethernet LAN. CSN technologies are generally used in the home environment – Multimedia over Coax (MoCa), Homeplug (Inhouse Powerline Communication), and Ultra Wide Band (UWB)/IEEE 802.15.4a are given as examples – and were thus considered necessary to integrate into the AVB concept [30, 32, 129]. CSNs do not play a role in automotive.

8 At the time, summer 2009, this was actually very progressive. Automotive Ethernet had not yet taken off. One car manufacturer, BMW, used Ethernet for flash updates and for a private link between HU and RSE (see Sections 3.1 and 3.2), the project SEIS (Security in Embedded IP-based Systems) funded by the German government had only just started [121, 135], and the interest of everyone else was very moderate. The bus system for automotive infotainment (AV) was MOST or the transmission was analog or LVDS. This was true also for BMW. It can be assumed that the involvement of Harman and Broadcom in starting the Avnu Alliance spurred the inclusion of the automotive industry. Harman had just provided the HU/RSE system for BMW and, together with Broadcom, BMW made promising progress toward the first use of 100 Mbps BroadR-Reach Ethernet.

9 The seven hops, 2 ms requirement has two explanations. One is historical, the other was derived from the harshest, i.e., smallest, maximum network delay requirement in the professional auto domain. A musician needs to hear the response to his/her action within 10 ms; 8 ms of these are needed for DSP delays and the delay of the sound traveling, e.g., between speaker monitor and the musician. This leaves a delay of 2 ms for the network. These 2 ms were split into a realistic number of hops, leaving some margin and taking into consideration that in the worst case a 100 Mbps AVB packet needs to wait at every switch behind a best-effort 1500-byte packet for about 122 μs [M. Johas Teener, email correspondence, 2013]. Seven is also the historical limit for the layer two network hop count, which is reused in AVB. This seven-hop limit was established early on in Ethernet development. While IEEE 802.3 was working on repeaters, DEC developed the first layer two bridge. The upper layer protocol DEC used for the bridge was time sensitive and did not support more than seven hops. A daisy chain of seven hops for a layer two bridged system seemed a reasonable worst case and, even though the DEC upper layer protocol is no longer used and technology has advanced to fast hardware supported switching, the number has stayed [Y. Kim, email correspondence, August 12, 2013].

10 Without IEEE 1722 AV data would be streamed using UDP. This allows that data is also routed through the network on layer three, instead of just being switched on layer two. Apparently, neither additional processing delays are that critical nor is much additional processing required in the end nodes (provided the UDP checksum is ignored) [B. Petersen, email correspondence, May 12, 2020].

11 [28] includes the IEC 61883-2, 4, 6, 7, 8 formats (i.e., Standard Definition Digital Video Cassette Recorder (SD-DVCR), MPEG2-Transport Stream of compressed video, uncompressed digital audio and music, satellite TV MPEG ITU-R BO.1294, digital video data ITU-R BT.601), IIDC 1394-based uncompressed industrial camera, and

formats that will be defined by the Musical Instrument Digital Interface (MIDI) manufacturers' association [118].

12 The 2016 IEEE 1722 version includes a Clock Reference Format (CRF), which allows for the distribution of event timing information within a system. This is of particular interest for Driver ASsist (DAS) functions, e.g., to be able to combine four camera pictures correctly in a surround view image. It would also allow the determination of the exact distance to an obstacle on the road from camera images. For autonomous driving cameras represent an important source of redundant information in addition to other sensors such as radar, ultrasound, etc.

13 This scenario is not as unlikely as it sounds. When, e.g., a car is parked in a garage where it has no GPS reception, the navigation system cannot use the clock of the GPS as a reference but must rely on another internal clock with potentially inferior qualities. Thus, while in the garage another ECU might have a better quality clock available that would be selected as the Grandmaster. As soon as the car leaves the garage and GPS becomes available again, this might now represent the most superior clock. Using the BMCA the Grandmaster would then change over to the GPS clock.

14 In practice the situation is the following. The PTP timestamp is supposed to occur at the leading edge of the first bit of a MAC header's destination address field. This is before the Ethertype field is available and can be evaluated. In consequence, every received packet is timestamped, and the timestamps are simply ignored for non-PTP packets.

15 The Multiple VLAN Registration Protocol (MVRP) provides for the registration into the correct VLAN(s) and Multiple MAC Registration Protocol (MMRP) provides for the registration and announcement of multicast addresses.

16 Note that when writing the first edition of this book, it was anticipated that Class C traffic would be specified to have a packet frequency of 1 kHz. This has not materialized. It was seen as more fitting to provide a traffic class based on a number of audio samples well aligned with processing capabilities, which would then have a varying packet frequency, depending on whether it was used with 44.1 kHz audio sampling or 48 kHz [35].

17 A driving force behind this initiative was actually the industrial automation community which, for years, has been trying to move the industry toward standard Ethernet networks. Because Ethernet supported only best effort traffic for a long time, the industrial automation industry created a number of solutions to support time-critical control traffic in an Ethernet network, solutions that generally diverged from the IEEE's solutions at layer two (see also Section 1.2.3). The TSN effort is thus a chance for the industry to converge further toward an IEEE solution. Note that industrial automation's requirements for time critical control traffic are, if anything, more stringent than automotive requirements [119].

18 The Avnu Automotive Profile actually specified two features that did not make it into any IEEE 802.1 specification (yet): The already discussed "Bridge Holdover" and "Exception Handling and Diagnostic Counters." "Exception Handling and Diagnostic Counters" addresses the fact that the IEEE specifications do not describe error cases or error handling. This makes interoperability testing of malfunctions and their recovery almost impossible.

19 As another element, a variation of the one-step clock synchronization, as had originally been available in IEEE 1588, was adopted with the IEEE 802.1AS-2020 revision. Two-step clock synchronization had been specified with IEEE 802.1AS-2011 because of perceived challenges of hardware implementations of the one-step version, and the possibility to implement the two-step version in software [M. Turner, Email correspondence, April 20, 2020]. With the reintroduction, the goal was to allow the reuse of hardware (developed in the meantime) that supported other 1588 profiles [D. Pannell, email correspondence, April 21, 2020].

20 IEEE 802.3 limits the Ethernet packet size to 1500 bytes payload. However, a concept called "jumbo frames" exists, which is used in various non-standardized variations and

allows payloads of up to 9000 bytes [125]. These packets are used to increase throughput by reducing per-packet processing and bandwidth overhead. The potential delay owing to packets blocking the egress port, however, is increased almost 6-fold. Today, this is one of the reasons not to use jumbo packets. Once IET and pre-emption find their way into products, the authors expect that jumbo packets will be used more frequently; including the automotive industry.

21 Other protocols of the TCP/IP protocol suite are the Address Resolution Protocol (ARP) and the Reverse Address Resolution Protocol (RARP), which translate layer 3 IP addresses to layer two Ethernet MAC addresses and vice versa (RFC 826, 1982 [131]). Furthermore, the Internet Control Message Protocol (ICMP, or IPMCv6) sends error messages or relays query messages (RFC 792, 1981 [133]). The Internet Group Management Protocol (IGMP) establishes IP multicast group membership for IPv4 (RFC 3376, 2002 [127]) on layer two MAC addresses, while Multicast Listener Discovery (MLD, RFC 3810, RFC 4804) does the same for IPv6. When using static multicast addresses, it is not necessary to provide IGMP or MLD for an Automotive Ethernet application. Last, but not least, the User Datagram Protocol (UDP) (RFC 768, 1980 [72]) is part of the TCP/IP protocol suite. See also Figure 7.1.

22 Subnetting allows the division of a single-class IP network into smaller networks. Means to simplify routing and network design are thus appealing. Subnetting was first specified in 1985 in RFC 950 [130]. Variable Length Subnet Mask (VLSM) allows a subnetted network to use more than one subnet mask and thus to use the assigned address space more efficiently. It was first addressed in 1987 in RFC 1009 [126]. Classless Inter-Domain Routing (CIDR) eliminated the concept of class addresses through the introduction of longest-prefix match lookups on IP addresses. With CIDR, the number of nodes in a network was selectable arbitrarily. CIDR also allowed the reduction of the number of entries in routing tables. It is said that, without this, the Internet would not have sustained [126]. CIDR was specified in 1993 RFC 1517, 1518, 1519, and 1520 [76]. With IPv6, neither VLSM nor CIDR are needed.

23 We also see routing in the Automotive Ethernet in-vehicle networks of today. Every ECU that receives more than one IP address has to perform routing within the ECU. Also, all IP packets coming in via mobile carriers are routed within the vehicle.

24 In the IT or telecom spaces (see also Section 1.2.2), layer two packets are often tunneled over layer three, so that customers can interconnect their server instances at layer two while benefiting from the functions of layer three [B. Petersen, email correspondence, May 20, 2020].

25 The service *CD_Player* is used as an example to explain the basic features of SOME/IP and RPCs. Every service has to be defined during the development process by its service interface. This is normally done with an Interface Description Language (IDL) and could look as follows:

```
Service CD_Player
{
track_number              //Field
{unsigned int track;      //the track number
set (track);              //Method for setting the track (uses a request/response method)()
get ();                   // Method for getting the actual track number played
}
tray.eject ();            // Event that is triggered if the eject button is pressed
Boolean tray_state;       // Status OPEN or CLOSED when tray is open or closed
                             respectively
tray_state: open_tray (); // Method that is used for open the tray, the return value of this
                             Method is
// the tray_state.
}
```

Following the above service interface definition, "A client would like to change the track to track number 10" would cause the command *CD_Player.track_number.set(10)* to be sent from, e.g., the Head Unit (HU, client) to the CD-player (server). The method of the service is *track_number.set*, the payload value is *10*, and the communication principle typically used for this would be a request/response method, meaning that a response for the set command is expected.

In the next case "A client would like to open the tray of the CD-Player and would like to know when the job is done." When using the above description, the command from any client (e.g., the Head Unit) to the CD-player would be *CD_Player.open_tray()*. In this example the client expects a response in the form of an acknowledgment. This is thus an example of the request & response communication principle. When the client receives *CD_Player.open_tray()==OPEN*, it knows that the command has been successfully completed.

There is more than one way to achieve the same result. The key in service-based communication is that, whichever way is chosen to achieve a result, the data types, data structures, methods, and communication principles must be clearly defined up front. In the above example the client could also send a read command ("get field") to receive information regarding the CD-player tray status. Or it could have subscribed to the CD-player, asking the CD-player to automatically inform the client every time the status of the tray changes (i.e., an "event"). The respective command would be *Subscribe. CD_Player.Eject()*. In the event of the tray opening, the CD-player would send *CD_Player.Eject()* to all subscribed clients. This is an example of an event from the server to the clients.

This example emphasizes the possibilities service-based communication offers in contrast to the CAN-like communication principle of fire & forget and simple messages only. The serialization of SOME/IP ensures that the information fits into the existing packet format like all other traffic. The content, however, ensures a type of contract for a service to be fulfilled between the communication parties.

26 IEEE 802.1 standardized a number of security related protocols, whose reuse for Automotive Ethernet is worth investigating. For example, IEEE 802.1x is very widely implemented for key management [67]. It defines Ethernet encapsulation for the Extensible Authentication Protocol (EAP) which, in turn, is a framework for the exchange of authentication messages. Another standard of interest is IEEE 802.1AR, the Secure Device Identity Standard, which was first published in 2009 and updated in 2015. It defines the device identity and cryptography to be used by the device and the operation within EAP-TLS/802.1x. It assumes hardware support for efficient operation.

27 The MACsec standard actually also allows end-to-end protection. However, this reduces many of the benefits compared to IPsec, like protecting all messages and multicast support. While this is suitable for small VPN-like scenarios, the disadvantages of the end-to-end support are rather significant inside a car.

References

[1] L. Völker, "One for All; Interoperability from AUTOSAR to GENIVI," in *1st Ethernet & IP @ Automotive Technology Day*, Munich, 2011.

[2] ITWissen, "QoS (Quality of Service)," 2020, continuously updated. [Online]. Available: www.itwissen.info/QoS-quality-of-service-Dienstguete.html. [Accessed August 15, 2016].

[3] ITU, *Definitions of Terms Related to Quality of Service*, Geneva: ITU, 2008.

[4] ITU-T, *Network Performance Objectives for IP-based Services*, Geneva: ITU, 2011.

[5] R. Brand, S. Carlson, J. Gildred, S. Lim, D. Cavendish, and O. Haran, "Residential Ethernet, IEEE 802.3 Call for Interest," July 2004. [Online]. Available: http://grouper.ieee.org/groups/802/3/re_study/public/200407/cfi_0704_1.pdf. [Accessed May 6, 2020].

[6] T. Lamont, "Napster: the Day the Music Was Set Free," February 24, 2013. [Online]. Available: www.theguardian.com/music/2013/feb/24/napster-music-free-file-sharing. [Accessed May 6, 2020].

[7] Gartner, "Gartner Says Declining Worldwide PC Shipments in Fourth Quarter of 2012 Signal Structural Shift of PC Market," January 14, 2013. [Online]. Available: www.bloomberg.com/press-releases/2013-01-14/gartner-says-declining-worldwide-pc-shipments-in-fourth-quarter-of-2012-signal-structural-shift-of-pc-market. [Accessed May 20, 2020].

[8] M. Johas Teener, "No-excuses Audio/Video Networking: the Technology Behind AVnu," August 24, 2009. [Online]. Available: http://avnu.org/wp-content/uploads/2014/05/No-excuses-Audio-Video-Networking-v2.pdf. [Accessed May 20, 2020].

[9] IEEE 802.3, "IEEE 802.3 Residential Ethernet Study Group Homepage," January 10, 2006 (closed). [Online]. Available: http://grouper.ieee.org/groups/802/3/re_study/. [Accessed May 6, 2020].

[10] Business Wire, "AVnu Alliance Launches to Advance Quality of Experience for Networked Audio and Video," August 25, 2009. [Online]. Available: www.businesswire.com/news/home/20090825005929/en/AVnu-Alliance-Launches-Advance-Quality-Experience-Networked. [Accessed May 6, 2020].

[11] Marvell, "Marvell Announces Industry-First Audio Video Bridging Family of SoCs with Integrated Switching, CPU and Endpoint Functionality," May 8, 2012. [Online]. Available: www.marvell.com/company/newsroom/marvell-announces-industry-first-audio-video-bridging-family-of-socs-with-integrated-switching-cpu-and-endpoint-functionality.html. [Accessed May 20, 2020].

[12] J. Urban, "Debunking Some Myths About AVB," May 17, 2013. [Online]. Available: http://blog.biamp.com/debunking-some-myths-about-avb/. [Accessed May 20, 2020].

[13] IEEE 802.1, "802.1 Plenary –11/2012 San Antonio Closing Slides," November 2012. [Online]. Available: www.ieee802.org/1/files/public/minutes/2012-11-closing-plenary-slides.pdf. [Accessed May 6, 2020].

[14] E. Hellerud, "Transmission of High Quality Audio over IP Networks," April 2009. [Online]. Available: https://core.ac.uk/download/pdf/52128195.pdf. [Accessed May 20, 2020].

[15] R. Kreifeld, "AVB for Professional A/V Use," July 30, 2009. [Online]. Available: http://avnu.org/wp-content/uploads/2014/05/AVnu-Pro__White-Paper.pdf. [Accessed May 20, 2020].

[16] H. Kaltheuner, "Das Universalnetz, Ethernet AVB: Echtzeitfähig und Streamingtauglich," *c't*, no. 13, pp. 176–181, 2013.

[17] M. Johas Teener, "Residential Ethernet Objectives, Requirements and Possible Solutions," May 9, 2009. [Online]. Available: www.ieee802.org/1/files/public/docs2005/liaison-mikejt-rese-objectives-requirements-0505.pdf. [Accessed May 20, 2020].

[18] R. Steinmetz, "Human Perception of Jitter and Media Synchronization," *IEEE Journal on Selected Areas in Communication*, vol. 14, no. 1, pp. 61–72, January 1996.

[19] K. Stanton, "AVB for Home/Consumer Electronics Use," August 11, 2009. [Online]. Available: http://avnu.org/wp-content/uploads/2014/05/AVB-for-Home-Consumer-Electronics-Use__White-Paper.pdf. [Accessed May 20, 2020].

[20] F. Held, "Digitale Audio-, Video- & Licht-Kommunikationsprotokolle in der Veranstaltungstechnik," October 12, 2012. [Online]. Available: https://entropia.de/wiki/images/b/b8/Avb_vortrag_felix.pdf. [Accessed May 20, 2020].

[21] R. Boatright, "Understanding New Audio Video Bridging Standards," May 10, 2009. [Online]. Available: www.embedded.com/understanding-ieees-new-audio-video-bridging-standards/. [Accessed October 24, 2013].

[22] M. Schettke, "AVB Audionetzwerke in der Praxis: Betriebssicherheit und deren automatisierte Überprüfung," June 10, 2014. [Online]. Available: http://schettke.com/files/Diplomarbeit_AVB-Praxis_Schettke_2014.pdf. [Accessed May 20, 2020].

[23] R. Kreifeld, "AVB for Automotive Use," July 20, 2009. [Online]. Available: http://avnu.org/wp-content/uploads/2014/05/2014-11-20_AVnu-Automotive-White-Paper_Final_Approved.pdf. [Accessed October 20, 2013, only updated version available].

[24] M. Kicherer (Turner) and T. Königseder, "BMW Proposal for an AVB Gen 2 Automotive Profile," *BMW White Paper*, Munich, 2013.

[25] K. Matheus, M. Kicherer (Turner), and T. Königseder, "Audio/Video Transmission in Cars using Ethernet," *BMW White Paper*, Munich, 2010.

[26] T. Hogenmüller, "Use Cases & Requirements for IEEE 802.3 RTPGE Ethernet," May 2012. [Online]. Available: http://grouper.ieee.org/groups/802/3/RTPGE/public/may12/hogenmuller_01_0512.pdf. [Accessed May 20, 2020].

[27] K. Matheus, M. Kaindl, S. Korzine, D. Goncalvez, J. Leslie, Y. Okuno, N. Kitajima, M. Gardner, R. Orosz, M. Jaenecke, D. Kim, R. Mei, and D. v. Knorre, "1 Pair or 2 Pairs for RTPGE: Impact on System Other than the PHY Part 1: Weight & Space," January 2013. [Online]. Available: www.ieee802.org/3/bp/public/jan13/matheus_3bp_01_0113.pdf. [Accessed May 14, 2020].

[28] IEEE Computer Society, *1722-2016 – IEEE Standard for a Transport Protocol for Time-Sensitive Applications in Bridged Local Area Networks*, New York: IEEE, 2016.

[29] H. Schulzrinne, S. Casner, R. Frederick, and V. Jacobson, "RTP: A Transport Protocol for Real-Time Applications," July 2003. [Online]. Available: http://tools.ietf.org/html/rfc3550. [Accessed May 20, 2020].

[30] IEEE Computer Society, *802.1AS-2011 – IEEE Standard for Local and Metropolitan Area Networks – Timing and Synchronization for Time-Sensitive Applications in Bridged Local Area Networks*, New York: IEEE SA, 2011.

[31] IEEE Computer Society, *1588-2008 – IEEE Standard for a Precision Clock Synchronization Protocol for Networked Measurement and Control Systems*, New York: IEEE, 2008.

[32] Hirschmann, "Precision Clock Synchronization, The Standard IEEE 1588," 2008. [Online]. Available: www.belden.com/docs/upload/precision_clock_synchronization_wp.pdf. [Accessed October 28, 2013, no longer available].

[33] EndRun Technologies, "Precision Time Protocol (PTP/IEEE-1588)," 2013. [Online]. Available: www.endruntechnologies.com/pdf/PTP-1588.pdf. [Accessed May 20, 2020].

[34] IEEE Computer Society, *802.1Q-2011 – IEEE Standard for Local and Metropolitan Area Networks – Media Access Control (MAC) Bridges and Virtual Bridge Local Area Networks*, New York: IEEE, 2011.

[35] G. Bechtel, B. Gale, M. Kicherer (Turner), and D. Olsen, *Automotive Ethernet AVB Functional and Interoperability Specification, Revision 1.4*, Beaverton: Avnu, 2015.

[36] S. Stein and R. Racu, "Ethernet Quality of Service @ Volkswagen," in 2nd Ethernet & IP @ Automotive Technology Day, Regensburg, 2012.

[37] J. Diemer, D. Thiele, and R. Ernst, "Formal Worst-Case Timing Analysis of Ethernet Topologies with Strict-Priority and AVB Switching," in 7th IEEE International Symposium on Industrial Embedded Systems (SIES12), Karlsruhe, 2012.

[38] IEEE Computer Society, 1733-2011 – IEEE Standard for Layer 3 Transport Protocol for Time-Sensitive Applications in Local Area Networks, New York: IEEE, 2011.

[39] J. Damori, "Are Layer 2 or Layer 3 protocols better? Yes," May 31, 2013. [Online]. Available: http://blog.biamp.com/are-layer-2-or-layer-3-protocols-better-yes/. [Accessed May 20, 2020].

[40] IEEE Computer Society, *802.1BA-2011 – IEEE Standard for Local and Metropolitan Are Networks – Audio Video Bridging (AVB) Systems*, New York: IEEE SA, 2011.

[41] IEEE Computer Society, *1722.1-2013 – IEEE Standard for Device Discovery, Connection Management, and Control Protocol for IEEE 1722 Based Devices*, New York: IEEE, 2013.

[42] J. Lane, "Digital Audio Listening in Car Is Increasing," October 11, 2010. [Online]. Available: http://audio4cast.com/2010/10/11/digital-audio-listening-in-car-is-increasing/. [Accessed May 20, 2020].

[43] M. Jochim and M. Osella, "The Need for IEEE Standardized Ethernet Mechanisms for Active Safety Applications," in *3rd Ethernet & IP @ Automotive Technology Day*, Leinfelden-Echterdingen, 2013.

[44] IEEE 802.1, "P802.1DG – TSN Profile for Automotive In-Vehicle Ethernet Communications Homepage," IEEE, 2019 (continuously updated). [Online]. Available: https://1.ieee802.org/tsn/802-1dg/. [Accessed April 14, 2020].

[45] H. Zinner and D. Hopf, "Time Sensitive Networking," in AVnu TSN Automotive Workshop and Tutorial, Munich, 2017.

[46] Wikipedia, "Time-Sensitive Networking," March 5, 2020. [Online]. Available: https://en.wikipedia.org/wiki/Time-Sensitive_Networking. [Accessed April 18, 2020].

[47] D. Pannell, "Choosing the Right Tools to Meet a Bounded Latency," in Automotive Ethernet Congress, *Munich*, 2020.

[48] D. Zebralla (Hopf), "Requirements on Future In-Vehicle Architectures for Automotive Ethernet," in Automotive Ethernet Congress, Munich, 2016.

[49] G. Bechtel, "P1722b PAR," October 28, 2019. [Online]. Available: http://grouper.ieee.org/groups/1722/contributions/2020/1722bPAR/P1722b_PAR_Detail-2020-01-14.pdf. [Accessed April 14, 2020].

[50] D. Pannell, "Audio Video Bridging Gen 2 Assumptions," July 16, 2013. [Online]. Available: www.ieee802.org/1/files/public/docs2013/avb-pannell-gen2-assumptions-0313-v15.pdf. [Accessed April 17, 2020].

[51] IEEE Computer Society, *802.1AS-2020 – IEEE Standard for Local and Metropolitan Area Networks – Timing and Synchronization for Time-Sensitive Applications*, New York: IEEE-SA, 2020.

[52] IEEE Computer Society, *802.1Qcc-2018 – IEEE Standard for Local and Metropolitan Area Networks – Bridges and Bridged Networks – Amendment 31: Stream Reservation Protocol (SRP) Enhancements and Performance Improvements*, New York: IEEE-SA, 2018.

[53] Wikipedia, "TTEthernet," April 10, 2020. [Online]. Available: http://en.wikipedia.org/wiki/TTEthernet/TTEthernet. [Accessed May 20, 2020].

[54] IEEE Computer Society, *802.1Qbv-2015 – IEEE Standard for Local and Metropolitan Area Networks – Bridges and Bridged Networks Amendment 25: Enhancements for Scheduled Traffic*, New York: IEEE-SA, 2015.

[55] D. Pannell, "IEEE TSN Standards Overview & Update," in *5th Ethernet & IP @ Automotive Technology Day*, Yokohama, 2015.
[56] IEEE Computer Society, *802.3br-2016 – IEEE Standard for Ethernet Amendment 5: Specification and Management Parameters for Interspersing Express Traffic*, New York: IEEE-SA, 2016.
[57] IEEE Computer Society, *802.1Qbu-2016 – IEEE Standard for Local and Metropolitan Area Networks – Bridges and Bridged Networks – Amendment 26: Frame Preemption*, New York: IEEE-SA, 2016.
[58] IEEE Computer Society, *802.1Qch-2017 – IEEE Standard for Local and Metropolitan Area Networks – Bridges and Bridged Networks – Amendment 29: Cyclic Queuing and Forwarding*, New York: IEEE-SA, 2017.
[59] IEEE Computer Society, *802.1Qcr-2020 – IEEE Standard for Local and Metropolitan Area Networks – Bridges and Bridged Networks – Amendment: Asynchronous Traffic Shaping*, New York: IEEE-SA, 2020 (est.).
[60] IEEE Computer Society, *802.1Qca-2015 – IEEE Standard for Local and Metropolitan Area Networks – Bridges and Bridged Networks – Amendment 24: Path Control and Reservation*, New York: IEEE-SA, 2015.
[61] IEEE Computer Society, *802.1CB-2017 – IEEE Standard for Local and Metropolitan Area Networks – Frame Replication and Elimination for Reliability*, New York: IEEE-SA, 2017.
[62] M. Jochim, "Ingress Policing," November 10–14, 2013. [Online]. Available: www.ieee802.org/1/files/public/docs2013/tsn-jochim-ingress-policing-1113-v2.pdf. [Accessed May 20, 2020].
[63] IEEE Computer Society, *802.1Qci-2017 – IEEE Standard for Local and Metropolitan Area Networks – Bridges and Bridged Networks – Amendment 28: Per-Stream Filtering and Policing*, New York, IEEE-SA, 2017.
[64] K. Matheus, "The Use of AVB and TSN in Automotive," in *TSNA*, Stuttgart, 2017.
[65] H. Zinner, "Recap on TSN Specs with Focus on Effort," in AVnu TSN Automotive Workshop and Tutorial, San Jose, CA, 2017.
[66] B. Gale, "Ethernet Security in the Car," in *IEEE-SA Ethernet & IP @ Automotive Technology Day*, Detroit, 2014.
[67] Y. Kim, "Ethernet Security," in *Automotive Ethernet Congress*, Munich, 2016.
[68] W. R. Stevens, *TCP/IP Illustrated, Volume 1, The Protocols*, Reading, MA: Addison Wesley Longman, 1994.
[69] Wikipedia, "Internet Protocol Suite," Wikipedia, April 27, 2020. [Online]. Available: http://en.wikipedia.org/wiki/Internet_protocol_suite. [Accessed May 6, 2020].
[70] Information Sciences Institute University of Southern California, "Transmission Control Protocol," September 1981. [Online]. Available: http://tools.ietf.org/html/rfc793. [Accessed May 20, 2020].
[71] Information Sciences Institute University of Southern California, "Internet Protocol," September 1981. [Online]. Available: http://tools.ietf.org/html/rfc791. [Accessed May 6, 2020].
[72] J. Postel, "User Datagram Protocol," August 29, 1980. [Online]. Available: http://tools.ietf.org/html/rfc768. [Accessed May 6, 2020].
[73] M. Kessler, "Ethernet in Small ECUs, Challenges and Chances," in *3rd Ethernet & IP @ Automotive Technology Day*, Leinfelden-Echterdingen, 2013.

[74] Wikipedia, "Motor Industry Software Reliability Association," March 2, 2020. [Online]. Available: https://en.wikipedia.org/wiki/Motor_Industry_Software_Reliability_Association. [Accessed April 19, 2020].

[75] Y. Rekhter, B. Moskowitz, G. D. Karrenberg, and E. Lear, "Address Allocation for Private Internets," February 1996. [Online]. Available: https://tools.ietf.org/html/rfc1918. [Accessed April 19, 2020].

[76] 3Com, "Understanding IP Addressing, Everything You Ever Wanted to Know," 2001. [Online]. Available: http://pages.di.unipi.it/ricci/501302.pdf. [Accessed May 20, 2020].

[77] Number Resource Organization, "Free Pool of IPv4 Address Space Depleted," February 3, 2011. [Online]. Available: www.nro.net/news/ipv4-free-pool-depleted. [Accessed May 6, 2020].

[78] S. Deering and R. Hinden, "Internet Protocol, Version 6 (IPv6)," December 1995. [Online]. Available: http://tools.ietf.org/html/rfc1883. [Accessed May 6, 2020].

[79] C. Bao, C. Huitema, M. Bagnulo, M. Boucadair, and X. Li, "IPv6 Addressing of IPv4/IPv6 Translators," October 2010. [Online]. Available: http://tools.ietf.org/html/rfc6052. [Accessed May 20, 2020].

[80] RIPE NCC, "The RIPE NCC Has Run out of IPv4 Addresses," November 25, 2019. [Online]. Available: www.ripe.net/publications/news/about-ripe-ncc-and-ripe/the-ripe-ncc-has-run-out-of-ipv4-addresses. [Accessed April 19, 2020].

[81] InfiniBand Trade Association, *Supplement to InfiniBand Architecture Specification Volume 1 Release 1.2.1*, Beaverton: InfiniBand Trade Association, 2010.

[82] A. Busnelli, "Car Software: 100M Lines of Code and Counting," June 26, 2014. [Online]. Available: www.linkedin.com/pulse/20140626152045-3625632-car-software-100m-lines-of-code-and-counting. [Accessed May 20, 2020].

[83] The Apache Software Foundation, "Apache Etch," 2020, continuously updated. [Online]. Available: http://etch.apache.org/. [Accessed May 20, 2020].

[84] A. Bouard, J. Schanda, D. Herrscher, and C. Eckert, "Automotive Proxy-based Security Architecture for CE Device Integration," in *Mobile Wireless Middleware, Operating Systems, and Applications*, Heidelberg: Springer, 2013, pp. 62–76.

[85] K. Weckemann, F. Satzger, L. Stolz, D. Herrscher, and C. Linnhoff-Popien, "Lessons from a Minimal Middleware for IP-Based In-Car Communication," in *Proceedings of the IEEE Intelligent Vehicles Symposium*, Alcala de Henares, 2012.

[86] Google Developers, "Protocol Buffers," Google Developers, April 2, 2012. [Online]. Available: https://developers.google.com/protocol-buffers/. [Accessed May 20, 2020].

[87] Apple, "Bonjour for Developers," Apple, 2020, continuously updated. [Online]. Available: https://developer.apple.com/bonjour/. [Accessed May 20, 2020].

[88] L. Völker, "Scalable Service-Oriented Middleware over IP (SOME/IP)," 2020, continuously updated. [Online]. Available: http://some-ip.com. [Accessed May 20, 2020].

[89] AUTOSAR, "SOME/IP Protocol Specification R19-11," November 28, 2019. [Online]. Available: www.autosar.org/fileadmin/user_upload/standards/foundation/19-11/AUTOSAR_PRS_SOMEIPProtocol.pdf. [Accessed May 5, 2020].

[90] Wikipedia, "Data Distribution Service," March 5, 2020. [Online]. Available: https://en.wikipedia.org/wiki/Data_Distribution_Service. [Accessed May 5, 2020].

[91] Object Management Group, *Data Distribution Service v.1.4*, Needham, MA: Object Management Group, 2015.

[92] Object Management Group, *RPC Over DDS v.1.0*, Needham, MA: Object Management Group, 2017.

[93] R. Moussalli, "Dealing with Security Threats in the Digital Age: Part 1," August 19, 2015. [Online]. Available: www.tatacommunications.com/blog/2015/08/dealing-with-security-threats-in-the-digital-age-part-1/. [Accessed May 20, 2020].

[94] C. Miller and C. Valasek, "Remote Exploitation of an Unaltered Passenger Vehicle," August 10, 2015. [Online]. Available: http://illmatics.com/Remote%20Car%20Hacking.pdf. [Accessed May 20, 2020].

[95] A. Greenberg, "Remote Exploitation of an Unaltered Passenger Vehicle," July 21, 2015. [Online]. Available: www.wired.com/2015/07/hackers-remotely-kill-jeep-highway/. [Accessed May 20, 2020].

[96] SANS Institute, "The Best Security Books to Have in Your Library," SANS Institute, 2020. [Online]. Available: www.sans.edu/research/book-reviews/article/security-books-best. [Accessed May 20, 2020].

[97] S. Singer, "IP Based Communication in Vehicles – Learnings from the IT Industry," in *Automotive Ethernet Congress*, Munich, 2016.

[98] H. G. Molter, "Introduction to Security," in *Automotive Ethernet Congress*, Munich, 2016.

[99] Information Security, "Who Is the Creator of the CIA Triad," December 27, 2013. [Online]. Available: https://security.stackexchange.com/questions/47697/who-is-the-creator-of-the-cia-triad. [Accessed April 23, 2020].

[100] Wikipedia, "SYN Flood," May 4, 2020. [Online]. Available: https://en.wikipedia.org/wiki/SYN_flood. [Accessed May 20, 2020].

[101] AUTOSAR, "Specification of Secure Onboard Communication, Release 4.3.1," December 8, 2017. [Online]. Available: www.autosar.org/fileadmin/user_upload/standards/classic/4-3/AUTOSAR_SWS_SecureOnboardCommunication.pdf. [Accessed May 5, 2020].

[102] SAE, "Cybersecurity Guidebook for Cyber-Physical Vehicle Systems, J3061," *SAE international*, Warrendale, 2016.

[103] ISO/SAE, *DIS 21434 Road Vehicles – Cybersecurity Engineering*, Geneva: ISO, 2020.

[104] H. Goto, "In-Vehicle Ethernet Technology Promoted by JASPAR and Industry Trends," in *4th Nikkei Electronics/Automotive Seminar*, Tokyo, 2016.

[105] R. Pallierer and M. Ziehensack, "Secure Ethernet for Autonomous Driving," in *Automotive Ethernet Congress*, Munich, 2016.

[106] E. Rescorla, "The Transport Layer Security (TLS) Protocol Version 1.3," August 2018. [Online]. Available: https://tools.ietf.org/html/rfc8446. [Accessed April 21, 2020].

[107] T. Dierks and E. Rescorla, "The Transport Layer Security (TLS) Protocol Version 1.2," August 2008. [Online]. Available: https://tools.ietf.org/html/rfc5246. [Accessed May 20, 2020].

[108] E. Rescorla and N. Modadugu, "Datagram Transport Layer Security," April 2006. [Online]. Available: https://tools.ietf.org/html/rfc4347. [Accessed April 21, 2020].

[109] M. Lindner, "Security Architecture for IP (IPsec)," 2007. [Online]. Available: www.ict.tuwien.ac.at/lva/384.081/infobase/L97-IPsec_v4-7.pdf. [Accessed May 20, 2020].

[110] Wikipedia, "IPsec," May 12, 2020. [Online]. Available: http://en.wikipedia.org/wiki/Ipsec. [Accessed May 20, 2020].

[111] IEEE Computer Society, *802.1AEcg-2017 – IEEE Standard for Local and Metropolitan Area Networks – Media Access Control (MAC) Security – Amendment 3: Ethernet Data Encryption devices*, New York: IEEE-SA, 2017.

[112] L. Völker, "Comparing Automotive Network Security for Different Communication Technologies," in *Automotive Ethernet Congress*, 2018.

[113] L. Völker, "Why Is Network Security in Vehicles so Hard?," in *Automotive Networks – IVN Technologies and EE Architecture*, Munich, 2018.

[114] L. Völker, *Choosing Network Security Solutions: Guidance for Automotive Use Cases*, Munich: Technica Engineering GmbH, 2020.

[115] D. Kleidermacher and M. Kleidermacher, *Embedded Systems Security – Practical Methods for Safe and Secure Software and Systems Development*, Oxford: Newnes, 2012.

[116] Wikipedia, "Books on Cryptography," April 6, 2020. [Online]. Available: https://en.wikipedia.org/wiki/Books_on_cryptography. [Accessed May 20, 2020].

[117] L. Apvrille, R. El Khayari, O. Henniger, Y. Roudier, H. Schweppe, H. Seudié, B. Weyl, and M. Wolf, "Secure Automotive On-board Electronics Network Architecture," May 30, 2010. [Online]. Available: www.evita-project.org/Publications/AEHR10.pdf. [Accessed May 20, 2020].

[118] Musical Instrument Digital Interface manufacturers association (MIDI), "IEEE Ethernet AVB, AVB and MIDI," 2013. [Online]. Available: www.midi.org/techspecs/avbtp.php. [Accessed November 2, 2013, no longer available].

[119] M. Johas Teener, "IEEE 802 Time-Sensitive Networking: Extending Beyond AVB," in *3rd Ethernet & IP @ Automotive Technology Day*, Leinfelden-Echterdingen, 2013.

[120] DLNA, "DLNA Fulfills Mission, Dissolves as non-Profit Trade Association," January 5, 2017. [Online]. Available: www.dlna.org/about/organization/. [Accessed April 13, 2020].

[121] M. Glaß, D. Herrscher, H. Meier, M. Piastowski, and P. Schoo, "SEIS – Sicherheit in Eingebetteten IP-Basierten Systemen," *ATZelektronik,* vol. 5, no. 1, pp. 50–55, 2010.

[122] DLNA, "About DLNA," DLNA, 2013. [Online]. Available: www.dlna.org/dlna-for-industry/about-dlna. [Accessed May 20, 2020].

[123] Open Connectivity Forum (OCF), "About UPnP," January 1, 2016. [Online]. Available: https://openconnectivity.org/developer/specifications/upnp-resources/upnp/. [Accessed May 20, 2020].

[124] Napster, "Company Info," 2020, continuously updated. [Online]. Available: http://us.napster.com//about?from=rhapsody. [Accessed May 20, 2020].

[125] Wikipedia, "Jumbo Frame," May 7, 2020. [Online]. Available: https://en.wikipedia.org/wiki/Jumbo_frame. [Accessed May 20, 2020].

[126] R. Braden and J. Postel, "Requirements for Internet Gateways," June 1987. [Online]. Available: www.ietf.org/rfc/rfc1009.txt. [Accessed May 20, 2020].

[127] B. Cain, S. Deering, I. Kouvelas, B. Fenner, and A. Thyagarajan, "Internet Group Management Protocol, Version 3," October 2002. [Online]. Available: http://tools.ietf.org/html/rfc3376. [Accessed May 20, 2020].

[128] T. Hümmler, "Router, Switches, Hub und Co. Ratgeber: Was ist im Netzwerk?," October 26, 2013. [Online]. Available: www.tecchannel.de/a/ratgeber-was-ist-was-im-netzwerk,2038788. [Accessed May 20, 2020].

[129] P. Klein, "Support for Coordinated Shared Network in 802.1AVB," January 2008. [Online]. Available: www.ieee802.org/1/files/public/docs2008/av-phkl-csn-0108-v1.pdf. [Accessed May 20, 2020].

[130] J. Mogul and J. Postel, "Internet Standard Subnetting Procedure," August 1985. [Online]. Available: www.ietf.org/rfc/rfc950.txt. [Accessed May 20, 2020].

[131] D. C. Plummer, "An Ethernet Address Resolution Protocol," November 1982. [Online]. Available: http://tools.ietf.org/html/rfc826. [Accessed May 20, 2020].

[132] B. Popper, "Rhapsody Rebrands Itself as Napster Because ¯_(ツ)_/¯," The Verge, June 14, 2016. [Online]. Available: www.theverge.com/2016/6/14/11936974/rhapsody-rebrands-as-napster. [Accessed May 20, 2020].

[133] J. Postel, "Internet Control Message Protocol," September 1981. [Online]. Available: http://tools.ietf.org/html/rfc792. [Accessed May 20, 2020].

[134] ISO, *ISO/IEC 29341- (1-13)-1:2008: Information Technology – UPnP Device Architecture*, Geneva: ISO, 2008.

[135] Universität Erlangen, Lehrtstuhl für Informatik 12, "SEIS – Sicherheit in Eingebetteten IP-Basierten Systemen," June 6, 2013. [Online]. Available: www12.informatik.uni-erlangen.de/research/seis/. [Accessed May 13, 2020].

8 Ethernet in Automotive System Development

The decision on the overall functionality to provide with a car is first of all a marketing decision and independent of the capabilities of in-vehicle networking technologies used and available. However, as soon as decisions have to be made on how to enable the functionality, in-vehicle networking becomes important. Aspects such as flexibility, scalability, or how to distribute functions are severely impacted by the properties of in-vehicle networking. This chapter thus discusses the opportunities and changes Automotive Ethernet brings to system development.

8.1 A Brief Overview on the System Development Process

The development process in the automotive industry follows the V-cycle. While Section 2.3.1 used the V-cycle to explain the responsibilities shared between car manufacturer and Tier 1 supplier, in this section the model is used to explain the changes the introduction of Ethernet-based communication brings to the development process. The important idea of the V-cycle is to follow a top-down approach on the development side and a bottom-up approach on the test side. On both sides, each new step requires the conclusion of the previous step. During the development, the later need for testing is directly supported with the provision of test cases. This ensures stringent test coverage.[1]

In the context of in-vehicle networking, it is not necessary to consider the complete car development process. Instead a focus on the development of the Electric Electronic (EE or E/E) architecture is sufficient (see also [1]). The task of an automotive EE architecture is to enable all required (electric and electronic) functions in a vehicle, while fulfilling the constraints given by cost targets and space limitations. Figure 8.1 gives an overview of the respective elements of the V-cycle. In the following, each step is briefly described, before the next sections concentrate on those steps affected by the introduction of Automotive Ethernet.

In the first step, product management and sales define the (EE enabled) system requirements, i.e., the customer functions a new car model should have. This includes the distinction between functions installed in every car and functions that can be bought by customers as options. Directly or indirectly, this definition also includes the interdependencies that might exist between some of the functions. In a complementary case, customers might only be able to buy a certain function if they bought another

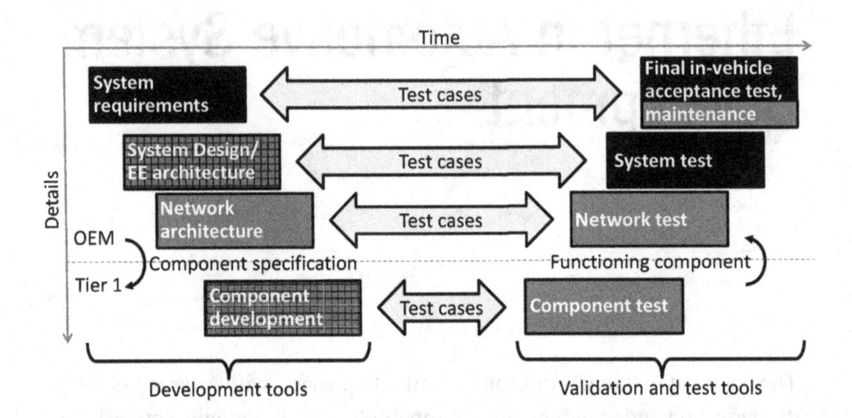

Figure 8.1 Overview of the elements of the automotive EE development V-cycle. The grey boxes indicate the areas with most changes, the checkered boxes indicate the areas in which there are some Ethernet induced changes concerning the software

function at the same time. An example is a rear view camera that can only be bought if the customer also selects a Head Unit (HU) with a suitable display. In an exclusive case, a customer cannot select two options simultaneously. For example, the rear view camera cannot be bought if the customer wants a HU without the appropriate display. There are various reasons to provide specific functions in a car. The most obvious is to offer functions the customers want to have. Others are profitability, image, regulatory requirements, or the wish to achieve a certain safety rating by, e.g., the European New Car Assessment Program (Euro NCAP) [2]. This first step of the development process does not define how the functions are enabled but represents the ideal result of the system [3]. With the next steps that look into the feasibility, it might turn out that this ideal result is not achievable. The system is then adapted accordingly.

The second step defines the EE architecture. In general, this means that the system engineers propose a solution on how the requirements can be implemented. First, all customer functions are broken down into smaller functional entities, so-called function blocks. Each function block describes one part of a customer function at a certain level. For the surround view system, for example, one function block starts the surround view function (e.g., the use of the reverse gear); one each records the images of the front, sides, and back; one combines the images; one identifies and marks pedestrians in those images; and, finally yet importantly, one displays the image to the driver.

Once all function blocks have been defined, the very crucial and complex task of partitioning the function blocks onto ECUs, sensors, and actuators follows. In an ideal world, this step is a cost optimization process, in which function blocks are clustered onto an optimal number of ECUs, sensors, and actuators depending on the basic/optional definition given by the requirements of step one. Input values to this process would be expected customer take rates (potentially including roll out plans over different car models), functional safety targets (ASIL levels, see ISO 26262 [4]), and preference toward a more integrated or a more distributed approach (see Section

8.3.2.1). The next steps would then define how the ECUs, sensors, and actuators are powered and communicate with each other and what the 3D routing layout would look like.

In a not so ideal world, however, there are lots of interdependencies with the following as well as the preceding steps. If, because of the spatial constraints of a certain car model, an optional function can be offered only when integrated into an ECU with basic functions, this basic ECU might become too expensive to meet the cost targets. Or, the proposed partitioning might stress the available data rate on the preferred networking technology and the targeted communication technology cannot be used. This in return affects the costs and the price target might not be met. Or, the function is too new to have a sound cost analysis and later developments show that changes to the price target are necessary. In the real world, partitioning functions onto ECUs is thus an iterative process in which the EE architecture is defined via various feedback loops. The output of this step is a description of ECUs, sensors, and actuators, as well as their interdependencies and communication requirements.[2]

Breaking up the customer functions and partitioning them onto ECUs as such is not affected by the introduction of Automotive Ethernet. Nevertheless, in combination with the selection of the networking technology, Automotive Ethernet might well impact the system design: The higher data rate might allow for a different partitioning of functions, e.g., onto fewer ECUs with more data exchange between them.

The EE architecture directly affects the network architecture. The network architecture defines the in-vehicle communication and the power supply. The in-vehicle communication describes which ECU is connected via which communication/IVN-technology to the rest of the system. Also, the inter-technology communication via gateways and their location in the network is defined. Supported data types and data rates, timing behavior, and quality are important criteria for the selection of the in-vehicle networking technology. Naturally, the introduction of Ethernet causes quite some changes to the in-vehicle communication. Specific implications are thus discussed in more detail in Section 8.3.2.

The network architecture has two further outputs. One is the spatial layout of the network architecture, i.e., the 3D position of the ECUs, sensors, and actuators; the routes for the harness elements that connect them; and the terminations for the in-vehicle networking technologies. Important considerations for the spatial layout are weight, installation space, harness diameter, maximum link lengths, inline connections between different areas of the car, pinning rules, whether the units are in dry or wet areas, in high or very high temperature areas, and diagnosability. After all, in the event of malfunctions, all units and elements of the harness need to be accessible in a garage. The changes Automotive Ethernet induces on connectors, wiring, or harness manufacturing have been addressed in Sections 4.2 and 5.1.

The second output is that the definition of the ECUs provided with the EE architecture also gives indications on the communication between function blocks within an ECU and thus of the design of the latter. While today the final design is done by the Tier 1 supplier of the ECU, car manufacturers can use the information for a first cost estimate and to provide some guidelines to the supplier.

Based on the provided information, the supplier then completes the design of the ECU and implements it. This includes hardware as well as software design, which is indeed impacted by Ethernet (see Section 8.2). The development of each ECU follows its own V-cycle. It is also possible to use Ethernet communication within an ECU: It is a design choice that can be made. However, it must be integrated like any other networking function. In the case of high data rates, generally Direct or Remote Direct Memory Access (DMA/RDMA) mechanisms need to be included for hardware acceleration. Intelligent approaches enable this without the use of special hardware. This is not limited to Ethernet, however, but is the same for other busses such as I^2C and therefore the topic does not require special attention in this chapter.

On the test side, first the ECUs as such are tested. Next, when connecting the ECUs, (only) the networking functions such as start-up and shut-down are tested. Only when these have been successful are the combined user functionalities tested before the final tests are performed inside the car. Last, but not least, the car and its network must be maintained throughout the lifetime of the car. The impact Automotive Ethernet has on component test, network test, and maintenance are described in Section 8.4. The system-level tests and in-vehicle tests are not really affected by the introduction of Automotive Ethernet.

8.2 Software Design

Two different software design approaches currently prevail in the automotive industry, AUTomotive Open System ARchitecture (AUTOSAR) and Portable Operating System Interface (POSIX). The following bullet points describe briefly how Automotive Ethernet affects the two approaches in the design process. For a detailed introduction to AUTOSAR, see e.g. [5].

- **AUTOSAR:** As with all operating systems, one of the core functions of AUTOSAR is to decouple the software from the hardware on which it is used. For this, AUTOSAR provides a set of Application Programming Interfaces (APIs) that very specifically fit automotive requirements and are very scalable. If desired, AUTOSAR can be used with very small 8-bit processors; although today in the automotive industry the use of 32-bit processors is standard. Furthermore, AUTOSAR divides the software into modules, which then, at least theoretically, can be developed independently by various companies. A mostly-automated, powerful configuration tool chain combines the function modules. The result is a specific software project, which is partially post-built configurable. AUTOSAR is the dominant architecture in the body, chassis, and power train domains within a car.

  AUTOSAR has supported Ethernet-based communication – first for DoIP – since 2009. The combination of AUTOSAR with SOME/IP (see Section 7.4) requires the extension of the configuration tool chain in order for it to cover SOME/IP plus the respective "Socket Adaptors" and Service Discovery (SD) configuration. The first AUTOSAR standard to describe these requirements is version 4.0.3. Further revisions of AUTOSAR incorporate more requirements

Table 8.1 Basic distinction between "Classic" and "Adaptive" AUTOSAR [6]

Classic AUTOSAR	Adaptive AUTOSAR
Deeply embedded ECUs	High-performance ECUs
μC ~1000 DMIPS	μC >20,000 DMIPS
C only	C++, others tbd
Static configuration	Run-time configuration
Monolithic updates	Partial updates and upgrades
Signal-oriented communication	Service-oriented communication
Low-speed communication	High-speed communication

coming from Automotive Ethernet (see Section 3.5.3 for more details). The Japan Automotive Software Platform and ARchitecture (JASPAR) also adopted the AUTOSAR standard.

In principle, the latest development "Adaptive AUTOSAR" supports Ethernet communication just like "Classic AUTOSAR" does. However, Adaptive AUTOSAR has been developed in order to be more suitable for new, more compute-intense functions as is needed, e.g., for autonomous driving or functions that connect with the backend outside the car. For supporting run-time configuration, over the air updates and upgrades, as well as a service-oriented communication, Ethernet is the communication technology to use with Adaptive AUTOSAR systems. The middleware functionality of Adaptive AUTOSAR thus uses SOME/IP (see Section 7.4). Table 8.1 gives a high-level overview of the differences between "Classic" and "Adaptive" AUTOSAR.

- **POSIX:** POSIX compatible Operating Systems (OS) such as QNX (see Section 3.2) or GENIVI (see Section 3.5.3) are mainly used within the infotainment domain. However, with the growing complexity in the driver assist domain, interest in deploying POSIX systems is increasing there, too. In contrast to traditional in-vehicle networking systems, Automotive Ethernet is the ideal complement to the POSIX approach; the socket-based communication of TCP/IP fits perfectly with the InterProcess Communication (IPC) solutions of the POSIX-compatible operating systems. Ethernet stacks are generally an integral part of these solutions. For the support of SOME/IP, extensions for code generators exist. Generally, an independent tool chain feeds these generators. An example of such a tool chain is based on an Interface Definition Language (IDL) named Franca, which is also used with development tools such as Eclipse.

8.3 Networking Architecture

8.3.1 EE Architecture Related Requirements

From a consumer industry perspective, it appears very complex to introduce a new in-vehicle networking technology. Why is it not possible to simply take an existing CE technology and (extensively) reuse it in cars? The simple answer is that CE technologies generally do not fulfill essential automotive requirements. One of the key differences between the IT/consumer and automotive industries emphasized in this book so far has been the physical environment the technology must cope with.

As a minimum requirement, ambient temperatures of –40°C to +105°C must be supported. For example, the inner side of the wing mirror of a car parked in the sun quite quickly heats up to this upper bound. In the engine compartment or the gearbox control, the temperature that needs to be supported increases to 125°C. This not only requires that the respective qualification programs are passed for the semiconductors, but also that the right type of housing is selected and that concepts for heat dissipation are included in the design. Next to the temperature range, mechanical strain, EMC, ESD, etc., also significantly differ in a car from an office environment (see also Chapter 4). As a consequence, Automotive Ethernet not only requires special semiconductor qualification but also optimized PHY technologies.

There are more differences than just temperature requirements. This is largely due to the very different use case experienced in a car. Just to give an idea: A typical car, and thus also its communication network and power system, drives 300,000 km in 15 years and faces 10,500 temperature changes [7]. Additionally, a user might start and park a car several times a day. The design of the communication network and power supply system thus has to consider the following effects:

- **Long lifetime:** Cars are used for many years, and around the world the average age of cars keeps increasing (see e.g. [8, 9]). This touches an important user requirement: The customer wants a reliable and robust car with minimal maintenance. As a consequence, a car manufacturer must **consider the aging effects** of all components inside the car, i.e., of active as well as passive parts such as, e.g., capacitors. The temperature changes and long activity cycles are particularly challenging. The PCB design, selection of electronic parts, and design of the communication system must take this into account. Furthermore, the long lifetime affects the supply chain. Suppliers must guarantee the **availability of (replacement) parts** for a significantly longer time than is usual in the consumer or IT industries. A minimum of 15 years is standard. This is not different for Automotive Ethernet than for any other in-vehicle networking technology. However, it might be new to traditional Ethernet suppliers.
- **Upgradeability:** The long lifetime of cars directly implies another aspect: Car owners generally keep their cars for several years (see e.g. [10]). It is therefore desirable to be able to update the car, especially in the infotainment and consumer interface domain. This might be done by exchanging ECUs to newer versions or by upgrading the software of a system. Furthermore, a buyer of a second-hand car might want to add features the previous owner did not care for. The possibility of adding functions later therefore increases the resale value of a car. In some cases, updates can be realized via software update or ECU exchange. In other cases, the desired function can only be added with an additional ECU. In order for this to be possible, the originally installed in-vehicle network would need to have been designed to allow for these later extensions. The implications are quite different for the different in-vehicle networking technologies. In the case of Automotive

Ethernet, one existing ECU in the network would need to have a switch and a currently unused PHY in order to allow a new ECU to be integrated into the network. To provide such an unused PHY in a network, however, adds costs (see also Section 8.3.2.2). Smart new concepts for ensuring the upgradeability must thus be developed for Automotive Ethernet.

- **Flexibility:** All major car manufacturers sell a range of different car models targeting a range of different market segments. For this to be feasible, it generally means some reuse but also changes between different car models in order to allow for differentiation and innovation. The reuse can be an ECU, the communication technology, or (parts of the) EE-architecture. This also requires the respective flexibility in the in-vehicle network. If a new ECU cannot be put on an existing shared bus, because the capacity of that bus is already at its limit, generally a new bus must be added, the gateways must be increased, etc. With a switched Automotive Ethernet architecture, the means to extend the architecture and thus the flexibility change (see also Section 8.3.2.2)
- **Long pauses between use:** Even if a car has been parked, i.e., was "off," for a long time, it must be able to start when the customer wants to use it again. With the increasing amount of electronics inside the car, this is a challenge as electronics consume power and thus drain the car's battery, even when the car is not in use. Some units need power because they must always stay alive, such as the ECU that enables keyless entry. Others simply consume quiescent current. One measure is to disconnect the battery from the ECUs when the car is not used (see also Section 6.1). Another is to **limit the quiescent current** of all electronic hardware (directly connected to the battery), i.e., to have (almost) **no power consumption when the car is not used**. The allowed quiescent current for a complete ECU that always stays on the power supply is in the area of a few 100 μA, which means the transceivers must be in the range of 10–20 μA quiescent current. In the consumer and IT industries, Ethernet semiconductors do not have to provide such low quiescent current values. Even if a device has not been used for a long time and the batteries are discharged, the user simply charges the batteries or connects the units to an electrical outlet. The consequence of automotive use of Ethernet is that ECUs with standard Ethernet components will consume too much quiescent current. Respective ECUs either cannot stay connected to the battery when the car is not in use, or other intelligent means must be provided for keeping just those parts that are needed in the off status connected to the power supply. This is one of the many criteria that must be considered in the EE architecture.
- **Low power (fuel, battery) consumption when the car is in use:** This requirement as such is not so different from the IT or consumer industries. All users would like small electricity bills, long battery life, or stand-by times. Ensuring low energy use in all devices is, therefore, a common concern in all electronics. From this perspective, the car manufacturer simply has a very complex product, which consists of a large number of parts from a large number of different suppliers and where each part must be efficient. With power saving, the car manufacturers have a rigid goal: They must meet government requirements in terms of CO_2 reductions for cars with

combustion engines. For electric cars, CO_2 emissions are not the concern, but customer satisfaction is. Satisfaction increases with the number of kilometers the car can drive, before it needs to be recharged. User awareness of power consumption will increase with electric cars, as drivers might be inclined to actively switch off the entertainment, or even air-conditioning system, in order to be able to drive some additional kilometers.

Chapter 6 discusses the various possibilities for reducing power consumption with the choices made in the in-vehicle network by either saving weight and/or by saving the power consumption (during runtime). Realizing "partial networking" [11] in particular also affects the Ethernet network architecture. If only those ECUs currently needed are alive, but all communicate in one vast Ethernet network, all switches in the communication paths must remain alive as well, even if the rest of the ECU is not required. This is not ideal because of the power budget of Ethernet components. So, determining which ECUs should include an Ethernet switch (as opposed to a simple endpoint connection) and serve as part of the fabric of the Ethernet network must be considered from this perspective as well. The simplest solution is, if a unit is not continuously needed during runtime, it does not contain a switch. If it does contain a switch, for other reasons, then, of course, the switch should have a separate power supply such that it can stay awake independently of the rest of the ECU.

It is by no means readily apparent which ECU can be deactivated in a partial networking situation. The acoustic park distance control is a good example to illustrate this. The acoustic signal is generally generated in the multimedia system, i.e., the HU. If this was completely taken off the power supply, because the user preferred not to have any entertainment, the park distance control would no longer work. For this reason, it makes sense to install coordination that controls the activation and deactivation of certain functions.

- **Fast availability:** In the consumer electronics domain, it is normal that consumers wait, e.g., for the availability of a mobile phone after complete switch off. Even to switch on a TV takes today almost as long as it did during the time of vacuum tubes. However, the user of a car expects all functions to be available immediately. This is the case when the car is being started, and it is also the case when ECUs have been put to sleep during the runtime of the car for the power saving reasons just discussed. Customers should not be aware that this is even happening. Once the car has been started, it must be able to go immediately. But, even to start the engine, the in-vehicle network must be available, as, e.g., the electronic immobilizer system requires communication with various ECUs to exchange and calculate the respective certificates before the engine can start. Because of the previously discussed requirement for low quiescent current, it is not possible to simply leave all needed ECUs connected to the power supply. This leads to the requirement of a **short start-up time**. After 100–200 ms the in-vehicle network must be ready to communicate, in order not to cause any noticeable delay in the wake up of the systems beyond. However, for the complete multimedia or navigation system these short start-up times can often not be reached and a user might have to wait a few seconds (see Table 7.2 in Section 7.1.2.4).

- **On–off changes:** An Ethernet-based IT network does not expect to face extreme temperatures or even a lot of temperature variations. Additionally, the whole network will be shut down only occasionally for maintenance reasons and equipment upgrades. In contrast, an in-vehicle network is started and shut down frequently, often several times during a single day. Short start-up times are thus crucial (see the previous bullet point and also Section 7.1.3). This requirement is the same for Automotive Ethernet as for all other in-vehicle networking technologies.
- **Quality and availability of power supply:** An Ethernet-based IT network is expected to be constantly and evenly powered. Even though a low electricity bill will also be of some concern in an IT environment, the risk of a power supply with significantly varying voltage or the risk of running out of power altogether is small and, if it happens, it is generally part of a more serious problem out of the network provider's control. In cars, this is different. The power supply is always limited. When the car is parked for a long time, the limit is the battery capacity and charge status. When the car is running, the limit is the amount of fuel in the tank (in combination with the battery charging capability). Additionally, low-voltage transients might occur in the case of old batteries, low temperatures, and engine starts. This can cause resets, which in return results in the necessity of fast recovery. Furthermore, the power needed will vary. Depending on the actual use of the car, more or fewer functions are connected to the network and the power supply. In the service case, just one ECU might be powered. Without giving any further details on how these issues are addressed in the EE architecture – all car manufacturers have different requirements and capabilities in this respect – they must be taken into account when designing in-vehicle networking and using Automotive Ethernet.

8.3.2 EE Architecture Related Choices

8.3.2.1 Centralized versus Distributed

During system design, there are two principal approaches on how to partition function blocks: An integrated/centralized approach and a distributed approach. In the integrated approach, many function blocks are designed into one ECU, while in the distributed approach only a few function blocks are designed into any one ECU.[3] Speaking in general terms, this means that to achieve the same functionality, the integrated approach results in fewer ECUs and a smaller communication network than the distributed approach. Provided the customer is willing to pay for the functionality, this generally makes the integrated approach less costly. At the same time, the integrated approach is less flexible and less scalable. Everyone is offered the same functions. Even if some of the functions are activated by software, so that it is still possible to sell functions as options, every customer will have the same hardware provisions in the car. Table 8.2 provides an overview of the differences.

Overall, one approach is not necessarily better than the other. The preference also depends on the market segment a car manufacturer wants to address, either with a

Table 8.2 Centralized/integrated versus distributed EE architecture

Centralized/integrated EE architecture	Distributed EE architecture
Fewer, more complex, and bigger ECUs	More, less complex, and smaller ECUs
Fewer connections in the in-vehicle network, which potentially makes it less complex	More connections in the in-vehicle network, which potentially makes it more complex
Does not scale well	Better scalability of overall car functionality (potentially over different models)
Fewer suppliers, which might not be the best for all integrated functions	For each ECU a different/optimum supplier can be chosen
Likely lower costs if customers are willing to pay for all functions	Likely lower costs if customers are only willing to pay for some functions

specific car model or in general. As a rule of thumb, integration is more common for basic functions, for functions that have been on the market for a long time, which have higher take rates, and/or for cars and functions in the lower price segment. Distribution is more likely if the function is an option, is new, and/or addresses the top end of the market. Naturally, there are other criteria for partitioning functions on ECUs than just the general preference on an integrated or distributed architecture, e.g., the physical or logical closeness of functionalities [1].

Furthermore, the choice for distributed or centralized is a matter of preference and trend. The upcoming automated driving is a good example. To start with, the intelligence will be centralized (and doubled for redundancy reasons). This centralized intelligence likes raw data from the sensors that have no intelligence themselves. Once there is more experience with this type of function, it might well happen that system optimization will lead to distributing some of the intelligence back to the sensors.

Automotive Ethernet supports both choices. In case of a more integrated architecture there might be fewer but more bandwidth hungry communication links, which Ethernet supports. In a distributed architecture there might simply be more communication that can be handled flexibly with Automotive Ethernet. Ethernet allows for the necessary scalability in the data rate by exchanging the PHY technology (only) and supports various different option configurations over time and different car models in a cost optimized way (see Section 8.3.2.5).

8.3.2.2 From a Bus to a Switched Network

The traditional in-vehicle network technologies (CAN, LIN, FlexRay, MOST, see Section 2.2) are all bus systems, meaning that the available bandwidth is shared between all connected units. Most often, a bus system has a logically linear topology; with MOST being an exception, using a ring topology. Owing to the simplex nature of the optical transmission MOST uses, closing the line to a ring saves having to use two POFs between neighboring units. Instead, it allows all units to communicate with each other, even though communication is in only one direction around the ring.

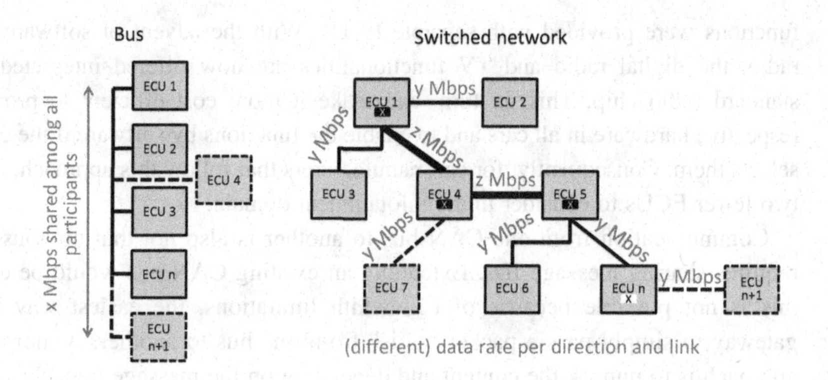

Figure 8.2 Important topology and extendibility differences between a bus and a switched network, added units are marked by dotted lines, the boxed "x" marks units with switches, the white box indicates that here a switch would need to be added

A fundamental property of a bus is that, in principle, all units can listen to and thus receive all data that is available on the channel. There are different methods for deciding whether a receiver actually does process the data, and traditional in-vehicle networking technologies do deploy a variety of such methods. There can be a message identifier, from which a receiver decides whether data is of interest (CAN). A predefined scheduled may be followed (LIN, FlexRay). An address can unambiguously identify the receiving unit (MOST).

If more units must be included in the communication on the bus, it seems straightforward to simply attach a new unit to the bus (see also Figure 8.2). Such a layer 1 connection requires that there is enough data rate available to support the traffic to and from the new unit. Repeaters work on layer 1 and might be used to extend the physical range of a bus. Hubs might be used to connect various users at one point. In the case of FlexRay, a star coupler is needed when the scale of the network increases such that the propagation delays of the signals become too large. Additionally, at layer two it must be ensured that the new unit is included in the channel access scheme, which might require reprogramming all of the units that share the same bus (e.g., in the case of FlexRay). In the case of MOST, the order of units must be observed.

Should there be reasons why a new unit cannot be connected to the same bus, e.g., because the data rate of the bus is not sufficient, a proxy/gateway is needed to transfer data from one bus to another. Traditionally, this gateway function between two busses of the same technology is performed on the Network or Transport Layers. However, because of the timing requirements and the availability of the star couplers, gateways between two FlexRay busses are not very common. For MOST, this is also unlikely not only because of the complexity (see also Section 3.1.2.1), but also because the use case is simply not probable. MOST focuses on the infotainment domain, which is a typical example of an application area in which many new functions evolve (causing new ECUs), but also in which these functions quickly become legacy (and are integrated with other ECUs). For example, to begin with, digital radio or digital TV functions were provided with separate ECUs. With the advent of software defined radio, the digital radio and TV functionalities are now offered integrated onto a standard radio chip. This, in turn, can make it more cost efficient to provide the respective hardware in all cars and to enable the functions by software if the customer selects them. Consequently, for car manufacturers that follow this approach, there are two fewer ECUs to consider in the infotainment domain.

Communication from one CAN bus to another is also not that obvious, as this requires sharing message IDs. Extending an existing CAN bus would be easier. If this is not possible because of bandwidth limitations, the easiest way is for a gateway to simply pass a packet "as is" from one bus to another. A more generic approach is to unpack the content and depending on the message, repackage it onto a new bus, which then might be FlexRay, LIN, MOST, or even Ethernet as well. Naturally, this second approach is more demanding of resources, causing more latency, too. In general, gateways require effort, which increases with the time criticality of the applications. Communication within one networking technology is thus preferable.

This motivates a so-called domain architecture; with the idea that most communication happens within one domain. However, the data rate provided by a certain bus might not be sufficient. Additionally, it is not always possible to avoid cross-domain traffic. When thinking of new concepts such as automated driving – which require a significant amount of redundancy – this is obvious. But there are many examples that are a lot simpler. For instance, displaying the picture of a rear view camera on the head unit requires cross-domain communication between the driver assist domain and the infotainment domain.

With Automotive Ethernet the basic concepts of the architecture are completely different. First, other than for 10BASE-T1S multidrop, which we can assume to be at the edges of the network, every connection is Point-to-Point (P2P), meaning that only two units are ever attached to the same link segment. The networking happens on layer two (or three, see Section 7.3.3). On layer two the switches pass data on, depending on addresses (in Figure 8.2 the switches are marked with an "X"). This is key for networking but completely new in the automotive industry. It supports all kinds of topologies and no gateways are needed from Ethernet to Ethernet, even if various different PHY technologies and speed grades are used (for examples see Section 5.2 to 5.6). Proxies/gateways are only needed between Ethernet and other in-vehicle networking technologies.

Extending the network requires adding another port to a switch or exchanging a PHY with a two-port switch (see also Figure 8.2). Because the links are always P2P this actually increases the overall capacity inside the network. Of course, the limits of the available data rate also need to be observed when extending an Ethernet-based network. It is better for the overall data rate to attach a unit directly or close to the main communication partner and not at the other end of the network. However, with the data rate so much higher to start with – 100 Mbps per link and direction – and technologies for even higher data rates being developed (see Section 5.6), the right architecture should always be able to provide enough bandwidth. This, in turn, gives more flexibility for network optimization and potentially allows prioritizing network optimization criteria differently.

8.3.2.3 Unicast, Multicast, and Broadcast

As was explained in Section 8.3.2.2, every unit attached to a bus will, in principle, see all packets transmitted on that bus and the bus therefore has native broadcast behavior. It is up to the receiving units whether the channel is used in a broadcast-like fashion, i.e., one where everybody reads all packets, or not. In a switched Ethernet network, this is different. Using P2P with only two units ever directly connected, the core transmission mode of a switched Ethernet network is unicast. However, an Ethernet system also supports multicast and broadcast for specific use cases and/or with the use of the specifically reserved MAC addresses.[4] Also, a switch performs broadcast-like flooding every time it receives a packet for which it does not know the packet's MAC destination address and where to send it on. The use of multicast and broadcast is frequently the case during start-up, when the network is used for the first time. Another example for multicast are IEEE 1722 audio video packets, which can be multicast within the AVB cloud (see Section7.1).

The obvious risk of broadcast and multicast messages is overloading the network, especially if the network has redundant paths (i.e., loops). In theory, any redundancy is taken care of with the Rapid Spanning Tree Protocol (RSTP) at initial start-up, but there is a remaining risk, not only of a high data load, but also of malfunctions. Even though there is a theoretical difference between broadcast and multicast, switches generally just flood both types of packets to all outgoing ports except the one via which the packet was received [12]. An exception are AVB capable switches, which make a hardware supported distinction between ports connecting to other units that are part of the AVB cloud and ports connecting to units which are not.

In Automotive Ethernet another effect is important. A PHY in an Ethernet-based network receives every packet it detects on the link and passes it to whatever processing capability it finds attached to its MII interface. In an end node this is likely to be a µC; in a middle node this is a switch. An end node µC will continue processing every packet that contains its own address, every broadcast message, and probably, depending on the implementation, most of the multicast messages. Because there is very limited filtering at the MAC level, it is up to the application to decide what to do with the content. If all packets in an Ethernet network were broadcast or multicast packets, e.g., because a CAN bus was emulated over the Ethernet network, the application on the µC would look at the message content of every packet before it could be dismissed. The µC of a small ECU, designed to process, e.g., 10 Mbps of data, would be congested just by the network functionality.

In an Automotive Ethernet network, it is thus better to avoid the use of broadcast and multicast. Under the assumption that an Automotive Ethernet network will always be small (in comparison with an IT network), fanouts, i.e., the use of multiple unicast transmissions instead of one multicast transmission, are likely to be preferable. If done with care, this is not only less strenuous for the network but also scales well, in case the system grows.

8.3.2.4 Throughput Optimization with Automotive Ethernet

Crucial for designing the communication network is that all data arrives at the receiver within the required timeframe. In order to ensure this, it is necessary to know all communication paths between units and the amount of data to be transmitted. Normally, this communication requirement is one of the outputs of the EE architecture. The communication architecture decides accordingly on the in-vehicle networking technology to use, how many gateways to include, and where/how to connect the units to the network in a way that it scales across option selection, car models, and functional updates. Functional updates are generally released once a year, but, of course, their requirements are not known when the car is being designed to start with. The in-vehicle network must therefore allow for future growth in data rate requirements.

Tools and simulation are used when designing the communication network, especially to determine how much of the available data rate is consumed. Data rate consumption is not always intuitively obvious, as the communication rarely follows deterministic and/or cyclic patterns but also contains event driven portions. Deterministic approaches do exist, but often result in the establishment of maximum possible traffic demand assuming the occurrence of bursts. Deterministic approaches thus allow the calculation of the maximum required buffer space in an ECU.

The rules behind communication network design that enable the determination of maximum loads, prioritization, and maximum delays, are not specific to Automotive Ethernet. Traditional in-vehicle networking technologies need them as well and they are thus well established in the automotive industry. Examples of existing tools evaluating in-vehicle network load performance include PREEvision (Vector), SystemDesk (dSpace), SymTA/S, and TraceAnalyzer (Symtavision) (see also [13]).

Naturally, the tools need(ed) to be extended with the communication paradigms of Ethernet. To compute reserves for the traditional technologies is very complex by comparison and "higher" data rates even require the inclusion of wave theory. The simulation of the bus physics of an Ethernet network must take the switched architecture into account. Between switches there are only P2P links, and the computation of the traffic load on those links is relatively straightforward. The critical design element is the load within the switches. The in-vehicle network design must ensure that a certain switch in the network does not become a bottleneck because of buffers that are too small. Depending on the tooling preferences within the car manufacturers, the respective tools have been/need to be extended accordingly, including AVB (see e.g. [14]).

8.3.2.5 Cost Optimization in a Switched Network

Costs inside a car are comprised of various elements (see Figure 8.3 or, e.g., [1]). The obvious cost elements are the costs of ECUs and harness as well as the costs associated with assembling the hardware into a car. The derivation of sound values for these elements is daily business within every car manufacturer and are an integral part when deciding on ECUs or the in-vehicle networking system to use. Additionally, every ECU or in-vehicle networking technology also causes costs in qualification and later warranty. However, those are difficult to nail down at the time of decision and are

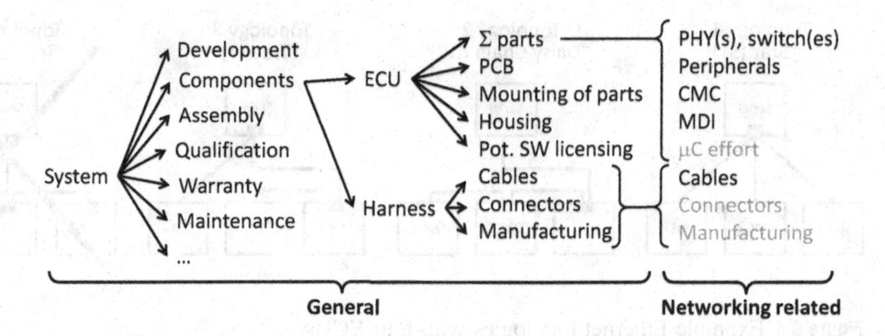

Figure 8.3 Overview of cost elements; light grey indicates the elements needed for the competitive comparison between technologies but not needed for the topology optimization [1]

thus generally not considered this early in the process. Even more difficult are those aspects that have monetary impacts but are difficult to measure such as the money saved by being more flexible or future-proof with a certain architecture or the marketing effect achieved by offering a certain function. These aspects, important as they are, are thus not (yet) included as concrete values in the cost assessment.

With respect to Automotive Ethernet, two types of cost-related evaluations can be expected. The first is a direct comparison between Automotive Ethernet and another networking technology. Most likely, Automotive Ethernet will be compared against technologies with similar use cases such as SerDes links for the higher speed grades and FlexRay or CAN(-FD/XL) for the lower speed grades. In this case, all costs directly related to the choice of the networking technology are added up (see also Figure 8.3): Cabling, connectors, harness manufacturing, transceiver chip costs, switches, peripherals, filters, CMC, SoftWare (SW) licensing fees,[5] and potentially additional effort in the μC, e.g., for compression or providing an MII interface.

The second evaluation is related to the actual Ethernet topology. Because Ethernet is not a bus but a switched network with P2P links, criteria new to the industry must be included in the optimization. This section will focus on the explanation of those, also, because a correctly optimized Ethernet network can make a big difference in the cost comparison between two competing technology solutions. As shown in Figure 8.3, the system costs consist of a significant number of other factors as well. The authors believe that Automotive Ethernet can impact those and more; up to changing the way cars are developed. To evaluate these scenarios is, however, not detailed further here. This section concentrates on the impact of the switch.

The switch is the main element that is new in the optimization. The switch allows for flexibility when selecting the topology, but also adds costs, and the location of switches thus needs to be considered carefully. The following describes the effects with a simple example of four "Ethernet ECUs," of which one is available in all cars and three are offered as options. For this example, it is assumed that the customers can select each option individually, i.e., there are no complement selections required or mutual exclusions of options. Figure 8.4 shows the three topologies that are, in principle, possible under this assumption. Key is that all topologies have enough

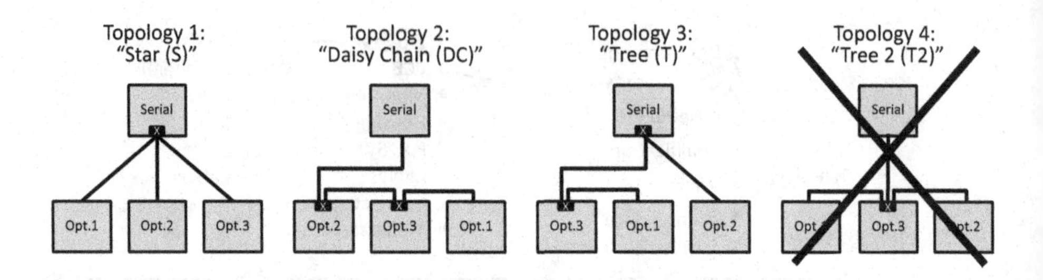

Figure 8.4 Example Ethernet topologies with four ECUs

PHYs and switch ports available to support all possible subsets of options selected. The "Tree 2 (T2)" topology does not allow for this.[6] If the customer chose only the outside two options and not the one with the switch, the serial ECU would not have enough ports available to connect them. For the Daisy Chain (DC) and the Tree (T) topologies, there are multiple variations of (optional) ECU orders possible. Figure 8.4 shows one example of each topology. When comparing the different types of topologies it thus makes sense to first select the optimal ECU order for each. In the example calculated in the following, the topology internal optimization was completed upfront and is not explained in the text. It is a fundamental requirement for all topologies that the required data rate is supported by all of them, which also needs to be confirmed before making the cost calculation. The numbering of the optional ECUs chosen in Figure 8.4 represents the optimum selection for the example discussed in the following.

There are some principal differences between the three topologies independent of the specifics of a certain car model: The number of PHYs and switches each solution needs, and the number of harness segments that must be defined per topology (see Table 8.3). Each harness segment describes one possible element of a harness that a car manufacturer must define for the harness manufacturing, depending on the options selected. If, in the star topology, Options 2 and 3 are not selected, nothing changes for the connection to Option 1. The overall number of link segments is three. If, in the daisy chain, Options 2 and 3 or only Option 3 are not selected, Option 1 requires a new link segment in order to be connected to the network. The overall number of segments is thus six. Table 8.3 compares the different topologies. The maximum length of cabling in the topologies requires the data from a specific car model. For the values shown in the left-hand part of Table 8.3 the distances between the units as defined in the right-hand part were assumed.

At first glance, the DC topology looks less favorable than the T topology, as both the summed up cable lengths as well as the number of link segments that must be defined are larger for DC than for T. However, in T, one of the switches is in the serial unit, i.e., in every car, independent from how many options the customer buys. Without knowing more details about the take rates of the options and their take rate combinations, it is not possible to say which topology is the most suitable. For the star scenario, it is important to know, for example, how often the customers choose how many of the options, i.e., how often only one, two, or three ports are needed in the

Table 8.3 Basic comparison values for the different topologies (left) using the example link distances on the right

Part	S	DC	T	[m]	Serial	Opt. 1	Opt. 2	Opt. 3
# PHYs	3	2	2	Serial	0	4	2	1
# 2port* switch	0	2	2	Opt. 1		0	6	3
# 3port* switch	1	0	0	Opt. 2			0	3
# segments	3	6	4	Opt. 3				0
Max. $\sum$ lengths	7	8	6					

Note: The switch will have (at least) one internal port. This is why a "2port switch" is sometimes called a "2+1port switch" or even a "3port switch"
* Two/three integrated PHYs.

Table 8.4 Example of average networking induced costs per vehicle depending on topology and car model take rates (the input data for this table is given at the end of the chapter)[7]

	Star	Daisy Chain	Tree
Model A (high end)	9.27	9.75	9.37
Model B (upper class)	7.70	7.34	7.30
Model C (middle class)	4.91	2.78	4.04
Volume dependent average	5.44	3.64	4.66

switch. This determines how often one, two, or all three ports of the switch are not needed and thus represent an unnecessary hardware provision. Note that it is not an option for the car manufacturers to simply make three variants of an ECU, each with a network interface of a different size (switch with three ports, two ports, or PHY only).

A car manufacturer can only handle a certain number of variants for each ECU, especially in logistics and assembly, but also in development, qualification, and purchasing as too many variants are problematic. Often, there are important functional reasons to have them. For example, a HU needs to have different variants for different countries/regions. On top of that, like other ECUs, a car manufacturer might want to offer a "light," "mid," and "full" version. If each of those versions then needed to be provided in versions with different Ethernet interfaces, this would quickly become unsustainable.

Table 8.4 shows the result of cost calculations for three different cars. Model A is a high-end car, Model B an upper class car, and Model C a middle class car. The table shows that, with a simple cost estimation, a different topology would be optimum for each of them. If the goal is to support only one topology in all car models, the daisy chain topology is the most cost efficient one in this example simply because the switch, which is a costly hardware element, is needed only when the respective option is selected. Note that a topology optimization might include other criteria, such as extendibility, discovery, start-up and flash, or power consumption. In a daisy chain, flash updates and start-up must be planned more carefully; especially if some of the

ECUs in the chain are options. Also, the costs for the additional harness segments have not been included. If too unfavorable from a cost perspective, it might be better to have a solution with as few link segments as possible.

Naturally, the example was very simplified and a real-life example is likely significantly more complex. Its purpose was to explain the additional elements needed to optimize an Automotive Ethernet Architecture, in comparison with traditional in-vehicle networking technologies.

8.3.2.6 Zonal Architecture

The in-vehicle EE architecture is continuously changing. While no two EE architectures are the same between two different car manufacturers (unless they cooperated on this, of course), certain overall trends can be observed in the architecture development.

At the beginning, ECUs were simply connected as needed. As the ECUs with software additionally had to be diagnosable and updatable, the respective ECUs needed to be connected to a (central) gateway that provided access. As most communication stayed within the functional domains, this automatically led to an EE architecture with a central gateway that has a number of domain-specific busses connected. Each ECU represents a specific application, connected by an application-specific bus, with a signal-based communication that mainly goes from one transmitting unit to one receiving unit [15, 16].

With increasing digitization (more functions, more ECUs, more software), a trend followed that saw an even stronger domain focus with the introduction of powerful domain ECUs. Common examples are the Head Unit (HU) for infotainment and a powerful ECU that "overlooks" the driver assist functions (especially with the trend toward autonomous driving). But other domain head ECUs are possible and are used in the industry. Between those powerful "domain head ECUs," most EE architectures foresee an Ethernet backbone that interconnects these ECUs (see e.g. [17]). At the time of writing, most EE architectures recently put on the street in series production cars were somewhere in between these two trends. This is possible, because the shift can happen gradually/per domain as the basic communication principles stay the same: The ECUs are still application-specific, within the domains the communication buses are also still application-specific, and the communication remains predominantly signal-based and 1-to-1. What changes is the communication to the domain head ECU (many-to-1) as well as the communication between the domain heads (many-to-many).

Which trend will persevere with respect to future EE architectures is open. Keywords are, e.g., cloud-based [18], centralized [15, 16], software defined [19], and zonal [19, 20]. Key to all these architectures is that they dissolve the functional domain boundaries on the physical level, all with a slightly different focus. Automotive Ethernet is the networking technology that supports all of them. VLANs allow for the separation of functional domains on a virtual level and eliminate the need for physical separation (see also Section 7.2). The inherent support of a service oriented architecture with the service oriented communication (see also Section 7.4) paves the path toward a software defined car that separates software from

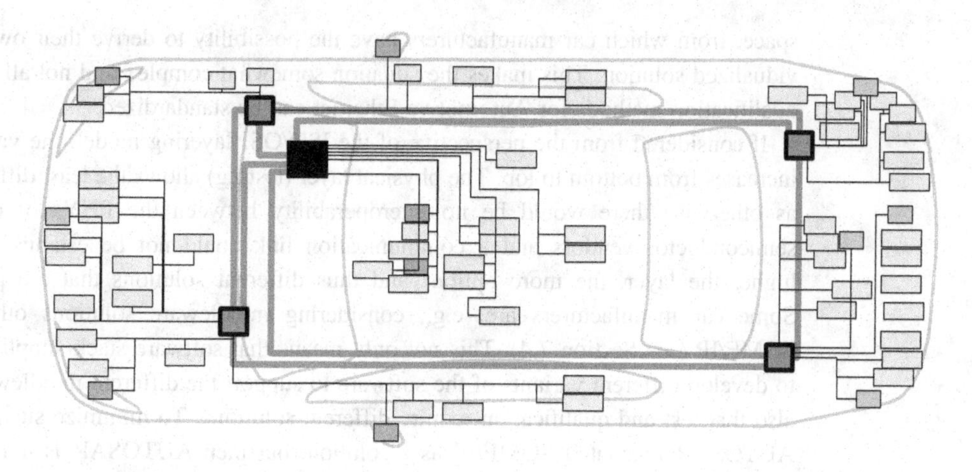

Figure 8.5 Basic principle of a zonal architecture

hardware and supports the software virtualization that is also needed for a cloud-based computing approach. The different PHY speeds scale the system so that any data rate needed can be supported efficiently. Switches allow the formation of whatever topologies are needed. The different QoS mechanisms available with TSN allow different data types to share the same link without loss of quality.

The core idea of the zonal architecture focuses on yet another aspect: The optimization of the wiring harness itself. The function of a car mandates the physical location of many of the ECUs inside the car. In traditional EE architectures, it thus happens frequently that ECUs located at the opposite ends of the car share the same bus and that the wiring is laid across the complete car. The length of a bus thus often reaches a length that is a multiple of the car length itself. Instead of (functional) domain head ECUs, the zonal architecture has "zone head ECUs." These zone heads are connected among each other via a high-speed Ethernet "core" bus. In every zone, local harness elements connect the ECUs in that zone (see also Figure 8.5). This significantly reduces the length of the wires that need to be put inside the car, while the zonal harness elements would allow for an automated harness production. ECUs from the same functional domain situated at different ends of the car would potentially have their communication tunneled over the Ethernet backbone, should the communication inside the zone use one of the legacy buses.

8.4 Test and Qualification

A car is a complex product. In order to ensure its correct function, test and qualification throughout the complete development and production process for all components are thus extremely important. Automotive Ethernet represents one of the many components to which this applies. What must be taken into consideration for Automotive Ethernet is that Automotive Ethernet in itself provides a large solution space, from which car manufacturers have the possibility to derive their own, individualized solution. This makes the situation somewhat complex and not all test and qualification methods for Automotive Ethernet can be standardized.

If considered from the perspective of the ISO/OSI layering model, the variability increases from bottom to top. The physical layer (testing) allows the least differences, as otherwise there would be no interoperability between the PHYs of different semiconductor vendors and a communication link could not be established. The higher the layer, the more options and thus different solutions that are possible. Some car manufacturers are, e.g., considering middleware solutions other than SOME/IP (see Section 7.4). This not only means that software stack suppliers have to develop different variants of the software to support the different middleware, but also that test and qualification require different solutions. To minimize such effects, AUTOSAR mandated SOME/IP as a solution, but then AUTOSAR is not used in every ECU.

This section will give a brief overview of test and qualification for Automotive Ethernet, approaching the topic from two sides: The tooling (see Section 8.4.1) and the procedures used to arrive at a tested and qualified product (see Section 8.4.2).

8.4.1 Tools

For any new in-vehicle networking technology, the availability of suitable tools is decisive, and it is necessary to ensure the availability of such tools very early in the development process. The tools needed during development are, ideally, available when development starts. However, these tools also need to be developed and tested before they can be used. It is highly recommended that any company creating a new standard, or creating their own solution that deviates from a standard, establishes partnerships with tool vendors very early in the process. In the authors' experience, the best tools are developed when the tools developers are directly involved in the system design or closely interact with the system developers. The Automotive Ethernet pilot project discussed in Section 3.4.2 was also used for the purpose of establishing the tooling infrastructure (see [13] for some of the output).

For Automotive Ethernet, not all tools need(ed) to be developed from scratch. Some reuse was possible from the IT industry, and some from the automotive legacy buses. All tools can exploit the ISO/OSI layering separation provided by Ethernet. In the following description, a distinction is made between the tools that support the development side of the v-cycle (see Figure 8.1) and tools that support the test side of the v-cycle. This includes tools used during the long time during which cars need to be maintainable once they have left the production site.

For development, e.g., tools are needed that enable the configuration of communication and the network, tools that simulate or calculate traffic volume, tools that support network design along QoS requirements, and tools that trace actual communication. An example of the latter is Wireshark, which allows the universal tracing of IP-based communication and was well established in the IT industry before being used in the automotive industry as well. For the test and validation side shown in Figure 8.1, each item is discussed separately in the following, highlighting the differences caused by Automotive Ethernet:

- **Component-level testing:** Tests at the component level evaluate the functionality of an ECU as a standalone component. However, many functions provided by ECUs require data from the in-vehicle communication network. In order to test the functionality of an ECU as a standalone component it is thus necessary that communication data is available during testing without needing the other ECUs. With parts being supplied by various different Tier 1s this is essential for the development of a car. In the automotive industry, tools for "rest/residual bus simulation" are used for this. These tools generate all data the ECU requires from the communication system. The introduction of Automotive Ethernet required substantial extensions of these tools. Not only are the communication patterns as such different, some of the communication paradigms change as well. It is no longer sufficient to just simulate the output data of other ECUs and provide it as input data to the ECU being tested. An Automotive Ethernet ECU needs to be registered as a new participant in the overall communication network. The ECU interacts with the partners in terms of service discovery or other method calls (see also Section 5.3). The rest-bus simulation tools must support all of this.

  In addition to enabling communication with the ECUs that are not physically available, the output of the ECU under test onto the communication network needs to be tracked and evaluated for correctness and completeness. Naturally, the respective tools need to support Automotive Ethernet communication. To aid testing at the ECU level, an Ethernet ECU test specification was made available by the OPEN Alliance [21].
- **Network-level testing:** These tests evaluate all functions that are directly related to the communication network. In the first step, the communication behavior of each unit is tested stand-alone for these functions with the help of the rest-bus simulation described in the previous bullet point. In the second step, several ECUs are connected. The component test of the previous step and the integration tests of the consecutive test both evaluate the (customer) functions of the ECU. These are excluded from the network-level testing unless they are necessary for locating erroneous behavior of an ECU in the network functions. This distinction is important because, without knowing that the communication network functions correctly, it makes no sense to test the communal functionality of the ECUs in the integration tests. Tests performed at the network level comprise, e.g., start-up and restart, shut down and sleep, load tests under expected circumstances, load test in the case of bursts, and behavior in the case of network overload.

  On the tool side this requires tools that generate the respective network traffic and loads and, naturally, these must support Automotive Ethernet in order to be useful with an Automotive Ethernet network. However, as Ethernet PHYs and switches need to be tested outside of the automotive world as well, the change and challenge is small with tools and experience already available. Furthermore, tools for data logging and evaluation are needed, which can be the same as for the

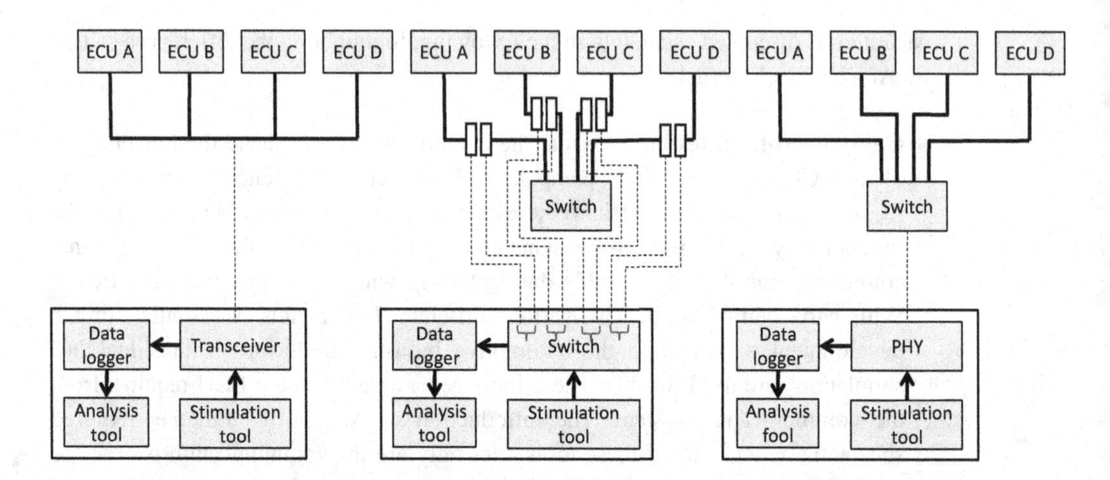

Figure 8.6 Data logging on a bus (left side) and in a switched network (middle and right side)

component-level tests. The change in the case of network tests for Automotive Ethernet is not the tools as such, but the way of adding the tools to the network (see Figure 8.6). When logging the data on a bus, the logger can be just another participant of that bus (left-hand diagram in Figure 8.6). In a switched Ethernet network, however, the traffic flow on every link can be different. To track all communication in a complex Ethernet network either each link needs to contain a transparent logger (middle diagram in Figure 8.6) or each switch has to provide an extra port that allows mirroring the communication to the data logger attached (right-hand diagram in Figure 8.6). With the costs of Ethernet ports, such a "mirroring port" adds costs to the ECUs. Breaking each link requires more complex test setups and impacts the timing behavior of the Ethernet communication. Whether the timing impact matters or not depends on the exact use case and is up to the tester to decide.

Tools for logging an Automotive Ethernet network have been made available. One other aspect that needs consideration in this context is the amount of data that is being logged. The considerations are two-fold. First, with video being analog or on distinct LVDS links, there was no temptation to log all video data in a network. With Ethernet, there suddenly is. Within a very short time this can lead to tremendous amounts of tracking data. It is important to have the ability to distinguish in the data logging between control traffic, which is likely the focus of the tests, and potentially dismiss the application data. Second, if the middle set-up shown in Figure 6.5 is used and, e.g., ECU A communicates with ECU D, the same content will be recorded twice. Once it is clear that no packets are dropped and that switching errors are no longer a concern, it might be desirable to avoid the redundancy.

- **System-level testing:** These tests investigate the functionality of the integrated system, either of the complete system or of subsystems, i.e., functional domains. The focus is on testing the customer functions against the requirement specifications with all (respective) ECUs available. To handle the complexity, test automation is important; independent of any in-vehicle networking technology used. Additionally, the above-discussed tooling can be used to log the communication data in case erroneous behavior must be evaluated.

- **Final in-vehicle acceptance test and maintenance:** The system tests are an important step toward the overall functionality, but, of course, all required functions have to pass the final in-vehicle acceptance test with all units integrated into the vehicle. This is, again, independent of the in-vehicle networking technology chosen. However, once a car has been qualified for Start of Production (SOP), the need for tests and tools has not ended. The difference is that after SOP, tests are performed only in the event of malfunctions in the field. Key for solving the customer issues in the field are the diagnostic capabilities designed into the ECUs to start with.

  These diagnostic functions follow two different approaches. First, if the component registers an anomaly considered during the development, the component will record the respective error code in its error memory. By reading the error code from this memory, the malfunction is known to the repair personnel. Additionally, there are various test functions available to better locate the source of a malfunction. As explained in Section 2.1 diagnostics was one of the very early use cases for in-vehicle networking.

  A good example is the electric window opener. If a customer complains of a defective window opener, the repair personnel will first read the error memory of the electronics inside the window opener. The error code might say that the motor of the window opener ceased functioning because the motor temperature was too high. The error code does not say whether the cause for it was a defect in the electric motor, or whether the mechanical run of the window was hampered. To find out, a diagnostic function of the external tester could stimulate the opening and closing of the window, while receiving data on the forces on the window. If the forces are in a defined value range, it will be necessary to change the electric motor. If the forces are outside of that range, it will be necessary to repair the mechanics and gaskets inside the door. This example describes a classic repair situation.

  Nevertheless, sometimes a problem arises that cannot be solved by the above methods. The specialists needing to investigate the issue must have access to all information sources available, including the in-vehicle network. So, even if this is a rare situation, it is essential that the in-vehicle network traffic is accessible in customers' cars and that the respective tools are capable of deciphering Ethernet traffic.

8.4.2 Test Concepts, Test Houses, and Test Suites

Good test concepts emerge during development, and not afterwards. The introduction of Automotive Ethernet is a good example of this. Having had a good test strategy as early as 2010 allowed the developers to introduce an almost error free system into series production in 2013. In particular, the physical layer was without errors at project start-up. This meant cost and time savings, not only for the car manufacturer who introduced the technology, but also for the chip vendor responsible for it. Good test concepts are thus crucial when developing complex systems and should always be part of the process. The goal is not only to find errors in the system, but to find them as early as possible. The later an error is found and the more fundamental the error is, the more expensive it is to resolve the error [22]. A good test concept thus means tremendous cost savings.

For the chip vendors the situation is the following. The newer the process technology used for the semiconductor production and the higher the degree of integration, the more costly the mask sets. Extensive upfront simulations and very good test solutions are essential to stay within the planned development costs and time. Without intensive upfront validation, the risk of costly errors found too late would be far too high. However, no matter where in the value chain the effort for test and validation is made, it does not come free, either, but is often overlooked in project plans.

Standardization helps! One problem is the availability and capability of good test and simulation resources. Normally, these are only needed in phases; namely, whenever there is a new development. Once this is completed, the resources are idle again. For this reason, it makes sense to professionalize the topic. And naturally, the business case for all parties involved is better when a standardized solution needs to be validated than when one vendor needs support for their individualized case. There are two ways to professionalize the topic:

1. **Test houses:** At a test house, testing is offered as a service. The offered tests are usually based on existing standards because, generally, only standards offer the chance of enough revenue to make the effort of implementing the tests worthwhile. For non-standard solutions, this is a lot more difficult. The worth of test houses is not always understood, especially by those who just see the costs. The alternative would be to implement the tests locally, which, when wanting to achieve the same quality, generally entails even higher costs. Not to perform the tests would be grossly negligent. As said in the introduction, if a fundamental error is discovered too late, the costs for resolving this increase manifold. As a consequence, customers often require respective test reports before integrating a product. For Automotive Ethernet these test reports might entail tests at the physical layer for compliance, interoperability, EMC (see [23][8]), at MAC level for TSN and (other) switch functions [24], but also at higher protocol and ECU levels [21, 25, 26], in hierarchical order. If the PHY does not function properly, it will not be possible to test the MAC or the ECU. If the ECU is tested first and the problem is in the PHY, then it is difficult to locate and eliminate the problem.
2. **Test suite:** In contrast to the test house, the test suite intends to automate a test system. This test system generally also follows the respective standards or can be adapted on behalf of the customer. A test suite is, therefore, a test tool that is given to the customer of the test suite manufacturer, so that the customers themselves can validate their implementations with this tool and can classify them as ok or not ok. The advantage of this solution is that the customers can use the tool directly at their facilities and can repeat the test as part of the development process. Thus, they can

continuously check the target behavior with the test tool. Such test suites are generally of interest to the suppliers, Tier 1 or Tier 2.

Often, a combination of test suite and test house makes sense. In this case, the supplier tests his product in-house with the help of the test suite before passing it to a test house for confirmation. This generally means that a single test run in the test house is sufficient. It might otherwise be very inefficient to have multiple runs in the test houses in order to find individual errors one by one during the tests there.

As already described in this chapter, the test sector is also based on market economy rules. It is also essential for test houses (and companies in test suite production) that there is a standard or thorough and accurate requirement document according to which the companies can develop their solutions. In the case of Automotive Ethernet, this is much easier thanks to the OSI ISO model and the optimal layer separation as compared with previous automotive solutions. The described advantages of standardizing solutions with respect to test and validation are, unfortunately, all too often forgotten.

8.5 Functional Safety and Ethernet

The objective of functional safety is "freedom from unacceptable risk of physical injury or of damage to the health of people either directly or indirectly (through damage to property or to the environment) by the proper implementation of one or more automatic protection functions (often called safety functions)" [4]. The goal of the respective ISO 26262 specifications is to reduce safety risks in cars to a minimum. The core of the ISO 26262 is the Automotive Safety Integrity Level (ASIL) classification of a function/ECU that is based on severity, exposure, and controllability of certain hazards and risks. If systems receive an ASIL classification, it ranges from ASIL level A (lowest, e.g., for rear lights) to ASIL level D (highest, e.g., for airbags [27]). Those functions/ECUs inside the car that are not safety relevant receive the rating "Quality Management (QM)."

Developing to safety requirements is seen as very critical and no one wants to be responsible for the violation of a safety target. The automotive industry therefore often applies the "end-to-end safety" principle. The "end-to-end safety" principle means that the application must ensure the safety target. The application thus must provide additional functions from the application context that allow for some errors on the communication link or in the communication protocol stack (e.g., in the middleware). The two most common principles for this are the so-called application-based CRC protection and the Alive Mechanism.

1. For **application-based CRC protection**, a CRC is calculated in the application context for a specific amount of data and is then transmitted with this data. The receiving unit will check the application-based CRC at application level. An error on any of the other layers that occurred during the communication will show in the application-based CRC, which means that no further reliability objective is required for all other layers.

2. The goal of the **Alive Mechanism** is that the recipient of data recognizes when the timing of the transmitting application is not correct. This could be that the transmitting application is delayed or even that it has stopped for longer than it should have. The recipient must recognize if the transmitter is, e.g., sending cyclically the same (now out of date) data instead of new information. In many driving situations, the use of out of date data is a safety-relevant error. To prevent this, an alive counter is sent directly out of the application context. With the alive counter, the recipient can recognize whether the transmitter is still "alive." Unfortunately, how (well) this works depends very much on the application and is therefore very difficult to be checked by communication partners in between, such as gateways or converters of data formats.

In some cases, systems are offered up to a security level of ASIL B, even without these mechanisms. But this is very limited and normally not very practical in normal practice. Communication systems, including Ethernet, are one part of the overall safety consideration. At the data link layer, Ethernet provides an end-to-end protection in the form of a CRC with every packet. In case a CRC error is found, Ethernet simply discards the packet and a higher layer (e.g., TCP, or the application) must handle the missing data (e.g., by initiating a retransmission).

End-to-end safety is particularly difficult to realize when the transmitting application uses a different communication system than the recipient. Such a scenario means that somewhere in the system a translation is necessary; generally in a gateway (see Figure 8.7). The gateway must convert the data and thus also evaluate the CRC and alive counter on the receive side of the gateway and then calculate a new CRC and

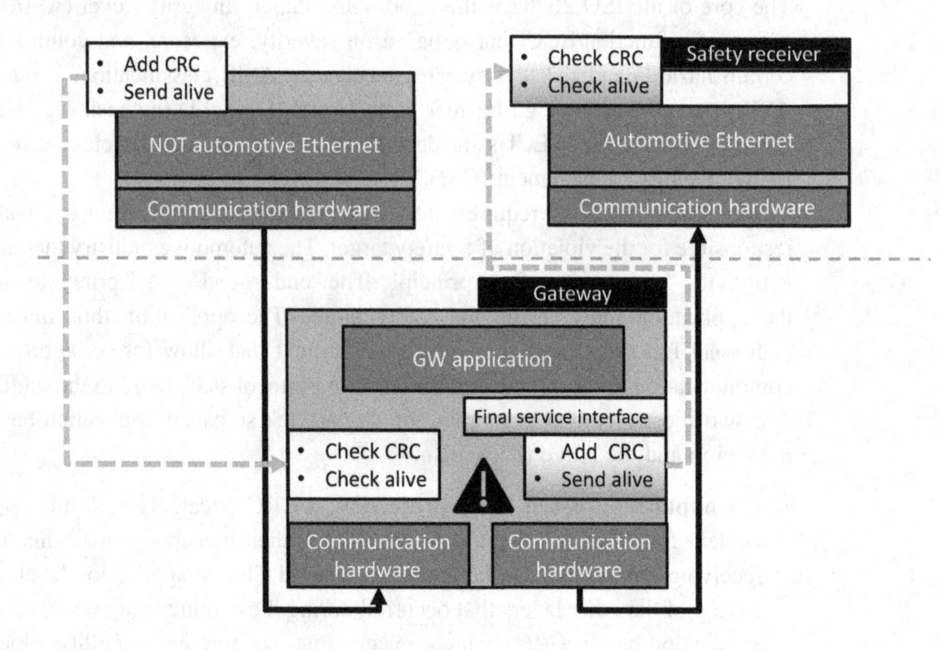

Figure 8.7 Translation of functional safety measures within a gateway

new alive counter for the send side in the gateway. Now the entire safety load is on the gateway and the gateway must have the respective application related knowledge in order to perform the task. This, in turn, results in a very expensive and, unfortunately, not scalable gateway solution. Furthermore, the expectation is that such a translation cannot go beyond a security level of ASIL B. The expectation is thus that such gateways will be used less and less in the future; this is also because many safety-critical use cases are on the rise due to the applications in the field of highly automated driving.

Thus, alternative solutions are required. One solution explored was to allow the cause of an error to be calculated back based on pure gateway algorithms. This did not persevere, because the behavior in the event of an error is strongly influenced by the function underlying this data, which is highly variable and individualized. To solve this generically was, at the time of writing, not possible. The authors therefore recommend the following two approaches:

1. **No data conversion**, but, instead, transfer of all data using the same communication principles from source to sink, e.g., in the form of an all Ethernet/IP transmission. This ensures that the actual data from the source to the sink does not have to be processed by a third party (see Figure 8.8). The challenge with this variant is to create a cost-effective network with which you can compete with currently inexpensive solutions such as LIN or CAN. This was one of the motivations for the development of 10BASE-T1S Ethernet (see Section 5.4).

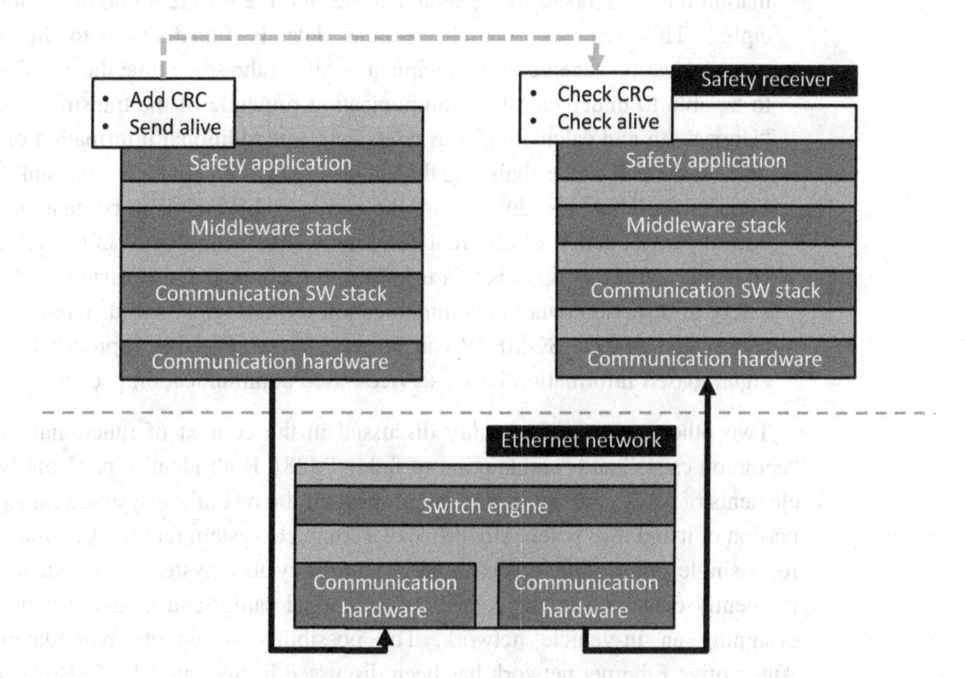

Figure 8.8 Gateway free deployment of functional safety measures in an all-Ethernet communication

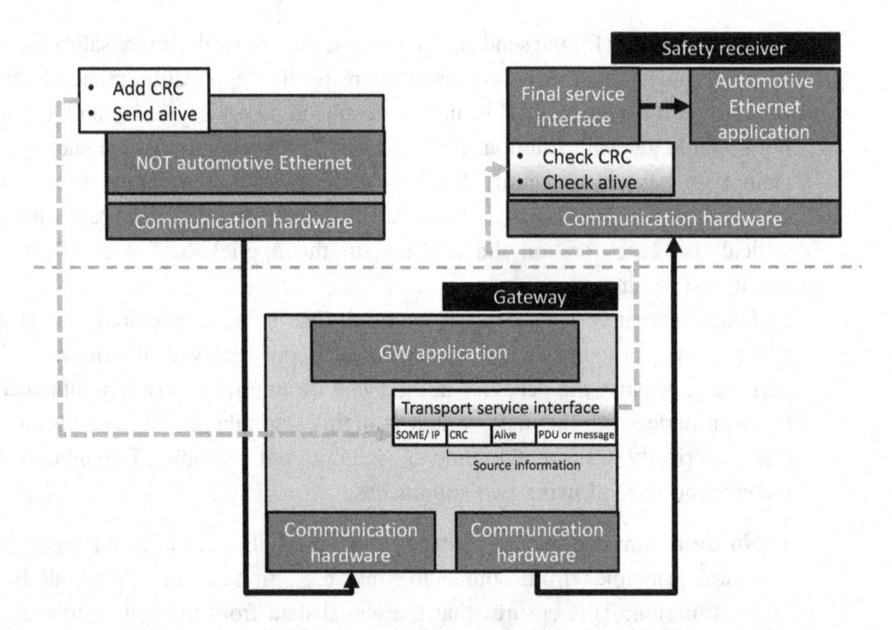

Figure 8.9 Effective tunneling of functional safety measures in a network using different technologies

2. **Tunneling of safety information**, meaning that the same source safety information data is relayed to the recipient, despite the different communication principles. This requires that the source data is transformed to fit into the communication scheme of the recipient, while at the same time the recipient needs to be able to understand the communication principle of the transmitter network. "Unpacking and calculating" the safety-relevant additional information only takes place at the end of the chain and thus in the application context of the sink (see also Figure 8.9). This procedure allows the end-to-end principle to be guaranteed even with the introduction of different communication technologies in the vehicle. The challenge with this approach is to ensure the efficient transmission of the source safety information data via communication technologies with different communication preferences. SOME/IP can be used to support this approach by packing signal-based information into a service-based communication packet.

Two other concepts frequently discussed in the context of functional safety are "common cause" and "single point of failure" [28]. Both identify particularly critical elements of a system that, if they fail, impact all the rest of the system or a significant portion of it and thus potentially provoke a complete system failure. A countermeasure for a single point of failure is to add redundancy to a system. Redundancy is thus frequently discussed in the context of functional safety and is also a subject when designing an in-vehicle network. The possibility of adding redundancy to an Automotive Ethernet network has been discussed in Section 7.1.4.5. However, for a system that includes communication, redundancy can come in various forms [29] and the following example gives an idea.

Assume the simple use case of a camera being connected to an ECU that processes the camera data as part of the surround view model. Having two communication lines laid in different routes through the car, mitigates the effect of a cable break on one of the routes. Should both wire pairs be using the same (multipin) connector at the camera or the processing ECU, the connector is the single point of failure. Redundancy thus also requires two connectors at each end. When the communication system is Ethernet, it would use switches at each end for the redundancy realization with IEEE 802.1CB. The switch duplicates the data at the transmitter and removes the redundant version of the data at the receiver. Now each switch is the single point of failure. Also, using Ethernet for both redundant routes can be a single point of failure if there is a design error in the communication system. This can be countered by using two different communication technologies – if two suitable ones are available – with the respective processing overhead now in the gateway instead of switch. So, as the gateway is now the single point of failure, why not use two cameras to start with, or one camera and one other sensor that allows the system to obtain the same information to start with? The next level is to use two different units that process the data.

8.6 Lessons Learned

Admitting mistakes is difficult. However, mistakes are also the best chance to learn. In this section the authors thus list some important lessons learned to offer the readers the chance not to repeat them. These lessons learned are numbered and formulated as boldly as possible, in order to make the point:

Lesson learned No. 1: Ethernet is not CAN (and should be considered differently from the start).

At the time of writing, CAN was the most prominent communication technology used. A large number of EE developers and EE architects grew in their profession with the CAN system. In the absence of any alternatives, they used this bus system as the basis for all activities. Many of them no longer knew the constraints associated with the design of CAN and took the existing CAN-specific limiting features as basic rules applicable to all communication. This resulted in some strange implementation proposals for Automotive Ethernet. The less experience the engineers had with other communication principles, the more this showed. Everything was dominated by the fieldbus CAN and its arbitration process, and the Ethernet system was designed accordingly.

Lesson learned No. 2: Ethernet is a networking technology (and not a bus).

As mentioned several times in this book, the use of Automotive Ethernet in the vehicle allows a transition from various distributed buses and bus systems to a (single) network. This means that EE architects have to deal with a network and have to learn how to design one. In order to be able to profit from a network, the network must be considered as a system for which basic rules should be set up at the beginning. Otherwise, the network complexity at a later stage might be overwhelming. This includes principles such as traffic shaping procedures or the securing of resources via various paths in the network through prioritization or other procedures to ensure the respective QoS.

At the beginning, it is possible to design the network without any complex procedures in order to "keep it simple." However, designers should be clear from the beginning where it makes sense to invest in complexity in order to master complexity. This means that you have to understand the problems and solutions in networks in order to be able to make the right decision.

Lesson learned No. 3: Bandwidth is not the issue.

As already described in Lessons learned No. 1, many EE architects think "CAN" when they think of in-vehicle communication. Since CAN, as a fieldbus, prioritizes the messages via the message IDs and their priority functions via arbitration, the delivery of a CAN message is not predictable in the classic sense. Attempts to solve this problem result in extrapolation and over-provisioning. Many EE architects and developers thought (and still think) that this can be solved with more bandwidth and fled to CAN FD, which promised higher data rates, or, at the time of writing, CAN XL, which promised even more bandwidth (see Section 2.2.2). The real problem controlling the data flow was thus overlooked. A simple example to visualize the problem is a street network without traffic lights in which the strongest car wins. What happens with such a road network in the case of more and more traffic? Except if you have the strongest car, who will get through in time?

The problem cannot be tackled with bandwidth, but only through a structural change. Ethernet offers many opportunities to break new ground here. Due to the explosive increase in communication volume in the automotive sector, it is extremely important to set the Ethernet network up properly and to establish a suitable set of rules early in the development process.

Lesson learned No. 4: The requirements have to be known.

As described with the previous lessons learned, it often happens that EE designers take the limits of a communication system and mistake them for requirements of the application. The authors have experienced in many discussions that the actual functional requirements of the application were not known or considered in detail, but that instead the limitations of the communication technology were considered as requirements of the application.

This in turn caused (causes) many application developers to make decisions based on the (available) communication schedule and not based on the actual requirements. In addition, there are many shortcomings in the software stacks which, in turn, ensure that entire schedules of ECUs units are based on the communication system and the ECUs' schedules are optimized accordingly. These solutions had (have) nothing to do with the actual functional requirements of the application.

Working out the true requirements of an application takes time and energy. However, this is time well spent because it provides the only sensible basis for further optimization. With these requirements developed from the context of the applications, new solutions can be developed that allow a significantly improved overall structure. To ensure optimization, the communication stacks of the software manufacturers also have to take this into consideration.

Lesson learned No. 5: Do not be overwhelmed by complexity.

EE developers from the automotive industry are now faced with an Ethernet solution space that has been created over decades, which they are supposed to deploy properly within a relatively short amount of time. In order to handle this, it first of all needs to be really clear what is needed (the requirements of the network). The developer should only choose what is really needed for the current problem; "keep it simple." It is extremely helpful to start small with prototypes or small series, in order to learn what complexity is needed or not. This might sound like a contradiction to the point made earlier, that some basic rules for the network should be in place. Yes, some complexity is needed in order to truly benefit from future scalability. However, some complexity can be added in subsequent systems following the familiarization with the system: "Make it as simple as possible but as complex as necessary."

Lesson learned No. 6: Not everything in a standard has to be used.

Many users of a standard start out believing that everything that is written in a standard must be used accordingly. But this is exactly where there is a latent danger of designing an over-engineered system. As with Lessons learned No. 5, it is important to know the requirements exactly and to weigh different solutions. It is often sensible to cherry pick. The EE architect often has the problem that the experts who developed a standard often do not know the exact benefits of the solution in practice. This is not to be read as finger pointing or blaming. New standards break new ground and no one can foresee what exactly will come out of it. But all those participating try to provide the future user with the best possible options. Some of those options might simply turn out to be unsuitable.[9] The EE architects should familiarize themselves or seek advice from experts that have experience with respective implementations. No one is born a master! The lessons learned described here also arose from some mistakes.

Lesson learned No. 7: "All Ethernet/IP" is not the only solution for all applications.

The Automotive Ethernet community is endeavoring to use Automotive Ethernet as comprehensively as possible. However, it is sometimes overlooked that there are communication approaches that are superior in their simplicity and that it does not necessarily make sense to address these local solutions with IP. In some cases, and especially to begin with, it might be valid to tunnel this data through an IP-based system safely in order to have the advantages of the IP architecture, but also not to lose the local simplicity and efficiency (nor to overwhelm the development with too many large changes at once). With the right tunneling solution, developers can create a very scalable system.

There are some discussions to simply have an "All IP" network by connecting different technologies at IP level. The development of solutions such as CAN XL might be seen in that context. However, many advantages of Ethernet (unambiguous addressing, security, topology flexibility, QoS, . . .) are lost with such an approach. The authors think that suitable tunneling mechanisms are better suited than to extend legacy busses to speak IP.

Lesson learned No. 8: Ethernet and SOME/IP do not automatically mandate services.

One of the key features of Ethernet based-communication is that it supports service-based communication and thus a service-based architecture. In the automotive industry, MOST (see Section 2.2.4), had been the first networking technology to allow for a service orientation, while all other legacy communication technologies were deploying signal-based communication. When BMW introduced Automotive Ethernet with the first application (see Section 3.1), BMW also wanted to implement the design patterns for a service-oriented architecture directly. We saw the introduction of this new networking solution as an opportunity to establish more modern, future-proof communication mechanisms. In hindsight, this was a challenge as it required quite some effort to include this in a predominantly signal-based architecture. From our experience, we would make the suggestion today to go with the service-oriented approach, especially for new clusters, and be careful with mixed service/signal clusters, especially if it is the first contact with service-based communication. When taking over existing systems or when just changing networking technology to Ethernet, which approach is most suitable and future-proof should be considered carefully. Note though that well-known and established design patterns, such as end-to-end safety mechanisms, can also be adopted in an Ethernet system (see Section 8.5).

Also, the service orientation of SOME/IP does not automatically mean that only service-oriented designs are possible when using SOME/IP. The solution provided by SOME/IP enables data containers to be transported without any difficulty and thus allows data to be tunneled from and to a non-IP-based system network such as CAN, LIN, or FlexRay. The service discovery feature of SOME/IP can be used independently.

Lesson learned No. 9: Do not base availability and redundancy on the network (only).

Redundancy provisions are essential when discussing functional safety, e.g., for use cases such as automated driving. However, data transmission with minimal errors is only one part of the solution. Furthermore, duplicating the transmission links is neither necessarily required, nor sufficient (see also Section 8.5). First of all, there are other methods in the design of the PHY with which many errors can be compensated for in the network; robust error handling, for example (see Section 4.1.4 and Chapter 5). Then the communication stack might contain a systematic error, for which a duplicated communication link also does not provide a solution. To use two different communication technologies, each with its own transmission link, mitigates the effect of systematic errors, provided that two technologies can be found that have the required capabilities. However, this is costly. More effective and simpler solutions for the design of redundant systems might be found with the risk analysis that do not rely on the communication system.

Lesson learned No. 10: It is not necessary to apply the same EMC requirements and solutions everywhere.

Last but not least, this last point of the lessons learned addresses the physical layer, which actually got the whole topic of Automotive Ethernet going. What can be observed here in particular is that it is often mandated that an EMC-related solution has to work for all use cases, taking the actual application into account as little as possible. This means that a large number of links in the vehicle are oversized just so that the same standard solution can be used everywhere, even though the extreme case is the exception. Please ask an EMC specialist about the limit values and you will see that it usually makes a big difference if the data line on which the EMC noise occurs is close to the vehicle antenna. However, if all links are treated as if they pass close to an antenna, even though the majority do not, an extremely important function/possibility of a network such as Ethernet is left unused. As the majority of an Ethernet network will be switched and thus consists of individual links that do not affect each other from an EMC perspective, EMC protection can actually vary from link to link. This results in a great cost saving potential. With the legacy bus systems, a separation at this level is generally not possible, since the basic principle here is that every participant must behave exactly the same on the bus. According to the authors, this cost advantage of Ethernet could even compensate for the additional costs of switches in the system network. However, it requires that EMC and networking specialists work together toward the same goal, and carefully redefine the respective limit values.

Notes

1 An example of a bottom-up test strategy in a narrower context was presented in Section 4.1.3 where EMC was tested at the semiconductor, ECU, and car levels.

2 Note that this is a very ECU/hardware centric approach that describes the prevailing car development process at the time of writing. With the increase of software and software enabled functions in the car, it is unlikely that this process will continue unchanged. The interrelations and interdependencies will be just too complex.

3 This applies not only to the EE architecture but to semiconductor design as well. The partitioning onto semiconductors and the partitioning onto ECUs are mutually dependent. If a semiconductor is available that integrates two function blocks, it is more likely that an ECU will, too, as the integrated chip might make it financially attractive. If there is no obvious market for integrating those two blocks in one ECU, e.g., because the take rate for one of the function blocks is low, such a chip is not necessarily offered. This makes it an interesting chicken and egg problem, as the integrated chip might reduce the price of the function block to a point where the take rate would increase. However, even though semiconductor availability is very important for automotive innovations, it is not the focus of this chapter.

4 The Ethernet broadcast address is FF:FF:FF:FF:FF:FF. Ethernet MAC addresses for multicast are identified by a "1" in the least significant bit of the first octet. A list of Ethernet multicast addresses can be found in [30].

5 SoftWare (SW) as a cost factor is comparatively new in the automotive industry as such. Generally, the costs for the SW development are seen as a one-time effort the Tier 1 vendor has to deal with during the development of the ECU. Implementing a different business model based on SW licensing is rare. However, it can be expected that the automotive

industry will see more of it in the future. Today, explicitly disclosed SW licensing costs are generally caused by a standard from the consumer industry being reused inside the vehicle (e.g., specific media compression formats). Note that car manufacturers expect hardware related licensing fees to be part of the hardware offer and to be dealt with by the Tier 2 (and thus Tier 1) supplier and to be transparent to the car manufacturer.

6 For the four-unit scenario shown, the T2 topology is actually a star topology with the center shifted to a different unit. However, the center is shifted to an optional unit. To emphasize the distinction to the S scenario, and to emphasize the possible addition of more optional units, this is called "Tree 2."

7 The following shows the input data used to derive the values of . While the take rates would represent high-end (Model A), upper class (Model B), and middle class (Model C) cars, neither they, nor the cost values have been taken from a real life example, but were selected to emphasize the differences. There is no point in including any low-end car in such a comparison as the number of options chosen for those is generally very small. For the cost values, the following, arbitrarily picked example values have been used:

- Standalone PHY: 1.00
- Switch logic: 0.90
- Integrated PHYs: 0.70
- CMC and Media Dependent Interface (MDI): 0.25
- Cable per [m]: 0.35

One-time costs such as the number of segments have not been included, as this would require absolute cost and volume values to do so. The table below shows the combination of take rates used for the example.

Opt. 3	Opt. 2	Opt. 1	Serial	Model A	Model B	Model C
no	no	no	yes	0%	15%	60%
no	no	yes	yes	3%	10%	20%
no	yes	no	yes	1%	4%	7%
no	yes	yes	yes	10%	28%	7%
yes	no	no	yes	1%	1%	3%
Yes	no	yes	yes	10%	2%	2%
Yes	yes	no	yes	5%	10%	1%
Yes	yes	yes	yes	70%	30%	0%
				0.1 Mio	1.0 Mio	5.0 Mio

8 At the time of writing, many of the OPEN Alliance test specifications were being transferred to ISO 21111 (see also Section 3.5.3). Once this is completed, they must be obtained from ISO.

9 Sometimes, standards even need to be violated in order to make a better product. However, this would be a very unfortunate outcome and risk interoperability. The violation of standards should thus always be accompanied with an industry consensus on why and how this is done.

References

[1] T. Streichert and M. Traub, *Elektrik/Elektronik-Architekturen im Kraftfahrzeug*, Berlin, Heidelberg: Springer Vieweg, 2012.

[2] Euro NCAP, "Euro NCAP," 2020, continuously updated. [Online]. Available: www.euroncap.com/en. [Accessed May 22, 2020].

[3] Wikipedia, "V-Model (Software Development)," November 21, 2019. [Online]. Available: http://en.wikipedia.org/wiki/V-Model_(software_development). [Accessed May 22, 2020].

[4] ISO, *ISO 26262:2018 – Road Vehicles – Functional Safety, Parts 1–12*, Geneva: ISO, 2018.

[5] O. Kindel and M. Friedrich, *Softwareentwicklung mit AUTOSAR: Grundlagen, Engineering, Management in der Praxis*, Heidelberg: dpunkt verlag, 2009.

[6] J. Neumüller, "Adaptive Platform," in *12th AUTOSAR Open Conference*, Lisbon, Portugal, 2020.

[7] VDE Verband der Elektrotechnik Elektronik Informationstechnik e.V., "ITG-Positionspapier: Kfz-Anforderungen an Elektronik-Bauteile," about 2005. [Online]. Available: https://docplayer.org/12978134-Itg-positionspapier-kfz-anforderungen-an-elektronik-bauelemente.html. [Accessed May 22, 2020].

[8] R. Charlton, "American Drivers Keeping Cars on the Road for Longer: Average Age Now 11.4 Years," August 9, 2013. [Online]. Available: www.huffingtonpost.com/reno-charlton/american-drivers-keeping-_b_3718301.html. [Accessed May 22, 2020].

[9] Green Car Congress, "IHS Markit: Average Age of Cars and Light Trucks in US Rises Again in 2019 to 11.8 Years," June 28, 2019. [Online]. Available: www.greencarcongress.com/2019/06/20190628-ihsmarkit.html. [Accessed April 25, 2020].

[10] kbb.com, "Average Length of U.S. Vehicle Ownership Hit an All-Time High," February 23, 2012. [Online]. Available: www.kbb.com/car-news/all-the-latest/average-length-of-us-vehicle-ownership-hit-an-all_time-high/2000007854/. [Accessed May 22, 2020].

[11] all-electronics.de, "Deutsche OEMs setzen Standards," June 14, 2011. [Online]. Available: www.all-electronics.de/deutsche-oems-setzen-standards/. [Accessed May 22, 2020].

[12] Wikipedia, "Multicast Address," April 29, 2020. [Online]. Available: http://en.wikipedia.org/wiki/Multicast_address. [Accessed May 22, 2020].

[13] M. Schaffert and D. Kim, "Automotive Ethernet Tooling Revision 13," June 8, 2018. [Online]. Available: http://opensig.org/tech-committees/tc4/. [Accessed May 22, 2020].

[14] N. Navet, H. H. Bengtsson, and J. Migge, "Early-stage Bottleneck Identification and Removal in TSN Networks," in *Automotive Ethernet Congress*, Munich, 2020.

[15] H. Zinner, J. Brand, and D. Hopf, "Automotive EE Architecture Evolution and the Impact on the Network," March 2019. [Online]. Available: http://ieee802.org/1/files/public/docs2019/dg-zinner-automotive-architecture-evolution-0319-v02.pdf. [Accessed April 28, 2020].

[16] D. Hopf and H. Zinner, "Automotive TSN Profile Based on Features, Architectures, or Requirements," July 2019. [Online]. Available: www.ieee802.org/1/files/public/docs2019/dg-hopf-features-architectures-requirements-0719-v02.pdf. [Accessed April 27, 2020].

[17] T. Hogenmüller, Single Twisted Pair Fast Ethernet, Call for Interest, Consensus Building Presentation v.0.4, 2014.

[18] D. Zebralla (Hopf), "Requirements on Future In-vehicle Architectures for Automotive Ethernet," in *Automotive Ethernet Congress*, Munich, 2016.

[19] M. Hiller, "Towards a Central Computing Architecture for In-Vehicle EE Systems," in *Automotive Ethernet Congress*, Munich, 2019.

[20] Y. Kaku, "Denso's Automotive Ethernet Technology for Future Architecture," in *Automotive Ethernet Seminar*, Osaka, 2019.

[21] T. Kirchmeier and G. Janker, *OPEN Alliance Automotive Ethernet ECU Test Specification, v2.0*, Irvine, CA: OPEN Alliance, 2017.

[22] T. Kirchmeier, "Design and Qualification of Automotive Ethernet," in *Automotive Ethernet Congress*, Munich, 2015.

[23] OPEN Alliance, "OPEN Alliance Automotive Ethernet Specifications," OPEN Alliance, 2020, continuously updated. [Online]. Available: www.opensig.org/Automotive-Ethernet-Specifications/. [Accessed April 30, 2020].

[24] F. Nikolaus, J. Angstenberger, M. Heinzinger, and M. Gosh, *Switch Semiconductor Test Specification, v1.0*, Irvine, CA: OPEN Alliance, 2018.

[25] T. Kirchmeier, "Automotive Ethernet: How to Handle the Difference between the Standard and its Implementation," in *IEEE-SA Ethernet & IP @ Automotive Technology Day*, Paris, 2016.

[26] T. Kirchmeier and G. Janker, *Test Process, TC 8 – ECU and Network Test, v1.0*, Irvine, CA: OPEN Alliance, 2016.

[27] embitel, "Understanding How ISO 26262 ASIL Is Determined for Automotive Applications," April 19, 2018. [Online]. Available: www.embitel.com/blog/embedded-blog/understanding-how-iso-26262-asil-is-determined-for-automotive-applications. [Accessed May 1, 2020].

[28] Wikipedia, "Single Point of Failure," April 13, 2020. [Online]. Available: https://en.wikipedia.org/wiki/Single_point_of_failure. [Accessed May 2, 2020].

[29] C. Humig, "Reliable Ethernet Communication for new Network Topologies," in *Automotive Ethernet Congress*, Munich, 2017.

[30] cavebear.com, "Multicast (Including Broadcast) Addresses," September 3, 1999. [Online]. Available: www.cavebear.com/archive/cavebear/Ethernet/multicast.html. [Accessed May 22, 2020].

9 Outlook

The car industry is at a turning point. On one side there is the (slowly, but surely) changing customer behavior, which affects the whole concept of mobility as well as the expectations of the timely availability of new functions. On the other side there is the increasing complexity of the car itself with growing amounts of software, electronics, and communication (from 100s of different suppliers) that must function together faultlessly.

For years the world has seen the number of vehicles sold grow; from about 66 million in 2005 to about 95 million in 2017 [1]. However, when not looking at the data in detail, the message of these numbers is deceiving. Not only has there been an overall decrease in recent years (to about 91 million in 2019),[1] but also the growth is very region-specific. In the EU and the U.S., e.g., almost the same amount of cars wee sold in 2019 as in 2005 [1, 2] (despite growing mobility [3, 4]). The market growth was solely driven by the emerging markets (especially in Asia/China). Table 9.1 gives an overview of the numbers.

The stalling of the car markets in the EU and the U.S. seems to be accompanied by a decreasing importance of owning a car. This can be deduced, e.g., from the fact that in these regions the age increases when people get their driver's license. In the U.S. in 1983, 46.2% of 16 year olds and 87.3% of 19 year olds had a license. In 2014 it was 24% of the 16 year olds and 69% of the 19 year olds. While this trend is most profound for young people, it is actually up to the age of 55 that the percentage of owners of driver's licenses has been decreasing in the U.S. [5, 6]. In Germany, a similar effect can be observed. The percentage of holders of driver's licenses in the age group 17–24 years reduced from 85.8% in 2010 to 79.2% in the few years to 2019 [7]. In Germany this is attributed to cars losing their nimbus as a status symbol, which – among other things – shows in fewer illegal modifications to cars [8] and fewer young people owning a car [9]. In the U.S., next to economic motivations, tougher regulations and an explosion of ride-hailing and ride-sharing services are given as reasons [10].

In a growing market such as China, being able to drive (your own car) yourself is a comparably new phenomena [11]. The percentage of people with a driver's license is thus still increasing (from about 13% in 2011 to about 26% in 2018 [12]), and can be expected to continue to grow – 26% of citizens having a driver's license is still low when compared with countries such as Germany or the U.S. However, car sales have also decreased in China since 2017 [1]. While these recent years have seen some

Table 9.1 Worldwide sales of passenger cars and commercial vehicles in 2005 and 2019 [1]; the numbers for Asia include the Middle East and Oceania; EU15 includes the 14 EU members that had joined the EU before 2004 plus the UK; EFTA represents Iceland, Norway, and Switzerland; NAFTA includes Canada and Mexico, as well as the U.S.

	2005		2019		CAGR [%]
	Sales [Mio]	Share [%]	Sales [Mio]	Share [%]	2005–2019
Overall	65.9	100	91.3	100	2.4
Asia (China)	20.4 (5.8)	30.95 (8.7)	44.0 (25.8)	48.2 (28.2)	5.6 (11.3)
NAFTA (U.S.)	20.2 (17.4)	30.7 (26.5)	20.8 (17.4)	22.8 (19.1)	0.2 (0.01)
Europe (EU15+EFTA)	21.1 (16.9)	31.95 (25.7)	20.8 (16.6)	22.8 (18.2)	–0.09 (–0.15)
Central & South America	3.1	4.7	4.5	4.9	2.7
Africa	1.1	1.7	1.2	1.3	0.4

economic instability, the consequences of another effect, as profound in China as in almost all other areas in the world, might be visible in this: Urbanization. Since 2008, more than half of the world's population lives in towns and cities and, while the rural population has stopped growing, the population in cities will continue to grow significantly (by 1.5 billion in 2030 and by another 1.5 billion in 2050 [13]). This growth is anticipated especially in Africa and Asia where, in turn, most cars are being sold today (see Table 9.1).

The (century long) trend of urbanization has always had economic reasons [14]. Towns and cities generate 80% of the gross national product [13], and attract with the promise of being able to partake. While economic success might allow more people to buy cars, cities are getting more crowded and owning a car in them becomes less attractive. Provided there are alternative means of transportation available, people use these alternatives; and even more so with the higher traffic congestion they face, the fewer parking spaces they find, and the more that legislation discourages ownership and use of cars. For city councils, noise reduction and lower CO_2 emissions are two of many reasons to implement car-free schemes (see e.g. [15, 16] for examples) and/or to encourage EV driving.[2] In 2019, even the BMW CEO conceded that people who live in cities might well do so without owning a car [17].

The huge proliferation of smart phones has a hand in this. In the U.S. in 2018, an average of 81% of adults owned a smartphone, 83% if they lived in an urban or suburban area, and 96% if they were aged between 18 and 29 [18]. In the developed countries in 2019, every adult had on average 1.2 mobile broadband subscriptions (0.75 in developing countries), and 93% of the population had LTE/WiMAX coverage (80% in developing countries) [19], which essentially means that everyone who lives in the city has/can have access. This gives a different edge to alternative means of transportation. For public transport, smartphone and Internet provide, e.g., access to timetables, up-to date route planning, and/or ticket purchasing. In the case the user wishes to use other means of transportation, they offer easy access to alternatives such as scooters or bicycles, car-sharing, ride-sharing, or ride-hailing.[3]

While today the revenues of these services seem small in comparison with those of the automotive industry (less than 1%),[4] they are what the generation of the millennials is growing up with. Smartphones influence and change user behavior with lasting effect; not only of people below the age of 30. One effect is that they assist the trend toward usership instead of ownership. In the consumer world, usership has become normal in areas such as music (e.g., Spotify), video (e.g., Netflix), storage (e.g., Dropbox), and many more [4]. This trend can be observed for mobility as well; and not only with respect to car-sharing. Car leasing, e.g., is just that, too, a car to use and not to own. The market for car leasing has continuously grown for years, also for non-commercial customers. In Germany, e.g., this might be a reason why the share of privately purchased cars decreases (from 40.3% to 34.4% between 2008 and 2019 [20]). There is even a new term describing uses of cars that do not require ownership and a respective term for mobility in general, "Car-as-a-Service (CaaS)" and "Mobility-as-a-Service (MaaS)" [4].

Leasing has another side effect: The lease has a termination date. Car owners are more likely to hold onto a car and to continue using it, even if they find out that they are not using the car enough or are missing some features [21]. A lease holder might simply not get a new lease. Leasing also impacts the expectations on the updateability of cars, as the potential resale value of a car is an important part of the business model of the leasing company (and for company-owned cars in general). With the customers being used to the fast changing world of smart phones, why is this so difficult with cars?

The smartphone allows easy access to the latest trends and developments. 2019 saw about 115 billion app downloads from Google Play and Apple's App Store [22].[5] Millennials in the U.S. use 66% of the digital media time with smartphone apps (the average over all age groups is 50%), and 70% of millennials say they are always looking for new and interesting apps [23]. In the statistics provided in [23], the download frequency for apps is measured in months. While only 1/3 of the users in the U.S. downloads more than one app per month, just imagine if you were to ask them how many apps they download in a year, or in three years; the average time it takes for a car facelift. If car manufacturers want to keep pace with technological and other changes customers expect, they must change and, potentially, speed up the market introduction of new features (see e.g. [24, 25], plus change their business model accordingly [26]).[6]

But, how? As said many times, the car is an extremely complex product. A 2008 publication speaks of 2000–3000 singular functions per vehicle that are combined into 250–300 functions with which car users operate the car [27]. As can be seen in Figure 2.2, the average number of ECUs per car has more than tripled since then. Just from 2010 to 2016, the amount of software in some cars increased from 10 Million lines of codes to 150 Million lines of codes [26] (while, in 1981, 50,000 lines of code apparently sufficed across GM's complete domestic car production [27]). The automotive semiconductor market has seen a tremendous growth (from around 380 USD per car in 2012 to more than 540 USD per car in 2018 [28]) across various types of semiconductors (analog, digital, special purpose, memory, etc.). Also, the number of IVN-technologies has increased [29], as have the communication relations.

Just consider a park distance control application that offers audible and visible feedback behind and in front of the car with, e.g., eight sensors and five ECUs involved (see e.g. [30]). If each sensor (or actuator) communicates with just one ECU but all ECUs have communication relations among them, this leads to $(8 + \Sigma(1{:}(5-1))) \times 2 = 18 \times 2$ theoretical communication relations (×2, because the communication is bi-directional). For a remote parking control function (which automatically parks cars into narrow parking spots) with 14 sensors, 11 ECUs, and 4 actuators this means $(14 + 4 + \Sigma(1{:}(11-1))) \times 2 = 74 \times 2$ theoretical communication relations. This is just an example of one customer function. When looking at the car overall, the number of different signals in the BMW EE architecture increased from 3000 in 1995 to 5500 in 2004 to 12,500 in 2015, and is expected to increase to about 80,000 in 2021 [31]. Every one of these signals must be maintained.

This complexity increase comes at a cost. Between 2009 and 2018, e.g., the overall worldwide spending for warranty claims by car manufacturers outgrew the car sales, with a solid Compound Annual Growth Rate (CAGR) of 3.2% [32] (compared with a CAGR of 2.4% in car sales).[7] In 2009, it was estimated that 50% of the automotive warranty costs were related to electronics and their embedded software [27]. That there is a connection might be conducted from Figure 2.2. European car manufacturers have, on average, the highest number of ECUs in their cars and, at the same time, spend an over-proportional amount on warranty claims: E.g., in 2018 their share of money spent on warranty claims was 50% [32], while they sold 29% of the cars [33].

And car manufacturers invest a lot to make the products their customers want. In 2017, eight of the 15 leading R&D investors were car manufacturers (the three German ones ranked within the top five) [34], with the R&D budget having grown continuously over the years [35]. The share spent on test and validation is not published. However, already by 2009, one third of all software in cars was devoted to just diagnosis [27]. It can thus be expected that the budget spent on test and validation is significant.

So, competition is fierce, customers' expectations are volatile, while complexity and costs for (developing) cars are increasing. A typical reaction is cutting costs, e.g., by squeezing money out of the supply chain, by reducing production costs,[8] or by becoming more efficient in development (which generally means doing more with smaller head count). Reducing functionality is generally not favored and seems to be a contradiction to progress and customer demands, especially for premium brands. Reducing functionality is thus not really an option. To enable and ensure more functionality in development, for less costs, seems contradictory. Even if economies of scales tend to reduce the costs of legacy ECUs, these ECUs must also be requalified in the context of the new functions surrounding them, while all new functions require the full effort anyway.

Not only in the authors' opinion will it be difficult for the automotive industry to continue to develop cars as they have done in the past 20 years (not only, but also because others are proving it can be done differently, see e.g. [36, 37]). Car manufacturers must take complexity out of the system, in a way which, at the same time, allows them to offer new functionality faster.

At this point we will make a small excursion into changes caused by innovations. There are, in principle, three types of changes: Incremental, revolutionary, and disruptive. Incremental changes improve key market features on the basis of existing methods and technologies. Revolutionary changes are based on new technological developments that improve key market features so profoundly that – among other consequences – they will leave behind those who did not adopt the change [38]. Disruptive innovations also change the market radically (which is why they are often confused with revolutionary innovations). The difference is that disruptive innovations induce a change because of features previously not available or not considered as important, while at the same time underperforming in key features [39]. Disruptive innovations thus change the key market requirements and definitely leave behind those not having adopted the change.

It can be a matter of perspective how an innovation is categorized. The development of CAN-FD, e.g., is an incremental step from CAN. The key market feature – data rate – is increased, while otherwise using the same communication principles. Automotive Ethernet is revolutionary. First of all, it further increases the data rate(s) but, additionally, introduces the concept of networking (instead of just sharing a bus) into cars.[9] Automotive Ethernet might become disruptive if it changes the way we develop cars. Electric cars, at the moment, are simply a market diversification in that they enable driving with a completely different engine. If lowest CO_2 emission suddenly became the new key market feature, electric cars would become disruptive, because it meant that lower range (and higher price) no longer mattered. Automated driving is a revolution and a technology leap is needed to achieve this. In its implications on many aspects of everyday life, it might be extremely disruptive [21].

So, should the car manufacturers implement incremental or revolutionary changes to handle the above challenges? The costs and improvements of incremental changes can generally be assessed more easily than those of revolutionary changes that might have unforeseen implications. However, are incremental changes enough to really make a difference? Or would revolutionary changes, in turn, simply be not manageable and break the manufacturers on the way to implementing them? We, the authors, do not claim to be the experts who know which way is best. To us, simply doing things the same way as in the past – just better – does not seem conducive to success, and the industry, in general, is aware of this. Many concepts of how to develop and build cars differently are being discussed. We would here like to highlight three keywords in this context that allow us to come back to in-vehicle networks later.

- **Software-defined cars**: Cars are becoming computers on wheels. Software is responsible for the customer experience, innovations, and the possibility to differentiate a car from the competitor's [40]. It requires that car manufacturers change from being hardware manufacturers to also/predominantly becoming software developers. This has huge implications.

  First of all, an EE architecture is required that allows for the separation of hardware and software and thus to decouple vehicle and vehicle–function development cycles [26]. New (software realized) functions will be treated as software releases with respective rollout structures. The possibility to reuse software/services will be an important asset. After having decided on the car's new features, the overall software comes first. How much has to be added? Where is it going to run? Is a hardware (or communication) upgrade needed? Software and its development would have to be treated as a product.

Naturally, software also means complexity and requires concept and strategy [41], e.g., emphasizing the need to implement an end-to-end automotive software platform with a unified software architecture, in order to provide native compatibility and to reduce complexity accordingly. The traditional car manufacturers are aware of this and are working on the respective changes. However, as the implications are huge, the changes are slow.

- **Service-Oriented Architecture (SOA)** is directly related to software-defined cars. Service orientation offers an abstraction that supports reusability (of services and software) and allows for a better formal specification of functions. When introducing hierarchical system layers with SOA, faster development of new customer functions is possible because not all layers need to be touched. New functions can be introduced frequently with updates to the high layers, while maintaining more stable modules and reusability at the lower layers. SOA allows one to think outside of the ECUs, i.e., in their functions and partitioning, as it establishes a new paradigm of services availability (anywhere in the system).

  It is controversially discussed in the industry whether all communication has to be service-based or whether some communication could or should stay signal-based. We showed in Section 8.6 that an Ethernet/SOME-IP system can also transport/tunnel signal-based information. This discussion is, however, beside the point. The classic signal-based communication in the automotive industry has simply reached its limits [31, 40] and to maintain a signal-based system (only) is not feasible.
- **Test automation:** The test and qualification effort in a car is huge and it will remain huge. Considering the test concepts alongside the development is essential to quality and competitiveness. The industry must move more toward test automation, again also in light of the increasing amount of software.

Automotive Ethernet supports all of the above aspects and it is the only in-vehicle networking technology currently in use that does so.[10] Ethernet embeds all necessary communication concepts perfectly. It was designed for distributed (software) systems, it inherently supports SOA (see Section 7.4 for SOME/IP), and it simply provides standard interfaces. With Automotive Ethernet, the in-vehicle network can be an enabler instead of a limiting factor.

However, the industry is still in its infancy in the adoption of Ethernet. While all major car manufacturers have Automotive Ethernet links in their vehicle EE architectures (or at least have committed projects to do so in the coming years) these steps are often small. In many cases Ethernet has been/is selected as merely yet another IVN technology when the required data rates mandate it. If used as such, Ethernet may be seen as just adding to complexity and not reducing it. Its potential to provide the right infrastructure to change the way cars are being developed (and to help tackle some of the problems discussed in this chapter) is not (yet) exploited.

Why is this so? Part of the reason surely is that traditional car manufacturers are still thinking primarily about the hardware [41]. They have an incredible knowhow in attaching price tags to the hardware. That the hardware costs of an Ethernet link are generally higher than that of, e.g., a CAN-FD link is more decisive than the impact this might have on complexity, flexibility, or sustainability; simply because no price tags have been attached to these aspects. Even the value of less abstract – but related – factors such as a universal security concept or the possibility of software reuse is generally not quantified and thus not really relevant for their decisions. If the industry was to take the route of making fundamental changes, a more comprehensive cost evaluation must be established, just as the industry must learn how to monetize software [27].

What we know is that Automotive Ethernet allows for it all: A mix and match IVN, where the technologies simply live side by side; a tunneling concept where an in principle service-based Ethernet backbone tunnels signal-based information from legacy busses; or the comprehensive all IP/Ethernet solution that supports more radical changes in the way we develop cars. Just this shows how powerful the technology is.

Meanwhile, many more supporting standardization activities increase the toolbox available to build the right Automotive Ethernet system and many more engineers in the whole chain learn and adapt. We are happy to have made a contribution.

Notes

1 It is difficult to say how this trend will continue. As this book was finalized amidst one of the world's biggest economic crises in 2020, it can definitely be expected that the numbers will drop significantly in the short term. If or how much they will recover in 2021 remains to be seen.

2 China started pushing green transportation more than a decade ago by offering tax cuts for consumers and government subsidies for car makers [43]. The motivators were to reduce the dependence on oil and to reduce the significant pollution problems in urban areas by establishing a profitable low emission industry in China [44]. The success showed in 2019: While Tesla, as a brand, sold the largest number of electric vehicles, it is the Chinese car manufacturers who, overall, sold almost three times as many electric cars as Tesla did, mainly into the local market [42].

3 The car industry is aware of this and is actively participating. For example, GM invested heavily in the ride-sharing company Lyft [45]. BMW and Daimler joined forces in Share Now, which offers not only car-sharing but addresses mobility overall [46].

4 For example, the car industry sold about 91 million cars in 2019 [1], while it is estimated that 2025 sees 0.5 million car sharing vehicles worldwide [47]. The revenue of the car industry in 2019 was estimated at 4 trillion USD [48], while Uber, Lyft, and Didi generated a revenue of about 14 billion USD [49], 3.8 billion USD [50], and an expected 9.5 billion USD [51], respectively.

5 While only 4% of the apps in Apple's App Store are travel related, 83% of the users have travel related apps on their phones [23]. Also, 4% of about 1.85 million [53] still adds up to 74,000 apps.

6 The key is to anticipate what customers will want in the future. The integration of new features into the car might be just one way to react to market requirements. Another way might be to simply integrate mobile devices and the cloud better so that (some) new feature are available to use in the car but are not provided by it.
7 The worldwide warranty claims payments peaked in 2017, where the largest load was carried by Volkswagen because of the Volkswagen Diesel scandal (apparently with a two year delay [32]). In the 2018 numbers used to calculate the CAGR, this exceptional situation thus only has a minor impact.
8 [25], e.g., suggests to cut production costs by further automating it, with the goal of compensating for the added R&D costs.
9 Allowing for a switched network is one of the most fundamental changes Automotive Ethernet introduces. Having a switched network instead of a shared bus allows the network to grow in bandwidth when links and nodes are added instead of just decreasing the available bandwidth for all nodes attached. Because of the switches, the size of the Automotive Ethernet network is not limited as is the case for the legacy technologies. Furthermore, the strict adherence to the ISO/OSI layering model not only allows for the reuse of protocols proven in other industries, it also allows for flexibility on the PHY speed grades attached. It also allows for a choice between switching and routing (see Section 7.3.3), and a number of QoS concepts as well as redundancy (see also the outlook of [52]).
10 Just for completeness: It goes without saying that Ethernet-based communication also supports the necessary safety and security.

References

[1] OICA, "2005–2019 Sales Statistics," 2020. [Online]. Available: www.oica.net/category/sales-statistics/. [Accessed May 9, 2020].

[2] ACEA, "The Automobile Industry Pocket Guide," 2020 (continuously updated). [Online]. Available: www.acea.be/publications/article/acea-pocket-guide. [Accessed May 9, 2020].

[3] Sustainable Mobility for All, "Global Mobility Report 2017," United Nations, Washington, DC, 2017.

[4] Roland Berger, "Embracing the Car-as-a-Service Model – The European Leasing and Fleet Management Market," January 2018. [Online]. Available: www.rolandberger.com/publications/publication_pdf/roland_berger_car_as_a_service_final.pdf. [Accessed May 17, 2020].

[5] J. Beck, "The Decine of the Driver's License," January 22, 2016. [Online]. Available: www.theatlantic.com/technology/archive/2016/01/the-decline-of-the-drivers-license/425169/. [Accessed May 9, 2020].

[6] M. Sivak and B. Schoettle, "Recent Decreases in the Proportion of Persons with a Driver's License across All Age Groups," January 2016. [Online]. Available: www.umich.edu/~umtriswt/PDF/UMTRI-2016-4_Abstract_English.pdf. [Accessed May 9, 2020].

[7] Kraftfahrt-Bundesamt, "Fahrerlaubnisbestand im Zentralen Fahrerlaubnisregister (ZFER)," 2020 (continuously updated). [Online]. Available: www.kba.de/DE/Statistik/Kraftfahrer/Fahrerlaubnisse/Fahrerlaubnisbestand/fahrerlaubnisbestand_node.html. [Accessed May 9, 2020].

[8] F. Rinke, "Immer weniger Menschen machen den Führerschein," May 3, 2019. [Online]. Available: https://rp-online.de/leben/auto/news/fuehrerschein-immer-weniger-junge-leute-besitzen-eine-fahrerlaubnis_aid-38535651. [Accessed May 9, 2020].

[9] H. Mortsiefer, "Keine Lust auf Auto," June 6, 2017. [Online]. Available: www.tagesspiegel .de/wirtschaft/autohersteller-und-junge-kunden-keine-lust-auf-auto/19893908.html. [Accessed May 6, 2020].

[10] T. Henderson, "Why Many Teens Don't Want to Get a Driver's License," March 6, 2017. [Online]. Available: www.pbs.org/newshour/nation/many-teens-dont-want-get-drivers-license. [Accessed May 10, 2020].

[11] P. Hessler, *Country Driving*, New York: HarperCollins Publisher, 2011.

[12] statista, "Number of Registered Car Drivers in China from 2011 to 2018," 2020. [Online]. Available: www.statista.com/statistics/288165/number-of-registered-car-drivers-in-china/. [Accessed May 9, 2020].

[13] United Nations Population Fund, "Urbanization," October 3, 2016. [Online]. Available: www.unfpa.org/urbanization. [Accessed May 9, 2020].

[14] Wikipedia, "Urbanization," May 5, 2020. [Online]. Available: https://en.wikipedia.org/wiki/Urbanization/Urbanization. [Accessed May 9, 2020].

[15] Wikipedia, "Car-free City," January 18, 2020. [Online]. Available: https://en.wikipedia .org/wiki/Carfree_city. [Accessed May 10, 2020].

[16] Wikipedia, "Car-free Movement," April 390, 2020. [Online]. Available: https://en .wikipedia.org/wiki/Car-free_movement#Urban_design. [Accessed May 10, 2020].

[17] Spiegel Online, "Das Auto ist zu Recht umstritten," October 25, 2019. [Online]. Available: www.spiegel.de/auto/aktuell/bmw-chef-zipse-hat-verstaendnis-fuer-kritik-am-auto-a-1293367.html. [Accessed May 9, 2020].

[18] Pew Research Center, "Mobile Fact Sheet," June 12, 2019. [Online]. Available: www .pewresearch.org/internet/fact-sheet/mobile/. [Accessed May 10, 2020].

[19] ITU, "Statistics," 2020 (continuously updated). [Online]. Available: www.itu.int/en/ITU-D/Statistics/Pages/stat/default.aspxStatistics/Pages/stat/default.aspx. [Accessed May 10, 2020].

[20] Kraftfahrtbundesamt, "Statistiken," 2020, continuously updated. [Online]. Available: www.kba.de/DE/Statistik/statistik_node.html. [Accessed May 17, 2020].

[21] R. A. Simons, *Driverless Cars, Urban Parking, and Land Use*, Milton Park: Routledge, 2020.

[22] statista, "Schätzung zur Anzahl der Downloads von Apps nach App-Stores weltweit in den Jahren 2017 bis 2019," January 2020. [Online]. Available: https://de.statista.com/statistik/daten/studie/993352/umfrage/anzahl-der-downloads-von-apps-nach-app-stores-weltweit/. [Accessed May 10, 2020].

[23] MindSea, "25 Mobile App Usage Statistics to Know in 2019," 2018. [Online]. Available: https://mindsea.com/app-stats/. [Accessed May 10, 2020].

[24] T. Grünweg, "Modellzyklen der Automobilhersteller: Eine Industrie kommt auf Speed," 10 February 2013. [Online]. Available: www.spiegel.de/auto/aktuell/warum-lange-entwicklungszyklen-fuer-autohersteller-zum-problem-werden-a-881990.html. [Accessed May 8, 2020].

[25] strategy&, "Automotive Trends 2019: The Auto Industry Must Find a Way to Balance Accelerating Innovation and FInancial Survival," 2019. [Online]. Available: www .strategyand.pwc.com/gx/en/insights/industry-trends/2019-automotive.html. [Accessed May 11, 2020].

[26] O. Burkacky, J. Deichmann, G. Doll, and C. Knochenhauer, "Rethinking Car Software and Electronics Architecture," February 14, 2018. [Online]. Available: www.mckinsey .com/industries/automotive-and-assembly/our-insights/rethinking-car-software-and-electronics-architecture#. [Accessed May 11, 2020].

[27] R. N. Charette, "This Car Runs on Code," February 1, 2009. [Online]. Available: https://spectrum.ieee.org/transportation/systems/this-car-runs-on-code. [Accessed May 11, 2020].

[28] J. Morra, "Automotive Semiconductor Sales to Keep Growing," November 26, 2018. [Online]. Available: www.sourcetoday.com/supply-chain/article/21867256/automotive-semiconductor-sales-to-keep-growing. [Accessed May 11, 2020].

[29] K. Matheus, "Evolution of Ethernet-based Automotive Networks: Faster and Cheaper," in *IEEE-SA Ethernet & IP @ Automotive Technology Day*, London, 2018.

[30] K. Matheus, "New Speed Grades: Automotive Ethernet Manifests Itself as a Future Proof In-Vehicle Networking System," in *Automotive Ethernet Congress*, Munich, 2017.

[31] J. Broy, "Service-oriented Architecture as a Mindset: Shaping the Next EE Architecture in a Digital Age," in *Automotive Networks*, Munich, 2019.

[32] Warranty Week, "Worldwide Automotive Warranty Expenses," August 22, 2019. [Online]. Available: www.warrantyweek.com/archive/ww20190822.html. [Accessed May 11, 2020].

[33] OICA, "World Ranking of Car Manufacturers 2017," 2018. [Online]. Available: www.oica.net/wp-content/uploads/World-Ranking-of-Manufacturers-1.pdf. [Accessed May 11, 2020].

[34] R. Ciechanski, "R&D in the Automotive Sector," March 21, 2018. [Online]. Available: https://home.kpmg/pl/en/home/insights/2018/03/r-and-d-in-the-automotive-sector.html. [Accessed May 11, 2020].

[35] R. Crossan, "Top 20 Carmakers R&D Spending Tops 70B Pounds in a Year as Transition to Electric and Autonomous Vehicles Accelerates," November 6, 2019. [Online]. Available: www.bdo.co.uk/en-gb/news/2019/top-20-carmakers-r-d-spend-tops-70bn-in-a-year. [Accessed May 11, 2020].

[36] L. Shipley, "How Tesla Sets Itself Apart," February 28, 2020. [Online]. Available: https://hbr.org/2020/02/how-tesla-sets-itself-apart. [Accessed May 13, 2020].

[37] S. Hansen, "Alle machen Auto; Fahrzeuge von Sony & Co," January 2020. [Online]. Available: www.heise.de/select/ct/2020/5/2001606351558364371. [Accessed May 12, 2020].

[38] V. Kotelnikov, "Radical Innovation versus Incremental Innovation," not kown. [Online]. Available: www.1000ventures.com/business_guide/innovation_radical_vs_incr.html. [Accessed May 13, 2020].

[39] C. M. Christensen, *The Innovator's Dilemma*, New York: HarperBusiness, 1997.

[40] M. Hiller, "Towards a Central Computing Architecture for In-Vehicle EE Systems," in *Automotive Ethernet Congress*, Munich, 2019.

[41] R. Fletcher, A. Mahindoro, N. Santhanam, and A. Tschiesner, "The Case for an End-to-end Automotive-software Platform," January 16, 2020. [Online]. Available: www.mckinsey.com/industries/automotive-and-assembly/our-insights/the-case-for-an-end-to-end-automotive-software-platform?cid=soc-web#. [Accessed May 12, 2020].

[42] M. Kane, "Global EV Sales for 2019 Now in: Tesla Model 3 Totally Dominated," Febuary 2, 2020. [Online]. Available: https://insideevs.com/news/396177/global-ev-sales-december-2019/. [Accessed May 10, 2020].

[43] E. Huang, "China Buys One out of Every Two Electric Vehicles Sold Globally," February 18, 2019. [Online]. Available: https://qz.com/1552991/china-buys-one-out-of-every-two-electric-vehicles-sold-globally/. [Accessed May 10, 2020].

[44] P. Marsters, "Electric Cars: The Drive for a Sustainable Solution in China," August 2009. [Online]. Available: www.wilsoncenter.org/publication/electric-cars-the-drive-for-sustainable-solution-china. [Accessed May 10, 2020].

[45] S. Trousdale, "GM Invests $500 Million in Lyft, Sets Out Self-driving Partnership," January 6, 2016. [Online]. Available: www.reuters.com/article/us-gm-lyft-investment/gm-invests-500-million-in-lyft-sets-out-self-driving-car-partnership-idUSKBN0UI1A820160105. [Accessed May 12, 2020].

[46] You Now, "Unsere Story," You Now, 2020. [Online]. Available: www.your-now.com/de/our-story. [Accessed May 8, 2020].

[47] statista, "Anzahl der Fahrzeuge auf dem weltweiten Carsharing-Markt von 2006 bis 2025," August 2, 2018. [Online]. Available: https://de.statista.com/statistik/daten/studie/388012/umfrage/anzahl-der-fahrzeuge-auf-dem-weltweiten-carsharing-markt/. [Accessed May 16, 2020].

[48] IBISWorld, "Global Car & Automo Industry Trends (2015–2020)," November 2019. [Online]. Available: www.ibisworld.com/global/market-research-reports/global-car-automobile-sales-industry/. [Accessed May 16, 2020].

[49] Uber Investor, "Uber Announces Results for Fourth Quarter and Full Year 2019," February 6, 2020. [Online]. Available: https://investor.uber.com/news-events/news/press-release-details/2020/Uber-Announces-Results-for-Fourth-Quarter-and-Full-Year-2019/. [Accessed May 16, 2020].

[50] Investor Lyft, "Lyft Announces Record Fourth Quarter and Fiscal Year Result," February 11, 2020. [Online]. Available: https://investor.lyft.com/news-releases/news-release-details/lyft-announces-record-fourth-quarter-and-fiscal-year-results/. [Accessed May 16, 2020].

[51] Forbes, "Is $80 Billion Valuation Achievable for Didi Chuxing's IPO?," December 24, 2018. [Online]. Available: https://investor.lyft.com/news-releases/news-release-details/lyft-announces-record-fourth-quarter-and-fiscal-year-results/. [Accessed May 17, 2020].

[52] K. Matheus and T. Königseder, *Automotive Ethernet*, 2nd edition, Cambridge: Cambridge University Press, 2017.

[53] statista, "Number of Apps Available in Leading App Stores as of 1st Quarter 2020," 2020. [Online]. Available: www.statista.com/statistics/276623/number-of-apps-available-in-leading-app-stores/. [Accessed May 13, 2020].

Index

Printed in the United States
by Baker & Taylor Publisher Services